Volume and Surface Area Formulas

Figure	Volume	Surface Area	Definitions
Rectangular Solid 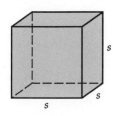	$V = lwh$	$S = 2lh + 2lw + 2wh$	l = length w = width h = height
Cube	$V = s^3$	$S = 6s^2$	s = side
Right Circular Cylinder	$V = \pi r^2 h$	$S = 2\pi rh + 2\pi r^2$	r = radius h = height
Sphere	$V = \dfrac{4}{3}\pi r^3$	$S = 4\pi r^2$	r = radius

Properties of Real Numbers

Associative Property of Addition

If a, b, and c are real numbers, then
$(a + b) + c = a + (b + c)$.

Commutative Property of Addition

If a and b are real numbers, then $a + b = b + a$.

Distributive Property

If a, b, and c are real numbers, then
$a(b + c) = ab + ac$.

Associative Property of Multiplication

If a, b, and c are real numbers, then
$(a \cdot b) \cdot c = a \cdot (b \cdot c)$.

Commutative Property of Multiplication

If a and b are real numbers, then $a \cdot b = b \cdot a$.

BEGINNING ALGEBRA

BEGINNING ALGEBRA

Robert G. Marcucci

San Francisco State University, California

Harold L. Schoen

University of Iowa, Iowa

HOUGHTON MIFFLIN COMPANY   Boston

Dallas   Geneva, Illinois   Palo Alto
Princeton, New Jersey

Cover design by Hannus Design Associates.
Interior design by George McLean.

IBM is a registered trademark of International Business Machines Corporation.
Apple is a registered trademark of Apple Computer, Inc.

Printed in the U.S.A.
Library of Congress Card Number: 89-80948

ISBN Number:
Text: 0-395-48152-X
Instructor's Manual: 0-395-52243-9
Answer Key: 0-395-52954-9
Printed Test Bank: 0-395-52955-7

ABCDEFGHIJ-D-9543210/89

Contents

7 *Linear Equations and Inequalities in Two Variables* *310*

8 *Systems of Linear Equations and Inequalities* *357*

9 *Radicals* *398*

10 *Quadratic Equations*   440

Preface

Beginning Algebra provides a balanced, mathematically sound approach to basic algebraic concepts and applications. Building in a natural way from the concrete ideas of arithmetic, we introduce algebraic symbolism as a convenient language for modern technology, real-world applications, and higher mathematics. Several features and a full set of instructional materials have been included to enhance conceptual understanding and the development and maintenance of procedural skills.

Features

Natural Transition from Arithmetic to Algebra

Much of Beginning Algebra can be viewed as a generalization of arithmetic. For example, the integers and real numbers are natural extensions of the positive rationals of arithmetic. Variables represent numbers, hence, their properties are the properties of numbers. Geometric measurement formulas, first introduced in arithmetic, include variables with restricted domains. Graphing ordered pairs of positive numbers is a topic of basic mathematics which extends easily to graphing in algebra. Students are reminded explicitly of these arithmetic ideas and techniques in the first chapter of *Beginning Algebra*, and their connections with algebraic ideas are emphasized in the remainder of the textbook.

Applications Integrated Throughout the Text

A rich variety of problems and applications are included. Some are straightforward applications of the previously learned content, but many are posed in realistic settings and require some nonroutine applications of the content. Applications are found throughout the textbook and are used to motivate the algebra topics, not just to apply them as an afterthought to the conceptual and skill development as is done in most textbooks. In fact, many sections of *Beginning Algebra* are introduced with an interesting application to establish the need for the content of the section.

Specific Problem-Solving Strategies

Problem solving is a very important goal of mathematics instruction at all levels, yet it is a difficult goal to achieve. Students must be taught strategies for solving problems; by and large, they do not learn to solve problems by mimicking the teacher or the examples in the textbook. In *Beginning Algebra*, specific strategies such as guess-and-test, direct translation of words to symbols, and use of the tables and diagrams are explained with worked-out examples and follow-up applications.

Early Use of Graphing

Since students have used graphs in their previous math studies and many understand visual representations of ideas better than symbolic ones, graphing is reviewed in the first chapter and used where appropriate throughout the textbook. For example, the solutions of linear equations (or the nonexistence of solutions) are examined graphically when they are first introduced. We believe this approach is both natural for the students and consistent with our philosophy of building the ideas of algebra from those of arithmetic.

Section Focus and Discussion Questions

Each section opens with a statement of focus. Less restrictive than behavioral objectives found in many texts, the focus statement serves as a guide for the students as they read the section. To further aid students reading, some questions at the end of each section are designed to test reading comprehension and to generate in-class discussion about the key ideas of the section.

Systematic Review

Review is built into the textbook in the following ways.

- *Preparing for the Next Section* The last five to ten exercises at the end of each end-of-section exercise set provide review of important ideas and procedures needed in the next section.
- *Chapter Summary* At the end of each chapter important terms and ideas from the entire chapter are represented for synthesis and review.
- *Chapter Review* Following each chapter summary, about ten representative exercises and problems are posed for each section of the chapter.
- *Chapter Test* A test containing 20 to 30 problems is also provided at the end of each chapter.

Boxed-in Procedures and Cautions

Steps required to carry out important procedures are summarized and highlighted in boxes for easy reference. Students are also warned of common errors in boxed cautions entitled "Be Careful."

Calculators

We believe that calculators are now so readily available to students that they are best treated as a tool for carrying out some of the more tedious computations needed in mathematics. Therefore, we have not flagged exercises or examples that use calculators. Rather, scientific calculators are discussed at points where they could be helpful. For example, keystrokes for finding roots and powers are described. While calculators would reduce the drudgery in some exercises, an instructor who prefers to restrict calculator use can easily do so skipping or revising the small number of examples that use calculators and requiring students to use tables instead of calculators in exercises.

Exercise Sets

In addition to the discussion questions and various types of review exercises described above, there are many (typically 50 to 100) exercises at the end of each section covering that section's topics. In keeping with the balanced approach in *Beginning Algebra*, there are large numbers of skill-oriented exercises ranging in each section from straightforward to quite complex and applications of the section's topics ranging from direct to nonroutine.

Supplements

The package of supplements to accompany *Beginning Algebra* includes the Student Solutions Manual, a computerized tutorial program, Instructor's Manual, an answer key for insertion in the instructor's textbook, a computerized test generator, and a printed test bank.

Student Solutions Manual

The Student Solutions Manual contains the complete worked-out-solution to every odd-numbered exercise in the textbook.

Expert Tutor

The Expert Tutor is an interactive instructional microcomputer program designed for student use. Each concept in the text is supported by a lesson on the Expert Tutor. Lessons on the tutor provide additional instruction, hints, examples, and worked-out solutions. All referenced by specific page numbers to the text, this instruction and practice can be used in several ways: (1) to cover material the student missed because of absence from class, (2) to repeat instruction on a skill or concept that the student has not yet mastered, or (3) to review material in preparation for examinations. No knowledge of computers is necessary on the part of the student or instructor. The Expert Tutor is available for the IBM^{R}PC and compatible computers with DOS 2.0 or higher and 256K of memory.

Instructor's Manual with Testing Program

The Instructor's Manual is one of three sources of testing material for *Beginning Algebra.* Four tests—two in free response format and two in multiple-choice format—and their solution keys are provided for each chapter. In addition, the Instructor's Manual contains documentation for the computerized test generator.

Answer Key

The Answer Key is a short booklet with answers to all of the even-numbered exercises in the text and to all problems on the chapter tests in the text. This booklet is designed to be placed inside the back cover of the instructor's copy of *Beginning Algebra.*

Computerized Test Generator

The Computerized Test Generator contains more than 1,500 test items. They are not repeats of items from the textbook or any other supplement. Organized according to the selections of *Beginning Algebra,* the Test Generator is designed to produce an almost unlimited number of tests for each chapter of the textbook or for any combination of chapters at the instructor's discretion. It is available for the Apple II family of computers, IBM PC or compatible computers, and Macintosh family of computers.

Printed Test Bank

For instructors who do not have computers to operate the Computerized Test Generator, the Printed Test Bank is a print-out of all the items in the database. Instructors may select items from the Printed Test Bank to be included on a test, but, of course, the test must be written or typed.

Acknowledgments

The authors would like to thank the following people who have reviewed the manuscript and provided many helpful suggestions.

Patricia Confort
Roger Williams College, RI

Gwen Dillard
Community College of Allegheny County, PA

Louise M. Ettline
Trident Technical College, SC

Frank Gunnip
Macomb Community College, MI

Marilyn Hamilton
Indiana Vocational Technical College, IN

John Kennedy
Santa Monica College, CA

Patricia C. McCann
Franklin University, OH

Michael J. Mears
Manatee Community College, FL

Gus Pekara
Oklahoma City Community College, KS

Ross Rueger
College of the Sequoias, CA

BEGINNING ALGEBRA

1 The Transition from Arithmetic to Algebra

1.1 Numbers and Operations

FOCUS

In this chapter, we review some important arithmetic concepts and procedures, as a foundation for your study of algebra. We begin here with the nonnegative numbers and percents, the four basic operations, and a fifth operation—exponentiation.

Algebra is concerned with the applications of numbers and other mathematical ideas to real-world problems like that described in Example 1.

EXAMPLE 1

Suppose you take a job after graduation at a salary of $18,000 per year. Your employer promises you raises of at least 6% each year for as long as you stay with the firm.

a. If you keep your job, what will be your salary in your fourth year?
b. How will the purchasing power of your salary in the fourth year compare with that in your first year, if there is a constant 4% inflation over the period?

Solution

a. Each year's salary is the previous year's salary plus 6% of that salary. Thus, in the second year, you will earn

$$18,000 + 6\% \text{ of } 18,000$$

To compute 6% of 18,000, we multiply $0.06 \times 18,000$. On most scientific calculators you may use the $\boxed{\%}$ key to compute 6% of 18,000 by

1

pressing the key sequence 18000 $\boxed{\times}$ 6 $\boxed{\%}$. From here on, we shall indicate such a calculation in the following form:

Expression	Key Sequence	Display
6% of 18,000	18000 $\boxed{\times}$ 6 $\boxed{\%}$	1080

where $\boxed{}$ indicates a calculator function key.

The table below shows your minimum salary in each of the first four years. (Salaries are rounded to the nearest dollar.)

Year	Raise	Salary
1	0	$18,000
2	6% of 18,000 = 1080	$19,080
3	6% of 19,080 = 1144.80	$20,225
4	6% of 20,225 = 1213.50	$21,439

b. A 4% inflation rate means that prices increase by an average of 4% each year. Then your 6% raise increases your purchasing power by only 2% per year. Computing as in part **a** but with 2% rather than 6%, you would find that your purchasing power in the fourth year is about $19,102, compared with $18,000 in your first year—about $1,102 greater. ∎

Example 1 uses three concepts from arithmetic—numbers, operations, and the idea of percent. We will review these concepts to see how they are used and extended in algebra.

Whole Numbers, Rational Numbers, Real Numbers

Which kinds of numbers will be helpful in a particular problem depends on the problem situation itself. In arithmetic, you used **whole numbers**. These are the numbers 0, 1, 2, 3, 4, ... The ". . ." means that this pattern continues indefinitely. There is no largest whole number, because we can always add 1 to any particular whole number to get the next whole number.

The whole numbers are **ordered**. That is, we can always determine whether a particular whole number is greater than, equal to, or less than another whole number. Moreover, we can use a **number line** to picture the whole numbers in the proper order. We construct the number line in Figure 1.1 as follows:

> **To construct a number line**
>
> 1. Draw a horizontal line, and add an arrowhead pointing to the right.
> 2. Choose and mark a convenient point on the line to represent 0. This point is called the **origin.**
> 3. Decide on a unit length, and repeatedly mark off this length on the line, beginning at 0 and moving to the right.
> 4. Label the resulting points 1, 2, 3, and so on, in order.

FIGURE 1.1

A number line marked in this way is a **graph,** or geometric representation, of the whole numbers.

A whole number a is **less than** a whole number b (written $a < b$) if a appears to the left of b on the number line. If $a < b$, we also say b is **greater than** a and write $b > a$. Furthermore, "$a \le b$" means that either $a < b$ or $a = b$; similarly, "$a \ge b$" means that either $a > b$ or $a = b$.

The statement "$a = b$" is an **equation.** It means that the expression on the left of "$=$" represents the same number as the expression on the right.

Numbers that can be written in the form $\frac{a}{b}$, where a and b are whole numbers with b not zero, are called nonnegative **rational numbers.** (In Chapter 2, we will introduce negative numbers as well.) The rational numbers include *fractions* such as $\frac{2}{3}$, $\frac{7}{4}$, and $\frac{200}{52}$, the whole numbers, and *decimals* like 0.15 and 143.7.

Any rational number can be written as a fraction. For example:

$$0.6 = \frac{6}{10}$$

$$4 = \frac{4}{1}$$

$$5.7 = \frac{57}{10}$$

$$0 = \frac{0}{1}$$

Like whole numbers, rational numbers can be placed on a number line, as you will see in Example 2.

Some useful numbers, such as $\sqrt{2}$ and π, cannot be written as fractions. So these are not rational numbers; they are, in fact, called *irrational numbers*. Yet they too can be placed on a number line. The positive irrational numbers, along with all the nonnegative rational numbers, make up the nonnegative **real numbers.** For that reason Figure 1.1 is sometimes called the *nonnegative* real number line.

Example 2 illustrates how to place rational numbers and other real numbers on a number line.

EXAMPLE 2   Draw a number line as in Figure 1.1. Find the points on the number line that correspond to

   a. $\dfrac{4}{5}$ **b.** 1.7   **c.** $\dfrac{22}{7}$ **d.** $\sqrt{2}$

Solution   **a.**   A number line is drawn in Figure 1.2. The number $\dfrac{4}{5}$ is between 0 and 1, so we begin by dividing the segment from 0 to 1 into five equal parts. The length of each part is $\dfrac{1}{5}$, so $\dfrac{4}{5}$ corresponds to the fourth point to the right of 0.

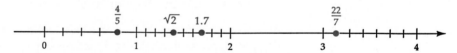

FIGURE 1.2

b.   The number 1.7 is between 1 and 2, so we divide the segment from 1 to 2 into 10 equal parts. The seventh point to the right of 1 corresponds to 1.7.

c.   Dividing 7 into 22 gives $3\dfrac{1}{7}$. Thus, $\dfrac{22}{7}$ is between 3 and 4. We divide the segment from 3 to 4 into 7 equal parts. The first point to the right of 3 corresponds to $\dfrac{22}{7}$.

d. $\sqrt{2}$ is about 1.4. You can verify this with a calculator by entering 2 followed by $\boxed{\sqrt{}}$. Thus, using the points we marked in part b, we can place $\sqrt{2}$ at (approximately) the fourth point to the right of 1.   ■

Operations with fractions

Here are the rules for performing the basic operations with fractions. The raised dot indicates multiplication; for example, $a \cdot n = a \times n$.

Operations with fractions

a. Simplifying: If n is not zero, then

$$\frac{a \cdot n}{b \cdot n} = \frac{a}{b}$$

b. Multiplying: $\dfrac{a}{b} \cdot \dfrac{c}{d} = \dfrac{a \cdot c}{b \cdot d}$

c. Dividing: $\dfrac{a}{b} \div \dfrac{c}{d} = \dfrac{a}{b} \cdot \dfrac{d}{c} = \dfrac{a \cdot d}{b \cdot c}$ (Changing $\dfrac{c}{d}$ to $\dfrac{d}{c}$ is called *inverting the divisor.*)

d. Adding: $\dfrac{a}{b} + \dfrac{c}{d} = \dfrac{a \cdot d + b \cdot c}{b \cdot d}$

e. Subtracting: $\dfrac{a}{b} - \dfrac{c}{d} = \dfrac{a \cdot d - b \cdot c}{b \cdot d}$

A fraction $\dfrac{a}{b}$ is said to be in **simplified or reduced form** if there are no whole numbers that are divisors of both the **numerator** a and the **denominator** b. For example, $\dfrac{2}{3}$ is in simplified form, but $\dfrac{6}{8}$ is not. To simplify $\dfrac{6}{8}$, we note that

$$\frac{6}{8} = \frac{3 \cdot 2}{4 \cdot 2} \qquad \text{Showing a common divisor}$$

$$= \frac{3}{4} \qquad \text{Simplifying}$$

EXAMPLE 3

Perform the indicated operations and write the result in simplified form.

a. $12 \div \dfrac{2}{3}$ b. $1\dfrac{1}{12} - \dfrac{1}{4}$

Solution

a. We begin by writing 12 as a fraction.

$$12 \div \frac{2}{3} = \frac{12}{1} \div \frac{2}{3}$$

$$= \frac{12}{1} \cdot \frac{3}{2} \qquad \text{Inverting the divisor}$$

$$= \frac{12 \cdot 3}{1 \cdot 2} \qquad \text{Multiplying numerators and denominators}$$

$$= \frac{36}{2} \qquad \text{Multiplying}$$

$$= 18 \qquad \text{Dividing 36 by 2}$$

b. We begin by noting that $1\frac{1}{12} = 1 + \frac{1}{12}$, and so we add

$$1 + \frac{1}{12} = \frac{1}{1} + \frac{1}{12} \qquad \text{Writing 1 as a fraction}$$

$$= \frac{1 \cdot 12 + 1 \cdot 1}{1 \cdot 12} \qquad \text{Using the addition rule}$$

$$= \frac{12 + 1}{12} \qquad \text{Multiplying}$$

$$= \frac{13}{12} \qquad \text{Adding}$$

Now we use $\frac{13}{12}$ instead of $1\frac{1}{12}$ in the subtraction:

$$1\frac{1}{12} - \frac{1}{4} = \frac{13}{12} - \frac{1}{4} \qquad \text{Substituting } \frac{13}{12} \text{ for } 1\frac{1}{12}$$

$$= \frac{13 \cdot 4 - 12 \cdot 1}{12 \cdot 4} \qquad \text{Using the subtraction rule}$$

$$= \frac{52 - 12}{48} \qquad \text{Multiplying}$$

$$= \frac{40}{48} \qquad \text{Subtracting}$$

$$= \frac{5 \cdot 8}{6 \cdot 8} \qquad \text{Showing a common divisor}$$

$$= \frac{5}{6} \qquad \text{Simplifying}$$

■

Percents

The idea of percent is important in real-world applications of arithmetic and algebra, as we saw in Example 1. *Percent* (abbreviated %) means simply "per hundred." Thus, if n is any number,

$$n\% = \frac{n}{100} = n \div 100$$

Percents can be rewritten as either decimals or fractions; conversely, every decimal and fraction is equivalent to a number in percent form.

EXAMPLE 4

Fill in the blanks. Round decimals to the nearest thousandth.

Fraction	$\dfrac{5}{6}$	?	?
Decimal	?	0.316	?
Percent	?	?	$16\frac{2}{3}\%$

Solution

First column:

To change $\dfrac{5}{6}$ to a decimal, we use the fact that the fraction bar means "divided by." Hence, we divide 6 into 5. The result is 0.8333333.... To round this to the nearest thousandth, we note that the digit in the ten-thousandths place (the fourth digit to the right of the decimal point) is less than 5. Hence, we remove it and every digit to its right, so, to the nearest thousandth, $\dfrac{5}{6} = 0.833$.

To change 0.833 to a percent, we multiply it by 100: $0.833 \cdot 100 = 83.3$. Thus, $0.833 = 83.3\%$.

Second column:

As a fraction, 0.316 is

$$\frac{316}{1000} = \frac{79 \cdot 4}{250 \cdot 4} = \frac{79}{250}$$

As a percent, 0.316 is $(0.316 \cdot 100)\% = 31.6\%$.

Third column:

As a fraction,

$$16\frac{2}{3}\% = \frac{50}{3}\%$$

$$= \frac{50}{3} \div 100$$

$$= \frac{50}{3} \cdot \frac{1}{100}$$

$$= \frac{50}{300}$$

$$= \frac{1}{6}$$

To find the equivalent decimal, we note that 1 divided by 6 is 0.1666666.... Since the digit in the ten-thousandths place is 6, which is greater than 5, we round up to 0.167. (We would do the same thing if the digit in the ten-thousandths place were 5.) The completed table is

Fraction	$\frac{5}{6}$	$\frac{316}{1000}$	$\frac{1}{6}$
Decimal	0.833	0.316	0.167
Percent	83.3%	31.6%	$16\frac{2}{3}\%$

∎

Algebraic Expressions

In arithmetic, we deal with strings of numbers connected by operations, called *arithmetic* or *numerical expressions;* examples are

$$7 + 8 \qquad 1.8 - 4 + 5.2$$

In algebra we use these along with letter symbols, called **variables,** that represent any number or sometimes particular unknown numbers. A string of numbers, operations, and variables is called an **algebraic expression.** Two examples of algebraic expressions are

$$5x + 10 \qquad r^2 - 2rs + 8s^2$$

where $5x = 5 \cdot x$ and $2rs = 2 \cdot r \cdot s$. Variables and algebraic expressions are the basis for the language of algebra, higher mathematics, and mathematics-related fields such as engineering and computer science. Example 5 illustrates the use of a variable n.

EXAMPLE 5   A solution is 86% acid. Make a table showing the amount of acid in 10, 20, and 30 gallons (gal) of the solution. Generalize the pattern to show the amount of acid in any number n of gallons.

Solution   To compute 86% of 10, multiply 0.86 by 10. On a calculator you may use the $\boxed{\%}$ key to compute 86% of 10 by pressing the following keys in the given sequence:

Expression	*Key Sequence*	*Display*
86% of 10	10 $\boxed{\times}$ 86 $\boxed{\%}$	8.6

We do the same for 20, 30, and n gal, obtaining the following table.

Gallons of Solution	Gallons of Acid
10	$0.86(10) = 8.6$
20	$0.86(20) = 17.2$
30	$0.86(30) = 25.8$
n	$0.86n$

In the table, $0.86n$ means the product of 0.86 and the number n. We may also write this product as $0.86 \times n$ or $0.86 \cdot n$ or $(0.86)(n)$. We would not usually write it as $n0.86$. ■

Exponentiation

An important operation in arithmetic and in algebra is **exponentiation,** or raising a number to a power. When the power is a positive whole number, exponentiation is shorthand for repeated multiplication, as you can see in the following definition:

The *n***th power of *a*,** written a^n, means

$$a^n = a \cdot a \cdot a \cdot \ldots \cdot a$$
$$n \text{ factors}$$

Here n represents a positive whole number and a represents any number; in a^n, n is called the **exponent,** and a is called the **base.** For example, 3^4 is read "the fourth power of 3"; 3 is the base, and 4 is the exponent.

For any number a, a^2 is read "a squared" or "the square of a," and a^3 is read "a cubed" or "the cube of a." Furthermore, a^1 is understood to be a. So $1.6^1 = 1.6$ and $45^1 = 45$. Example 6 further illustrates the meaning of exponents.

EXAMPLE 6   Write each of the following as a whole number. Use paper and pencil for parts a and b, and use a calculator for parts c and d.

   a. 26^2 **b.** 12^3 **c.** 5^{10} **d.** 13^6

Solution   **a.** In 26^2, 26 is the base and 2 is the exponent. So, by definition,

$$26^2 = 26 \cdot 26 = 676$$

   b. In 12^3, 12 is the base and 3 is the exponent. So, by definition,

$$12^3 = 12 \cdot 12 \cdot 12 = 1728$$

The $\boxed{y^x}$ key on your calculator raises a number to a power. (On some calculators the exponentiation key is $\boxed{x^y}$.) Use the key sequences below to solve parts c and d.

	Expression	Key Sequence	Display
c.	5^{10}	5 $\boxed{y^x}$ 10 $\boxed{=}$	9765625
d.	13^6	13 $\boxed{y^x}$ 6 $\boxed{=}$	4826809

■

DISCUSSION QUESTIONS 1.1

1. Classify each of the following numbers as a real number, a rational number, or a whole number (each number may be classified as two or more of these). Use your calculator for help in classifying $\sqrt{5}$ and $\sqrt{16}$.

$$31.7, \; 4{,}509, \; \sqrt{5}, \; \sqrt{16}, \; \frac{45}{9}, \; \frac{2}{3}, \; 145\%, \; 0, \; \frac{15}{10}$$

2. In Figure 1.3, think of circles 1, 2, and 3 as each containing either all whole numbers, all nonnegative rational numbers, or all nonnegative real numbers. Match each of these kinds of numbers to the correct circle, noting that some circles fall completely within others.

3. What kind of number must n be in Example 5? What is the least value n could have? Why?

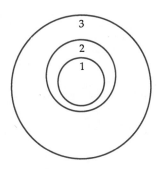

FIGURE 1.3

4. Write "five to the sixth power." Name the base and the exponent.

5. Write "four cubed" using numbers and find its value.

EXERCISES 1.1

In Exercises 1–5, match the given number with a point on the number line of Figure 1.4.

1. $\dfrac{3}{2}$ **2.** $\dfrac{3}{10}$ **3.** 2.15 **4.** 0.3 **5.** 300%

FIGURE 1.4

In Exercises 6–11, round to the nearest hundred.

6. 239 **7.** 470 **8.** 818.72 **9.** 1049.5 **10.** 4650 **11.** 68,358

In Exercises 12–16, round to the nearest thousandth.

12. 0.8256 **13.** 0.27511 **14.** 5.2396783 **15.** 1.33333333 **16.** 0.9995

In Exercises 17–33, perform the indicated operations and give results in simplified form.

17. $\dfrac{6}{7}\cdot\dfrac{2}{3}$ **18.** $2\dfrac{2}{3}\cdot\dfrac{3}{4}$ **19.** $\dfrac{5}{6}\cdot4\dfrac{2}{3}$ **20.** $1\dfrac{1}{4}\cdot3\dfrac{3}{8}$ **21.** $\dfrac{1}{2}\div\dfrac{2}{3}$ **22.** $\dfrac{2}{5}\div\dfrac{3}{4}$

23. $2\dfrac{1}{4}\div3$ **24.** $4\div1\dfrac{1}{4}$ **25.** $\dfrac{1}{5}+\dfrac{3}{5}$ **26.** $\dfrac{2}{4}+\dfrac{3}{4}$ **27.** $6\dfrac{1}{4}+8\dfrac{1}{3}$ **28.** $9\dfrac{2}{3}+7\dfrac{5}{6}$

29. $\dfrac{5}{8}-\dfrac{2}{5}$ **30.** $\dfrac{1}{3}-\dfrac{1}{4}$ **31.** $2\dfrac{1}{2}-1\dfrac{4}{5}$ **32.** $18-11\dfrac{1}{8}$ **33.** $5\dfrac{1}{2}-3\dfrac{2}{3}$

In Exercises 34 and 35, complete the table. Round decimals to the nearest thousandth.

34.

Common fraction	$\dfrac{1}{2}$	$\dfrac{1}{3}$	?	$\dfrac{3}{4}$	?	?	$\dfrac{3}{5}$
Decimal	0.5	?	?	0.75	?	0.04	?
Percent	50%	$33\dfrac{1}{3}\%$	$66\dfrac{2}{3}\%$	?	20%	?	?

35.

Common fraction	?	$\dfrac{1}{8}$	?	?	$\dfrac{7}{8}$	?	?
Decimal	?	?	0.375	?	?	?	1.0
Percent	$16\dfrac{2}{3}\%$	?	?	$62\dfrac{1}{2}\%$	?	$83\dfrac{1}{3}\%$	?

In Exercises 36–46, convert the number to a percent.

36. 0.07 **37.** 2.1 **38.** 0.14 **39.** $\dfrac{3}{10}$ **40.** $\dfrac{7}{20}$ **41.** 0.195

42. $\dfrac{18}{50}$ **43.** $\dfrac{7}{5}$ **44.** 2.65   **45.** $\dfrac{21}{30}$ **46.** 0.763

In Exercises 47–53, convert the percent to a decimal.

47. 36%   **48.** 2%   **49.** $\dfrac{1}{2}$%   **50.** 295%   **51.** 0.9%   **52.** 5.4%   **53.** 5.23%

In Exercises 54–61, find the percentage.

54. 9% of 120   **55.** 140% of 1.3   **56.** 100% of $\dfrac{1}{4}$ **57.** 77.4% of 270.9

58. 9.5% of 2000   **59.** $10\dfrac{1}{2}$% of 180   **60.** $1\dfrac{1}{2}$% of 378   **61.** 256% of 15

In Exercises 62–69, write the expression as a single rational number. A calculator will be needed for at least some.

62. 3^4 **63.** 4^3 **64.** 0.8^3 **65.** 0.7^2 **66.** 4^{10} **67.** 10^5 **68.** 2^{15} **69.** 26^5

70. A druggist has 10,000 milliliters (ml) of cough syrup in a large container. He sells cough syrup in 25-ml bottles for $2.50, and in 125-ml bottles for $9.
   a. How many 25-ml bottles of cough syrup can the druggist make?
   b. How many 125-ml bottles of cough syrup can the druggist make?
   c. How much money can the druggist make if he sells 125-ml bottles of cough syrup?
   d. Does he make more money selling the cough syrup in 25-ml bottles or in 125-ml bottles? How much more?

71. The table below relates the setting of the cruise control on a car (which keeps it moving at a constant speed), the time spent traveling on a freeway, and the distance traveled. Use the equation, $d = rt$, where d is the distance traveled, r is the rate of travel, and t is the time spent traveling to complete the table.

Cruise Setting (miles per hour)	Time (hours)	Distance (miles)
55	8	?
62	?	223.2
?	10.5	525
65	t	?
r	12	?
68	?	d
?	9	d
r	t	?

72. The required textbook for a course is available hardbound for $36.00 and paperbound for $21.95. At the end of the year, used hardbound texts in good condition can be resold to the bookstore for 50% of the purchase price, and used paperbound books for 30% of the purchase price. Assuming that you take good care of your books and resell them at the end of the year, which is the more economical to buy?

Preparing for Section 1.2

In Exercises 73–76, perform each computation.

73. $2.61 + 0.765 + 1.83$ **74.** $7.64 - 0.87$ **75.** $0.782 \cdot 1.2$ **76.** $52.48 \div 2.5$

In Exercises 77–80, complete each statement.

77. When numbers are added, the answer is called the __?__.

78. When two numbers are subtracted, the answer is called the __?__.

79. In $15 \div 5 = 3$, the divisor is __?__ and the quotient is __?__.

80. Since $46 = 23 \cdot 2$, we say that 2 and 23 are __?__ of 46 and that 46 is the __?__ of 23 and 2.

1.2   Order of Operations

FOCUS

We continue our review, discussing the order in which operations are to be carried out in an arithmetic or algebraic expression.

The meaning of an arithmetic or algebraic expression can be changed by the order in which we perform indicated operations. And the order of operations can be changed by adding or removing any of the following *grouping symbols*.

Name	Parentheses	Brackets	Braces
Symbol	( )	[ ]	{ }
Example	$8 - (2 + 4)$	$2 + 3[5 - 2]$	$7\{15 - 4\}$

To make sure that every expression means the same thing to everyone, mathematicians have agreed on the following order for performing operations.

Order in which operations should be performed

1. Perform the operations within the innermost grouping symbol, using the order prescribed in steps 2 to 4 to remove that grouping symbol. Remove all other grouping symbols, working from inner to outer, in the same way. Then continue with steps 2 to 4.
2. Perform all indicated exponentiations, working from left to right.
3. Perform all indicated multiplications and divisions, working from left to right.
4. Perform all indicated additions and subtractions, working from left to right.

The next two examples are applications of these rules.

EXAMPLE 1   Compute:

a. $6 + 7^3$ **b.** $(6 + 7)^3$ **c.** $5 + 3^5 - 6 \cdot \dfrac{5}{2}$

Solution

a. There are no grouping symbols, so first raise 7 to the third power:

$$6 + 7^3 = 6 + 343$$

Then add: $6 + 343 = 349$

b. First do the operation indicated in the parentheses:

$$(6 + 7)^3 = (13)^3$$

Then raise the result to the third power:

$$(13)^3 = 2197$$

c. $5 + 3^5 - 6 \cdot \dfrac{5}{2} = 5 + 243 - 6 \cdot \dfrac{5}{2}$ Raising 3 to the fifth power

$\qquad\qquad = 5 + 243 - 15$ Multiplying 6 times $\dfrac{5}{2}$

$\qquad\qquad = 248 - 15 = 233$ Adding and subtracting, left to right   ∎

When an expression has more than one level of grouping symbols, they are removed in order from the inside out.

EXAMPLE 2

Compute: $\quad 6 \cdot 3^2 + [5(7 - 4)] \div \dfrac{5}{2}$

Solution

Since the parentheses are the innermost grouping symbols, we remove them first by subtracting 4 from 7, getting

$$6 \cdot 3^2 + [5 \cdot 3] \div \frac{5}{2}$$

The brackets remain, so we next multiply $5 \cdot 3$ to eliminate them:

$$6 \cdot 3^2 + 15 \div \frac{5}{2}$$

There are no more grouping symbols, so we perform the exponentiation 3^2, getting

$$6 \cdot 9 + 15 \div \frac{5}{2}$$

Multiplication and division come before addition. Recall that

$$15 \div \frac{5}{2} = 15 \cdot \frac{2}{5}$$

so we have

$$6 \cdot 9 + \frac{30}{5} = 54 + 6$$

Finally, the sum of 54 and 6 is 60. ∎

BE CAREFUL!

$5 \cdot 6^4 = 5 \cdot 1296 = 6480$, **but** $(5 \cdot 6)^4 = 30^4 = 810{,}000$.
$5 + 6^4 = 5 + 1296 = 1301$, **but** $(5 + 6)^4 = 11^4 = 14{,}641$.
Doing the operations in the proper order will keep you from making errors.

Estimating Results

One way to guard against errors, whether on a calculator or with paper and pencil, is to get into the habit of estimating the result either before or after you perform the computation. One way to estimate is as follows:

1. Round the numbers in the problem to numbers that you can manipulate.
2. Use the rounded numbers to solve the problem mentally.

EXAMPLE 3   A woman traveled 672 miles (mi) by car in 14 hours (h) at approximately constant speed. Without computing, which choice would you estimate to be closest to her speed?

20 mph   50 mph   80 mph   400 mph

Solution   The best choice is 50, since 672 is about 700 and $700 \div 14 = 50$. (The exact solution is 48.) ■

Averages

Another type of estimate is the average of a collection of numbers. The **average** or **arithmetic mean** of n numbers is the sum of the n numbers divided by n. The average of a collection of numbers always lies between the smallest and the largest of the numbers. It is a rough estimate for every number in the collection. For example, we use the average income of a country to roughly describe the level of income of all its people.

EXAMPLE 4   A student scored 82, 75, 93, 84, and 71 on five quizzes in English class. What is her average quiz score?

Solution   First we add: $82 + 75 + 93 + 84 + 71 = 405$

Since there are five numbers ($n = 5$), we divide the sum by 5:

$$405 \div 5 = 81$$

Her average quiz score is 81. This also means that her five-quiz total, 405, is the same as if she had scored 81 on all five quizzes. ■

DISCUSSION QUESTIONS 1.2

1. What operation should be done first in each of the following?
   a. $10 + 2 \times 9$ **b.** $(10 + 2) \times 9$ **c.** $2 \times (9 + 10)$

2. Which of the following pairs of expressions are equal?
   a. $\left(\frac{1}{2} \cdot \frac{2}{3}\right) \cdot \frac{1}{4}$ and $\frac{1}{2} \cdot \left(\frac{2}{3} \cdot \frac{1}{4}\right)$ **b.** $(10 + 8) - 2$ and $10 + (8 - 2)$

   c. $(12 - 6) - 4$ and $12 - (6 - 4)$ **d.** $\left(\frac{2}{5} \div \frac{1}{2}\right) \div \frac{1}{4}$ and $\frac{2}{5} \div \left(\frac{1}{2} \div \frac{1}{4}\right)$

3. How is the average of a set of numbers computed? Can the average be greater than the largest number in the set? Smaller than the smallest number in the set? Explain.

EXERCISES 1.2

In Exercises 1–31, perform the indicated operations, giving the result as a single rational number.

1. 4^5

2. $\left(\dfrac{1}{2}\right)^3$

3. $(2 - 0.3)^2$

4. $6 \div 3 \cdot 4 - 1$

5. $8 \div 2 + 6 \cdot 3$

6. $\dfrac{1}{5} \div \left(\dfrac{2}{3} \cdot \dfrac{1}{2}\right)$

7. $\dfrac{5}{8} \cdot \left(\dfrac{1}{4} \div \dfrac{25}{12}\right)$

8. $\left(\dfrac{3}{2} \cdot 6\right) \div \dfrac{9}{20}$

9. $\left(0 \div \dfrac{1}{4}\right) \cdot 5$

10. $5 \cdot \left(\dfrac{1}{2} \div \dfrac{1}{3}\right) \cdot 2$

11. $\left(\dfrac{1}{4} \cdot \dfrac{3}{4}\right) \div \left(\dfrac{1}{3} \div \dfrac{2}{3}\right)$

12. $(2.1 \cdot 0.2) + (0.8 \div 0.2)$

13. $(4.12 \div 4) + (0.06 \cdot 3)$

14. $2.7 + 40.8 \div 8$

15. $0.375 \div 0.5 + 8$

16. $12.8 - (3.9 \cdot 2) + 7.6$

17. $5^4 + 4^5 \cdot 3^4$

18. $32 + 6^3 \, (3 \cdot 2^4)$

19. $1.2(3.6 + 5.7 \cdot 1.5) - 3.4$

20. $8.9 + 3.5(6.9 + 4.2)$

21. $(10 + 2)^2 \div 36 + 4$

22. $4 \cdot 30 - 3(7 - 3)^2$

23. $2[4 - 6 \div (17 - 3 \cdot 5)]$

24. $120 \div 2[14 - 2(5 - 3)]$

25. $\dfrac{2}{3}\left[\dfrac{1}{4} + \dfrac{3}{4}(12 - 4)\right] \div \dfrac{1}{2}$

26. $0.4\left[\dfrac{1}{3} + \dfrac{2}{3}\left(\dfrac{1}{6} + \dfrac{5}{12}\right)\right] \div \dfrac{1}{3}$

27. $3\{[4 - (6 - 5)] + 5 \cdot 6\}$

28. $\{5[6 - (8 - 5)] + 13\} \div 7$

29. $10.6 - \{2.4 + 5.2[4.6 - (1.8 + 2.8)]\}$

30. $\{1.5 \div 0.5 \cdot 1.4 + [2.3 - 1.5(4.1 - 3.8)] \div 0.2\}$

31. $3.6 \div \{1.6[4.2(1.8 - (4.9 - 4.6))]\}$

In Exercises 32–37, compute the average of the given collection of numbers.

32. 8, 12, 13

33. 14, 19, 21

34. 1.3, 0.24, 3.7, 0.26

35. 0.29, 0.11, 0.33, 0.47

36. 30.4, 45.2, 92.7, 26.9, 14.6, 23.5

37. 126, 87, 72, 21, 183, 207, 44, 65, 150, 45

In Exercises 38–41, choose from a, b, or c the number you *estimate* to be nearest to the average of the given collection of numbers. Your results in Exercises 36–41 may be of help.

38. 80, 120, 130   **a.** 90   **b.** 110   **c.** 150

39. 1.4, 1.9, 2.1   **a.** 2.1   **b.** 1.5   **c.** 1.9

40. 13, 2.4, 37, 2.6   **a.** 25   **b.** 18   **c.** 35

41. 12.6, 8.7, 7.2, 2.1, 18.3, 20.7, 4.4, 6.5, 15, 4.5

   a. 11   **b.** 20   **c.** 50

42. A student took eight math quizzes during the semester. The sum of all his quiz scores was 112. Which of the following is the average of his quiz scores?
   a. $112 \cdot 8$ **b.** $112 - 8$ **c.** $8 + 112$ **d.** $112 \div 8$

43. A teacher earns $22,400 a year. Which of the following is his average monthly salary?

a. $22,400 - 12,000$ **b.** $\dfrac{22,400}{12}$

c. $(22,400 \cdot 12) \div 12$ **d.** $22,400 \cdot 12$

In Exercises 44–46, match the problem with the one of the following computations that would solve it:

a. $(2 + 4) \cdot \dfrac{3}{4}$ **b.** $4 + 2 \cdot \dfrac{3}{4}$ **c.** $2 + 4 \cdot \dfrac{3}{4}$ **d.** $2 \cdot \left(4 + \dfrac{3}{4}\right)$

44. For a holiday, Jill bought 3 boxes of candy, one each for her mother and her two brothers. Her mother's box weighed 4 pounds (lb) and her brothers' boxes weighed $\dfrac{3}{4}$ lb each. How much did the 3 boxes weigh altogether?

45. A firefighter stood on the second rung of a ladder. He went up 4 more rungs. The distance between consecutive rungs is $\dfrac{3}{4}$ foot (ft) and the first rung is $\dfrac{3}{4}$ ft above the ground. How far is the firefighter above the ground?

46. Alan bought six 1-lb bags of apples on Thursday. By Saturday, his roommates had eaten one-fourth of each of four of the bags. How many pounds of apples were left?

47. A company orders 12 cases of candy bars. Each case contains equal amounts of 3 different kinds of candy bars: brand A, brand B, and brand C. If there are a total of 72 candy bars in each case, how many of each kind did the company order?

In Exercises 48–54, estimate by rounding and then choose the closest given answer. Do *not* compute the exact answer.

48. The atomic weight of copper is 64 and the atomic weight of gold is 196. How much more is the atomic weight of gold than the atomic weight of copper?

a. 260   **b.** 60   **c.** 160   **d.** 140

49. The Mathematics Department secretary can type 82 words per minute. It takes him 48 minutes of typing to finish a certain report. How many words does the report contain?

a. 130   **b.** 4000   **c.** 3200   **d.** 4800

50. It is 279 mi from Chicago to Davenport and 164 mi from Davenport to Des Moines. How many miles is it from Chicago to Des Moines via Davenport?

    a. 600   **b.** 330   **c.** 400   **d.** 440

51. Last month a used car dealer made a profit of $38,703. She made an average profit of $1843 on each car sold. How many cars did she sell last month?

    a. 20   **b.** 30   **c.** 35   **d.** 50

52. The local library has 53,614 books. It plans to buy 16,295 new books next year. How many books will be in the library after next year?

    a. 66,000   **b.** 70,000   **c.** 61,000   **d.** 78,000

53. A student pays $287 per month rent. How much rent will she pay in 1 year?

    a. $4000   **b.** $3600   **c.** $2400   **d.** $1200

54. Rosa had $1418 in her bank account. She withdrew $659 from her account to pay a dentist bill. How much remained in Rosa's bank account?

    a. $600   **b.** $700   **c.** $900   **d.** $460

Preparing for Section 1.3

55. Draw a 90° angle.

56. Draw a square.

57. Draw a rectangle that is not a square.

58. What are perpendicular lines?

59. Draw a right triangle.

1.3   Geometric Formulas

FOCUS

We discuss and use formulas for the area, perimeter, and volume of several geometric figures. These formulas provide some early practice in the use of variables, as well as a tool for solving problems involving geometry.

Areas and Perimeters

The **area** of a closed two-dimensional geometric figure is a measure of the size of the enclosed region. The greater the area of a wall, for example, the more paint will be needed to paint it, and the more plaster to cover it. Given two pieces of cloth from the same bolt, the one with the greater area will cost more. Area is measured in square units, such as square inches or square meters.

The **perimeter** of a closed two-dimensional geometric figure is the total distance around it. A circle is an exception; the distance around a circle is called its **circumference.** A perimeter or circumference is measured in units of length, such as inches or meters.

Formulas for the area A and perimeter P (or circumference C) of some important figures are shown in Table 1.1. The letters in these formulas are *variables,* but because they are measures of lengths or areas, they may be assigned only positive values. When two variables are placed together with no operation sign between them, they are to be multiplied. Thus, in these formulas, $lw = l \cdot w$ and $bh = b \cdot h$.

EXAMPLE 1

Find the area and perimeter of a rectangle with

a.   Length = 9 ft, width = 6 ft
b.   Length = 15 ft, width = 4 yards (yd)

Solution

a.   Substituting 9 for l and 6 for w in the formulas for area and perimeter of a rectangle, we obtain

$$A = lw = 9 \text{ ft} \cdot 6 \text{ ft}$$
$$= 54 \text{ square feet (written ft}^2)$$

$$P = 2l + 2w = 2 \cdot 9 \text{ ft} + 2 \cdot 6 \text{ ft}$$
$$= 18 \text{ ft} + 12 \text{ ft}$$
$$= 30 \text{ ft}$$

Note that computing the area involves multiplying two numbers with the dimension feet (that is, feet × feet), and the result has the unit square feet (that is, ft^2). The computation of the perimeter involves doubling numbers with the dimension feet and then adding them, and the result is in feet.

b.   In this case, the length and width are given in different units. We must convert one of them to the unit of the other before we can use the formulas. There are 3 ft in 1 yd, so

$$15 \text{ ft} = (15 \text{ ft}) \div (3 \text{ ft/yd}) = 5 \text{ yd}$$
$$4 \text{ yd} = (4 \text{ yd}) \times (3 \text{ ft/yd}) = 12 \text{ ft}$$

There are two ways to compute the area:

$$A = lw \qquad\qquad A = lw$$
$$= 15 \text{ ft} \cdot 12 \text{ ft} \qquad = 5 \text{ yd} \cdot 4 \text{ yd}$$
$$= 180 \text{ ft}^2 \qquad\qquad = 20 \text{ yd}^2$$

TABLE 1.1   Area and Perimeter Formulas

Figure	Area	Perimeter	Definitions
Rectangle 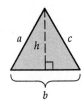	$A = lw$	$P = 2l + 2w$	l = length w = width
Triangle	$A = \dfrac{1}{2}\,bh$	$P = a + b + c$	b = base h = height to b a, c = sides
Parallelogram	$A = bh$	$P = 2a + 2b$	b = base h = height a = side
Circle	$A = \pi r^2$	$C = 2\pi r$	r = radius C = circumference

Similarly, there are two ways to compute the perimeter:

$$P = 2l + 2w \qquad\qquad P = 2l + 2w$$
$$= 2 \cdot 15 \text{ ft} + 2 \cdot 12 \text{ ft} \qquad = 2 \cdot 5 \text{ yd} + 2 \cdot 4 \text{ yd}$$
$$= 30 \text{ ft} + 24 \text{ ft} \qquad = 10 \text{ yd} + 8 \text{ yd}$$
$$= 54 \text{ ft} \qquad\qquad = 18 \text{ yd} \qquad\blacksquare$$

EXAMPLE 2   Find the area of the triangle in Figure 1.5.

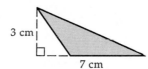

3 cm

7 cm

FIGURE 1.5

Solution   In this case, the height $h = 3$ centimeters (cm) is outside the triangle, and the base $b = 7$ cm is not at the bottom of the triangle. Nevertheless, the area formula $A = \dfrac{1}{2} bh$ applies:

$$A = \frac{1}{2} bh$$
$$= \frac{1}{2} \cdot 7 \text{ cm} \cdot 3 \text{ cm}$$
$$= 10.5 \text{ cm}^2 \qquad\blacksquare$$

There are several important points to recall about circles. First, the term **radius** is used both to name a segment joining the center of a circle to any point on the circle *and* to name the length of such a segment. Second, the **diameter** of a circle is the length of a segment with endpoints on the circle that contains the center of the circle, and the term diameter is also used to name such a segment. Interpreted as a length, the diameter of a circle is twice its radius. Third, the formulas for the area and circumference of a circle contain the irrational number π, which we approximate as $\dfrac{22}{7}$ or 3.14 in pencil-and-paper computations. (We write this approximation as $\pi \approx \dfrac{22}{7}$ or 3.14.) If your calculator has a $\boxed{\pi}$ key, use it instead of one of these estimates. When you use the $\boxed{\pi}$ key, your calculator approximates to 8 or even 14 decimal places, depending on the calculator type.

EXAMPLE 3

Find the circumference of a circle whose radius is 70 inches (in.).

Solution

Substitute 70 for r in the circumference formula. If you are computing with paper and pencil, use $\pi \approx \dfrac{22}{7}$:

$$
\begin{aligned}
C &= 2\pi r \\
  &= 2 \cdot \pi \cdot 70 \text{ in.} \\
  &\approx 2 \cdot \frac{22}{7} \cdot 70 \text{ in.} \\
  &= 440 \text{ in.}
\end{aligned}
$$

If you are using a calculator that has a $\boxed{\pi}$ key, enter this key sequence:

Expression	*Key Sequence*	*Display*
$2\pi \cdot 70$	2 $\boxed{\times}$ $\boxed{\pi}$ $\boxed{\times}$ 70 $\boxed{=}$	439.82297

Your display may differ slightly from this, depending on how your calculator approximates π. However, to the nearest whole number the circumference is 440 in. ∎

EXAMPLE 4

A circle is cut out of a parallelogram as shown in Figure 1.6. Find the area that remains.

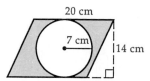

20 cm

7 cm

14 cm

FIGURE 1.6

Solution

The area of the shaded region in Figure 1.6 is equal to the area of the parallelogram less the area of the circle:

$$
\begin{aligned}
\text{Area of parallelogram} &= bh \\
&= 20 \text{ cm} \cdot 14 \text{ cm} \\
&= 280 \text{ cm}^2
\end{aligned}
$$

$$
\begin{aligned}
\text{Area of circle} &= \pi r^2 \\
&\approx \frac{22}{7} \cdot (7 \text{ cm})^2 \\
&= 154 \text{ cm}^2
\end{aligned}
$$

$$
\begin{aligned}
\text{Area of shaded region} &\approx 280 \text{ cm}^2 - 154 \text{ cm}^2 \\
&= 126 \text{ cm}^2
\end{aligned}
$$
∎

The next example requires some thought rather than just substituting numbers into formulas.

EXAMPLE 5 A real estate company recently bought an entire rectangular city block in Chicago. It plans to divide the block into five lots. The two end lots will be of equal size, but each will be twice as large as each of the three equal-sized inner lots. What is the area of each lot if the area of the block is 49,630 ft^2?

Solution At first reading, this appears to be a complex problem. In such cases, a diagram can help to sort out the data. We know that the block is in the form of a rectangle and is to be divided into five lots. Each end lot has twice the area of each inner lot, as in the sketch in Figure 1.7.

FIGURE 1.7

We see that if the end lots, I and V, are divided in half (at the dashed lines), the entire block is then made up of seven equal-sized parts. To find the area of each of the seven lots, we must use division. Since the total area of the block is 49,630 ft^2, each equal part has area

$$49,630 \text{ ft}^2 \div 7 = 7090 \text{ ft}^2$$

This means that the area of each of the three inner lots is 7090 ft^2. Since the end lots have twice the area, each end lot contains 14,180 ft^2. ∎

BE CAREFUL!

Don't confuse the *area* of a figure with its *perimeter* or *circumference*. The area is the measure in square units of the enclosed region. The perimeter or circumference is the distance around the figure in units of length.

A *square* of side s is a rectangle in which $l = w = s$. Substituting s for both l and w, we obtain $A = lw = s^2$ for the area of a square of side s. This is why the symbol s^2 is read ''s squared.''

FIGURE 1.8

Now suppose the area of a square is 16, as in Figure 1.8, and we want to find s, the length of a side. Here s is the positive number for which $s^2 = 16$; and because $4^2 = 16$, we know that $s = 4$. This suggests the idea of a "square root."

> For a number N, such that $N \geq 0$, we say that x is a **square root** of N if $x^2 = N$. We write $x = \sqrt{N}$.

Thus, a square root of 16 is 4, so we write $\sqrt{16} = 4$. In general, the positive square root of a number A is the length of the side of a square with area A. (Later, we will see that $(-4)^2$ is also 16, and so numbers will also have negative square roots.)

EXAMPLE 6

For each of the following values of A, find the square root of A to the nearest tenth. Then draw a square in which each side is $\sqrt{A}$ inches. Use your calculator, if necessary, by keying $A \boxed{\sqrt{\ }}$.

a. $A = 9$ **b.** $A = 7$ **c.** $A = 12$

Solution

Part a can be solved by inspection, but we shall use a calculator for parts b and c.

a.   We note that $3^2 = 9$, so $\sqrt{9} = 3$.

	Expression	Key Sequence	Display	Rounded Result
b.	$\sqrt{7}$	$7 \boxed{\sqrt{\ }}$	2.6457513	2.6
c.	$\sqrt{12}$	$12 \boxed{\sqrt{\ }}$	3.4641016	3.5

Squares with these sides are drawn in Figure 1.9.

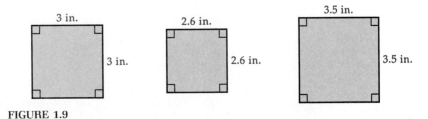

FIGURE 1.9

Volumes and Surface Areas

The **volume** V of a closed solid (three-dimensional) figure is a measure of the amount of space in the interior of the figure and is given in cubic units (such as feet $\times$ feet $\times$ feet, or ft^3). The *surface area S* of a rectangular solid is a measure in square units of the outer surface of the solid. Formulas for the volume V and surface area S of some important figures are shown in Table 1.2.

The formula $S = 2\pi rh + 2\pi r^2$ in the table gives the **total surface area** of a cylinder, including the two circular bases. The area of the curved surface, called the **lateral surface area,** of a right circular cylinder with radius r and height h is given by

$$LA = 2\pi rh$$

We apply these formulas in the next two examples.

EXAMPLE 7

Find the volumes and surface areas of the given geometric solids. (Use the $\boxed{\pi}$ key and round to the nearest hundredth.)

a. Right circular cylinder: $r = 5$ in.; $h = 8$ in.
b. Sphere: $r = 5$ in.

Solution

Substitute the appropriate dimensions into the formulas in each case. Note that V is in cubic units and S is in square units.

a. $V = \pi r^2 h$
$= \pi \cdot (5 \text{ in.})^2 \cdot 8 \text{ in.}$
$\approx 628.32 \text{ in.}^3$

We will compute the lateral surface area LA and the total surface area S.

$LA = 2\pi rh$
$= 2 \cdot \pi \cdot 5 \text{ in.} \cdot 8 \text{ in.}$
$\approx 251.33 \text{ in.}^2$

We use this value for $2\pi rh$ in the formula for S.

$S = 2\pi rh + 2\pi r^2$
$= 251.33 \text{ in.}^2 + 2 \cdot \pi \cdot (5 \text{ in.})^2$
$\approx 251.33 \text{ in.}^2 + 157.08 \text{ in.}^2$
$= 408.41 \text{ in.}^2$

TABLE 1.2 Volume and Surface Area Formulas

Figure	Volume	Surface Area	Definitions
Rectangular Solid 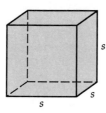	$V = lwh$	$S = 2lh + 2lw + 2wh$	l = length w = width h = height
Cube 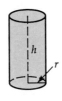	$V = s^3$	$S = 6s^2$	s = side
Right Circular Cylinder	$V = \pi r^2 h$	$S = 2\pi r h + 2\pi r^2$	r = radius h = height
Sphere	$V = \dfrac{4}{3}\pi r^3$	$S = 4\pi r^2$	r = radius

    b. $V = \dfrac{4}{3}\pi r^3$

$\qquad\qquad = \dfrac{4}{3} \cdot \pi \cdot (5 \text{ in.})^3$

$\qquad\qquad \approx 523.60 \text{ in.}^3$

$\qquad\quad S = 4\pi r^2$

$\qquad\qquad = 4 \cdot \pi \cdot (5 \text{ in.})^2$

$\qquad\qquad \approx 314.16 \text{ in.}^2$

∎

DISCUSSION QUESTIONS 1.3

1. State the formula for the area of a rectangle.

2. In some books, $P = 4s$ is given as a formula for the perimeter of a square of side s. Explain why this formula gives the same result for the square as $P = 2l + 2w$.

3. If the radius of a circle is given in centimeters, in what units will the circumference be? The area?

4. At one time there was a bill before the Indiana state legislature that would have made the value of π exactly 3 by law. Discuss the effect such a law could have had on commerce in the state.

EXERCISES 1.3

In Exercises 1–8, find to the nearest tenth the area and perimeter of the rectangle with the given dimensions ($l =$ length and $w =$ width). (*Note:* 1 m = 100 cm.)

1. $l = 7$ in. $w = 5$ in.
2. $l = 2$ m $w = 400$ cm
3. $l = 5$ m $w = 400$ cm

4. $l = 8.3$ ft $w = 6.8$ ft
5. $l = \dfrac{3}{4}$ in. $w = \dfrac{3}{4}$ in.
6. $l = 2\dfrac{1}{2}$ yd $w = 1\dfrac{3}{4}$ yd

7. $l = 1.2$ m $w = 0.8$ m
8. $l = 36.4$ mi $w = 15.9$ mi

In Exercises 9–14, find to the nearest tenth the area and perimeter of the parallelogram with the given dimensions.

9. $a = 11$ in. $b = 23$ in. $h = 9$ in.
10. $a = 9$ cm $b = 8$ cm $h = 7$ cm

11. $a = 7.2$ ft $b = 8.6$ ft $h = 5.9$ ft
12. $a = 6.1$ in. $b = 5.3$ in. $h = 4.6$ in.

13. $a = 3\dfrac{1}{2}$ in. $b = 5\dfrac{1}{4}$ in. $h = 3\dfrac{1}{3}$ in.
14. $a = 114$ ft $b = 98$ ft $h = 100$ ft

In Exercises 15–20, find to the nearest tenth the area of the triangle with the given dimensions.

15. $b = 12.6$ ft $h = 9.4$ ft   **16.** $b = 3.94$ m $h = 12.47$ m   **17.** $b = 2.7$ cm $h = 1.9$ cm

18. $b = 3$ ft $h = 18$ in.   **19.** $b = 1\frac{1}{2}$ yd $h = 2\frac{3}{4}$ ft   **20.** $b = \frac{5}{3}$ ft $h = \frac{2}{3}$ ft

In Exercises 21–24, find the diameter of a circle with the given radius.

21. 10 in.   **22.** 8.5 cm   **23.** $3\frac{1}{4}$ ft   **24.** 5.5 cm

In Exercises 25–28, find the radius of a circle with the given diameter.

25. 26 cm   **26.** 7.68 cm   **27.** $8\frac{1}{5}$ ft   **28.** $4\frac{3}{4}$ in.

In Exercises 29–38, find to the nearest tenth the circumference and the area of the circle with the given dimensions.

29. $r = 4$ cm   **30.** $r = 21$ ft   **31.** $d = 6.5$ in.   **32.** $d = 4.7$ yd   **33.** $r = \frac{7}{5}$ yd

34. $d = 129$ in.   **35.** $r = 6\frac{2}{3}$ ft   **36.** $d = 9.6$ m   **37.** $d = 27$ in.   **38.** $r = 32.5$ cm

In Exercises 39–46, find to the nearest tenth the volume and the surface area of the rectangular solid with the given dimensions. (1 m = 100 cm)

39. $l = 6$ ft $w = 5$ ft $h = 4$ ft   **40.** $l = 1.1$ m $w = 29$ cm $h = 1.7$ m

41. $l = 20.5$ in. $w = 18.6$ in. $h = 12.6$ in.   **42.** $l = 6\frac{1}{2}$ ft $w = 3\frac{1}{3}$ ft $h = 4\frac{3}{5}$ ft

43. $l = 3$ ft $w = 19$ in. $h = 2\frac{1}{4}$ ft   **44.** $l = 1\frac{1}{3}$ yd $w = 4\frac{1}{2}$ ft $h = 2$ ft, 4 in.

45. $l = 2.7$ m $w = 230$ cm $h = 425$ cm   **46.** $l = 16.3$ in. $w = 1.7$ ft $h = 0.8$ yd

47. Find to the nearest tenth the volume and surface area of a sphere with $r = 10$ cm.

48. Find to the nearest tenth the volume, total surface area, and lateral surface area of a right circular cylinder with $r = 14$ in. and $h = 2$ ft. (Compute first using inches and then using feet.)

In Exercises 49–54, the area of a square is given. Find to the nearest tenth the length of its side. (You will need a calculator.)

49. 49 in.^{2} **50.** 36 in.^{2} **51.** 58 ft^2 **52.** 93 ft^2 **53.** 1.9 ft^2 **54.** $42\frac{3}{4}$yd^2

In Exercises 55–60, find the area of the shaded region in the given figure. Assume that the appearance of the figure is accurate.

55.

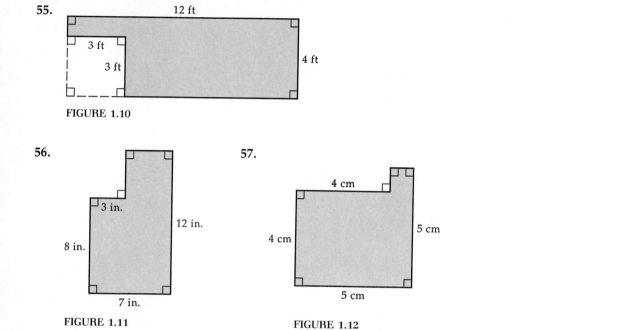

FIGURE 1.10

56.

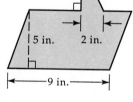

FIGURE 1.11

57.

FIGURE 1.12

58.

FIGURE 1.13

59.

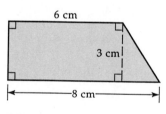

FIGURE 1.14

60.

(*ABCD* is a square.)

FIGURE 1.15

In Exercises 61–70, read the problem carefully, draw a figure illustrating the problem, if needed, and then solve the problem.

61. A city needs to replace part of the synthetic turf on its football field. The field is 100 yd long and 45 yd wide, and each end zone is 10 yd long and as wide as the field. How many square yards of new turf will be required to cover the entire field including the end zones?

62. An airplane crash-landed in a farmer's wheat field and ruined the crop in an approximately rectangular region 21 ft wide by 270 ft long. The pilot agreed to pay damages to the farmer at the rate of $10 per square yard of damaged wheat. How much should the pilot pay the farmer? (*Hint*: Change the dimensions of the damaged region to yards before finding its area.)

63. A painting is 38 cm wide and 55 cm long. The painting is put in a rectangular frame that is 8 cm wide, as shown in Figure 1.16. Find the perimeter of the outside edge of the frame, the area of the painting, and the area of the frame.

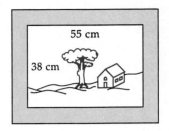

FIGURE 1.16

64. A fence is to be built around a 12-ft by 16-ft rectangular garden.
  a. If the posts are 4 ft apart, how many posts will be needed?
  b. If each post costs $5, how much will the posts cost in all?
  c. The fence will surround all but 4 ft of the garden, which will be closed by a gate. How much fence is needed?
  d. If the fence costs $2 per foot, what is its total cost?
  e. If the gate costs $32.95, what is the total cost of the gate, fence, and posts?

65. A college dormitory is 100 ft wide and 140 ft long. On the fifth floor there are staircases, rest rooms, and a hall that together account for 5000 ft^2 of floor space. The remainder of the fifth floor consists of 30 double rooms of equal size. What is the approximate area (floor space) of each room?

66. A basketball floor is an 84-ft by 50-ft rectangle. The coach has the basketball players run 40 laps around the floor after practice. About how far do they run? Is it more than 1 mi? More than 2 mi? (1 mi = 5280 ft)

67. A basketball is 9.5 in. in diameter. Find its volume and its surface area.

68. A baseball is 8.5 in. around at its greatest circumference. Find its volume and its surface area. (*Hint*: First use $C = 2\pi r$ and your calculator to find r.)

69. A tool box is 0.7 m long, 43 cm wide, and 39 cm high. Find its total surface area.

70. A cabinet maker is making a box that must be 16 in. long and 9 in. wide and must hold 720 in.^{3}. How high should he make it?

71. A box in the shape of a cube has a volume of 34 cm^3. Use your calculator to find the length of the box to the nearest tenth.

72. A circle is drawn inside a parallelogram as in Figure 1.17. State which of the given dimensions are needed, and which are not needed, to find each of the following. Compute each.
   a. The area of the circle
   b. The circumference of the circle
   c. The area of the parallelogram
   d. The perimeter of the parallelogram
   e. The area of the shaded region

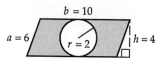

FIGURE 1.17

In Exercises 73 and 74, answer each question that can be answered. If not enough information is given, tell what else is needed.

73. A gallon of paint covers about 480 ft^2 of wall surface. A woman wants to paint four walls in her house. Each wall is 15 ft long and 8 ft high. (Ignore windows and doors.)
   a. What is the area of one wall?
   b. How much paint is needed to paint two walls?
   c. How much paint is needed to cover the four walls with two coats of paint? (Assume that the same amount is needed for each coat.)
   d. How long will it take the woman to paint the four walls?

74. Land in Mingo Junction costs $25 per square meter. A rectangular house with dimensions 35 m by 23 m is to be built in Mingo Junction on a square lot with sides of 54 m.
   a. What is the area of the lot?   **b.** What is the cost of the lot?
   c. What is the area of the land taken up by the house?
   d. How much of the lot will not be covered by the house?

Preparing for Section 1.4

75. What is a variable?

In Exercises 76–82, perform the indicated computations.

76. $6 \cdot (4 + 8)$ **77.** $6 \cdot 4 + 6 \cdot 8$ **78.** $13 \cdot 5 - 13 \cdot 2$

79. $13 \cdot (5 - 2)$ **80.** $2 \cdot (5 \cdot 6)$ **81.** $(2 \cdot 5) \cdot 6$

82. Which of your answers in Exercises 76–81 are equal? Give general rules or number properties that these equal answers suggest.

1.4   Number Properties

FOCUS

We discuss the commutative, associative, and distributive properties of numbers, and then use them to combine and simplify algebraic expressions.

Two numbers can be added in any order, so that, for example,

$$7 + 10 = 10 + 7$$

In symbols, we have the following:

Commutative property for addition

$a + b = b + a$

where a and b are any nonnegative real numbers.

Since two numbers can be multiplied in any order, there is also a commutative property for multiplication.

Commutative property for multiplication

$ab = ba$

where a and b are any nonnegative real numbers.

For now we restrict these properties to *nonnegative* real numbers, but in Chapter 2 we will see that they hold true for *all* real numbers.

It is important to recognize that subtraction and division are *not* commutative, since, for example,

$$5 - 3 \neq 3 - 5 \quad \text{and} \quad 8 \div 4 \neq 4 \div 8$$

(The symbol " $\neq$ " means "is not equal to.")

The commutative properties are illustrated in a geometric context in our first example.

EXAMPLE 1 A student in a geometry class was confused about which side to use as l and which as w when computing the area and perimeter of rectangle $ABCD$ in Figure 1.18. He considered two possibilities: (1) $l = AD$ and $w = AB$, and (2) $l = AB$ and $w = AD$. Explain, using the commutative properties and the area and perimeter formulas, why (1) and (2) would give the same area and perimeter.

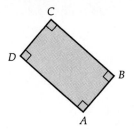

FIGURE 1.18

Solution

For the area, (1) would give

$$A = lw = (AD)(AB)$$

and (2) would give

$$A = lw = (AB)(AD).$$

[Note that $(AD)(AB)$ means the product of the lengths AD and AB.] The commutative property for multiplication ensures that $(AD)(AB) = (AB)(AD)$.

For the perimeter, (1) would give

$$P = 2l + 2w = 2(AD) + 2(AB)$$

and (2) would give

$$P = 2l + 2w = 2(AB) + 2(AD)$$

In this case, the commutative property for addition ensures that the two ways of computing P give the same result.   ■

Next we consider another pair of properties for addition and multiplication. The first tells us that when three or more numbers are added, they can be grouped in any way; thus, for example,

$$(6 + 9) + 12 = 6 + (9 + 12) = 27$$

because   15 $+ 12 = 6 +$ 21 $= 27$

Associative property for addition

$(a + b) + c = a + (b + c)$

where a, b, and c are any nonnegative real numbers.

Similarly, when three or more numbers are multiplied, they can be grouped in any way; for example,

$$(2 \cdot 5) \cdot 0.5 = 2 \cdot (5 \cdot 0.5) = 5$$

because

$$10 \ \cdot \ 0.5 \qquad 2 \ \cdot \ 2.5 \qquad = 5$$

Associative property for multiplication

$(ab)c = a(bc)$

where a, b, and c are any nonnegative real numbers.

These properties are illustrated in a geometric context in the next example.

EXAMPLE 2

Find the volume of the rectangular solid in Figure 1.19 using the following as length, width, and height.

	Length	Width	Height
a.	10	3	7
b.	3	10	7
c.	7	3	10

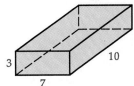

FIGURE 1.19

Solution

The volume of a rectangular solid is given by the formula

$$V = lwh$$

Since the same three numbers will be multiplied in each of the three cases (although in different orders and groupings), the commutative and associative properties for multiplication ensure that the volume will be the same each time. For the three approaches to finding V, we have

a. $V = 10 \cdot 3 \cdot 7$ **b.** $V = 3 \cdot 10 \cdot 7$ **c.** $V = 7 \cdot 3 \cdot 10$

$= (10 \cdot 3) \cdot 7$ $= (3 \cdot 10) \cdot 7$ $= (7 \cdot 3) \cdot 10$

$= 30 \cdot 7$ $= 30 \cdot 7$ $= 21 \cdot 10$

$= 210$ $= 210$ $= 210$

In general, the volume of a rectangular solid is the same no matter which dimensions are called the length, width, and height. Geometrically, this means that the volume of a rectangular solid does not change when it is rotated in space.   ■

When we multiply 3×82, we actually multiply 3×80 and 3×2 and then add the products. In symbols,

$$3 \times 82 = 3 \times (80 + 2)$$
$$= 3 \times 80 + 3 \times 2$$
$$= 240 + 6$$
$$= 246$$

This procedure is based on the distributive property.

Distributive property

$$a(b + c) = ab + ac$$

and $(b + c)a = ba + ca$

where a, b, and c are any nonnegative real numbers.

This property is applied in the next example.

EXAMPLE 3

Multiply using the distributive property.

a. $6(y + 3)$ **b.** $7(2 + n^2)$ **c.** $(2x + 1)5$

Solution

The letters are variables; they simply represent any numbers.

a. $6(y + 3) = 6 \cdot y + 6 \cdot 3$ **b.** $7(2 + n^2) = 7 \cdot 2 + 7 \cdot n^2$
$\qquad\qquad = 6y + 18$ $\qquad\qquad\qquad = 14 + 7n^2$

c. $(2x + 1)5 = (2x) \cdot 5 + 1 \cdot 5$
$\qquad\qquad\quad = 10x + 5$

We also use the associative and commutative properties to write

$$(2x) \cdot 5 = 2 \cdot 5 \cdot x = 10x$$ ■

> ### BE CAREFUL!
>
> **Be sure to multiply both terms in a sum in parentheses by the multiplier. For example, $6(y + 3)$ *is not* $6y + 3$; rather, $6(y + 3) = 6 \cdot y + 6 \cdot 3 = 6y + 18$.**

Algebraic Expressions

An algebraic expression consisting of a number or a variable or the product of a number and one or more variables is called a *term*. For example, each of the following is a term: 0.75, $6x$, and $9m^2n^5$. The expression $2x + 7$ consists of two terms, $2x$ and 7.

Terms in which the same variables are raised to the same powers are called **like terms.** In general, like terms can be combined to form a new term by using the distributive property, but unlike terms cannot be so combined.

EXAMPLE 4

Name the terms in each of the following expressions, and determine which are like terms. Add all like terms.

a. $3m + 4m + m$ **b.** $2xy + 3xy^2 + 5x^2y + 7xy$

Solution

a. The terms are $3m$, $4m$, and m (or $1m$). All three are like terms, because the only variable in each is m, raised to the first power. By the distributive property applied to all three terms,

$$3m + 4m + m = (3 + 4 + 1)m$$

$$= 8m$$

b. The terms are $2xy$, $3x^2y$, $5xy^2$, and $7xy$. The like terms are the first and last. The other terms have the same variables but raised to different powers, so they cannot be combined.

$$2xy + 3xy^2 + 5x^2y + 7xy = (2 + 7)xy + 3xy^2 + 5x^2y$$

$$= 9xy + 3xy^2 + 5x^2y$$ ∎

Number properties can be used together to remove parentheses and combine like terms, giving an equivalent expression that is in a simpler form. This process is sometimes called *simplifying the expression*.

EXAMPLE 5   Remove parentheses and combine like terms in each expression.

a. $7(x + 3) + x$ **b.** $(36a^2 + 2b) + 4a^2$

Solution   **a.** In this part, we include each step and indicate the property that was used.

$$7(x + 3) + x = 7x + 7 \cdot 3 + x \qquad \text{Distributive property}$$
$$= (7x + 21) + x \qquad \text{Order of operations}$$
$$= 7x + (21 + x) \qquad \text{Associative property for addition}$$
$$= 7x + (x + 21) \qquad \text{Commutative property for addition}$$
$$= (7x + x) + 21 \qquad \text{Associative property for addition}$$
$$= (7 + 1)x + 21 \qquad \text{Distributive property}$$
$$= 8x + 21 \qquad \text{Adding}$$

b. In this part, our work is briefer, at about the level that will be expected of you in your homework.

$$(36a^2 + 2b) + 4a^2 = (36a^2 + 4a^2) + 2b$$
$$= 40a^2 + 2b \qquad ■$$

If a term contains one or more variables, then the **numerical coefficient,** or simply the coefficient, is the number that multiplies the variable(s). Example 5 suggests this shortcut for adding like terms.

Adding like terms

Like terms can be added by simply adding the coefficients of the terms.

The next example illustrates the use of this shortcut.

EXAMPLE 6   Remove parentheses and combine like terms.

a. $8(t^2 + t) + 2t^2$ **b.** $2(x^3 + xy) + 4[(2xy + 1) + 8]$

Solution

a. $8(t^2 + t) + 2t^2 = 8t^2 + 8t + 2t^2$
$$= 10t^2 + 8t$$

Notice that $10t^2$ and $8t$ are not like terms because the exponents of t are not the same.

b. This expression involves grouping within grouping. Remove the inner grouping first.

$$2(x^3 + xy) + 4[(2xy + 1) + 8]$$

$= 2(x^3 + xy) + 4[2xy + 9]$ Removing inner grouping

$= 2x^3 + 2xy + 8xy + 36$ Removing other grouping

$= 2x^3 + 10xy + 36$ Adding like terms   ∎

BE CAREFUL!

a. $2 + 3x$ *is not* $5x$ because 2 and $3x$ are not like terms.

b. $4x + 5x$ *is not* $9x^2$; rather, $4x + 5x = (4 + 5)x = 9x$.

As we have seen, algebraic expressions arise in geometric formulas and in word problems. An important skill in solving such problems is translating English expressions into algebraic expressions. Table 1.3 shows some translations of phrases that apply to numbers.

TABLE 1.3 Translations from English to Algebra

English Expression	Algebraic Expression
4 more than a number x	$x + 4$
The sum of any two numbers	$x + y$ or $a + b$, and so on
Twice s added to twice t	$2s + 2t$
A number x added to one more than x	$(x + 1) + x$
The product of any three numbers	abc
The sum of n and the next two whole numbers	$n + (n + 1) + (n + 2)$
Three times the square of a number y	$3y^2$

In the next example, we translate an English expression to an algebraic expression.

EXAMPLE 7

Suppose you have $50. You spend m dollars on a movie and p dollars on popcorn. Write an algebraic expression for the amount of money you have left.

Solution

The money you spend must be subtracted from $50. Since you spend m dollars and p dollars, the amount you have left is $50 - m - p$.   ∎

DISCUSSION QUESTIONS 1.4

1. According to the dictionary, "to commute" means to exchange. How does that apply to the commutative properties for addition and multiplication?

2. "To associate" means to bring into relationship as a partner. How does that apply to the associative properties for addition and multiplication?

3. "To distribute" means to spread out or to deal out in shares. How does that apply to the distributive property?

EXERCISES 1.4

1. Give the numerical coefficient of each of the following terms: $3x$, $7x$, $5x^2$, xy, x^2y, $8xy$, $2xy^2$, $7x^2y$.

2. Identify the like terms among the following: $3x$, $7x$, $5x^2$, xy, x^2y, $8xy$, $2xy^2$, $7x^2y$.

In Exercises 3–12, evaluate each expression if $x = 5$ and $y = 2$.

3. $x + y$

4. $6x + 4$

5. $\dfrac{1}{4} + \dfrac{2}{3}y$

6. $6y + 2y^2$

7. $\dfrac{1}{2} + \dfrac{5}{6}x^2$

8. $0.3(x + 0.4y)$

9. $(7 + 3y^3)2 + 3$

10. $2[3(x + 4) + 1] + 2x$

11. $x + 3.2[y + 1.3(x + 2.1)]$

12. $[7 + (x^2 + 4)] - (2x^2 - x^2)$

In Exercises 13–30, fill in the blank and give the property that you apply to do so.

13. $8a = a \cdot \underline{\ ?\ }$

14. $5x + 3 = 3 + \underline{\ ?\ }$

15. $3(x + 2) = 3x + \underline{\ ?\ }$

16. $2(4y^2) = \underline{\ ?\ }\, y^2$

17. $0.5 + 0.9n = \underline{\ ?\ } + 0.5$

18. $0.5 + 0.9n = 0.5 + n \cdot \underline{\ ?\ }$

19. $4x + 3x = \underline{\ ?\ }\, x$

20. $(3r + 8) + 1 = 3r + \underline{\ ?\ }$

21. $3(x + 2) = (x + 2) \cdot \underline{\ ?\ }$

22. $5x + 3 = \underline{} \cdot 5 + 3$

23. $5(y + \underline{}) = 5y + 20$

24. $3(2 \cdot \underline{}) = 6x$

25. $(4m + \underline{}) + 5 = 4m + 12$

26. $\underline{}(2 + x) = 6 + 3x$

27. $(p + \underline{}) \cdot 6 = 6p + 18$

28. $\underline{} \cdot z = z \cdot 4.2$

29. $7.1x + 1.3x = \underline{} x$

30. $\underline{}(4r) = 12r$

In Exercises 31–54, use number properties to remove any grouping symbols, and combine like terms.

31. $4(x + 6)$ **32.** $3(y + 5)$ **33.** $2(5 + 3x)$ **34.** $8(2 + 3y)$ **35.** $(2x + 1)4$

36. $\left(\frac{2}{7} + \frac{1}{2}x\right)\frac{1}{8}$ **37.** $\frac{3}{10}\left(x^2 + \frac{2}{3}x\right)$ **38.** $12(2y + y^3)$ **39.** $8(xy + x^2)$ **40.** $4(a^3 + ab)$

41. $8x + 3x + 2$ **42.** $7 + 2x + 12x$ **43.** $8 + 5y + 7y$

44. $3(x + 4) + 20$ **45.** $1.5(2x + 6) + 2.5$ **46.** $2.8(10y + 2.4) + 3.5$

47. $x^2 + 3x + 2x^2$ **48.** $y^3 + 3y^2 + 7y + y^2$ **49.** $\frac{1}{2}\left(x + \frac{1}{5}\right) + \frac{1}{4}\left(\frac{1}{2}x + \frac{1}{3}\right)$

50. $3(5 + 7xy) + 4(xy + 3)$ **51.** $4[5 + 2(3x^5 + 1)]$ **52.** $7 + 3[4(2 + 3t^2) + 9t]$

53. $4\{3 + 2[x + 3(x^2 + x + 2)] + 2x\}$ **54.** $3\{2 + 3[(y + x + 6)4 + y] + x\}$

In Exercises 55–58, the length and width of a rectangle are given. Find the area in terms of x, and simplify your result.

55. $l = 8$
$w = 2x + 5$

56. $l = x + 8$
$w = 6$

57. $l = 7 + 2x + x^2$
$w = 5$

58. $l = 10$
$w = x^3 + 6x + 3$

In Exercises 59–62, the length and width of a rectangle are given. Find the perimeter in terms of x, and simplify your result.

59. $l = x + 3$
$w = 2 + 7x$

60. $l = 3(x + 1)$
$w = 5x$

61. $l = 5x^2 + 1$
$w = x^2 + x$

62. $l = 6 + 3x^3$
$w = x^2 + x^3$

In Exercises 63–66, the dimensions of a rectangular solid are given. Find the surface area in terms of x, and simplify your result.

63. $l = 5$
$w = x + 3$
$h = 9$

64. $l = 3$
$w = 3(x + 2)$
$h = 7$

65. $l = x^2 + x$
$w = 10$
$h = 8$

66. $l = 8$
$w = 3$
$h = x^2 + 2x + 1$

In Exercises 67–74, x represents any whole number. Write an algebraic expression for each English expression.

67. A number that is 5 times as large as x

68. A number that is 10 more than x

69. The next whole number after whole number x

70. A number that is three times the sum of x and 20

71. A number that is 6 more than twice x

72. A number that is 5 more than twice the product of 5 and x

73. The product of two more than x and 7 more than x

74. The sum of 9 times x and 4 more than x

75. Suppose a hamburger costs h cents, french fries cost f cents, and an orange drink costs d cents. What is the cost of two hamburgers, one order of french fries, and an orange drink? If you paid for this order with a $5 bill, what was your change?

76. A piece of wood is 15 ft long. What length remains after x ft are cut off?

77. If your car averages 28 mi to the gallon of gasoline, about how far could you travel on a tank that holds t gal?

78. How many inches are in x ft? How many feet are in y in.?

Preparing for Section 1.5

In Exercises 79–82, construct a number line and plot the number on it.

79. 4.2   **80.** 3.8   **81.** 0.56   **82.** 0.68

In Exercises 83–86, match the given number to a point on the number line in Figure 1.20.

83. 1.8   **84.** $\dfrac{13}{4}$

85. 2.17   **86.** $\dfrac{1}{6}$

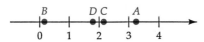

FIGURE 1.20

1.5 **Graphs**

FOCUS

Our final topic for review is the graphing of pairs of non-negative numbers on a rectangular coordinate system. We then use graphs to solve certain types of problems.

Equations like the area and perimeter formulas we developed earlier in this chapter represent relationships between two or more variables. One way to better understand and make use of such relationships is to construct their graphs.

The graph of a relationship between two variables is constructed on a **rectangular coordinate system.**

To construct a rectangular coordinate system *(see Figure 1.21)*

1. Draw two perpendicular number lines, called the **axes.** (The number lines are usually drawn as a horizontal and a vertical line.)
2. The point of intersection of the axes is called the **origin.** Assign zero on each axis to the origin.
3. Assign positive numbers to points to the right of the origin on the horizontal axis.
4. Assign positive numbers to points above the origin on the vertical axis.

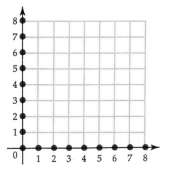

FIGURE 1.21

Each point on a rectangular coordinate system corresponds to a pair of numbers. For example, point *A* in Figure 1.22 corresponds to the numbers 3 and 2, and we write *A*(3, 2). The first number in the pair is the number of units *to the right* from the origin to the point *along the horizontal axis;* the second number is the number of units *above the origin* to the point *along the vertical axis.* Hence, (3, 2) corresponds to the point that is 3 units to the right of the origin in the horizontal direction and 2 units above the origin in the vertical direction. The process of locating the point that corresponds to the pair (3, 2) is called "plotting (3, 2)," and the point is said to have the *coordinates* 3 and 2.

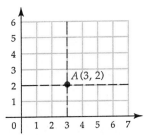

FIGURE 1.22

We write the pair of numbers (the coordinates) in parentheses, separated by a comma, to show that they are an *ordered pair*. That is, their order is important, because (3, 2) and (2, 3) represent different points. For now, we will plot only zero and positive coordinates, to represent points above and to the right of the origin. In later chapters, we will include negative coordinates as well.

EXAMPLE 1

Plot these ordered pairs.
a. (0, 4)   **b.** (2.5, 1.8)

Solution

a. Since the first coordinate is zero, we do not move to the right of the vertical axis; we remain on it. Then we count 4 up from the origin. The result is point *A* in Figure 1.23.

b. We count 2.5 to the right from the origin along the horizontal axis, and 1.8 up from the horizontal axis. The point with coordinates (2.5, 1.8) is point *B* in Figure 1.23.

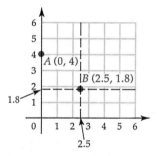

FIGURE 1.23

A formula that relates two variables can be graphed by plotting ordered pairs that satisfy the relationship. A formula like $P = 4s$ relates the perimeter *P* of a square to the length *s* of its sides. We can form ordered pairs (*s*, *P*) by choosing values for *s* and using the formula to find the related values for *P*. Such ordered pairs are said to *satisfy* the formula. The graphs of these ordered pairs lie on the graph of the formula.

EXAMPLE 2

Plot six ordered pairs of the form (*s*, *P*) that satisfy the formula for the perimeter of a square, $P = 4s$.

Solution

We assign six positive but relatively small values to *s* and compute the corresponding values of *P*, as in Table 1.4. These are our six ordered pairs. To plot them, we first draw a set of axes. Since the ordered pairs are in the form (*s*, *P*), we label the horizontal axis *s* and the vertical axis *P*, as in Figure 1.24. Then we actually plot each of the ordered pairs (*s*, *P*) from Table 1.4.

TABLE 1.4

s	$p = 4s$
1	$4 \cdot 1 = 4$
1.5	$4 \cdot 1.5 = 6$
2	8
2.5	10
3	12
3.5	14

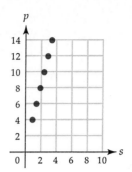

FIGURE 1.24

Notice that the points appear to lie on a straight line. In fact, if all the ordered pairs of numbers that satisfy $P = 4s$ could be plotted, they would form the line shown in Figure 1.25.

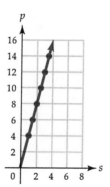

FIGURE 1.25   ∎

The line that is suggested by the points we plotted in Example 2 is called the *graph* of the equation $P = 4s$. An equation is related to its graph in the following way.

Connection between an equation and its graph

1. The graph of every ordered pair that satisfies an equation (that is, makes an equation true) lies on the graph of the equation.
2. The coordinates of any point on the graph of an equation satisfy the equation.

Because of this close connection, the graph can be used to help interpret the relationship given by the equation.

EXAMPLE 3

In Example 2, we plotted some ordered pairs that satisfied the formula for the perimeter of a square, $P = 4s$. Use the graph in Figure 1.25 to answer these questions:

a.   What is the length of the side of a square that has perimeter 5?
b.   What is the perimeter of a square with sides of length π?
c.   When the side of a square increases by 1 unit, say from 1 to 2 or from 2 to 3, by how much does its perimeter increase?

Solution

a.   We need to find s in the ordered pair $(s, 5)$. To do so, locate 5 on the P axis. Sketch a horizontal line through 5 that intersects the graph of $P = 4s$ at some point (point A in Figure 1.26). Then sketch a vertical line through A that intersects the s axis. The intersection gives the s coordinate of point A, here about $1\frac{1}{4}$. So a square with perimeter 5 has sides of length $1\frac{1}{4}$.

b.   We need to find P in (π, P). Locate π (about $3\frac{1}{7}$) on the s axis. Sketch a vertical line through π that intersects the graph at some point (B in Figure 1.26). Then sketch a horizontal line through B that intersects the P axis. The intersection gives the P coordinate of point B, about 12.6. So the ordered pair $(\pi, 12.6)$ satisfies the perimeter formula, $P = 4s$. The perimeter of a square with sides of length π is 12.6.

c.   The triangles in Figure 1.26 show that when the side s increases by 1 (say, from 4 to 5), the perimeter P increases by 4 (from 16 to 20 on the graph).

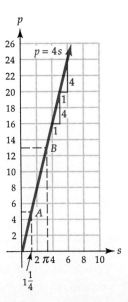

FIGURE 1.26

Not all graphs are straight lines, as we see in the next example.

EXAMPLE 4

Graph the formula for the area of a square, $A = s^2$, with s on the horizontal axis and A on the vertical axis.

Solution

As usual, we assign several values to s and use the formula to compute the value of A corresponding to each value of s. The results are shown in Table 1.5. This generates ordered pairs that lie on the graph. When we plot these points, we see that they do not lie in a line. Rather, they seem to fit on the curve that is sketched in Figure 1.27.

TABLE 1.5

s	$A = s^2$
1	$1^2 = 1$
2	$2^2 = 4$
3	9
4	16

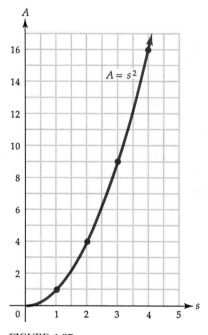

FIGURE 1.27

Another geometric application is given in the next example. For this example, recall that rectangles with a given perimeter, say 36 cm, may have different lengths and widths. For example, a rectangle in which $l = 13$ and $w = 5$ and one in which $l = 10$ and $w = 8$ both have perimeter 36. Note, too, that the areas of these rectangles are not equal. The area of the first rectangle is $lw = 13 \times 5 = 65$, and the area of the second is $lw = 10 \times 8 = 80$.

EXAMPLE 5

Figure 1.28 is the graph of area A versus width w for rectangles with perimeter 36 cm. Use the graph to determine which of the possible rectangles with perimeter 36 cm has the greatest area.

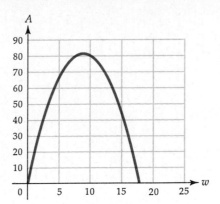

FIGURE 1.28

Solution   We look for the highest point on the graph (that is, the point for which the area A is greatest). By reading numbers off the graph, we see that it occurs at (9, 81), which corresponds to $w = 9$ cm and $A = 81$ cm^2. Since $A = lw$, it follows that

$$81 = l \cdot 9$$

Thus, $l = 9$ cm, too (because $9 \cdot 9$ is 81); hence, the rectangle with perimeter 36 cm that has the greatest area is a 9-cm by 9-cm square.   ∎

DISCUSSION QUESTIONS 1.5

1. What are the coordinates of the origin in a rectangular coordinate system?

2. If we know a point is on the horizontal axis, what can be said about the value of one of its coordinates?

3. If we know a point is on the vertical axis, what can be said about the value of one of its coordinates?

4. Is it true that the graph of every equation is a line? Explain.

EXERCISES 1.5

In Exercises 1–8, find the point in Figure 1.29 that corresponds to each ordered pair.

1. (2, 4)	**2.** (0, 3)	**3.** (5, 0)	**4.** (3, 3)
5. (1.3, 5.1)	**6.** (4.2, 1.6)	**7.** (4, 4)	**8.** (5, 3)

In Exercises 9–16, find five ordered pairs of nonnegative numbers that satisfy each equation; then draw a set of axes, label the horizontal axis x and

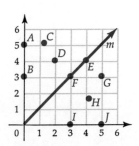

FIGURE 1.29

the vertical axis y, plot the five ordered pairs, and sketch the graph that these points suggest.

9. $y = x$ 10. $y = x + 2$ 11. $y = 2x$ 12. $y = 2x + 3$

13. $y = 10 - x$ 14. $y = 12 - 2x$ 15. $y = x^2$ 16. $y = x^2 + 1$

In Exercises 17–22, match the given equation to one of Figures 1.30 to 1.35.

17. $y = x - 2$ 18. $x = 2$ 19. $y = 4 - x$

20. $y = 3$ 21. $y = 4 - 2x$ 22. $y = 2x + 1$

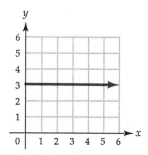

FIGURE 1.30

FIGURE 1.31

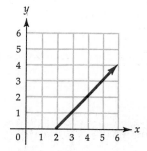

FIGURE 1.32

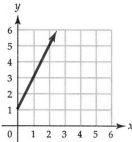

FIGURE 1.33

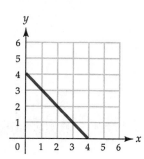

FIGURE 1.34

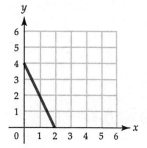

FIGURE 1.35

23. Half-line (or ray) m in Figure 1.29 divides the rectangular coordinate system into three sets of points, those that lie on m, those that lie above and to the left of m, and those that lie below and to the right of m. Give a rule that describes the coordinates of points that are on ray m.

24. Give a rule that describes the coordinates of points that lie above and to the left of m in Figure 1.29.

25. Give a rule that describes the coordinates of points that lie below and to the right of m in Figure 1.29.

26. The area of a rectangle with length 6 is given by $A = 6w$. Graph this equation with w on the horizontal axis and A on the vertical axis. (*Note: w* may be greater than *l*.)

27. Use the graph you constructed in Exercise 26 to estimate the width of a rectangle if its length is 6 and its area is 9.3.

28. The perimeter of a rectangle whose width is 6 is given by $P = 2l + 12$. Graph this equation with l on the horizontal axis and P on the vertical axis.

29. The area of a triangle with base 6 is given by $A = \frac{1}{2}(6)h$ or $A = 3h$.

    Graph this equation with h on the horizontal axis and A on the vertical axis.

30. Two graphs are given in Figure 1.36, one of the formula for the area of a circle, $A = \pi r^2$, and one of the formula for the circumference of a circle, $C = 2\pi r$. Which is the area graph and which is the circumference graph? How can you tell?

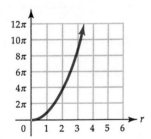 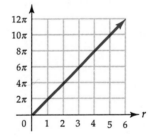

FIGURE 1.36

31. Setting the cruise control on a car keeps it moving at a constant speed on the open highway. If we set the cruise control at r mph, then in 2 h we will travel a distance $d = 2r$.

    a. With r on the horizontal axis and d on the vertical axis, graph this equation.
    b. Use your graph to estimate the rate r at which we must set the cruise control to go a distance of 114 mi in 2 h.

Exercises 32–36 refer to this situation.

   Michelle threw a baseball into the air. Figure 1.37 shows the graph of its height h in feet versus the time t in seconds after she threw it. She threw the ball at time $t = 0$.

32. For how many seconds was the baseball going up?

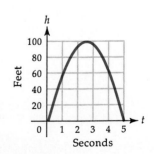

FIGURE 1.37

33. What was the maximum height that the ball reached?

34. How many seconds after Michelle threw the ball did it reach its maximum height?

35. For how many seconds was the baseball coming down?

36. How many seconds after Michelle threw the ball did it come back to the ground?

In Exercises 37–39, suppose that the height of a parallelogram is twice its base.

37. Find an equation for the area of the parallelogram in terms of the base. (*Hint:* Substitute $2b$ for h in the formula $A = bh$.)

38. Graph the equation you found in Exercise 37.

39. Use the graph in Exercise 38 to find the base of the parallelogram if its area is 6, 10, 14, and 20.

In Exercises 40–42, suppose that a piece of wood 12 ft long is cut into two pieces.

40. If x is the length of one piece and y is the length of the second piece, find an equation for y in terms of x.

41. Make a table showing all even-number values of x from 0 to 12, and the corresponding values of y.

42. Plot the ordered pairs (x, y) on a rectangular coordinate system with x on the horizontal axis and y on the vertical axis, and join the points.

Preparing for Section 2.1

43. Recall that on a Celsius thermometer, $0°$ is the freezing point of water. Writing $+$ before a temperature that is above the freezing point and $-$ before one that is below the freezing point, write a number to represent these Celsius temperatures.
    a. $1°$ above freezing
    b. $1°$ below freezing
    c. $12°$ above freezing
    d. $6°$ below freezing

44. Refer to Figure 1.38, which is a Celsius thermometer that is marked off every $2°$. Determine the temperature indicated by each of the following letters, and note whether it is above or below freezing by preceding it with $+$ or $-$.
    a. A b. B c. C d. D

45. Draw a thermometer like that in Figure 1.38, but horizontal. Mark each of the following temperatures on your thermometer. By how many degrees does each temperature differ from $0°$?
    a. $-8°$ b. $+8°$ c. $-12°$ d. $+12°$

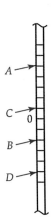

FIGURE 1.38

Chapter Summary

Important Ideas

To perform operations with fractions

a. *Simplifying:* If n is not zero, then $\dfrac{a \cdot n}{b \cdot n} = \dfrac{a}{b}$

b. *Multiplying:* $\dfrac{a}{b} \cdot \dfrac{c}{d} = \dfrac{a \cdot c}{b \cdot d}$ c. *Dividing:* $\dfrac{a}{b} \div \dfrac{c}{d} = \dfrac{a}{b} \cdot \dfrac{d}{c} = \dfrac{a \cdot d}{b \cdot c}$ (Changing $\dfrac{c}{d}$ to $\dfrac{d}{c}$ is called *inverting the divisor.*)

d. *Adding:* $\dfrac{a}{b} + \dfrac{c}{d} = \dfrac{a \cdot d + b \cdot c}{b \cdot d}$ e. *Subtracting:* $\dfrac{a}{b} - \dfrac{c}{d} = \dfrac{a \cdot d - b \cdot c}{b \cdot d}$

The **nth power of a,** written a^n, means

$$a^n = \underbrace{a \cdot a \cdot a \cdot \ldots \cdot a}_{n \text{ factors}}$$

Order in which operations should be performed:

a. Perform the operations within the innermost grouping symbol, using the order prescribed in steps b to d to remove that grouping symbol. Remove all other grouping symbols, working from inner to outer, in the same way. Then continue with steps b to d.

b. Perform all indicated exponentiations, working from left to right.

c. Perform all indicated multiplications and divisions, working from left to right.

d. Perform all indicated additions and subtractions, working from left to right.

Important geometric formulas

Rectangle:	$A = lw$	$P = 2l + 2w$
Triangle:	$A = \dfrac{1}{2}bh$	$P = a + b + c$
Parallelogram:	$A = bh$	$P = 2a + 2b$
Circle:	$A = \pi r^2$	$C = 2\pi r$
Rectangular solid:	$V = lwh$	$S = 2lw + 2lh + 2wh$
Cube	$V = s^3$	$S = 6s^2$
Right circular cylinder	$V = \pi r^2 h$	$S = 2\pi rh + 2\pi r^2$
Sphere	$V = \dfrac{4}{3}\pi r^3$	$S = 4\pi r^2$

Commutative properties

(Addition) $a + b = b + a$
(Multiplication) $ab = ba$

where a and b are any nonnegative real numbers.

Associative properties

(Addition) $(a + b) + c = a + (b + c)$
(Multiplication) $(ab)c = a(bc)$

where a, b, and c are any nonnegative real numbers.

Distributive property

$$a(b + c) = ab + ac \qquad (b + c)a = ba + ca$$

where a, b, and c are any nonnegative real numbers.

Chapter Review

Section 1.1

1. Draw a number line and plot $\frac{2}{9}$, $\frac{8}{5}$, π, and 1.26 on it.

In Exercises 2–9, perform the indicated operation. Give fraction results in simplified form.

2. 5^3 3. 0.2^5 4. $\frac{2}{5} \cdot 1\frac{1}{3}$ 5. $1\frac{4}{5} \div 2\frac{2}{3}$ 6. $5\frac{1}{3} + 4\frac{3}{4}$

7. Take 15% of 80   8. Take 125% of 200

9. Take 53% of 186 (to the nearest hundredth)

In Exercises 10–13, fill in the blanks.

10. $\frac{7}{5} = \underline{\ ?\ } \%$ 11. $0.591 = \underline{\ ?\ } \%$

12. $183\% = \underline{\ ?\ }$ (as a decimal)   13. $0.346 = \underline{\ ?\ }$ (as a reduced fraction)

14. The range for an A on a mathematics test was 85% to 100%. If there were a possible 150 points on the test, what is the least whole number of points that will be worth an A?

15. Long distance telephone rates between two cities are $1.08 for the first three minutes and $0.24 for each additional minute over three. Complete the following table.

Length (min.)	Cost, First 3 min.	Cost, Over 3 min.	Total Cost
3	$1.08	$(3 - 3)(0.24) = \$0$	$1.08
4	$1.08	$(4 - 3)(0.24) = \$0.24$	?
5	$1.08	?	?
?	?	$1.20	?
?	$1.08	?	$3.24
?	$1.08	$(m - 3)(0.24)$	?
x	$1.08	?	?

Section 1.2

In Exercises 16–20, compute and give your answers as a single rational number.

16. $16 - (8 - 3)$ **17.** $6 + 7 \cdot 4$ **18.** $2 + 6^2$ **19.** $18 \div (14 - 5) + 6^2$ **20.** $7\left(\dfrac{2}{3} + \dfrac{3}{4}\right) + 3 \cdot 2^4$

21. Five professors in the Mathematics Department have published 31, 14, 10, 6, and 4 articles, respectively, in mathematics journals. What is the average number of articles they have published?

22. Dandy Record Company received shipments containing 2695, 1413, and 180 records. They distribute these records equally to their 16 stores. How many records will each store receive?

23. A sporting goods company manufactures 26 baseball gloves per hour. In an average 8-h working day, 18 defective gloves are produced. The number of acceptable gloves produced in 1 day would be:
   a. $8 \cdot 18$ **b.** $(26 - 18) \cdot 8$
   c. $(26 \cdot 8) - 18$ **d.** $(26 + 8) - 18$

Section 1.3

In Exercises 24–29, fill in the blank.

	Figure	Base	Height	Area
24.	Parallelogram	5 cm	4 cm	?
25.	Triangle	14 in.	6 in.	?
26.	Triangle	6 ft	42 in.	?
27.	Parallelogram	3 yd	7 ft	?
28.	Parallelogram	?	8 ft	36 ft^2
29.	Triangle	10 m	?	40 m^2

30. Find the area and perimeter of a 12-in. by 2 ft rectangle.

31. The length of a cube is 3.1 cm. Find its volume and surface area.

32. The length, width, and height of a rectangular solid are 4 in., 6 in., and 9 in., respectively. Find the volume and the surface area of the solid.

33. Find the length of the side of a square if its area is 100 in.^{2}.

34. One side of a parallelogram is 8 cm and the height to that side is 6 cm. Find the area of the parallelogram.

35. Find the areas of the shaded regions in Figure 1.39*a* and *b*.

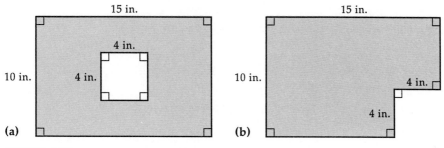

FIGURE 1.39

36. In Figure 1.40, a square is attached to the lower corner of a rectangle. What is the perimeter of the entire closed region? What is its area?

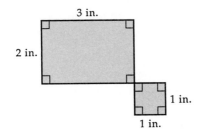

FIGURE 1.40

37. The area of a rectangular room is 800 ft^2. If the length of the room is 40 ft, what is its width?

38. Find the area of the triangle in Figure 1.41.

39. Jan has a rectangular piece of tin 136 cm by 78 cm. She makes a box by cutting a square with sides of 12 cm from each corner of the piece of tin and then folding up the flaps. What is the area of the bottom of the box? What is the volume of the box?

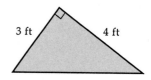

FIGURE 1.41

Section 1.4

In Exercises 40–42, evaluate each expression if $r = 2$ and $s = 4$.

40. $2r + s^2$ **41.** $s - (r + 1)$ **42.** $4[5 + 3(s - 2) + 6r] + 2s^2$

In Exercises 43–45, fill in the blanks and give the property that you apply to do so.

43. $2y + 6 = \underline{\ ?\ } + 2y$ **44.** $5x^2 + 7x^2 = \underline{\ ?\ } x^2$ **45.** $4 + 3x = \underline{\ ?\ } + 4$

In Exercises 46–48, simplify the expression.

46. $x^2 + x^3 + 2x^2 + 5$ **47.** $\dfrac{1}{3}\left(6x - \dfrac{3}{4}\right) + \dfrac{1}{2}(x + 4)$ **48.** $8 + 3\{2[4rs + 3(r^2s + 2rs) + 3] + r^2s + 1\}$

49. If r represents a number, what is twice the number divided by one more than the number?

50. Find the volume and surface area of a rectangular solid whose length, width, and height are 6, x, and $2x$, respectively.

51. If n represents any odd whole number, what is the product of n and the next odd number?

Section 1.5

In Exercises 52–55, plot the ordered pairs on the same rectangular coordinate system.

52. $(0, 1)$ **53.** $(2, 5)$ **54.** $\left(\dfrac{3}{4}, 7\dfrac{1}{2}\right)$ **55.** $(1.4, 2.9)$

56. If x represents the first coordinate in each of the ordered pairs in Exercises 52–55, and y represents the second coordinate, find an equation that gives y in terms of x for the given pairs.

In Exercises 57–60, draw a coordinate system and label the horizontal axis x and the vertical axis y. Plot five ordered pairs of nonnegative numbers that satisfy the equation, and sketch in the graph of the equation.

57. $y = 3x$ **58.** $y = 3x - 2$ **59.** $y = x^2 + 3$ **60.** $y = x^2 + x$

For Exercises 61–63, indicate whether each ordered pair is on the graph constructed in Exercise 60. Do not extend the graph.

61. $(20, 420)$ **62.** $(56, 7)$ **63.** $(25, 650)$

64. The surface area of a sphere is given by $S = 4\pi r^2$. Graph this equation with r on the horizontal axis and S on the vertical axis.

65. Use the graph you constructed in Exercise 64 to estimate the radius of a sphere whose surface area is 6 square units.

66. Use the graph you constructed in Exercise 64 to estimate the radius of a sphere whose surface area is π square units.

Chapter Test

1. A student's scores on math tests during a semester were 63, 87, 80, 71, 95, and 72. Find the student's average score.

2. Convert $\dfrac{3}{8}$ to a percent. **3.** What is 15% of 24?

4. A professor accepted a new position at a 40% increase in salary. If her salary in her old position was $36,000, what is her new salary?

5. Compute: $4\dfrac{1}{2} + 12 \div 6 \cdot 2\dfrac{2}{3}$

6. Fill in the blank and name the property that allows you to do so:
a. $6 + 3y = \underline{\ ?\ } + 6$ **b.** $7(3n) = \underline{\ ?\ }n$

7. Remove parentheses and combine like terms:
a. $5(2x + y) + 2x + y$ **b.** $5 + 2\{3 + 2[2x^2 + 1] + 3x^2\} + x$

8. If $x = 1$, then $4\{2 + [3(3 + 2x) + 1]\} - 8 = \underline{\ ?\ }$.

9. On a rectangular coordinate system, label the horizontal axis as x and the vertical axis as y. Graph five ordered pairs that satisfy the equation $y = x^2 + 1$, then draw the curve that appears to go through these points.

10. Tom is making a box 10 in. long and 5 in. wide. The volume of the box must be 10 in.^{3}. How high should Tom make the box?

11. A rectangular vegetable garden is 15 m by 20 m. The potato crop occupies a 5-m by 6-m rectangular region of the garden, and the string beans occupy 25 m^2. What is the combined area of the parts of the garden occupied by the potatoes and string beans?

12. Find the area and perimeter of the parallelogram in Figure 1.42.

13. Find the area and perimeter of the triangle in Figure 1.43.

14. A rectangular yard measures 9 m long and 5.5 m wide. The owner decides to put two circular flower gardens in the yard, each with a radius of 0.75 m. The remaining part of the yard will be a vegetable garden.
a. How many square meters of the yard will be used for flowers? For vegetables?
b. To protect the flowers from roving animals, the gardener put a wire fence around the flower gardens. How much fence was needed?

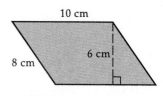

FIGURE 1.42

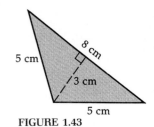

FIGURE 1.43

15. The wooden box in Figure 1.44 has no top. The wood used to make boxes costs $2.50 per square foot. How much would it cost to buy the wood needed to make this box?

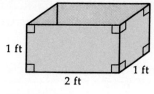

1 ft

2 ft   1 ft

FIGURE 1.44

16. A car cost $12,000 in 1988 and is losing value at the rate of 10% per year. Complete this table.

Number of Years Since 1988	Value of Car at End of Year
0	$12,000
1	$90\% \cdot 12{,}000 = 10{,}800$
2	?
3	?
4	?

17. Setting the cruise control on a car keeps it going at a constant speed on the open highway. Write an equation for the distance d, in miles, you would travel in t hours at a cruise control setting of 60 mph.

18. Graph the equation you found in Problem 17 with t on the horizontal axis and d on the vertical axis. Count each unit on the vertical axis as 20 mi.

Problems 19–23 refer to the graph of the population of Conesville, given in Figure 1.45.

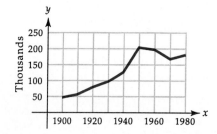

FIGURE 1.45

19. About what was the population of Conesville in 1900?

20. In about what year had the 1900 population doubled?

21. During what ten-year periods was the population decreasing?

22. In about what year was the population greatest?

23. About what was the greatest population of Conesville?

2 *Real Numbers*

2.1 Real Numbers and Absolute Value

FOCUS

In this chapter, we apply the number properties and operations to both positive and negative real numbers. We begin here by introducing the negative real numbers and the idea of the absolute value of a number.

These headlines might well appear in your local newspaper:

RAMS' NET YARDAGE IS −14
DOW JONES INDUSTRIALS: −14 POINTS FOR THE DAY
HIGH TEMPERATURE IN MINNEAPOLIS IS −14°

The number −14, which is read "negative fourteen," appears in each head-line. In all three cases, the minus sign indicates that we are dealing with a quantity that is almost like 14 but also is somehow different from 14. The difference is one of direction—a loss rather than a gain, down rather than up, backward rather than forward.

This difference in direction can be shown on a number line that has been extended to the left of zero. On it, we can plot 14 and −14 as in Figure 2.1.

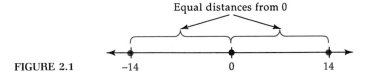

FIGURE 2.1

Extended in this way, the number line suggests that for each whole num-ber a, we can define a number "negative a" that is the same distance from 0 but in the opposite direction. (Several are shown in Figure 2.2.) Thus, with 1 we can associate -1, with 2 we can associate -2, with 14 we can associate -14, and so on. (Zero is considered to be neither positive nor negative, and $-0 = 0$.) Together, the whole numbers and their negatives are called the **integers.** Hence, we have the following names.

Integers:	$\ldots -3, -2, -1, 0, 1, 2, 3, \ldots$
Whole numbers (nonnegative integers):	$0, 1, 2, 3, \ldots$
Positive integers:	$1, 2, 3, \ldots$
Negative integers:	$\ldots -3, -2, -1$

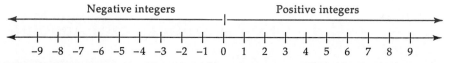

FIGURE 2.2

In a similar way, for any nonnegative real number r, we can define its negative $-r$. The nonnegative real numbers together with their negatives form the collection of *real numbers*. The **positive real numbers** are all the nonnegative real numbers except 0; the negatives of these are called the **negative real numbers.** Some examples are given in Figure 2.3.

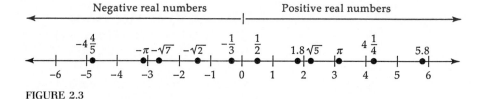

FIGURE 2.3

Here are some uses of positive and negative real numbers:

	Negative	Positive
In football	-5 means 5 yards lost	5 means 5 yards gained
On the stock market	-5 means $5 lost	5 means $5 gained
For altitudes	-500 means 500 feet below sea level	500 means 500 feet above sea level

Sometimes a "$+$" is placed before a positive number, as in $+5$ for 5, to emphasize that it is positive, but we will rarely use that notation. A number that carries a plus or minus sign is called a **signed number.**

EXAMPLE 1

Suggest numbers to represent a savings account deposit of $25, no transaction at all, and a withdrawal of $168.

Solution

If we decide to call a deposit positive, then a withdrawal is negative, and we have the following results:

	Number
Deposit of $25	25
No deposit or withdrawal	0
Withdrawal of $168	−168

    The **opposite** of a number is the number that is the same distance from 0 on the number line but in the opposite direction. Thus, the opposite of 168 is −168, and the opposite of −168 is 168. In symbols, we write the opposite of a real number a as $-a$; then the real number $-a$ has the opposite $-(-a)$. Now, because both $-(-a)$ and a are the same distance from 0 as $-a$, but in the opposite direction, it follows that $-(-a) = a$. This is shown in Figure 2.4.

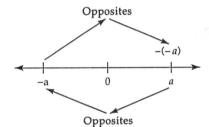

FIGURE 2.4

EXAMPLE 2

Give the opposite, $-x$, of each number x.

a. $x = 3.5$ **b.** $x = -\dfrac{2}{5}$ **c.** $x = 0$

Solution

a. Replace x by 3.5 in the expression $-x$.

$$-x = -3.5$$

b. Replace x by $-\dfrac{2}{5}$ in the expression $-x$.

$$-x = -\left(-\frac{2}{5}\right) = \frac{2}{5}$$

c. Since 0 is neither positive nor negative, we have

$$-x = -0 = 0$$

Absolute Value

Recall that the set of real numbers consists of all numbers that can be placed on a number line. This includes zero, all positive and negative integers, terminating decimals (decimals with a finite number of places), and fractions, as well as many positive and negative numbers that cannot be written as fractions, that is, *irrational numbers*. (The decimal forms of some fractions and all irrational numbers are infinite, or nonterminating.)

An important idea about real numbers is that of absolute value. The **absolute value** of a real number is its distance from zero on the number line. If *n* represents a real number, then the absolute value of *n* is written $|n|$. Because distances are always positive, absolute values are positive (except that $|0| = 0$).

EXAMPLE 3	Write each of the following as a real number with no absolute value sign or parentheses. **a.** $|1.3|$ **b.** $\left|-\dfrac{1}{3}\right|$ **c.** $-|3|$ **d.** $-(-\pi)$
Solution	The number line in Figure 2.5 will help us find these values.

FIGURE 2.5

a. By definition, $|1.3|$ is the distance between 0 and 1.3. Hence $|1.3| = 1.3$.

b. $\left|-\dfrac{1}{3}\right|$ is the distance between $-\dfrac{1}{3}$ and 0. Hence $\left|-\dfrac{1}{3}\right| = \dfrac{1}{3}$.

c. $-|3|$ means the opposite or negative of $|3|$. We must first find $|3|$ and then take its opposite or negative. Since $|3| = 3$, we have $-|3| = -3$.

d. $-(-\pi)$ means the opposite of $-\pi$. So we must first find $-\pi$ on the number line and then find its opposite. Now $-\pi$ is π units (about 3.14 units) to the *left* of zero. Then $-(-\pi)$ is the number that is π units from zero but to the right, as in Figure 2.5. Note that $-(-\pi)$ means "the opposite of the opposite of π," but that's just π. ∎

Example 3 suggests the following rule for finding the absolute value of any number.

Finding absolute values

If a represents a real number, then

$$|a| = \begin{cases} a \text{ if } a \text{ is positive or zero} \\ \text{the opposite of } a \text{ if } a \text{ is negative} \end{cases}$$

In Chapter 1, we defined variables and algebraic expressions. We also found the values of algebraic expressions when the variables were assigned nonnegative values. In algebra, variables may also represent negative numbers.

EXAMPLE 4

Evaluate if $x = -3$ and $y = 2$:

a. $-x$ **b.** $-|x|$ **c.** $5 - |-y|$ **d.** $2[|x| + 3(1 + |2y|)]$

Solution

To *evaluate* an expression means to replace the variables (letters) with their numerical values and simplify.

a.   We substitute -3 into the expression in place of x:

$$-x = -(-3) = 3$$

Note that we place parentheses around -3 to separate the two minus signs. The symbol $--3$ could be misread as -3.

b.   If $x = -3$, then $-|x| = -|-3| = -3$.

c.   If $y = 2$, then

$$5 - |-y| = 5 - |-2|$$
$$= 5 - 2$$
$$= 3$$

d.   We substitute -3 for x and 2 for y and use the rules in Chapter 1 for order of operations.

$$2[|x| + 3(1 + |2y|)] = 2[|-3| + 3(1 + |2 \cdot 2|)]$$
$$= 2[3 + 3(1 + 4)]$$
$$= 2[3 + 3 \cdot 5]$$
$$= 2[3 + 15]$$
$$= 2 \cdot 18$$
$$= 36$$ ∎

> ### *BE CAREFUL!*
> a. $|-5| = 5$, but $-|5| = -5$.
> b. $-(-5) = 5$, but $-|-5| = -5$.
> c. x may represent a negative number, say, $x = -2$. Then $-x$ is $-(-2) = 2$, a positive number.

Negative Numbers on a Calculator

The $\boxed{+/-}$ key on a calculator allows you to enter a negative number, as is illustrated below.

Expression	Key Sequence	Display
-7.8	$7.8 \boxed{+/-}$	-7.8
$-(-6)$	$6 \boxed{+/-} \boxed{+/-}$	6
$-(-(-6))$	$6 \boxed{+/-} \boxed{+/-} \boxed{+/-}$	-6

DISCUSSION QUESTIONS 2.1

1.   Use the $\boxed{+/-}$ key to display each of the following numbers on your calculator:
   a. -67 **b.** -0.0009 **c.** $-(-43)$

2.   Enter $-(-(-(-6)))$ and $-(-(-(-(-6))))$ on your calculator by using the $\boxed{+/-}$ repeatedly. Give a general rule for the value of such numbers in terms of the number of minus signs.

3.   Explain how $|-6|$, $-|6|$, and $-(-6)$ are different.

EXERCISES 2.1

1.   Order the absolute values of the following numbers from smallest to largest:

$$-3, -\frac{1}{4}, -1.5, 3, \frac{1}{4}, 1.5$$

In Exercises 2–7, find the absolute value of the given number, then plot the given number and its absolute value on the same number line.

2. 5   **3.** -3.5 **4.** 1.6   **5.** 0   **6.** -4 **7.** $-\pi$

In Exercises 8–15, use a positive or negative number to represent each situation.

8. A profit of \$75   **9.** A temperature of 12°F below zero

10. Three over par in golf   **11.** A gain of 4 yards (yd) in football

12. A loss of $3\frac{1}{8}$ points (dollars) on the stock market

13. An elevation of 43.6 meters (m) below sea level

14. Twenty dollars overdrawn on a checking account

15. Ten points "in the hole" in a card game

In Exercises 16–28, write each expression as a real number with no parenthesis or absolute value signs.

16. $|2.6|$ **17.** $-\left(-8\frac{1}{3}\right)$ **18.** $|-156|$ **19.** $-\left|61\frac{2}{9}\right|$ **20.** $-|-19.75|$

21. $-|0|$ **22.** $-(-36)$ **23.** $-(-|-16|)$ **24.** $|17| + |-6|$ **25.** $|17| - |-6|$

26. $2 \cdot |-5|$ **27.** $\left|-\frac{2}{3}\right| \div \left|\frac{1}{4}\right|$ **28.** $|-2| \cdot [8 + |-4| \cdot (|1| + |-4|)]$

In Exercises 29–56, evaluate each expression if $a = -3$ and $b = 2$.

29. $-a$ **30.** $|a|$ **31.** $-(-b)$ **32.** $-|b|$ **33.** $|-a|$

34. $-|-b|$ **35.** $|b|$ **36.** $-(-a)$ **37.** $-(-|b|)$ **38.** $-[-(-a)]$

39. $5|a|$ (Hint: $5|a|$ means $5 \cdot |a|$.)   **40.** $|5b|$ **41.** $|a| + |b|$

42. $|a| - |b|$ **43.** $|a|\,|b|$ (Hint: $|a|\,|b|$ means $|a| \cdot |b|$.)   **44.** $3|a| + |b|$

45. $|-4|\,|b|$ **46.** $|a| + b$ **47.** $b + 2|a|$ **48.** $\|a\|$ **49.** $\|b\|$

50. $\dfrac{|a|}{|b|}$ **51.** $\dfrac{3|b|}{4|a|}$ **52.** $2|a| - 3|b|$ **53.** $2|a| + 2|b|$ **54.** $(4 + |a| + |b|) \div 3$

55. $100 - 2[3|a| + 2|4b| - (|a| - |b|)]$ **56.** $|a|\{b + 2[6|a| - 3(|b| + 1)]\}$

Preparing for Section 2.2

In Exercises 57–59, perform the indicated operation.

57. $5\frac{1}{5} - 3\frac{9}{10}$ **58.** $5.14 + 3.78$ **59.** $5.03 - 2.98$

In Exercises 60–63, draw a number line and use it to show the first number and the result of the indicated operation. For example, for $2 + 4$, you would show 2 and the number that is 4 units to the right of 2, namely 6, on a number line.

60. $3 + 3$ **61.** $6 - 2$ **62.** $4\frac{1}{2} + 2\frac{1}{2}$ **63.** $5.3 - 2.8$

2.2   Addition of Real Numbers

FOCUS

We discuss how to add real numbers—both positive and negative.

If a football team runs two plays, gaining 5 yd on the first play and 3 yd on the second, they are 5 + 3 or 8 yd beyond where they started. Similarly, if a team gains 12 yd on the first play and loses 5 yd on the next, we must "put together" or add 12 yd and −5 yd to determine how far the team is beyond their starting point. This calls for addition involving both positive and negative numbers.

Adding Real Numbers

The addition of real numbers can easily be illustrated on a number line. Think of addition as moving along the line. If we are adding a positive number, we will move to the right. If we are adding a negative number, we will move to the left.

EXAMPLE 1   Compute by using a number line:

a. $7 + 5$ **b.** $7 + (-5)$ **c.** $(-7) + 5$ **d.** $(-7) + (-5)$

Solution   **a.** On the number line in Figure 2.6, we begin by locating the first number, 7. We then move (or count) 5 units to the right (since 5 is positive). The sum, of course, is 12.

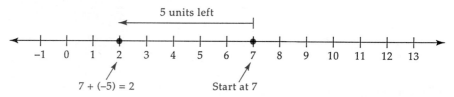

FIGURE 2.6

b. On the number line in Figure 2.7, we first locate 7. We then move 5 units to the left, since we are adding −5. The sum is $7 + (-5) = 2$.

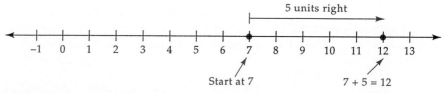

FIGURE 2.7

c. On the number line in Figure 2.8, we locate the first number, -7. We then move 5 units to the right, since we are adding 5. The sum is $(-7) + 5 = -2$.

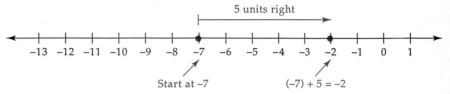

FIGURE 2.8

d. On the number line in Figure 2.9, we first locate -7. We then move 5 units to the left, since we are adding -5. The sum is $(-7) + (-5) = -12$.

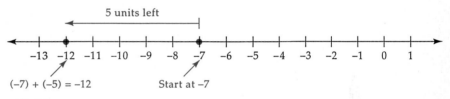

FIGURE 2.9   ■

For many people, it is helpful to think of a number line when adding real numbers. However, there are also computational rules that you may want to use.

Adding real numbers

1. To add two positive numbers, add their absolute values (which is the same as adding the numbers as you did in arithmetic).
2. To add two negative numbers, add their absolute values and give the result a minus sign.
3. To add a positive number and a negative number, find the difference of their absolute values. If the positive addend has the greater absolute value, the result is positive. If the negative addend has the greater absolute value, give the result a minus sign.

The next examples use these rules along with the commutative and associative properties, which apply to *all* real numbers. Recall that the commutative property states that two (or more) numbers may be added in either order, that is,

$$a + b = b + a$$

The associative property states that three (or more) numbers may be paired in any way for addition, that is,

$$(a + b) + c = a + (b + c)$$

When positive and negative numbers are to be added, it is simplest to add all the positive numbers first, then add the negative numbers separately, and finally combine the two totals.

EXAMPLE 2   Add:

a.   35
        −18
         70
        −25

b. $14 + (−12) + (−17) + 8 + (−36)$

Solution

a.   We add the positive numbers first, then add the negative numbers separately.

$$
\begin{array}{rcrcr}
35 & \text{and} & -18 & \text{combining} & 105 \\
70 & & -25 & & -43 \\
\overline{105} & & \overline{-43} & & \overline{62}
\end{array}
$$

b.   We use the same method to find this sum.

$$
\begin{array}{rcrcr}
14 & \text{and} & -12 & \text{combining} & 22 \\
8 & & -17 & & -65 \\
\overline{22} & & -36 & & \overline{-43} \\
& & \overline{-65} & &
\end{array}
$$

Obviously, adding 0 to a real number does not change it.

Additive identity

For any real number a,

$$a + 0 = 0 + a = a$$

For this reason, 0 is called the **additive identity.**

The sum of any real number and its opposite is 0. For example, $9 + (-9) = 0$ and $12 + (-12) = 0$. Thus,

Additive inverse

For any real number a,

$$a + (-a) = (-a) + a = 0$$

For this reason, $-a$ is called the **additive inverse** of a.

Absolute Value of a Sum

The absolute value of a sum is found by *first* computing the sum and then finding the absolute value.

EXAMPLE 3

Evaluate each expression if $r = -2$ and $s = 6$.

a. $|r + 2s|$ **b.** $r + 2(s + 4|r + (-s)|)$

Solution

In each part, we first substitute -2 for r and 6 for s. Then we add and take the absolute value.

a. $|r + 2s| = |(-2) + 2(6)|$

$= |(-2) + 12|$

$= |10|$

$= 10$

b. $r + 2(s + 4|r + (-s)|) = (-2) + 2(6 + 4|(-2) + (-6)|)$

$$= (-2) + 2(6 + 4|-8|)$$

$$= (-2) + 2(6 + 4(8))$$

$$= (-2) + 2(6 + 32)$$

$$= (-2) + 2(38)$$

$$= (-2) + 76$$

$$= 74$$ ∎

BE CAREFUL!

$|a + b|$ is not always the same as $|a| + |b|$. **For example,**

$$|10 + (-4)| = |6| = 6$$

but $$|10| + |-4| = 10 + 4 = 14$$

Calculator Addition

Positive and negative numbers can be added on a calculator, as the next example shows.

EXAMPLE 4   Use your calculator to add:

a. $3.14 + (-13.9)$ **b.** $-45.97 + (-67.8)$

Solution   For now, we shall use the following key sequences. Later we shall see that shorter sequences will also work.

a.   *Expression*   *Key Sequence*   *Display*
  $3.14 + (-13.9)$ 3.14 $\boxed{+}$ 13.9 $\boxed{+/-}$ $\boxed{=}$ -10.76

b.   *Expression*   *Key Sequence*   *Display*
  $-45.97 + (-67.8)$ 45.97 $\boxed{+/-}$ $\boxed{+}$ 67.8 $\boxed{+/-}$ $\boxed{=}$ -113.77 ∎

Ordering Real Numbers

Which is a higher temperature, $-10°$ or $-7°$? Would it make sense to say that -7 is greater than -10? In fact, we do just that.

> **Order for real numbers**
>
> If a and b represent any two real numbers, $a > b$ (read "a is greater than b") provided a is to the right of b on a number line. Moreover, $a < b$ ("a is less than b") means $b > a$.

Thus, *any* negative number is less than 0 and every positive number. For example, $-100 < 1$ even though 100 is much greater than 1.

EXAMPLE 5

Place these numbers in order, smallest to largest:

$$5 \quad -8 \quad 2.1 \quad -9 \quad -3.6 \quad 3.2 \quad -4 \quad -3 \quad 2.5$$

Solution

We plot these numbers on the number line in Figure 2.10. As you can see, the least number is -9, since it is furthest to the left; next is -8; then -4; and so on. Since it is furthest to the right, 5 is the greatest number in the list. From least to greatest, we have:

$$-9, -8, -4, -3.6, -3, 2.1, 2.5, 3.2, 5$$

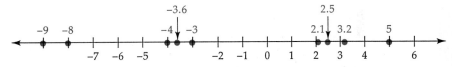

FIGURE 2.10

Next we consider an application of real-number addition.

EXAMPLE 6

During one week the price of a certain stock changed as follows: Monday, down $\frac{1}{8}$ dollar; Tuesday, down $\frac{5}{8}$ dollar; Wednesday, unchanged; Thursday, up $1\frac{3}{4}$ dollars; Friday, down $2\frac{1}{8}$ dollars. Find the stock's net change in value for the week.

Solution

Use positive numbers for rises and negative numbers for falls. The net change is the sum of the daily changes:

$$\left(-\frac{1}{8}\right) + \left(-\frac{5}{8}\right) + 0 + 1\frac{3}{4} + \left(-2\frac{1}{8}\right)$$

You can find the net change, $-1\frac{1}{8}$, by computing this sum.

In the solution of Example 7, we added positive and negative fractions. To do so, we used the rules in Section 1.1 for operations on nonnegative fractions with the rules for adding real numbers in this section. For example,

$$1\frac{3}{4} + \left(-2\frac{1}{8}\right) = -\left(\left|-2\frac{1}{8}\right| - \left|1\frac{3}{4}\right|\right)$$ Rules for adding real numbers

$$= -\left(2\frac{1}{8} - 1\frac{3}{4}\right)$$ Finding absolute values

$$= -\left(\frac{17}{8} - \frac{7}{4}\right)$$ Adding $2 + \frac{1}{8}$ and $1 + \frac{3}{4}$

$$= -\frac{(17 \cdot 4 - 8 \cdot 7)}{8 \cdot 4}$$ Rule for subtracting fractions

$$= -\frac{(68 - 56)}{32}$$ Multiplying

$$= -\frac{12}{32}$$ Subtracting

$$= -\frac{3 \cdot 4}{8 \cdot 4}$$ Showing common divisors

$$= -\frac{3}{8}$$ Simplifying

DISCUSSION QUESTIONS 2.2

1. What property implies that $-6 + 8 = 8 + (-6)$?

2. What property implies that $(-5 + 7) + (-8) = -5 + (7 + (-8))$?

3. Suppose $x + (-9) = -9$. What value must x have? What general rule does this illustrate?

4. Suppose $8 + x = 0$. What value must x have? Suppose $y + (-6) = 0$. What value does y have? What general rule is illustrated by these equations?

5. Suppose x and y represent numbers on the number line in Figure 2.11. Which is greater, x or y? Explain.

FIGURE 2.11

EXERCISES 2.2

In Exercises 1–5, use a number line to find the indicated sum. Check your answers with a calculator.

1. $(-3) + 5$ **2.** $10 + (-13)$ **3.** $(-7) + (-2)$ **4.** $(-4) + (-3)$ **5.** $2.6 + (-1.4)$

In Exercises 6–23, find each sum.

6. $9 + (-19)$ **7.** $\left(-\dfrac{2}{7}\right) + \left(-\dfrac{3}{7}\right)$ **8.** $3.6 + (-2.1)$ **9.** $0 + (-12)$

10. $(-8.3) + 8.3$ **11.** $\dfrac{4}{5} + \left(-\dfrac{9}{5}\right)$ **12.** $(-86) + 90$ **13.** $(-13) + (-5)$

14. $(-25) + (-25)$ **15.** $4.32 + (-6.32)$ **16.** $\dfrac{13}{3} + \left(-\dfrac{13}{3}\right)$ **17.** $(-62) + 50 + (-18)$

18. $3.7 + (-4.9) + (-16.1)$ **19.** $\dfrac{1}{2} + \left(-\dfrac{5}{8}\right) + \dfrac{1}{4}$ **20.** $(-5.6) + (-6.3) + (-4.4)$

21. $(-245) + (-467)$ **22.** $(-2583) + 3988$ **23.** $(-5.86) + 7.89$

In Exercises 24–40, evaluate each expression if $x = 5$, $y = -4$, and $z = 3$.

24. $(-x) + (-y)$ **25.** $-(x + y)$ **26.** $(-z) + y$ **27.** $(x + y) + z$

28. $x + (y + z)$ **29.** $x + |y + z|$ **30.** $(-y) + x + y$ **31.** $|x + y| + z$

32. $-|z + (-z)|$ **33.** $2.2x|6.8 + 1.3|y\,||$ **34.** $|3.2x| + 4.5|y|$ **35.** $3z(|y| + 2|y + z|)$

36. $x(4|x + y| - 3) + 2z$ **37.** $|-5.8| + 3.1|z| + 8.7|x|$ **38.** $|x|(6.73 - 1.2|z|) + |3.64x|$

39. $46.3 - 3.6[4.2|2.5z + y| + 3.2z(y + 3.2x)]$ **40.** $8.6[1.4|y| + 0.5xz(3.4z + y)] - 1.1|-y|$

41. Copy the number line in Figure 2.12. On it, plot the opposites of a and b.

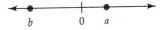

FIGURE 2.12

42. Copy the number line in Figure 2.13. On it, plot a number b that is less than a and a number d that is less than c.

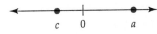

FIGURE 2.13

43. Copy the number line in Figure 2.14. On it, plot a positive number b such that $|b| > |a|$. Then plot a negative number c such that $|c| > |a|$.

FIGURE 2.14

44. Copy the number line in Figure 2.15. On it, plot a positive number b such that $|b| > |a|$. Then plot a negative number c such that $|c| > |a|$.

FIGURE 2.15

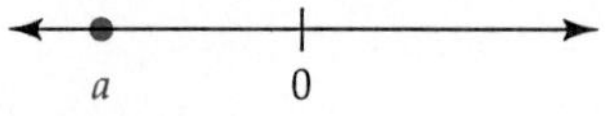

In Exercises 45–52, write the series of steps as an addition problem. Then find the sum and state it in terms of the given situation.

45. Going up 500 feet (ft) in a balloon, then going down 120 ft

46. Withdrawing $40, then depositing $50

47. Withdrawing $50, then depositing $40

48. The temperature rising 5°, then falling 13°, then rising 6°

49. The temperature falling 13°, then rising 5°, then falling 8°

50. The temperature falling t°, then falling 8°, then rising 10°

51. The temperature falling 12°, then rising t°, then rising $2t$°

52. Climbing a 250-ft hill and then descending 321 ft into a mine shaft

In Exercises 53–56, use positive and negative numbers to solve the problem.

53. The Chicago Bears gained 7 yd on their first play in a game against the Eagles. On the next play they lost 2 yd, and on their third play they gained 12 yd. What was the Bears' net yardage on these three plays?

54. Solid frozen oxygen at a temperature of −218.4°C must be heated 31.6°C in order to evaporate. At what temperature does oxygen evaporate?

55. Jason Electronics made a profit of $6541 during the first quarter of the year. In the remaining quarters it lost $8042, gained $3512, and lost $1322, respectively. Did Jason Electronics show a profit for the year? If yes, how much? If not, how much did it lose?

56. The temperature in Billings at 6 a.m. was −7°. By noon the temperature had risen 22°, and by 6 p.m. it was down to 5°. What was the temperature at noon? How many degrees did the temperature fall between noon and 6 p.m.?

57. A helium balloon launched from Death Valley (280 ft below sea level) rose to a height of 1350 ft above sea level. To find how far the balloon rose during its flight, the balloonist did the following computation:

$$-280 \text{ ft} + 1350 \text{ ft} = 1070 \text{ ft}$$

Is this solution correct? Explain your answer, and draw a diagram showing (a) sea level, (b) the point at which the balloon was launched, and (c) the point to which the balloon rose. Mark distances on the diagram.

58. In Exercise 57, in order to reach a height of 2500 ft above sea level, how many feet above the ground in Death Valley must the balloon rise?

Preparing for Section 2.3

In Exercises 59–62, perform both operations and compare the results.

59. $10 - 5$ and $10 + (-5)$

60. $34 - 15$ and $34 + (-15)$

61. $\dfrac{3}{4} - \dfrac{1}{3}$ and $\dfrac{3}{4} + \left(-\dfrac{1}{3}\right)$

62. $53.4 - 39.6$ and $53.4 + (-39.6)$

In Exercises 63–68, draw a number line and find x, the number that must be added to the first number to get the number on the right.

63. $5 + x = 9$

64. $10 + x = 6$

65. $6 + x = -3$

66. $-4 + x = 0$

67. $-2 + x = -7$

68. $-3 + x = 5$

2.3 Subtraction of Real Numbers

FOCUS

We discuss how to subtract real numbers, and see that any subtraction can be completed by using a related addition.

A relationship between addition and subtraction of whole numbers is illustrated by the following examples.

$$10 - 7 = 3 \text{ because } 10 = 3 + 7$$
$$12 - 4 = 8 \text{ because } 12 = 4 + 8$$

In general,

$$a - b = c \text{ whenever } a = c + b$$

This relationship between addition and subtraction holds for any real numbers a, b, and c. We now use it to subtract. For example, suppose we wanted to find the difference $7 - 9$. From the above we know that

$$7 - 9 = n \text{ provided } 7 = n + 9$$

Subtracting on a Number Line

Let us use the number line in Figure 2.16 to see what number we must *add* to 9 to obtain 7. We start at 9 and must move 2 to the left to reach 7; thus, $7 = -2 + 9$. This tells us that n is -2, so $7 - 9 = -2$.

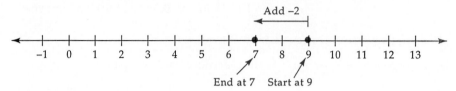

FIGURE 2.16

EXAMPLE 1

Subtract using a number line.

  a. $-6 - 7$ **b.** $9 - (-2)$ **c.** $-4 - (-5)$

Solution

  a. $-6 - 7 = n$ provided $-6 = n + 7$. On the number line in Figure 2.17, start at 7. You must move 13 *to the left* to reach -6; thus, $-6 = -13 + 7$. Then n is -13 and $-6 - 7 = -13$.

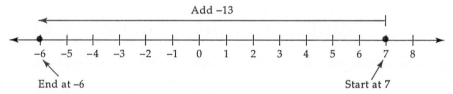

FIGURE 2.17

  b. $9 - (-2) = n$ provided $9 = n + (-2)$. On the number line in Figure 2.18, start at -2. You must move 11 *to the right* to reach 9; thus, $9 = 11 + (-2)$. Then n is 11 and $9 - (-2) = 11$.

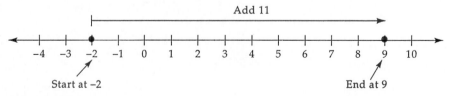

FIGURE 2.18

  c. $-4 - (-5) = n$ provided $-4 = n + (-5)$. On the number line in Figure 2.19, start at -5. You must move 1 *to the right* to reach -4; thus, $-4 = 1 + (-5)$ and $-4 - (-5) = 1$.

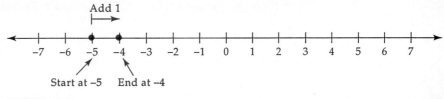

FIGURE 2.19

Subtracting as a Related Addition

It is not very convenient to use the number line to subtract. Fortunately, there is another way. Note that, for the problems in Example 1, we have

$$-6 - 7 = -13 \text{ and } -6 + (-7) = -13$$

$$9 - (-2) = 11 \text{ and } 9 + 2 = 11$$

$$-4 - (-5) = 1 \text{ and } -4 + 5 = 1$$

Hence, we can convert subtractions into equivalent additions. We have the following general rule:

Changing subtraction to addition

To subtract a real number b from a real number a, add the opposite of b to a. In symbols:

$$a - b = a + (-b)$$

Remember: If b is negative, then $-b$ is positive.

This rule is illustrated with a calculator in the next example.

EXAMPLE 2

Solution

Use your calculator to verify that $87.4 - (-56.7) = 87.4 + 56.7$.

Expression	Key Sequence	Display
$87.4 - (-56.7)$	87.4 $\boxed{-}$ 56.7 $\boxed{+/-}$ $\boxed{=}$	144.1
$87.4 + 56.7$	87.4 $\boxed{+}$ 56.7 $\boxed{=}$	144.1

The two key sequences give the same result. Which takes fewer keystrokes?

In the remaining examples, we apply the rule with paper and pencil.

EXAMPLE 3

Subtract by first converting to addition:

a. $7.4 - 9.2$ **b.** $(-36) - (-40)$

Solution

a. $\begin{aligned}7.4 - 9.2 &= 7.4 + (-9.2) \qquad &\text{Converting to addition}\\ &= -[|-9.2| - |7.4|] \qquad &\text{Rules for addition}\\ &= -[9.2 - 7.4] \qquad &\text{Taking absolute values}\\ &= -1.8 \qquad &\text{Subtracting}\end{aligned}$

b. $\begin{aligned}(-36) - (-40) &= (-36) + 40 \qquad &\text{Converting to addition}\\ &= |40| - |-36| \qquad &\text{Rules for addition}\\ &= 40 - 36 \qquad &\text{Taking absolute values}\\ &= 4 \qquad &\text{Subtracting}\end{aligned}$ ∎

EXAMPLE 4

Evaluate the following if $x = -3$ and $y = 4$:

a. $y - x + 5$ **b.** $18 - (y + x)$

Solution

In each case, replace the letters by their numerical values. Then change all subtractions to additions and add.

a. $\begin{aligned}y - x + 5 &= 4 - (-3) + 5\\ &= 4 + 3 + 5\\ &= 12\end{aligned}$

b. $\begin{aligned}18 - (y + x) &= 18 - (4 + (-3))\\ &= 18 - 1\\ &= 17\end{aligned}$ ∎

We use subtraction of negative numbers in the following application.

EXAMPLE 5

After three hands in a card game, Phil was 8 points in the hole. In other words, his score was -8. Before the next hand, Phil discovered that the scorer had made a mistake and given him 12 negative points too many. Phil found his corrected score by taking away (that is, subtracting) the -12 points. What is Phil's corrected score?

Solution   Phil's corrected score is gotten by subtracting -12 points from his original score:

$$-8 - (-12) = -8 + 12$$
$$= +4$$

His corrected score is $+4$.   ∎

Distance Between Two Numbers

We defined $|a|$ to be the distance from a to 0 on a number line. Absolute values can also be used to define the distance between any two numbers a and b.

Distance between a and b

The *distance between two real numbers a and b* is $|a - b|$.

EXAMPLE 6   Find the distance between r and s.

    a. $r = 4, s = -9$ **b.** $r = -12.9, s = 7.6$

Solution   In each case, the distance is $|r - s|$. We substitute the given numbers for r and s, and then subtract *before* taking the absolute value of the result.

    a. $|r - s| = |4 - (-9)|$ Substituting

               $= |13|$ Subtracting

               $= 13$ Taking the absolute value

    b. $|r - s| = |-12.9 - 7.6|$ Substituting

               $= |-20.5|$ Subtracting

               $= 20.5$ Taking the absolute value   ∎

DISCUSSION QUESTIONS 2.3

1. Suppose your bank account showed a balance of $100 at the end of February. Suppose that the bank then discovered that a withdrawal of $10 (i.e., $-\$10$) had been charged against your balance by mistake, and so they "took away" or subtracted the erroneous withdrawal. What would be your corrected balance?

2. Using Question 1 as an example, discuss why "taking away" a negative number actually increases a number (such as the bank account balance).

3. Compute $(10 - 3) - 4$ and $10 - (3 - 4)$. Compare your answers. Do you think subtraction is associative?

EXERCISES 2.3

In Exercises 1–6, determine whether $-x$ is a positive or negative number when x has the given value.

1. 6   **2.** -5 **3.** 42   **4.** $-\dfrac{1}{2}$ **5.** $-(-10)$ **6.** -3.25

In Exercises 7–12, use a number line to perform the indicated computation.

7. $-2 + 5$

8. $6 - (-3)$

9. $-4 - 3$

10. $-5 - (-6)$

11. $5 + 2 - (-2)$

12. $5 - 2 + (-2)$

In Exercises 13–15, find the distance from a to b. Then plot a and b on a number line to show that your computed distance is correct.

13. $a = -3, b = -5.5$ **14.** $a = 3\dfrac{1}{4}, b = -7\dfrac{1}{2}$ **15.** $a = 4.2, b = -4.3$

In Exercises 16–33, perform the indicated computation.

16. $7.3 - 4.9$

17. $\left(-\dfrac{3}{4}\right) - \dfrac{1}{4}$

18. $(-12) - (-8)$

19. $(-45) - 10$

20. $17.3 - 21.6$

21. $\dfrac{4}{3} - 2$

22. $20 - (-4)$

23. $0 - \dfrac{9}{17}$

24. $325 - (-476)$

25. $(-457) - 895$

26. $10.9 - (-8.6) + (-6.3)$

27. $(-6) + 28 - (-6)$

28. $(-584) + 997 - (-352)$ **29.** $(-1235) - (-5476) + (-1658)$ **30.** $6 - |16 - (-3)|$

31. $2|3 - (-5)|$

32. $5 + [-8 - 7] - (-12)$

33. $14.5 - \{2.6 - [10.4 - (-2.6)]\}$

In Exercises 36–56, evaluate each expression if $x = 2$, $y = 4$, and $z = -3$.

34. $z - x$

35. $(x - y) + z$

36. $x + (y - z)$

37. $x - (y + z)$

38. $x - y + (-z)$

39. $(-x) + (z - y)$

40. $|(-x) + y| + z$

41. $y - |x + z|$

42. $|(-x) - y| - z$

43. $z + z + z$ **44.** $y - (x - z)$ **45.** $2x - 3y - (x - z)$

46. $z - 2(3x - 2y)$ **47.** $5(-z) - (2x - z)$ **48.** $(3 + z) - (2y - x)$

49. $4.6(-z) - 6.2x + 5.1y$ **50.** $9.8x - 4.6(1.2x - 3.1y)$ **51.** $100x - 452|z|$

52. $572y - 39(271x + 896|z|)$ **53.** $(452 - 127x) - (672 + 398|z|)$

54. $890 + [100(-765x + z) - 345y]$ **55.** $x - \{z - [y - (x + z)]\}$

56. $2x - \{[z - (2y - z)] - x\}$

57. On a test, both Peter and Maria scored 88. However, Professor Fermat had made some scoring errors. To correct them, the professor gave Peter 5 more points for problem 7, and he took away -5 that he had given Maria on problem 6. What was Peter's corrected score? What was Maria's?

In Exercises 58–64, use positive and negative numbers to solve the problem.

58. A skin diver swimming at 175 ft below sea level sees a shark 95 ft above him. How far is the shark below sea level?

59. If liquid hydrogen at $-252°C$ is cooled $7.14°C$, it freezes. What is the freezing point of hydrogen?

60. International Telepathy stock started the day selling for $62\frac{1}{8}$. At noon it sold for $58\frac{3}{4}$. It finished the day selling for $61\frac{1}{2}$. How many points did the stock drop by noon? What was the net change in the price of the stock for the day?

61. A Polaris submarine is 326 m below sea level. It fires a missile that travels a total of 3847 m straight up and then explodes. At what elevation does the missile explode?

62. A wonder drug kills 18 strains of virus every hour. If there are 296 strains of virus now, how many were there 6 h ago?

63. Lori's checkbook showed a balance of $26. She then wrote a check for $83. What was her new balance? Later she deposited $130. What was her balance after the deposit?

64. Complete this table of batting averages (BAs).

Player	BA on July 4	BA at season's end	Change
Boggs	.398	.365	−.033
Mattingly	.304	.349	?
Puckett	.333	?	−.004
Allen	?	.324	+.018

Preparing for Section 2.4

In Exercises 65–67, evaluate each pair of expressions when:

a. $x = 2$ **b.** $x = -1$

65. $-(x - 3); -x + 3$ **66.** $-(7 - x); -7 + x$ **67.** $-(-6 - x); 6 + x$

68. Use Exercises 65–67 to propose a rule for removing parentheses when they are preceded by a minus sign.

69. Use the distributive property (which applies to all real numbers) to rewrite $5(2 + (-2))$.

70. Since $2 + (-2) = 0$ and $(5)(0) = 0$, what is $5(2 + (-2))$?

2.4 Multiplication and Division of Real Numbers

FOCUS

We discuss how to multiply and divide both positive and negative real numbers.

Multiplication

The multiplication of whole numbers can be interpreted as repeated addition; for example,

$$3 \cdot 5 = 5 + 5 + 5 = 15$$

and

$$4 \cdot 2 = 2 + 2 + 2 + 2 = 8$$

We apply this same interpretation to the product of a positive integer and a negative integer.

$$3(-5) = (-5) + (-5) + (-5) = -15$$

and

$$4(-2) = (-2) + (-2) + (-2) + (-2) = -8$$

The commutative property for multiplication tells us that $ab = ba$, and it applies to all real numbers, whether they are positive or negative. We can use this property to find the product of a negative number times a positive number, for it allows us to interchange the factors. Hence, we have, for example,

$$(-5)(3) = 3(-5) = (-5) + (-5) + (-5) = -15$$

and $$(-2)(4) = 4(-2) = (-2) + (-2) + (-2) + (-2) = -8$$

We must also consider the product of two negative numbers, as in $(-3)(-5)$. Let us assume that the distributive property (which says that $a(b + c) = ab + ac$) holds for all real numbers (it does), and use it to find the sum $(-3)(-5) + (-15)$. We have

$$(-3)(-5) + (-15) = (-3)(-5) + (-3)(5) \qquad \text{Since } (-3)(5) = -15$$

$$= (-3)[(-5) + 5] \qquad \text{By the distributive property}$$

$$= 3(0) = 0$$

Thus, when -15 is added to $(-3)(-5)$, the result is 0; hence, $(-3)(-5)$ is the additive inverse of -15 and must therefore be equal to 15.

The general rules suggested by these examples are summarized below. They apply to all real numbers, not just to the integers in the examples.

Multiplying real numbers

1. The product of two positive or two negative numbers is the product of their absolute values.
2. The product of a positive number and a negative number (in either order) is the opposite of the product of their absolute values.
3. The product of 0 and any number (in either order) is 0.

EXAMPLE 1

Multiply:

a. $18(-7)$ b. $\left(-\dfrac{1}{2}\right)(-6)$ c. $(-3.2)(5)$

Solution

a. One number is positive and the other is negative. Hence, we multiply their absolute values and make the product negative.

$$18(-7) = -(18)(7) = -126$$

b. Both numbers are negative, so the product is positive. Just multiply $\left(\frac{1}{2}\right)(6)$:

$$\left(-\frac{1}{2}\right)(-6) = \left(\frac{1}{2}\right)(6)$$
$$= \frac{6}{2}$$
$$= 3$$

c. As in part a, one number is positive and the other is negative:

$$(-3.2)(5) = -16.0 \qquad\blacksquare$$

Obviously, multiplying any real number by 1 does not change it.

Multiplicative identity

For any real number a,

$$a \cdot 1 = 1 \cdot a = a$$

For this reason, 1 is called the **multiplicative identity**.

The **reciprocal** of a nonzero real number a is $\frac{1}{a}$. Note that the product of a number and its reciprocal is 1, since

$$(a)\left(\frac{1}{a}\right) = \left(\frac{a}{1}\right)\left(\frac{1}{a}\right) \qquad \text{Writing } a \text{ as a fraction}$$
$$= \frac{a \cdot 1}{1 \cdot a} \qquad \text{Multiplying numerators and denominators}$$
$$= \frac{a}{a} \qquad \text{Multiplying}$$
$$= 1$$

The *reciprocal of a nonzero fraction,* $\frac{a}{b}$, is $\frac{b}{a}$.

Multiplicative inverse

For any nonzero real number a,

$$(a)\left(\frac{1}{a}\right) = \left(\frac{1}{a}\right)(a) = 1$$

For this reason, $\dfrac{1}{a}$ is also called the **multiplicative inverse** of a.

Division

In Section 2.3, we referred to a relationship between addition and subtraction, first for whole numbers and then for real numbers. A similar relationship between multiplication and division for whole numbers is illustrated by the following examples.

$$10 \div 5 = 2 \text{ because } 10 = 2 \cdot 5$$

$$21 \div 7 = 3 \text{ because } 21 = 3 \cdot 7$$

This relationship holds as well for the real numbers—both positive and negative. For example,

$$(-10) \div 5 = -2 \text{ because } -10 = -2 \cdot 5$$
$$21 \div (-7) = -3 \text{ because } 21 = -3 \cdot (-7)$$
$$(-12) \div (-4) = 3 \text{ because } -12 = 3 \cdot (-4)$$
$$0 \div (-5) = 0 \text{ because } 0 = 0 \cdot (-5)$$

The division $5 \div 0$ is not defined, nor is any division by zero.

The general rules suggested by these examples are summarized below. They apply to all real numbers, not just to the integers in the examples.

Dividing real numbers

1. The quotient of two positive or two negative numbers is the quotient of their absolute values.
2. The quotient of a positive number and a negative number (in either order) is the opposite of the quotient of their absolute values.
3. For any real number n,

$$0 \div n = 0$$

but $n \div 0$ is not defined.

The rules for operations with real numbers are illustrated in the next three examples.

EXAMPLE 2   Divide:

  a. $18 \div (-3)$ **b.** $[(-18) \div 3] \div 2$ **c.** $[(-4) + (-18)] \div (-2)$

Solution **a.** One number is positive, and the other is negative. Hence, their quotient is negative.

$$18 \div (-3) = -[(18) \div (3)]$$
$$= -6$$

  b. First, compute $(-18) \div 3$, noting that, as in part a, one number is positive. Then divide the result by 2.

$$[(-18) \div 3] \div 2 = (-6) \div 2$$
$$= -3$$

  c. First, add the numbers in parentheses, and then divide the result by -2.

$$[(-4) + (-18)] \div (-2) = (-22) \div (-2) \qquad \text{Because both } -22$$
$$\qquad\qquad\qquad\qquad\qquad\qquad\qquad \text{and } -2 \text{ are negative}$$
$$= 11 \qquad\qquad\qquad\qquad\qquad\qquad\blacksquare$$

EXAMPLE 3   Suppose the daily low temperatures (in degrees Celsius) for the last week were as follows: Sunday, $-10°$; Monday, $-2°$; Tuesday, $-4°$; Wednesday, $0°$; Thursday, $-1°$; Friday, $-3°$; and Saturday, $1°$. Find the average low temperature for the week.

Solution   Recall that the average of a set of numbers is found by dividing the sum of the numbers in the set by the number of numbers in the set. In this case, the sum is

$$(-10°) + (-2°) + (-4°) + 0° + (-1°) + (-3°) + 1° = -19°$$

There are seven numbers in all. The average is $(-19°) \div 7 \approx -2.7°$. $\blacksquare$

EXAMPLE 4   Evaluate $(-2)(x + 4)$ and $(-2)x + (-8)$, if $x = -4$. Compare the results in each part.

Solution $(-2)(x + 4) = (-2)[(-4) + 4]$
$$= (-2) \cdot 0 = 0$$

$$(-2)x + (-8) = (-2)(-4) + (-8)$$
$$= 8 + (-8) = 0$$

Notice that the two expressions have the same value.   ∎

BE CAREFUL!

Don't confuse the rules for determining the signs of the results of operations on two negative numbers. Two negatives always make a positive only for multiplication and division. For example,

Addition:	$-6 + (-3) = -9$
Subtraction:	$-6 - (-3) = -3$
	but $-3 - (-6) = 3$
Multiplication:	$(-6)(-3) = 18$
Division:	$(-6) \div (-3) = 2$

DISCUSSION QUESTIONS 2.4

1. Which property of numbers is illustrated by each of the following? Explain.
   a. $(-3)x = x(-3)$ **b.** $17(y + 2) = 17y + 34$
   c. $2(25n) = 50n$ **d.** $(x + 3) + (-3) = x$

2. Indicate which of the following are true and which are false, and explain why. Correct the false statements, and explain why your statement is correct.
   a. $5 + (2 + x) = 7 + 5x$ **b.** $5(x + 2) = 5x + 10$
   c. $5(2x) = 10 \cdot 5x = 50x$ **d.** $6(x - 3) = 6[x + (-3)] = 6x - 3$

3. Which property of numbers is illustrated by Example 4?

4. Explain why $0 \div 3 = 0$, but $3 \div 0$ is not a real number.

5. Use your calculator to perform the following operations (remember, a number like -6.4 is entered by keying 6.4 $\boxed{+/-}$):
   a. $(7.1)(-6.4)$ **b.** $(-3.5)(-4.6)$ **c.** $(-4.6) \div (-0.7)$
   Are the results on your calculator consistent with the rules in this section for multiplying and dividing positive and negative numbers?

EXERCISES 2.4

In Exercises 1–42, perform the indicated computation.

1. $12(-10)$ **2.** $(-8)(-3)$ **3.** $(-9)7$ **4.** $18 \cdot 5$

5. $\left(-\dfrac{4}{9}\right)\left(-\dfrac{3}{2}\right)$ **6.** $(-4.9)(0)$ **7.** $3.3(-4.2)$ **8.** $(-1.5)(-3.6)$

9. $\left(\dfrac{1}{4}\right)(-5.6)$ **10.** $\left(2\dfrac{1}{3}\right)\left(-3\dfrac{2}{5}\right)$ **11.** $\left(-1\dfrac{1}{2}\right)\left(-3\dfrac{1}{4}\right)$ **12.** $(-24.5)\left(-\dfrac{1}{5}\right)$

13. $8(-3)(6)$ **14.** $(-3)(-7)(-2)$ **15.** $(2.4)(4.6)(-1.9)$ **16.** $(-3.5)(-2.7)(-3.8)$

17. $28 \div (-7)$ **18.** $(-200) \div 8$ **19.** $(-12) \div 0.3$ **20.** $(-15) \div (-0.5)$

21. $14.4 \div (-2)$ **22.** $60.6 \div (-6)$ **23.** $(-108) \div (-9)$ **24.** $(-80) \div (-16)$

25. $\dfrac{9}{5} \div \dfrac{3}{10}$ **26.** $\left(-\dfrac{3}{4}\right) \div \dfrac{5}{8}$ **27.** $168 \div (-12)$ **28.** $248 \div (-24)$

29. $\left(-\dfrac{12}{7}\right) \div \dfrac{3}{16}$ **30.** $\left(-\dfrac{2}{7}\right) \div \left(-\dfrac{5}{14}\right)$ **31.** $0 \div (-9)$ **32.** $0 \div (-2.6)$

33. $(-4.23) \div (-9)$ **34.** $(-928) \div (-232)$ **35.** $(-963)(529) \div (-42)$ **36.** $(14) \div (-0.7)$

37. $2\dfrac{1}{2} \div \left(-4\dfrac{3}{4}\right)$ **38.** $(-5)(-8) \div 3$ **39.** $-10 \div 3(-6)$ **40.** $|(-7)(13)|$

41. $\left|\dfrac{1}{2} \div \left(-\dfrac{1}{2}\right)\right|$ **42.** $|(-16)(-1)|$

In Exercises 43–57, evaluate each expression if $p = -2$, $q = 4$, and $r = -6$.

43. $p(q + r)$ **44.** $pq + r$ **45.** $pq + pr$ **46.** $q(p - r)$ **47.** $qp - qr$

48. $p + qr$ **49.** $p \div q$ **50.** $q \div p$ **51.** $pr \div q$ **52.** $p \div rq$

53. $(-p) \div (q + r)$ **54.** $[p + (-q)] \div r$ **55.** $(q + p) \div (-r)$ **56.** $|pq| \div |r|$ **57.** $p \cdot |qr|$

58. Yearly earnings for Dynamic Plastics Corp. (in thousands of dollars) were:

1984	\$2,940 profit	1988	\$17,805 loss
1985	\$8,275 profit	1989	\$3,420 loss
1986	\$19,600 profit	1990	\$1,700 profit
1987	\$11,350 loss		

Use a calculator to find the average annual earnings for Dynamic Plastics Corp. for the years 1984 to 1990.

59. An electron carries an electric charge of -1. An atom of sodium has 11 electrons, and an atom of chlorine has 17 electrons. What is the total electric charge of the electrons in one molecule of salt? (A salt molecule is made up of one atom of sodium and one atom of chlorine.)

60. A train is traveling at 73 mph. For every second the engineer applies the brakes, the train slows down by 4 mph. To slow the train for a dangerous curve, the engineer applied the brakes for 13 seconds (s). How fast was the train traveling at the end of 13 s?

61. The low temperatures in Duluth were $-12°F$ on Sunday and Tuesday, $3°F$ on Monday, $0°F$ on Wednesday, $5°F$ on Thursday, and $-8°F$ on Friday. What was the low temperature on Saturday if the average of the low temperatures for the week was:
a. $-4°F$?   **b.** $0°F$?   **c.** $2°F$?

Preparing for Section 2.5

In Exercises 62–65, remove grouping symbols and combine like terms.

62. $6(3x + 2)$ **63.** $5(x + 3) - 10$ **64.** $3[2(y + 2) + 1] + y$ **65.** $2\{1 + 3[s + 2(s + 1)]\} + 2s$

In Exercises 66–67, write an algebraic expression for each English expression.

66. The next even number after n, if n is an even number

67. Three times the sum of x and y

2.5   # Simplifying Algebraic Expressions

FOCUS

We use order-of-operations rules and the properties of numbers to combine and simplify algebraic expressions that involve real numbers—both positive and negative.

In Chapter 1, we worked with algebraic expressions, which are expressions that involve variables. The associative, commutative, and distributive properties and the rules governing the order of operations gave us guidelines for operating on such expressions. However, there we were limited to the positive real numbers and zero. In this section, we consider algebraic expressions that may involve or represent any real numbers.

Combining Like Terms

The variables in expressions like $(-5)(2x - y)$ represent numbers; hence, the expressions themselves represent numbers—the numbers that result from the indicated operations. And, since algebraic expressions represent numbers, they can be added, subtracted, multiplied, and divided.

In Section 1.4, we defined *like terms* as terms in which the same variables are raised to the same powers. We also saw that, because of the distributive property, like terms can be added. Since the distributive property holds for all real numbers, we can add like terms with any real-number coefficients and with variables that may represent any real numbers, positive or negative. We can also subtract like terms, since any subtraction can be changed to an addition.

EXAMPLE 1

Combine like terms:

a. $5x + (-9x)$ **b.** $7xy + 3y - 5x + 9xy$

Solution

a. These are like terms, so we can write this expression in a simpler form without changing its value. We use the distributive property:

$$5x + (-9x) = (5 + (-9))x$$

$$= -4x$$

No matter what number x is assigned, $5x + (-9x)$ has the same value as $-4x$. We verify this by letting $x = -3$, for instance. Then

$$5x + (-9x) = 5(-3) + (-9)(-3)$$

$$= (-15) + 27 = 12$$

and $-4x = (-4)(-3) = 12$

b. Although there are four terms in this expression, the two terms containing xy are the only like terms. Hence, we have

$$7xy + 3y - 5x + 9xy = 16xy + 3y - 5x \qquad \text{Adding } 7xy \text{ and } 9xy \qquad ■$$

Removing Parentheses

Parentheses preceded by a minus sign can sometimes cause difficulties. Recall that for any number x,

$$-x = (-1)x$$

Now suppose that a minus sign precedes two or more terms in parentheses, as in the expression $-(a + b - c)$. Using the distributive property, we have

$$
\begin{aligned}
-(a + b - c) &= (-1)(a + b - c) &&\text{Since } -x = (-1)x \\
&= (-1)a + (-1)b + (-1)(-c) &&\text{By the distributive property} \\
&= -a - b + c &&\text{Since } -x = (-1)x
\end{aligned}
$$

In general, the following rule holds:

Parentheses preceded by a minus sign

To remove parentheses that are preceded by a minus sign, change the signs of *all* terms within the parentheses.

This rule is illustrated in the next example.

EXAMPLE 2

Remove the parentheses and combine all like terms:

a. $5y - (7 - 2y)$ **b.** $5z^3 + (7 - 2z^3)$

Solution

a. We can remove the parentheses by changing all signs within them. Then we can combine like terms:

$$
\begin{aligned}
5y - (7 - 2y) &= 5y - 7 + 2y \\
&= 7y - 7
\end{aligned}
$$

b. In this case, the parentheses are preceded by a plus sign. They can simply be removed:

$$
\begin{aligned}
5z^3 + (7 - 2z^3) &= 5z^3 + 7 - 2z^3 \\
&= 3z^3 + 7
\end{aligned}
$$
■

EXAMPLE 3

Here is a trick you can use to puzzle your friends. It is especially surprising if several people are involved. Have each person follow these steps:

Steps	Example
1. Write down any whole number between 50 and 99. Do not show it to anyone.	63
2. Add 74 to your number.	$63 + 74 = 137$
3. Strike off the hundreds digit and add it to the resulting number.	137 $37 + 1 = 38$
4. Subtract the result of step 3 from your original number. Your result is 25.	$63 - 38 = 25$

The result is always 25, no matter what number is picked originally. Can you explain why this trick works?

Solution

Operations involving variable expressions are very helpful in explaining a trick like this one. Suppose the number chosen in step 1 is n. (Remember, n represents a whole number between 50 and 99.) In step 2, we add 74 to n. This gives us $n + 74$, and we note that $n + 74$ is between 124 and 173. In step 3, the hundreds digit is struck off and then added to the resulting number. Of course, the hundreds digit is always 1. (Do you see why?) To strike 1 off the hundreds place is to subtract 100, so we now have $(n + 74) - 100$. But the 1 we just struck off is now added, giving us $[(n + 74) - 100] + 1$. Finally, this expression is subtracted from n. Here is a step-by-step solution:

Steps in Words	Steps in Symbols
1. Choose a number	n
2. Add 74.	$n + 74$
3. Strike off 1 in the hundreds place and add 1 to the result.	$(n + 74) - 100 = n - 26$ $(n - 26) + 1 = n - 25$
4. Subtract the result from n.	$n - (n - 25) = 25$

Notice that the result is 25 no matter what n is.   ∎

Further Examples

The next example shows how to simplify some rather complicated expressions.

EXAMPLE 4

Simplify: $2[x^2 + 3(x^2 - 5)] - 5(x^2 + 3)$

Solution

We start by removing the innermost grouping:

$2[x^2 + 3(x^2 - 5)] - 5(x^2 + 3)$

$\quad = 2[x^2 + 3x^2 - 15] - 5(x^2 + 3)$ Removing innermost grouping

$\quad = 2[4x^2 - 15] - (5x^2 + 15)$ Combining like terms; multiplying

$\quad = 8x^2 - 30 - 5x^2 - 15$ Removing grouping

$\quad = 3x^2 - 45$ Combining like terms   ∎

Remember: After you remove groupings and combine like terms, the new expression has the same numerical value as the original expression for any given values of the variables.

EXAMPLE 5

In Example 4, we showed that

$$2[x^2 + 3(x^2 - 5)] - 5(x^2 + 3) = 3x^2 - 45$$

Find the value of each side of this equation when $x = -2$.

Solution

The left side takes many steps to evaluate. First replace x by -2.

$$2[x^2 + 3(x^2 - 5)] - 5(x^2 + 3) = 2\{(-2)^2 + 3[(-2)^2 - 5]\} - 5[(-2)^2 + 3]$$
$$= 2[4 + 3(-1)] - 5(7)$$
$$= 2[4 + (-3)] - 35$$
$$= 2[1] - 35$$
$$= 2 - 35$$
$$= -33$$

We also get -33 when we evaluate $3x^2 - 45$, but it is much simpler to do so:

$$3x^2 - 45 = 3(-2)^2 - 45$$
$$= (3)(4) - 45$$
$$= 12 - 45$$
$$= -33 \qquad \blacksquare$$

DISCUSSION QUESTIONS 2.5

1. Why can't y be zero in the expression $2x \div y$?

2. List three terms that are *like* each of these, and explain why they are like terms. (There are many correct answers. Name any three.)
   a. x^5 **b.** $3ab$

3. How many terms are there in each expression? Name them.
   a. $5xy - 17z + 3$ **b.** $12\dfrac{x}{y} - 14$ **c.** xyz

4. What are the numerical coefficients of the terms in Question 3? (Hint: See Section 1.4.)

5. Does $a(b - c) = ab - ac$ for all real numbers? Explain.

EXERCISES 2.5

In Exercises 1–8, use the variable x (and y if needed) to transform each phrase into mathematical symbols.

1. The opposite of a number

2. The sum of two numbers

3. The opposite of the sum of two numbers

4. Four times a given area

5. Two more than three times a given temperature

6. Six less than two times a given rate of speed

7. Eight times one number minus three times another number

8. One-half times the expression in Exercise 7

In Exercises 9–14, evaluate each pair of expressions when $x = -1$.

9. $3x + x$; $4x$

10. $3x - x$; $2x$

11. $-(x + 3)$; $-x - 3$

12. $-(x - 3)$; $-x + 3$

13. $6x^3 + 8x^3$; $14x^3$

14. $6x^2 - 8x^2$; $-2x^2$

In Exercises 15–52, simplify each expression.

15. $11x + 5x$

16. $8y - y$

17. $4d + 3 - 2d$

18. $9w - 8 + w + 6$

19. $12 - 3z + 4z - 7$

20. $7x - 3y - 9x + 8y$

21. $6.3a + 5.9b - 5.9a - 3.2b$

22. $-12g + 3h + 11g + 4h$

23. $\frac{4}{3}u + \frac{1}{6}v - \frac{5}{9}u + \frac{1}{3}v$

24. $-10.3c - 8d + 13.7c + 5.3d$

25. $3x^2 - (2x + 3)$

26. $7y - (4y - 9)$

27. $6n - (6 - 6n)$

28. $8p + (p - 8)$

29. $2k - (5k + 3) + 5k^3$

30. $2x - 5y - (8x - 3y)$

31. $12q^2 - (4p + 3q) - 2q$

32. $2(3b + 4) - 2b$

33. $12w + 4(2 - 6w)$

34. $-5(z + 1) + 3z + 2$

35. $9g + 2(6 - 2g) - 3g$

36. $4x - 15(1 + x) + 10$

37. $mn - (2mn - 1) + n$

38. $6m^3 - (3m^2 + 5) - m^2$

39. $(n + 3) - (2n - 5)$

40. $(2n - 1) - (3n - 4)$

41. $3.12(x^2 - 1.33) - 5.32(1.52x^2 + 3.33)$

42. $1.6(4.3x - 6.2) - 7.4(1.2x - 5.1)$

43. $1.4(3 + 4.2z) - (6z + 3.1)$

44. $2.2h - 3(5.4 - h) + 4.7(2h - 8.5) + 1.2$

45. $2[4 + 3(xy - 5)]$

46. $5[-2xy + 6(2 - xy)]$

47. $-[3z - (2 - 10z)]$

48. $-4[3(2k - 5) - (4 - 3k)]$

49. $7(w + 5) + 18 + 2[4(w - 6) + 5]$

50. $4(2n + 7) - 5 - [3(n + 3) + 2n]$

51. $3[8(2x - 5) - 4x] - [2(2x + 3) + 6]$

52. $8(4y^2 - 3) - 3(y - 1) + 6y^2$

Exercises 53–55 refer to this situation.

The following sequence of questions describes a trick similar to the one described in Example 3:

a. Choose a number.
b. Double the number.
c. Subtract 100 from the expression in b.
d. Take half of the expression in c.
e. Subtract the expression in d from the number you chose.

53. Using the variable n to stand for the chosen number, give the cumulative algebraic expression corresponding to each step.

54. Follow the five steps for $n = -11$, $n = 24$, and $n = 250$. What do you notice?

55. Show that the trick works by simplifying the variable expression you obtained in step e of Exercise 53.

56. The length l of a rectangle is $x - 2$, and its width w is $x - 7$.
a. Use the formula $p = 2l + 2w$ to write an expression for the perimeter of the rectangle. Simplify the expression.
b. If $x = 10$, find the perimeter of the rectangle by first finding the length and width and computing $2l + 2w$.
c. Find the perimeter of the rectangle when $x = 10$ by substituting into the simplified expression you obtained in part a. Should your answer be the same as that in part b?
d. Find the area of the rectangle if $x = 10$.

Exercises 57–60 refer to the following situation:

The Sparkle Jewelry Company mass-produces earrings and necklaces. However, one-eighth of the earrings produced and one-twelfth of the necklaces produced are defective. The company pays $5000 per month for fixed costs (rent, insurance, and the like). Let the variable x stand for the number of earrings produced in 1 month and the variable y stand for the number of necklaces produced in 1 month.

57. Write an expression involving x and/or y for (a) the number of defective earrings produced per month and (b) the number of defective necklaces produced per month.

58. Write an expression for (a) the number of nondefective earrings produced per month and (b) the number of nondefective necklaces produced per month. If possible, simplify these expressions.

59. The company sells a *pair* of earrings for $7.25 and a necklace for $16. They sell only their nondefective earrings and necklaces. Write an expression for (a) the company's revenue per month from earring sales and (b) the company's revenue per month from necklace sales. (Revenue is the amount of money received from sales.)

60. Write an expression for the company's total revenue per month.

Preparing for Section 3.1

61. Construct a rectangular coordinate system. (See Section 1.5 for guidelines.)

62. In the system you constructed in Exercise 61, label the origin *O*. What are the coordinates of *O*?

For Exercises 63–70, plot the ordered pair on a rectangular coordinate system.

63. (2, 3)   **64.** (1, 4)   **65.** (0, 5)   **66.** (4, 0)

67. (2, 5)   **68.** (5, 2)   **69.** (3, 3)   **70.** (4, 4)

Chapter Summary

Important Ideas

Absolute value
If *a* is any real number, then the *absolute value* of *a* is written $|a|$, and

$$|a| = \begin{cases} a \text{ if } a \text{ is positive or zero} \\ \text{the opposite of } a \text{ if } a \text{ is negative} \end{cases}$$

Additive identity
For any real number *a*,

$$a + 0 = 0 + a = a$$

Additive inverse
For any real number *a*,

$$a + (-a) = (-a) + a = 0$$

Multiplicative identity
For any real number *a*,

$$a \cdot 1 = 1 \cdot a = a$$

Multiplicative inverse
For any nonzero real number *a*,

$$(a)\left(\frac{1}{a}\right) = \left(\frac{1}{a}\right)(a) = 1$$

Chapter Review

Section 2.1

1. Plot these numbers on a number line: -5, $-(-3)$, $|-1.6|$, $-|-2|$.

In Exercises 2–10, evaluate each expression if $x = -2$ and $y = -5$.

2. $|x| + 4$

3. $-(-x)$

4. $|y| + |x|$

5. $\frac{1}{8} - \frac{1}{2}|y|$

6. $|x|\,|y| \div \frac{2}{5}$

7. $\frac{|x|}{|y|} + 4.1|x|$

8. $5|y| + |-231|$

9. $1.6|y| + 4.2|x|$

10. $|-2|[|x| + 3(1 + |y|)]$

11. If a gain in weight of 6 pounds (lb) is written $+6$, what does -4 mean? What does 0 mean?

Section 2.2

In Exercises 12–15, find each sum.

12. $-\frac{1}{2} + \left(-\frac{3}{4}\right)$

13. $7.3 + (-5.4)$

14. $-3 + (-2) + 7$

15. $8.93 + (-6.98) + (-5.04)$

In Exercises 16–23, evaluate each expression if $p = 3$ and $r = -10$.

16. $-(p + r + 7)$

17. $\frac{1}{2}p + r + \left(-\frac{4}{5}\right)$

18. $4.6p + r + 7.8$

19. $7|5p + r| + 4|r|$

20. $|5p + (-34)| + 3|r + (-48)|$

21. $1.2[2.6|r + 5.6| + 4.3(2p - 1.4)]$

22. $p\{5[r + 4(5p - 12)] + 14\} + (-42)$

23. $10 - 0.2\{r + 5[p - (r + 12)]\}$

24. The price of a certain microcomputer decreased by \$125. If the price before the decrease was \$1200, write an expression involving *addition* to represent the new price. Find the new price.

Section 2.3

In Exercises 25–30, perform the indicated computations.

25. $17 - (-6)$

26. $3.7 + (-6.2) - 5.3$

27. $-\left(\frac{1}{6}\right) - \left(\frac{1}{5}\right)\left(\frac{1}{7}\right)$

28. $2[14 - 3(1 - (-2))] - (-5)$

29. $62[123 - (-136)] + (-315)$

30. $12\{9 - \left(\frac{1}{4}\right)[48 - (-19)]\}$

In Exercises 31–35, evaluate each expression if $n = -3$.

31. $n + |n| - 6$

32. $2(5 - n)$

33. $-(n - 4)$

34. $2|n| + 3[(-2) + 2(3 - (-1))]$ **35.** $1.4[(5|n|)^2 - 3.6(-n)] - [n + (-2.7)]$

36. Find $|(-3) - (-9)|$. Show on a number line that your result is the distance between -3 and -9.

37. A low temperature of $-8°$ is predicted for tomorrow. If the present temperature is $4°$, by how much must it change?

Section 2.4

In Exercises 38–46, perform the indicated operations.

38. $(-5)(-1.5)(3)$

39. $\left(-\dfrac{1}{2}\right)\left(\dfrac{3}{4}\right) \div \left(-\dfrac{2}{3}\right)$

40. $0 \div (-3)$

41. $[4 + (-6)](-3) - (-2)(-3)$

42. $(-7.2)[8.1 + (-9.8) - (-4.6)] \div (-2)$

43. $18 \div [21 - (-72)(13 + (-15))]$

44. $20|(-12) \div (-4)| \div (-16)$

45. $|(-3)\{(-5) - (-4)[(-3) + (-5)(6 + (-10))]\}|$

46. $-(275)(-189)[712 + (-368)(421 - 512)]$

In Exercises 47–50, evaluate each expression if $x = -2$ and $y = 3$.

47. $5xy - y$ **48.** $\dfrac{3x}{y} + 2xy$ **49.** $|xy - 1.6| - (x - y)$ **50.** $\left(\dfrac{1}{2}\right)\{-3x - [y + (-2)(2x - 3y)]\}$

51. A quarterback threw three passes for gains of 6, 19, and 32 yd, one pass for no gain, and one pass for a loss of 7 yd. What was his average yardage for the five passes?

Section 2.5

In Exercises 52–62, remove grouping symbols and combine like terms.

52. $3(x - 5) + 14$

53. $2x - 12(x - 4) - (6x - 2)$

54. $3x - (3x - 2y + xy)$

55. $2[3(c - 2) - (c - 3)]$

56. $3k - (14 - 12k)$

57. $-[3z - (2 - 10z)]$

58. $2.3(1.7 - 5.2x^2) + 4.2x - x^2$

59. $\dfrac{2}{3}\left(\dfrac{3}{4}x - \dfrac{5}{6}\right) - \left(\dfrac{1}{4} - 2x\right)$

60. $18m^3 + (-12m^2) - (-2)[13m + (-m^2)]$

61. $124 - 206[x + (-163)] - (-528x)$

62. $7x + (-3)\{4[6x - 5] - (-3x)\}$

63. The width of a rectangle is w. Its length is 8 less than four times its width. Write an algebraic expression for the length of the rectangle.

64. Recall that the perimeter of a rectangle is twice its width plus twice its length. Write an algebraic expression for the perimeter of the rectangle in Exercise 63.

65. Write an algebraic expression for the average of four numbers: -6, x, 5, and $3x$. Remove parentheses and combine like terms in your expression.

Chapter 2 Test

1. Write an algebraic expression for "six less than twice a number n."

In Exercises 2–5, perform the indicated computation.

2. $[3 - (-8)](-5)$

3. $(-7) - (-3) - (-4)$

4. $\dfrac{48}{-6} - 5(-3)$

5. $|(-2) \cdot 3| - |3(-9)|$

6. Evaluate $3x^2 - 2xz + 3(x + y)$ if $x = 3$, $y = -1$, and $z = 2$.

7. Evaluate $x(|y| - |z|)$ if $x = 7$, $y = -3$, and $z = -\dfrac{3}{2}$.

In Exercises 8–11, remove parentheses and combine like terms.

8. $(x + 2) - (2x - 4)$

9. $3[x + 2(x - 8)] + 8(x - 2)$

10. $12k - (3k - 14)$

11. $2[-(1 + s) + 3(4 - 2s)]$

12. Evaluate $3(x + y) - 2|t|$ if $x = 1$, $y = -2$, and $t = 2$.

13. The average daily low temperature in Waterloo during one week was $1°$. The low temperature was $-2°$ on Sunday, $-4°$ on Monday, $0°$ on both Tuesday and Thursday, $3°$ on Wednesday, and $6°$ on Friday. What was the low temperature on Saturday?

14. In the first five hours of a day, the value of a certain stock showed these hourly changes:

$$\text{down } \frac{1}{8}, \text{ up } \frac{1}{4}, \text{ down } \frac{1}{2}, \text{ down } \frac{1}{4}, \text{ up } \frac{3}{8}$$

After the sixth hour the stock had returned to its beginning value. By how much did the value of the stock change during the sixth hour?

15. A sky diver jumped from an airplane flying 9500 ft above the ground. In the first 8 s, the diver fell at an average rate of 315 ft/s. Find her distance above the ground after 8 s.

16. The length of a rectangular room is 1 ft less than twice its width. If x represents the width, find an expression for the perimeter of the room. Remove parentheses and combine like terms in your expression.

3 Linear Equations and Inequalities

Linear Equations and Their Graphs

FOCUS

In the last chapter we extended the number line to include the negative real numbers; here we do the same with the rectangular coordinate system. Then we use graphs to estimate solutions of linear equations.

Graphing Points and Lines

In Section 1.5, we constructed a rectangular coordinate system on which we could graph ordered pairs of nonnegative real numbers. In Section 2.1, we saw that the number lines (which serve as the axes in a rectangular coordinate system) can be extended to include negative real numbers. By extending the axes, then, we can produce a coordinate system on which we can graph ordered pairs of real numbers—both positive and negative.

> **To construct a rectangular coordinate system for all real numbers** *(Figure 3.1)*
>
> 1. Draw two perpendicular number lines, called the *axes*. (The number lines are usually drawn as a horizontal and a vertical line.)
> 2. The point of intersection of the axes is called the *origin*. Assign zero on each axis to the origin.
> 3. On the horizontal axis, assign positive numbers to points to the right of the origin, and negative numbers to points to the left of the origin.
> 4. On the vertical axis, assign positive numbers to points above the origin, and negative numbers to points below the origin.

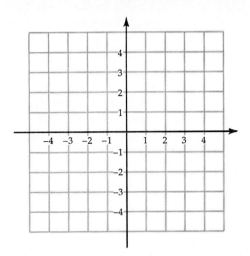

FIGURE 3.1

On these axes, we can plot ordered pairs in which either or both coordinates are negative, as you will see in the next example.

EXAMPLE 1

Plot these ordered pairs.

a. $(-5, 3)$ **b.** $(-2.4, -3.1)$

Solution

a. The first coordinate is -5, so we count 5 units to the left of the origin on the horizontal axis. Then we count 3 up from that axis. The result is point A in Figure 3.2.

b. Move 2.4 units to the left of the origin on the horizontal axis, and 3.1 down on the vertical axis. The result is point B.

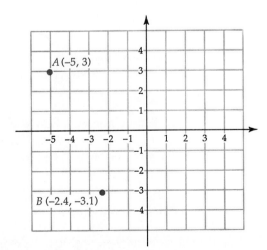

FIGURE 3.2

Recall from Section 1.5 that the graph of an equation consists of all points whose coordinates satisfy the equation. Equations in the two variables x and y are almost always graphed using pairs of the form (x, y). Hence, the first coordinate in an ordered pair is also called the **x coordinate,** and the second coordinate is called the **y coordinate.** Since first coordinates correspond to points on the horizontal axis, that axis is also called the x *axis.* Similarly, the vertical axis is called the y *axis.* Sometimes, the first coordinate in an ordered pair is called the **abscissa,** and the second coordinate, the **ordinate.**

EXAMPLE 2   Plot six ordered pairs in the form (x, y), three with positive x coordinates and three with negative x coordinates, that satisfy the equation $y = x + 2$.

Solution   To generate ordered pairs that lie on the graph, we first assign three positive and three negative numbers to x. Then we use the given equation to compute the corresponding values of y. The results are listed in Table 3.1.

Next, we draw a set of axes. Since the ordered pairs are in the form (x, y), we label the horizontal axis x and the vertical axis y. Finally, we plot the ordered pairs (x, y) from Table 3.1, as is done in Figure 3.3.

TABLE 3.1

x	$y = x + 2$
1	$1 + 2 = 3$
2	$2 + 2 = 4$
4	6
-1	$-1 + 2 = 1$
-1.5	0.5
-3	-1

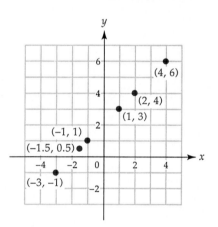

FIGURE 3.3   ∎

A **first-degree equation** is an equation in which the variable (or each variable) has the exponent 1. For example, the equation in Example 2, $y = x + 2$, is a first-degree equation in the two variables x and y, whereas

$$3x - 4 = 6$$

is a first-degree equation in one variable, x. On the other hand,

$$x^2 + 10 = 15$$

is not of degree one; it is a second-degree (or quadratic) equation. In this chapter, we consider only first-degree equations.

In Figure 3.3, the six ordered pairs that satisfy $y = x + 2$ appear to lie on a straight line, suggesting that the graph of this first-degree equation is a line. In fact, the graph of any first-degree equation in one or two variables is a line. For this reason, a first-degree equation is also called a **linear equation.** Since a line is determined by any two of its points, we can graph a first-degree equation as follows:

To graph a linear equation

1. Find three ordered pairs that satisfy the equation by assigning small values to one variable (usually x) and computing the corresponding values of the other variable (usually y).
2. Plot these three ordered pairs on a rectangular coordinate system.
3. Draw the line through these three points. This is the graph of the equation. (If the points do not all lie in a line, you have made an error.)

These guidelines are used in the next example.

EXAMPLE 3

Graph each equation.

a. $y = -2x + 4$ **b.** $y = 5$

Solution

a. Find three ordered pairs (x, y) that satisfy $y = -2x + 4$. We arbitrarily choose x to be 0, -1, and 3 and solve for y. The results are shown in Table 3.2. We then plot the three ordered pairs and draw the line containing them, as is done in Figure 3.4.

TABLE 3.2

x	$y = -2x + 4$
0	$-2(0) + 4 = 4$
-1	6
3	-2

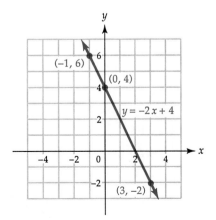

FIGURE 3.4

b. *Any* ordered pair with y coordinate 5 satisfies the equation $y = 5$. (And x can be anything!) We choose 0, 1, and -2 for x and produce Table 3.3, keeping in mind that $y = 5$ no matter what the value of x. We then plot the three ordered pairs and draw the line through these points. Notice in Figure 3.5 that the graph is a horizontal line.

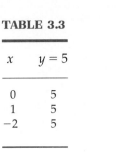

TABLE 3.3

x	$y = 5$
0	5
1	5
-2	5

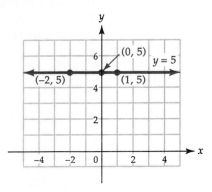

FIGURE 3.5   ■

Graphical Solutions of Linear Equations

An equation that is true for all real-number values of the variable or variables is called an **identity.** Thus,

$$x + 5 = 5 + x$$

is an identity. (Try it.) An equation that is not true for all values of the variables, but may be true for some values, is called a **conditional equation.** Hence, $0 = x + 2$ is a conditional equation.

Any replacement of the variable that makes a conditional equation true is called a **solution** or **root** of the equation. *Solving* an equation means finding all the roots (or verifying that there are none). Graphs can be used to find, or at least to estimate, the roots of an equation.

EXAMPLE 4   Graph $y = x + 2$. Use the graph to find the solution of $0 = x + 2$.

Solution   In Figure 3.3, we plotted six points that satisfy $y = x + 2$. We obtain the graph of this linear equation by drawing the line containing the six points. This is done in Figure 3.6. Any point that satisfies the equation $y = x + 2$ is on this line. Notice that the equation $0 = x + 2$ is the result of replacing y with 0 in $y = x + 2$. Hence, the solution of $0 = x + 2$ is the x coordinate of the pair $(x, 0)$ that satisfies $y = x + 2$. Now any point with y coordinate 0 is on the x axis. Thus, the point $(x, 0)$ that we are looking for is the point at which the graph of $y = x + 2$ crosses the x axis. That point is P in Figure 3.6, and its coordinates are $(-2, 0)$. Thus, $x = -2$ is the root of $0 = x + 2$. You can verify this easily by substitution: $0 = -2 + 2$.

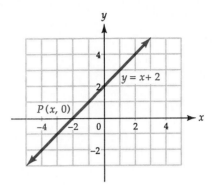

FIGURE 3.6

Obviously, if $0 = x + 2$, then it is also correct to write $x + 2 = 0$. This **symmetric property** holds for all equations.

Symmetric property of equality

If $a = b$, then $b = a$, where a and b are real numbers or expressions representing real numbers.

EXAMPLE 5

In Example 3 (Figure 3.4), we graphed the line $y = -2x + 4$. Use the graph to find the solutions of the following equations.

a. $-2x + 4 = -3$ **b.** $-2x + 4 = 1$

Solution

a.   Notice that $-2x + 4 = -3$, or, equivalently, $-3 = -2x + 4$, is the result of replacing y by -3 in $y = -2x + 4$. So we are looking for the x coordinate of the point $(x, -3)$ that is on the graph of $y = -2x + 4$. All points with y coordinate -3 lie on the line $y = -3$. In Figure 3.7 (page 106), we have graphed $y = -3$ on the same axis as $y = -2x + 4$. The point at which the two lines intersect is A $(3.5, -3)$. Hence, $x = 3.5$ is the root of $-3 = -2x + 4$. You can check by substituting 3.5 for x in this equation.

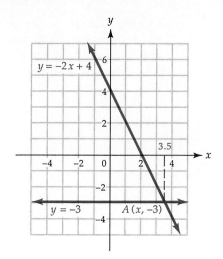

FIGURE 3.7

b.   Replacing y with 1 in the equation $y = -2x + 4$ gives

$$1 = -2x + 4$$

which is equivalent by symmetry to

$$-2x + 4 = 1$$

So we are looking for the x coordinate of the point $(x, 1)$ that is on the graph of $y = -2x + 4$. All points with y coordinate 1 lie on the line $y = 1$, which is shown in Figure 3.8 with the graph of $y = -2x + 4$. The point at which the line $y = 2x + 4$ crosses the horizontal line $y = 1$ is B (1.5, 1). Hence, $x = 1.5$ is the root of $1 = -2x + 4$. You can check by substituting 1.5 for x in that equation.

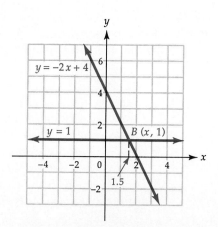

FIGURE 3.8   ■

DISCUSSION QUESTIONS 3.1

1. In a rectangular coordinate system, what is the horizontal axis called? What is the vertical axis called? What is the point (0, 0)?

2. In the ordered pair (a, b), what is the ordinate? What is the abscissa?

3. Which of the following equations have straight-line graphs? Which is the equation of a horizontal line? A vertical line?
   a. $y = a$ **b.** $2x - 3y = b$ **c.** $x^2 + y^2 = c$ **d.** $x = 0$

EXERCISES 3.1

In Exercises 1–16, plot the ordered pairs on a rectangular coordinate system. Mark each point with the correct letter.

1. $A(2, 4)$	**2.** $B(0, -3)$	**3.** $C(8, 0)$	**4.** $D(-7, 5)$
5. $E(-3, -3)$	**6.** $F(4, -9)$	**7.** $G(-7, -1)$	**8.** $H(-4, 6)$
9. $I(0, 5)$	**10.** $J(6, 10)$	**11.** $K(-0.5, 2.5)$	**12.** $L(-5, 0)$
13. $M(-1, -4)$	**14.** $N(2.25, -1.5)$	**15.** $P(2, 2)$	**16.** $Q(-4, -4)$

In Exercises 17–26, find the point in Figure 3.9 that corresponds to each ordered pair.

17. (0, 2)

18. (8, 1)

19. (−6, 3)

20. (−8, 0)

21. (−2, −3)

22. (0, −4)

23. (6, −7)

24. (−7, −7)

25. (3, 5)

26. (7, 0)

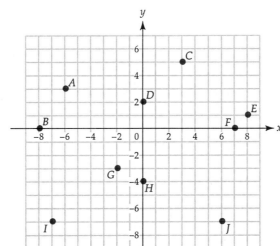

FIGURE 3.9

In Exercises 27–38, graph the first equation. Use the graph to find the solution of the second equation. Check by substituting your graphical solution into the second equation for x.

27. $y = -3x$; $6 = -3x$ **28.** $y = 4x - 1$; $0 = 4x - 1$ **29.** $y = 7 - x$; $-3 = 7 - x$

30. $y = x - 5$; $x - 5 = 1$ **31.** $y = x - 3$; $-2 = x - 3$ **32.** $y = 2x + 8$; $2x + 8 = 3$

33. $y = 7 - 2x$; $7 - 2x = 5$ 34. $y = 6 - x$; $6 - x = -9$ 35. $x - 6 = y$; $x - 6 = -4$

36. $y = 4x - 9$; $4x - 9 = -6$ 37. $y = 8x$; $-12 = 8x$ 38. $y = 5x$; $-6 = 5x$

39. **a.** Plot the points $(-3, -2)$, $(-2, -1)$, $(-1, 0)$, $(0, 1)$, $(1, 2)$, and $(2, 3)$.
     b. Write an equation in x and y that represents the relationship between the x and y coordinates of these points.

40. **a.** Plot the points $(-6, 4)$, $(-4, 2)$, $(-2, 0)$, $(0, -2)$, $(1, -3)$, and $(2, -4)$.
     b. Write an equation in x and y that represents the relationship between the x and y coordinates of these points.

41. **a.** Plot the points $(-2, 10)$, $(-2, -3)$, $(4, -3)$, and $(4, 10)$.
     b. Draw line segments between these points in the order in which they are given. Finally, join $(4, 10)$ and $(-2, 10)$. What kind of figure do you get?
     c. Find the area and perimeter of the figure.

42. **a.** Plot the points $(-7, 0)$, $(3, 0)$, $(-6, -4)$, and $(4, -4)$.
     b. Draw line segments between these points in the order in which they are given. Finally, join $(4, -4)$ and $(-7, 0)$. What kind of figure do you get?
     c. Find the area of the figure.

Exercises 43–49 refer to this situation.

A bicyclist kept a record of the distance d (in kilometers) she bicycled in t hours of riding. The record is shown as Table 3.4.

43. With t on the horizontal axis, make a graph of her record.

44. Draw the line through the given points.

45. Some points on the line you drew in Exercise 44 have no meaning in this situation. Explain.

46. Describe the location in the rectangular coordinate system of the points that do make sense for this situation. Explain why they do.

47. How far would the bicyclist ride in 1.25 hours (h)?

48. How far would she ride in 3.5 h?

49. Write an equation that gives her bicycling distance d in t hours of riding.

TABLE 3.4

t	d
0	0
0.5	8
1	16
1.5	24

Preparing for Section 3.2

In Exercises 50–52, check by substitution to see whether the given number is a root of the equation.

50. $-6 = 5x + 3$; $x = -1.8$ **51.** $2x + 6 = 10 - 3x$; $x = 5$ **52.** $6y - 3 = \dfrac{2y}{5}$; $y = 0.5$

53. Place these numbers in order from smallest to largest.

-3, -3.2, 2.8, π, $2\dfrac{3}{4}$, -2.9, 0, 3.1

3.2 The Addition and Subtraction Properties

FOCUS

We discuss two closely related properties that are used to solve equations.

In Section 3.1, we estimated the roots of conditional linear equations by graphing. In this section and in Section 3.3, we will use certain properties of equality to find exact solutions of equations.

To understand the first of these properties, the **addition property,** think of an equation as a balanced scale (see Figure 3.10). The scale is balanced because the left and right sides are equal. If the same weight is added to both sides of the scale, it will still be balanced. This suggests the addition property.

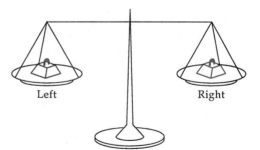

Left   Right

FIGURE 3.10

Addition property of equality

Any number or expression can be added to both sides of an equation without changing the relationship between the sides.

Two equations that have the same solutions are called **equivalent equations.** The addition property says that a given equation $a = b$ is equivalent to $a + c = b + c$. In symbols, the addition property states that

$$\text{if } a = b, \text{ then } a + c = b + c$$

EXAMPLE 1

For practice, write an equation that is equivalent to $x - 3 = 12$ by adding (a) -12, (b) 3, and (c) 10, respectively, to both sides. Which gives the simplest result?

Solution

a.   Adding -12 to both sides yields

$$(x - 3) + (-12) = 12 + (-12)$$
$$x - 15 = 0$$

b.   Adding 3 to both sides gives

$$(x - 3) + 3 = 12 + 3$$
$$x = 15$$

c.   Adding 10 to both sides gives

$$(x - 3) + 10 = 12 + 10$$
$$x + 7 = 22$$

The equation in part b is the simplest, since it shows clearly that $x = 15$ makes the equation true and hence is a root of $x - 3 = 12$. Notice that 15 is also a root of the equations we found in parts a and c. We may always add *any* number or expression to both sides of an equation, but in this case adding 3 produced the solution.   ∎

In many equations (and inequalities), a number is shown to be added to the variable term. A first step in solving the equation is to add the opposite of that number to both sides of the equation.

EXAMPLE 2

Use the addition property to solve:

a. $x - 10 = 18$ b. $x + 0.9 = -0.4$

Solution

a. $x - 10$ may be written as $x + (-10)$, so -10 has been added to the variable term. Hence, we should add $-(-10) = 10$ to both sides of the equation.

$$x + (-10) = 18$$

$$[x + (-10)] + 10 = 18 + 10 \qquad \text{Adding 10 to both sides}$$

$$x + [(-10) + 10] = 18 + 10 \qquad \text{By the associative property}$$

$$x = 28 \qquad \text{Simplifying both sides}$$

To check, we substitute 28 for x in the original equation and simplify. We see that the left side does equal the right side:

$$28 - 10 = 18$$

b. In $x + 0.9$, 0.9 has been added to the variable term. Hence, we should add -0.9 to both sides of the equation.

$$x + 0.9 = -0.4$$

$$[x + 0.9] + (-0.9) = -0.4 + (-0.9) \qquad \text{Adding } -0.9 \text{ to both sides}$$

$$x + [0.9 + (-0.9)] = -0.4 + (-0.9) \qquad \text{By the associative property}$$

$$x = -1.3 \qquad \text{Simplifying both sides} \qquad \blacksquare$$

Note that in each equation in Example 2, the number we added to both sides was the opposite of the number shown to be added to the variable term.

But what if there are variable terms on both sides? In that case, first add to both sides the opposite of one variable term. (We usually choose the variable term on the right side in order to get all variable terms on the left.) Then proceed as above.

EXAMPLE 3

Solve:

a. $3x + 9 = 2x - 6$ **b.** $5x - 4 = 4x + 6$

Solution

a. We note that the equation has an x term on each side. We can add either $-2x$ or $-3x$ to both sides to eliminate one of these terms. Adding $-2x$ gives us the following steps:

$$3x + 9 = 2x - 6$$

$$(3x + 9) + (-2x) = (2x - 6) + (-2x) \qquad \text{Adding } -2x \text{ to both sides}$$

$$(3x - 2x) + 9 = (2x - 2x) - 6 \qquad \text{Grouping like terms}$$

$$x + 9 = -6 \qquad \text{Simplifying both sides}$$

Now we solve by adding -9 to both sides:

$$(x + 9) + (-9) = (-6) + (-9) \qquad \text{Adding } -9 \text{ to both sides}$$
$$x = -15 \qquad \text{Simplifying both sides}$$

Check this value by substituting it in the original equation.

b.   We add $-4x$ to both sides:

$$5x - 4 = 4x + 6$$
$$(5x - 4) + (-4x) = (4x + 6) + (-4x) \qquad \text{Adding } -4x \text{ to both sides}$$
$$(5x - 4x) - 4 = (4x - 4x) + 6 \qquad \text{Grouping like terms}$$
$$x - 4 = 6 \qquad \text{Simplifying both sides}$$

Now we add 4 to both sides:

$$(x - 4) + 4 = 6 + 4 \qquad \text{Adding 4 to both sides}$$
$$x = 10 \qquad \text{Simplifying both sides}$$

Check by substituting 10 for x in the original equation. ∎

Of course, adding a number is like subtracting its opposite. Hence, we also have the **subtraction property.**

Subtraction property of equality

Any number or expression can be subtracted from both sides of an equation without changing the relationship between the sides.

Number of Solutions

The graphing technique we used in Section 3.1 gives us the same solution as the use of the addition and subtraction properties. (But it is not always as accurate.)

EXAMPLE 4   Solve $x + 1.3 = 3$ in two ways.

a.   Using the graphs of $y = x + 1.3$ and $y = 3$.
b.   Using the addition or subtraction property.

Solution   **a.**   We shall graph the two equations on the same axes. We first find three ordered pairs that satisfy $y = x + 1.3$; they are listed in Table 3.5.
    The graph of $y = 3$ is the horizontal line three units above the x axis. Hence, we obtain the graphs in Figure 3.11. The root of the equation $x + 1.3 = 3$ is the x coordinate of the point P at which the two graphs

intersect. It appears to be between 1 and 2, but closer to 2; let's call it 1.6 or 1.7.

TABLE 3.5

x	$y = x + 1.3$
0	1.3
3	4.3
−2	−0.7

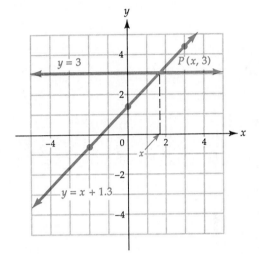

FIGURE 3.11

b. $x + 1.3 = 3$

$(x + 1.3) - 1.3 = 3 - 1.3$ Subtracting 1.3 from both sides

$x = 1.7$ Simplifying both sides

This is the exact value of the root that the graphing method only estimated for us in part a.   ■

Example 4 and the examples in Section 3.1 suggest that the solution of a conditional linear equation in one variable can be represented graphically as the intersection of two lines. Two lines can be related in one of three ways: (1) One line may fall exactly on top of the other, so that the two are not really different. Then they intersect in infinitely many points. (2) If the two lines are different, they may not intersect at all; they are then *parallel lines*. (3) If the two lines are different and not parallel, they intersect in exactly one point. This suggests the following result.

Roots of a linear equation in one variable

The number of roots of a linear equation in one variable is either zero, exactly one, or infinitely many (if the equation is an identity).

EXAMPLE 5

Solve:

a. $-2(x - 3) - (4 - 5x) = 2(x - 5)$
b. $x + 3 = x - 2$
c. $3x + 6 = 3(x - 1) + 9$

Solution

a. $-2(x - 3) - (4 - 5x) = 2(x - 5)$

$$-2x + 6 - 4 + 5x = 2x - 10 \qquad \text{Removing parentheses}$$

$$3x + 2 = 2x - 10 \qquad \text{Combining like terms}$$

$$(3x + 2) - 2x = (2x - 10) - 2x \qquad \text{Subtracting } 2x$$

$$x + 2 = -10 \qquad \text{Simplifying both sides}$$

$$x = -12 \qquad \text{Subtracting 2}$$

This equation has exactly one root, -12.

b. $$x + 3 = x - 2$$

$$(x + 3) - x = (x - 2) - x \qquad \text{Subtracting } x \text{ from both sides}$$

$$3 = -2 \qquad \text{Simplifying both sides}$$

Of course, 3 does not equal -2, and no value of x will change that. Hence, this equation has no solution

c. $3x + 6 = 3(x - 1) + 9$

$$3x + 6 = 3x - 3 + 9 \qquad \text{Removing parentheses}$$

$$3x + 6 = 3x + 6 \qquad \text{Combining like terms}$$

We see that the left and right sides of this last equation are identical, so it is obviously satisfied by *any* real number x. Since the original equation is equivalent to the last one, the original equation is also an identity. ■

DISCUSSION QUESTIONS 3.2

1. What is a conditional equation? Give two examples.

2. What are equivalent equations? Give two equations that are equivalent to $3x + 4 = x - 5$.

3. Describe the three types of linear equations in one variable with regard to the number of solutions. Give an example of each type.

EXERCISES 3.2

1. Perform the following operations, one at a time, on the equation $2 + x = 10$. Which results in the simplest equation?
   a. Add $(-x)$ to both sides.   **b.** Subtract 10 from both sides.
   c. Subtract $2x$ from both sides.   **d.** Add -2 to both sides.

In Exercises 2–5, estimate the solution of each equation by graphing. Then use the properties in this section to find the exact solution.

2. $0.3 - x = -1.6$ 3. $x + \dfrac{1}{3} = 7\dfrac{1}{6}$ 4. $-5 = x - 2$ 5. $2(x - 3) - (5 - x) = -3$

In Exercises 6–34, solve each equation and check by substitution.

6. $3 - x = 5$ 7. $w - 9 = 13$ 8. $6 = 16 - x$ 9. $9 + y = 4$

10. $8 + y = 5$ 11. $-8.3 = w - 1.5$ 12. $-4.2 = w + 13.6$ 13. $a + 12 = 15$

14. $m - \dfrac{1}{3} = 6$ 15. $1\dfrac{1}{3}a - \dfrac{1}{2} = \dfrac{1}{3}a$ 16. $2\dfrac{1}{6}m = 1\dfrac{1}{6}m + 3\dfrac{2}{3}$ 17. $7y = 3 + 6y$

18. $10 + 4z = 3z$ 19. $5u - 8 = 4u$ 20. $6u + 3 = 5u$ 21. $9 + 2v = v$

22. $11 + 7v = 6v$ 23. $5t + 4 = -5 + 4t$ 24. $4 + 3t = 2t + 9$ 25. $0.19 - s = 0.36 - 2s$

26. $7.4 - 1.6d = 8 - 2.6d$ 27. $7t - 5 = 10t + 8 - 4t$ 28. $\dfrac{3}{8}z - \dfrac{3}{4} = 5 - \dfrac{5}{8}z$

29. $12k - 4 = 11k + 1$ 30. $19 - 3k = 2k + 5 - 6k$ 31. $8n - 5 = (7n - 6) + 11$

32. $3(x + 1) = (2x - 1) + 4$ 33. $2(3x - 1) - (5x + 4) = 3$ 34. $-6(2y - 1) = 7 - 13(y + 1)$

In Exercises 35–43, write and solve the proper equation.

35. Eleven more than y is twenty-five. Find y.

36. Three is equal to the opposite of x. What is x?

37. Twice a number is equal to eight less than the number. Find the number.

38. Six minus four times a number is five times the number subtracted from three. Find the number.

39. Nine times a number plus one is equal to eight times the number minus one. Find the number.

40. A number is increased by 225.8. The result is 348.7. What is the number?

41. After losing 12 pounds (lb) on a diet, Lou weighed 142 lb. What was Lou's weight before the diet?

42. A savings account contained $352.34 after $24.53 in interest for the year was added. What was in the account at the beginning of the year?

43. A pitcher's earned run average increased by 1.13 in July. At the end of July his earned run average was 4.21. What was his earned run average at the beginning of July?

Preparing for Section 3.3

In Exercises 44–49, evaluate each expression if $x = -4$.

44. $5x$ **45.** $12x$ **46.** $-3x$ **47.** $-6x$ **48.** $4x + 3$ **49.** $-2x - 5$

In Exercises 50–51, solve for $3x$.

50. $3x + 1 = 22$ **51.** $-9 = 3x + 6$

3.3 The Multiplication and Division Properties

FOCUS

We introduce two more properties and use them, sometimes with the addition and subtraction properties, to solve linear equations in one variable.

To solve an equation, we must get it into the form:

$$x = \,?$$

The addition and subtraction properties are a big help in doing this, and they can be used to solve many equations. However, these properties do not help in solving an equation like $3x = 12$. For that we need the **multiplication property.**

> **Multiplication property of equality**
>
> Both sides of an equation can be multiplied by the same number without changing the relationship between the sides.

In the next example, this property is used to solve two equations.

EXAMPLE 1

Solve:

a. $5x = -32$ **b.** $-\dfrac{1}{2}x = 18$

Solution

a.   Our goal is to obtain the form $x = ?$, and for that we need to eliminate the multiplier 5. (Recall that such a number is called the *coefficient* of x.) Since $x = 1 \cdot x$, we would like to change the coefficient 5 to the coefficient 1. To do this, we should multiply both sides by $\dfrac{1}{5}$, the *reciprocal* or *multiplicative inverse* of 5:

$$5x = -32$$

$$\frac{1}{5}(5x) = \frac{1}{5}(-32) \qquad \text{Multiplying both sides by } \frac{1}{5}$$

$$\frac{5x}{5} = \frac{-32}{5} \qquad \text{Removing parentheses}$$

$$x = -6\frac{2}{5} \qquad \text{Simplifying}$$

b.   We choose -2 as our multiplier because $(-2)\left(-\dfrac{1}{2}\right) = 1$. Then

$$-\frac{1}{2}x = 18$$

$$(-2)\left(-\frac{1}{2}x\right) = (-2)18 \qquad \text{Multiplying both sides by } -2$$

$$\frac{2}{2}x = -36 \qquad \text{Removing parentheses}$$

$$x = -36 \qquad \text{Simplifying} \qquad \blacksquare$$

Both the addition and multiplication properties must be used, sometimes several times, to solve most linear equations. The following procedure shows how to decide which operations to use to reach a solution in the desired form. The operations should be performed in the given order.

Solving linear equations

1. If there are parentheses, apply the rules to remove them.
2. If there are like terms on the same side, combine them.
3. If there is a variable term on the right side but not on the left, interchange the entire left and right sides. (This is not a necessary step, but it makes the procedure easier.)
4. If there is a variable term on each side, add the opposite of the variable term on the right side to both sides. (Or, subtract the variable term on the right from both sides.)
5. If a number is shown to be added to the variable term, add the opposite of that number to both sides. (Or, subtract that number from both sides.)
6. If the coefficient of the variable is not 1, multiply both sides by the reciprocal of the coefficient.
7. Check your solution by substituting it into the original equation.

Let's use this procedure to solve a few equations. Not all the steps will be needed in every problem, of course.

EXAMPLE 2

Solve:

a. $6(2 - r) + 7(r - 2) = \dfrac{3}{4}$ **b.** $6(x - 2) = 4(x + 3) - 32$

c. $2(x - 3) = 2(x - 12) + 18$

Solution

a. $6(2 - r) + 7(r - 2) = \dfrac{3}{4}$

$12 - 6r + 7r - 14 = \dfrac{3}{4}$ Removing parentheses

$r - 2 = \dfrac{3}{4}$ Combining like terms

$r - 2 + 2 = \dfrac{3}{4} + 2$ Adding 2 to both sides

$r = 2\dfrac{3}{4}$ Simplifying

Check by substituting $2\dfrac{3}{4}$ for r in the original equation:

$$6\left(2 - 2\frac{3}{4}\right) + 7\left(2\frac{3}{4} - 2\right) = 6\left(-\frac{3}{4}\right) + 7\cdot\frac{3}{4}$$

$$= -\frac{18}{4} + \frac{21}{4}$$

$$= \frac{3}{4} \qquad \text{(It checks!)}$$

b. $6(x - 2) = 4(x + 3) - 32$

$6x - 12 = 4x + 12 - 32$ Removing parentheses

$6x - 12 = 4x - 20$ Combining like terms

$2x - 12 = -20$ Subtracting $4x$ from both sides

$2x = -8$ Adding 12 to both sides

$x = -4$ Multiplying both sides by $\frac{1}{2}$

The check is left to you.

c. $2(x - 3) = 2(x - 12) + 18$

$2x - 6 = 2x - 24 + 18$ Removing parentheses

$2x - 6 = 2x - 6$ Combining like terms

$-6 = -6$ Subtracting $2x$ from both sides

Since this equation is true no matter what the value of x, the original equation is an identity. Every real number is a solution. ∎

Of course, multiplying by a nonzero number is like dividing by its reciprocal. Hence, we also have the **division property.**

Division property of equality

Both sides of an equation can be divided by the same nonzero number without changing the relationship of the sides.

As the next example shows, linear equations can be used to solve word problems. Later in this chapter, we will present further strategies for solving such problems.

EXAMPLE 3

How should 464 marbles be divided between two children so that one child receives eight less than three times as many marbles as the other child?

Solution

The problem can be summarized as follows:

(Child 1's marbles) + (Child 2's marbles) = 464

We let n represent the unknown number of marbles of the first child. Then the second child should receive $3n - 8$ marbles. (Do you see why?) The equation that is given above in words becomes, in symbols,

$$n + (3n - 8) = 464$$

We solve it as follows:

$4n - 8 = 464$	Removing parentheses
$4n = 472$	Adding 8 to both sides
$n = 118$	Dividing both sides by 4

In the terms of the word problem, the first child should receive 118 marbles How many marbles should the second receive? ∎

DISCUSSION QUESTIONS 3.3

1. How would you check whether -4 is a solution of the equation in Example 2, part b?

2. What is the coefficient of n in each of these terms?

   a. $-6n$ **b.** $\dfrac{n}{6}$ **c.** $-n$

3. What is the reciprocal of each of the numbers?

   a. -6 **b.** 2.1 **c.** -1

EXERCISES 3.3

1. Perform the following operations one at a time on the equation $8x = -48.$ Which results in the simplest equation?

   a. Add 48 to both sides.   **b.** Multiply both sides by $\dfrac{1}{8}$.

   c. Subtract 8 from both sides.   **d.** Divide both sides by -48.

In Exercises 2–4, explain each step in the solution as is done in the examples.

2. $11x - 5 = 39$

   $11x = 44$

   $x = 4$

3. $5 - 2b = b - 2.5$

   $5 - 3b = -2.5$

   $-3b = -7.5$

   $b = 2.5$

4. $12c + 3 = 4c - 5 + 3c$

   $12c + 3 = 7c - 5$

   $5c + 3 = -5$

   $5c = -8$

   $c = -\dfrac{8}{5}$

In Exercises 5–33, solve each equation and check by substitution.

5. $z - 35 = 6z$

6. $2y + 6 = 10$

7. $4y + 5 = 21$

8. $\dfrac{1}{2}x + \dfrac{1}{3} = -1$

9. $-\dfrac{1}{3}w + \dfrac{3}{8} = -\dfrac{1}{6}$

10. $-1.2a + 11 = 35$

11. $2p - 1 = 5 + p$

12. $8r + 1 = 12r + 27$

13. $3m - 5 = 4m + 3 - m$

14. $2y - 3 + 6y = 10 - 3y$

15. $2w - (w + 1) = 4w + 6$

16. $\dfrac{5}{6}p + \dfrac{1}{4}(2 - p) = -\dfrac{1}{8}p - \dfrac{1}{6}$

17. $8(x - 2) = 50 + 5(12 - 2x)$

18. $1.9t + 9 - 2.8t = 0.6t$

19. $10 - 7m = 4(11 - 6m)$

20. $2(7r - 1) - 9 = 41 + 10r$

21. $6g - (3g + 8) = 16$

22. $5n - (7n - 4) = 7 - 3n$

23. $8 - (4n + 2) = 6n - 10n$

24. $\dfrac{2}{5}(3x + 5) = -\dfrac{3}{5}(x - 1)$

25. $2x - \dfrac{4}{5}(3x - 2) = \dfrac{3}{5}(6 + x)$

26. $2(5y + 3) - 3 = 6(y - 3)$

27. $5 + 4(y - 2) = 2(y - 7) + 1$

28. $8(q - 2) - 5(q - 12) = 50$

29. $5(3y - 6) = -8 + 7y$

30. $7z - 2(5 + 4z) = 8$

31. $5 - (2 - z) = 3z + 5$

32. $2(u - 4) - 5 = 7 + 3(2u - 1)$

33. $3(u + 1) - 2(u - 4) = 5(2u - 3) + 2$

In Exercises 34–37, graph the first equation. Use the graph to find the solution of the second equation. Check by substituting your graphical solution into the second equation for x.

34. $2x + 3y = 8$; $2x + 6 = 8$

35. $2x + 3y = 8$; $10 + 3y = 8$

36. $5x - 2y = 3$; $5x - 8 = 3$

37. $5x - 2y = 3$; $10 - 2y = 3$

In Exercises 38–42, translate each given situation into an equation and solve it by using the methods in this section.

38. If 24 is subtracted from eight times a certain number, the result is 96. What is the number?

39. If 6 is added to one-fourth of the number of degrees in the January minimum temperature in Solon, the result is zero. What was the January minimum temperature in Solon?

40. A 360-foot (ft) pipe is cut into three pieces so that the second piece is three times as long as the first piece and the third piece is twice as long as the second piece. How long is each piece of pipe?

41. The length of a rectangular city lot is 55 ft more than twice the width. Find the length and width of the lot if the perimeter of the lot is 350 ft.

42. Charles has five times as much money as Debbie. If Charles spends $40 and Debbie earns $20, they will have equal amounts. How much money does each have now?

Preparing for Section 3.4

In Exercises 43–45, place the given numbers in increasing order from least to greatest.

43. $-4, -6, 7, 0, 3$ **44.** $4.62, 3.989, 4.572, 4.61999, 4.7$ **45.** $\dfrac{2}{3}, \dfrac{3}{4}, \dfrac{4}{5}, \dfrac{4}{3}, \dfrac{3}{2}$

In Exercises 46–48, find three numbers that you could substitute for x so that the resulting inequalities hold.

46. $x + 3 < 5$ **47.** $x - 2 \geq 0$ **48.** $2x < 12$

3.4   Linear Inequalities

FOCUS

We introduce and discuss linear inequalities and properties for solving them.

Inequalities

An **inequality** is a statement saying that two numbers are, or may be, unequal. There are four kinds of inequalities, each with its own symbol:

$$a < b \quad \text{means} \quad a \text{ is less than } b$$
$$a \leq b \quad \text{means} \quad a \text{ is either equal to or less than } b$$
$$a > b \quad \text{means} \quad a \text{ is greater than } b$$
$$a \geq b \quad \text{means} \quad a \text{ is either equal to or greater than } b$$

More specifically, $5 > 1.6$ and $2x \leq 15$ are both inequalities.

We graph inequalities in one variable on a number line by marking on the line the interval or intervals in which the inequality is satisfied. If an endpoint of an interval is included, we mark it with a solid circle; if not, we use an open circle.

EXAMPLE 1

Graph the following inequalities on a number line.

a. $x < -2$ **b.** $m > -1$ **c.** $x \geq 5$ **d.** $y \leq 0$

Solution

a.   All numbers to the left of -2 on the number line satisfy this inequality, but -2 does not. Hence, we mark the number line as in Figure 3.12, with an open circle around -2.

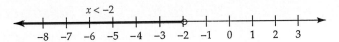

FIGURE 3.12

b. All numbers to the right of -1 on the number line satisfy this inequality, but -1 does not. Hence, we mark the number line as in Figure 3.13, with an open circle around -1.

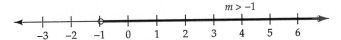

FIGURE 3.13

c. All numbers to the right of 5 on the number line satisfy this inequality, and 5 does, too. Hence, we mark the number line as in Figure 3.14, with a solid circle around 5.

FIGURE 3.14

d. All numbers to the left of 0 on the number line satisfy this inequality, and 0 does, too. Hence, we mark the number line as in Figure 3.15, with a solid circle around 0.

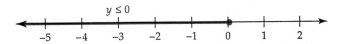

FIGURE 3.15 ∎

Inequalities do not exhibit a symmetric property. On the contrary, inequalities exhibit the **asymmetric property.** Interchanging the sides of an inequality reverses its meaning, or sense.

Asymmetric property of inequality

If $a < b$, then $b > a$.
If $a > b$, then $b < a$.
If $a \leq b$, then $b \geq a$.
If $a \geq b$, then $b \leq a$.

The Addition and Subtraction Properties

To solve an inequality, we must get it into one of these forms:

$$x > ? \text{ or } x < ? \text{ or } x \geq ? \text{ or } x \leq ?$$

The addition and subtraction properties apply in precisely the same form for inequality as they do for equality.

Addition and subtraction properties of inequality

Any number or expression can be added to both sides of an inequality without changing the relationship between the sides.

Any number or expression can be subtracted from both sides of an inequality without changing the relationship between the sides.

EXAMPLE 2

Use the addition or subtraction property to solve these inequalities.

a. $n + \dfrac{1}{6} < \dfrac{1}{3}$ **b.** $5x - 4 \geq 9 + 4x$

Solution

a. Here $\dfrac{1}{6}$ has been added to the variable term n, and so we should subtract $\dfrac{1}{6}$ from both sides:

$$n + \frac{1}{6} < \frac{1}{3}$$

$$\left(n + \frac{1}{6}\right) - \frac{1}{6} < \frac{1}{3} - \frac{1}{6} \qquad \text{Subtracting } \frac{1}{6} \text{ from both sides}$$

$$n + \left(\frac{1}{6} - \frac{1}{6}\right) < \frac{1}{3} - \frac{1}{6} \qquad \text{By the associative property}$$

$$n < \frac{1}{6} \qquad \text{Simplifying both sides}$$

We have reduced the inequality to an equivalent one that has the variable alone on one side and a single number on the other. We say that we have solved the inequality. Of course, the inequality does not have just one root. *All* numbers to the left of $\dfrac{1}{6}$ on the number line satisfy the inequality.

b.

$$5x - 4 \geq 9 + 4x$$

$$(5x - 4) - 4x \geq (9 + 4x) - 4x \qquad \text{Subtracting } 4x \text{ from both sides}$$

$$(5x - 4x) - 4 \geq 9 + (4x - 4x) \qquad \text{Grouping like terms}$$

$$x - 4 \geq 9 \qquad \text{Combining like terms}$$

$$(x - 4) + 4 \geq 9 + 4 \qquad \text{Adding 4 to both sides}$$

$$x \geq 13 \qquad \text{Simplifying}$$

We sometimes graph the solution of an inequality on a number line.

EXAMPLE 3

Solve for n and graph the solution on a number line.

$$\frac{1}{2} - (n + 1) \geq \frac{4}{5} - 2n$$

Solution

$$\frac{1}{2} - (n + 1) \geq \frac{4}{5} - 2n$$

$$\frac{1}{2} - n - 1 \geq \frac{4}{5} - 2n \qquad \text{Removing parentheses}$$

$$-n - \frac{1}{2} \geq \frac{4}{5} - 2n \qquad \text{Combining like terms}$$

$$\left(-n - \frac{1}{2}\right) + 2n \geq \left(\frac{4}{5} - 2n\right) + 2n \qquad \text{Adding } 2n \text{ to both sides}$$

$$(-n + 2n) - \frac{1}{2} \geq \frac{4}{5} + (-2n + 2n) \qquad \text{Grouping like terms}$$

$$n - \frac{1}{2} \geq \frac{4}{5} \qquad \text{Simplifying both sides}$$

$$\left(n - \frac{1}{2}\right) + \frac{1}{2} \geq \frac{4}{5} + \frac{1}{2} \qquad \text{Adding } \frac{1}{2} \text{ to both sides}$$

$$n \geq 1\frac{3}{10} \qquad \text{Simplifying both sides}$$

The graph of $n \geq 1\frac{3}{10}$, given in Figure 3.16, includes $1\frac{3}{10}$ and all points to the right of it.

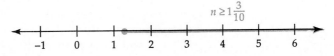

FIGURE 3.16

The Multiplication and Division Properties

We can multiply both sides of an equation by the same number, but can we do the same for an inequality? Consider the inequality $10 > 8$. If we multiply both sides by -1, we get -10 and -8. This gives the inequality $-10 > -8$, which is false. We can see why in Figure 3.17: Multiplying a number by a negative number puts the product on the opposite side of zero from the original number. Multiplying both sides of an inequality by a negative number thus reverses the relationship. Here,

$$10 > 8 \text{ becomes } -10 < -8$$

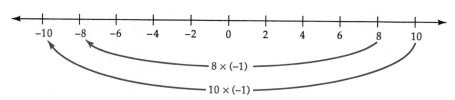

FIGURE 3.17

Multiplication property of inequality

Both sides of an inequality can be multiplied by the same *positive* number without changing the relationship of the sides.

  If $b > c$ and a is positive, then $ab > ac$.

If both sides of an inequality are multiplied by the same *negative* number, the relationship of the sides is reversed.

  If $b > c$ and a is negative, then $ab < ac$.

Since multiplying by a nonzero number is like dividing by its reciprocal, we also have a *division property of inequality*.

Division property of inequality

Both sides of an inequality can be divided by the same *positive* number without changing the relationship of the sides.

  If $b > c$ and a is positive, then $b \div a > c \div a$.

If both sides of an inequality are divided by the same *negative* number, the relationship of the sides is reversed.

  If $b > c$ and a is negative, then $b \div a < c \div a$.

The multiplication and division properties for inequality hold for $\leq$ and $\geq$ as well as for $<$ and $>$.

EXAMPLE 4   Solve:

a. $6x \leq -30$ **b.** $-\dfrac{2}{3}x > 18$ **c.** $-3y - 2 > 12 + 2y$

Solution   **a.** We must divide both sides by 6:

$$6x \leq -30$$
$$6x \div 6 \leq -30 \div 6 \qquad \text{Dividing both sides by 6}$$
$$x \leq -5 \qquad \text{Simplifying both sides}$$

b. Multiply both sides by $-\dfrac{3}{2}$, the reciprocal of $-\dfrac{2}{3}$. Since $-\dfrac{3}{2}$ is a negative number, the inequality is reversed from $>$ to $<$. Thus, we have

$$-\frac{2}{3}x > 18$$
$$\left(-\frac{3}{2}\right)\left(-\frac{2}{3}x\right) < \left(-\frac{3}{2}\right)18 \qquad \text{Multiplying both sides by } -\frac{3}{2}$$
$$x < -27 \qquad \text{Simplifying}$$

c. $-3y - 2 > 12 + 2y$

$$-5y - 2 > 12 \qquad \text{Subtracting } 2y \text{ from both sides}$$
$$-5y > 14 \qquad \text{Adding 2 to both sides}$$
$$y < -\frac{14}{5} \text{ or } -2\frac{4}{5} \qquad \text{Dividing both sides by } -5$$

To check an inequality, we can substitute a solution into the original inequality. Here, since any number less than $-2\dfrac{4}{5}$ is a solution, we can try -3:

$$-3(-3) - 2 > 12 + 2(-3)$$
$$9 - 2 > 12 - 6$$
$$7 > 6$$

Since $7 > 6$, our solution seems reasonable. As a further check, we substitute -2 for y. It should *not* be a root:

$$-3(-2) - 2 > 12 + 2(-2)$$
$$6 - 2 > 12 - 4$$
$$4 > 8$$

Since 4 is not greater than 8, we have found that -2 is not a root, as we predicted. This check does not ensure that we are right, but it should give us confidence. ∎

DISCUSSION QUESTIONS 3.4

1. Give an example of an inequality that is an identity.

2. State the asymmetric property of inequality. Apply it to $17 \leq 2x - 5$.

3. If both sides of $-2x < 4$ are divided by -2, what is the result? Must the sign be reversed?

EXERCISES 3.4

In Exercises 1–18, solve each inequality, if necessary, and graph the solution on a number line.

1. $x < -2$

2. $\frac{1}{6} \geq y$

3. $x - 5 \leq 12$

4. $-2 > y + 3$

5. $10n < 9n + 10$

6. $2k - 3 > k - 1$

7. $13w \geq 12w + 7$

8. $10y \leq 15 - (-9y)$

9. $3u - 16 > 5$

10. $4u + 15 < 9$

11. $8y < 15 + 5y$

12. $3x > 7 - 4x$

13. $\frac{1}{5}v - \frac{1}{6} \geq \frac{1}{3}v + \frac{5}{6}$

14. $-\frac{5}{6}v + \frac{1}{3} \leq 5 + \frac{1}{6}v$

15. $1.2s - 4.7 < 3.4s + 6.08$

16. $0.7 - 3.4d > 2.3d - 1.9d + 4.5$

17. $11h + 25 < 9h$

18. $124x - 114 > 12 - 35x$

In Exercises 19–20, explain each step in the solution as is done in the examples.

19. $3x - 2(2 - x) > 4(1 - 2x) + 5$
$3x - 4 + 2x > 4 - 8x + 5$
$5x - 4 > -8x + 9$
$13x - 4 > 9$
$13x > 13$
$x > 1$

20. $10 - 3x > 6(2x - 5) - 5$
$10 - 3x > 12x - 30 - 5$
$10 - 3x > 12x - 35$
$10 - 15x > -35$
$-15x > -45$
$x < 3$

In Exercises 21–32, solve each inequality.

21. $12k - 4 > 11k + 1$

22. $19 - 3k \leq 2k + 5 - 6k$

23. $2y - 3 + 6y < 10 - 3y$

24. $1.9t + 9 - 2.8t > 0.6t$ **25.** $3(2t - 1) + 2 < 4(2 - t)$ **26.** $2x - \dfrac{4}{5}(3x - 2) \geq \dfrac{3}{5}(6 + x)$

27. $2(5y + 3) - 3 < 6(y - 3)$ **28.** $5 + 4(y - 2) > 2(y - 7) + 1$ **29.** $8(q - 2) - 5(q - 12) \leq 50$

30. $5(3y - 6) > -8 + 7y$ **31.** $7z - 2(5 + 4z) \leq 8$ **32.** $5 - (2 - z) < 3z + 5$

Preparing for Section 3.5

For Exercises 33–38, recall that the area A of a rectangle with length l and width w is $A = lw$.

33.   Let $A = 42$ and $l = 7$. Find w.

34.   Let $A = 39$ and $l = 13$. Find w.

35.   Let $A = 22$ and $w = 7$. Find l.

36.   Let $A = 54$ and $w = 15$. Find l.

37.   Find a formula for l in terms of A and w.

38.   Find a formula for w in terms of A and l.

3.5   Transforming Formulas

FOCUS

We use the addition, subtraction, multiplication, and division properties to rewrite formulas in equivalent forms.

The perimeter formula for a rectangle $P = 2l + 2w$ is in a useful form for finding P if we are given l and w. However, if P and w are given, it would be more useful to have an equivalent equation with l by itself on the left side. The properties and procedures of Sections 3.2 and 3.3 can be used to write equivalent formulas.

EXAMPLE 1   The formula for the perimeter of a rectangle is $P = 2l + 2w$. Find a formula that gives l, the length of the rectangle, in terms of P and w, its width.

Solution   In a sense there are three variables, P, l, and w, in the formula $P = 2l + 2w$. To solve the formula for l, we let P and w be fixed for the time being, as if they were numbers. We then solve for l just as if we were solving a linear equation. Remember, as we apply the procedure of Section 3.3, that l is the variable we wish to isolate.

$$P = 2l + 2w$$
$$2l + 2w = P \qquad \text{Interchanging sides}$$
$$2l = P - 2w \qquad \text{Subtracting } 2w \text{ from both sides}$$
$$l = \frac{1}{2}(P - 2w) \qquad \text{Multiplying both sides by } \frac{1}{2}$$

As a check, we shall use this formula to find *l* when $P = 30$ and $w = 5$, and then check that result in the original formula $P = 2l + 2w$. First,

$$l = \frac{1}{2}(P - 2w)$$

$$= \frac{1}{2}(30 - 2 \cdot 5)$$

$$= \frac{1}{2}(20) = 10$$

Now, with $l = 10$ and $w = 5$, we should compute $P = 30$:

$$P = 2l + 2w$$

$$= 2 \cdot 10 + 2 \cdot 5 = 30 \qquad \blacksquare$$

BE CAREFUL!

We always assume, without explicitly stating so, that:
1. A formula does not apply for any value of a variable that makes a denominator zero.
2. When we divide both sides of an equation by an expression containing a variable, that expression is not zero.

EXAMPLE 2

A baseball player's batting average, call it *A*, can be found by using the formula

$$A = \frac{h}{b}$$

where *h* is the number of hits and *b* is the number of official times at bat. (The formula does not apply for $b = 0$.)

a. Find a formula for *h* when *A* and *b* are known.
b. Use the formula to find the number of hits by Ty Cobb in his lifetime, if he officially batted 11,420 times and had a batting average of 0.367 (the highest of any player in history).

Solution

a. We must solve $A = \frac{h}{b}$ for *h*:

$$A = \frac{h}{b}$$

$$\frac{h}{b} = A \qquad \text{Interchanging sides}$$

$$h = Ab \qquad \text{Multiplying both sides by } b$$

b. To find Cobb's total hits, we substitute $A = 0.367$ and $b = 11,420$ in the formula $h = Ab$:

$$h = 0.367 \cdot 11,420 = 4191.14$$

So Cobb got 4191 hits. (*Note:* Our calculation gave 4191.14, because 0.367 is approximate.) ∎

EXAMPLE 3

Solve for s in

$$r = (2s + 5) \div t$$

Solution

The variable—that is, the variable we wish to isolate—is s. We notice that $(2s + 5) \div t$ can be written in several ways:

$$(2s + 5) \div t = \frac{2s + 5}{t} = \frac{1}{t}(2s + 5)$$

The last form will be easiest for us to use.

$$r = \frac{1}{t}(2s + 5)$$

$$= \frac{1}{t} \cdot 2s + \frac{1}{t} \cdot 5 \qquad \text{Removing parentheses}$$

$$\frac{2}{t} \cdot s + \frac{5}{t} = r \qquad \text{Interchanging sides; simplifying}$$

$$\frac{2}{t} \cdot s = r - \frac{5}{t} \qquad \text{Subtracting } \frac{5}{t} \text{ from both sides}$$

$$\frac{t}{2} \cdot \frac{2}{t} \cdot s = \frac{t}{2} \cdot r - \frac{t}{2} \cdot \frac{5}{t} \qquad \text{Multiplying both sides by } \frac{t}{2}$$

$$s = \frac{rt}{2} - \frac{5}{2} \qquad \text{Simplifying}$$

This equation may also be written

$$s = \frac{1}{2}rt - \frac{5}{2} = \frac{1}{2}(rt - 5) = \frac{(rt - 5)}{2}$$

∎

EXAMPLE 4   Solve $\dfrac{x}{a} - \dfrac{y}{b} = 1$ for y. Use your result to find y if $x = -5$, $a = 4$, and $b = \dfrac{1}{3}$.

Solution

$$\dfrac{x}{a} - \dfrac{y}{b} = 1$$

$$-\dfrac{y}{b} = 1 - \dfrac{x}{a} \qquad \text{Subtracting } \dfrac{x}{a} \text{ from both sides}$$

$$y = -b\left(1 - \dfrac{x}{a}\right) \qquad \text{Multiplying both sides by } -b$$

We substitute the given values into this equation:

$$y = -\dfrac{1}{3}\left[1 - \left(-\dfrac{5}{4}\right)\right]$$

$$= -\dfrac{1}{3}\left(\dfrac{9}{4}\right)$$

$$= -\dfrac{3}{4}$$

∎

DISCUSSION QUESTIONS 3.5

1. A bright student solved Example 3 in this way:

$$r = \dfrac{2s + 5}{t}$$

$$rt = 2s + 5 \qquad \text{Multiplying both sides by } t$$

$$2s + 5 = rt \qquad \text{Interchanging sides}$$

$$2s = rt - 5 \qquad \text{Subtracting 5 from both sides}$$

$$s = \dfrac{1}{2}(rt - 5) \qquad \text{Multiplying both sides by } \dfrac{1}{2}$$

Is her method correct? Explain your answer.

2. By what must each side of $5rp = 10$ be divided in order to get p by itself?

3. When we divide both sides of an equation by $x - 3$, to what number are we assuming x is *not* equal? When we divide both sides by $2x^2$, to what number are we assuming x is not equal?

EXERCISES 3.5

In Exercises 1–24, solve each of the equations for the indicated variable.

1. $x + y = 10$ for x

2. $x + y = 10$ for y

3. $2x - y = 41$ for x

4. $2x - y = 41$ for y

5. $a - \frac{1}{4}b = \frac{1}{5}$ for a

6. $a - \frac{1}{4}b = \frac{1}{5}$ for b

7. $0.2x - 1.5y = 2.1$ for y

8. $0.2x - 1.5y = 2.1$ for x

9. $ax + by = c$ for x

10. $ax + by = c$ for y

11. $y = mx + b$ for x

12. $y = mx + b$ for m

13. $3x^2y - 5x = 7x + 2$ for y

14. $\frac{x}{a} + \frac{y}{b} = 1$ for y

15. $\frac{x}{a} + \frac{y}{b} = 1$ for a

16. $y - a = m(x - b)$ for y

17. $y - a = m(x - b)$ for m

18. $y - a = m(x - b)$ for x

19. $y - a = m(x - b)$ for b

20. $y - a = m(x - b)$ for a

21. $120x \div 24y = \frac{r}{s}$ for x

22. $120x \div 24y = \frac{r}{s}$ for y

23. $120x \div 24y = \frac{r}{s}$ for r

24. $120x \div 24y = \frac{r}{s}$ for s

25. **a.** Solve $P = 4s$ for s.
    b. Find s when $P = 144$.

26. **a.** Solve $P = 2l + 2w$ for w.
    b. Find w when $P = 96$ and $l = 42$.

27. **a.** Solve $A = lw$ for w.
    b. Find w when $A = 75$ and $l = 15$.

28. **a.** Solve $F = \frac{9}{5}C + 32$ for C.
    b. Find C when $F = 212$.

29. **a.** Solve $C = 2\pi r$ for r.
    b. Find r when $C = 80$.

30. **a.** Solve $A = \frac{1}{2}bh$ for b.
    b. Find b when $A = 10.25$ and $h = 22$.

31. **a.** Solve $A = \frac{1}{2}h(a + b)$ for a.
    b. Find a when $A = 120$, $b = 8$, and $h = 6$.

32. **a.** Solve $V = \pi r^2 h$ for h.
    b. Find h when $V = 770$ and $r = 7$.

33. **a.** Solve $E = IR$ for R.
    b. Find R when $E = 342$ and $I = 9$.

34. **a.** Solve $y = mx + b$ for m.
    b. Find m when $y = 0$, $x = \frac{2}{3}$, and $b = 48$.

35. **a.** Solve $E = md^2$ for m.
    b. Find m when $E = 20.48$ and $d = 3.2$.

36. **a.** Solve $h = k(1 + \frac{1}{273}t)$ for k.
    b. Find k when $h = 63$ and $t = 546$.

37. **a.** Solve $A = 2\pi r^2 + 2\pi rh$ for h.
    b. Find h when $A = 200$ and $r = 4$.

In Exercises 38–39, use the time-rate-distance formula $d = rt$ as needed to solve the problem.

38. Jody plans to take a 175.5-kilometer (km) bicycle trip. She can pedal at an average speed of 13 km/h. How long will it take Jody to complete her trip?

39. An imported car averages 32 miles per gallon (mi/gal) of gasoline when traveling at a speed of 45 mph. For every 5-mph increase in speed there is a decrease of 3 mi/gal. How many miles per gallon will the car average at a speed of 70 mph?

Exercises 40–43 refer to this situation.

Sleez Rentals rents cars for $26 per day plus 36 cents per mile. Acme Rentals rents cars for $23 per day plus 25 cents per mile. John plans to rent a car for 3 days and drive 561 mi.

40. How many miles of driving does John average per day?

41. How much will it cost John to rent a car from Sleez Rentals?

42. How much will it cost John to rent a car from Acme Rentals?

43. Which company provides the better deal for John? How much money does he save by taking the better deal?

Preparing for Section 3.6

44. What is 0.35 as a percent?   **45.** What is 48.3% as a decimal?

46. What is $\frac{3}{5}$ as a percent?   **47.** What is 26% of 120?

In Exercises 48–50, solve for x.

48. $\dfrac{3}{5} = \dfrac{x}{20}$ **49.** $\dfrac{2}{3} = \dfrac{9}{x}$ **50.** $\dfrac{x}{5} = \dfrac{3}{8}$

51. A 12-ounce (oz) can of soda sells for 75 cents, whereas a 16-oz can sells for 90 cents. Which size is less expensive per ounce?

3.6   Ratio and Proportion

FOCUS

We use the ideas of ratio and proportion to write equations and solve real-world problems.

Ratio

The **ratio** of a number a to a number b is the fraction $\dfrac{a}{b}$ (provided that b is not zero). All the rules for simplifying fractions also apply to ratios, since a ratio is essentially a fraction. For example, the ratio of 10 to 4 just means $\dfrac{10}{4}$, which can be simplified:

$$\frac{10}{4} = \frac{5 \cdot \cancel{2}}{2 \cdot \cancel{2}}$$

$$= \frac{5}{2}$$

Thus, the ratio 10 to 4 is the same as the ratio 5 to 2.

There are several ways of writing the ratio 10 to 4, all with essentially the same meaning:

As a fraction: $\qquad\qquad \dfrac{10}{4}$

As a decimal: $\qquad\qquad \dfrac{2.5}{1}$ or just 2.5

With the division sign: $\quad 10 \div 4$

With a ratio sign: $\qquad 10{:}4$

In this book, we will most often use the fraction form for ratios.

EXAMPLE 1

There are 16,234 men and 14,568 women enrolled in a certain midwestern university. Write the ratio of men to women in each of the above forms.

Solution

To write the ratio *of* men *to* women, we write a fraction with the number of men in the numerator and the number of women in the denominator and simplify:

$$\text{As a fraction: } \frac{16{,}234}{14{,}468} = \frac{8117}{7234}$$

To write the ratio as a decimal, we divide numerator by denominator:

$$\text{As a decimal: } \frac{8117}{7234} \approx \frac{1.12}{1} \text{ or } 1.12$$

To write it with the division sign, we replace the fraction bar with the division sign:

With the division sign: 8117 ÷ 7234

Finally, we have:

In ratio form: 8117:7234   ∎

Proportion

An important idea in mathematics, and in many other fields, is that of proportion. A **proportion** is an equation stating that two ratios are equal. The idea of equal ratios is very important in converting from one size or amount to another while maintaining an essential characteristic, such as strength of mixture or relative price.

The general form of a proportion is

$$\frac{a}{b} = \frac{c}{d}$$

The **cross products** of a proportion are found by multiplying the numerator of each ratio by the denominator of the other. Here, the cross products are *ad* and *bc*. If a proportion is true, then its cross products are equal.

Cross multiplication

1. If $\dfrac{a}{b} = \dfrac{c}{d}$, then $ad = bc$.

2. If $ad = bc$, then $\dfrac{a}{b} = \dfrac{c}{d}$, $\dfrac{b}{a} = \dfrac{d}{c}$, $\dfrac{a}{c} = \dfrac{b}{d}$, and $\dfrac{c}{a} = \dfrac{d}{b}$, provided neither a, b, c, nor d is zero.

Cross multiplication can be used in practical situations to check for proportionality.

EXAMPLE 2

A recipe for fruit punch called for pineapple juice and lemon soda in the ratio of 3 to 2. Laura mixed 9 pints (pt) of pineapple juice and 8 pt of lemon soda. Brian's punch contained 16 pt of pineapple juice and 12 pt of lemon soda. Will Laura's punch taste the same as Brian's? Did either one follow the recipe correctly?

Solution   If Laura and Brian used ingredients in the same ratio, their fruit punch will taste and look the same. In other words, the ratio of pineapple juice to lemon soda must be the same in each punch. For Laura, we have the ratio

$$\frac{\text{Pineapple}}{\text{Lemon}} = \frac{9 \text{ pt}}{8 \text{ pt}} = \frac{9}{8}$$

For Brian, $$\frac{\text{Pineapple}}{\text{Lemon}} = \frac{16 \text{ pt}}{12 \text{ pt}} = \frac{16}{12} = \frac{4}{3}$$

To decide whether the two ratios are equal, we compare the cross products.

$$\frac{9}{8} = \frac{4}{3} \text{ if } (9)(3) = (4)(8)$$

The cross products are not equal. Therefore, the ingredients were not used proportionally. Laura's punch will not taste the same as Brian's.

Can you answer the second question? Did either follow the recipe?   ■

Cross multiplication also allows us to find one of the four terms in a proportion when the other three are known.

EXAMPLE 3   Solve for x to the nearest hundredth:

a. $\dfrac{x}{5} = \dfrac{9}{16}$ b. $\dfrac{2.3}{x} = \dfrac{1.6}{5}$ c. $\dfrac{2\frac{1}{2}}{3\frac{3}{4}} = \dfrac{x}{10}$

Solution   These proportions can all be solved in essentially the same way, regardless of where the unknown is located.

a. $\dfrac{x}{5} = \dfrac{9}{16}$

$16x = 5 \cdot 9$ Cross multiplying

$16x = 45$ Simplifying

$x = \dfrac{45}{16}$ Dividing both sides by 16

≈ 2.81

b. $\dfrac{2.3}{x} = \dfrac{1.6}{5}$

$1.6x = 2.3 \cdot 5$ Cross multiplying

$1.6x = 11.5$ Simplifying

$x = \dfrac{11.5}{1.6}$ Dividing both sides by 1.6

≈ 7.19

c. $\dfrac{2\frac{1}{2}}{3\frac{3}{4}} = \dfrac{x}{10}$

$3\dfrac{3}{4} \cdot x = 2\dfrac{1}{2} \cdot 10$ Cross multiplying

$\dfrac{15}{4}x = 25$ Simplifying

$x = 25 \div \dfrac{15}{4}$ Dividing both sides by $\dfrac{15}{4}$

$= 25 \cdot \dfrac{4}{15}$ Inverting and multiplying

$= \dfrac{100}{15}$ Completing the division

≈ 6.67 ∎

Applications

Many word problems, like the next one, can be solved by finding a term in a proportion.

EXAMPLE 4   A 10-oz can of tomato sauce costs 49 cents. At the same price per ounce, how much would a 16-oz can cost?

Solution   This problem can be solved without proportions, but the use of a proportion is easiest. First note that if the prices per ounce are the same, the ratio of price to weight will be the same for the two cans. For the small can we have

$$\frac{\text{Price}}{\text{Weight}} = \frac{49}{10}$$

Letting P be the price of the large can, we have, for that can,

$$\frac{\text{Price}}{\text{Weight}} = \frac{P}{16}$$

We equate the two ratios to form the proportion, cross multiply, and solve the resulting equation:

$$\frac{49}{10} = \frac{P}{16}$$

$10P = 784$ Cross multiplying

$P = 784 \div 10$ Dividing both sides by 10

$= 78.4$

The 16-oz can should cost about 78 cents.   ■

Similar figures are geometric figures that have the same shape and appearance but not necessarily the same size. For example, a photograph and its enlargement are similar. The enlargement is bigger, but it has the same proportions as the photograph. Lengths of line segments in a figure are always proportional to the corresponding segments in a similar figure, a fact we use in the next example.

EXAMPLE 5

The two triangles in Figure 3.18 are similar. Find the lengths of the sides marked x and y.

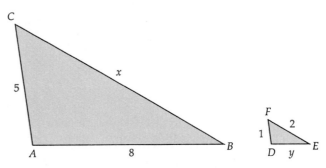

FIGURE 3.18

Solution

The pairs of corresponding sides are AC and DF, AB and DE, and BC and EF. Since the triangles are similar, the ratios of corresponding sides are equal:

$$\frac{AC}{DF} = \frac{AB}{DE} = \frac{BC}{EF}$$

Substituting the values marked on the figures, we obtain

$$\frac{5}{1} = \frac{8}{y} = \frac{x}{2}$$

To find y, we use the first two terms:

$$\frac{5}{1} = \frac{8}{y}$$

$$5y = 8 \qquad \text{Cross multiplying}$$

$$y = \frac{8}{5} \qquad \text{Dividing both sides by 5}$$

$$= 1\frac{3}{5}$$

To find x, we use the first and last terms:

$$\frac{5}{1} = \frac{x}{2}$$

$$1 \cdot x = 5 \cdot 2 \qquad \text{Cross multiplying}$$

$$x = 10 \qquad \text{Simplifying} \qquad\qquad \blacksquare$$

In Chapter 1, we reviewed percent ideas. A percent is actually a special ratio; namely, $n\%$ means $\dfrac{n}{100}$. Therefore, ratio and proportion ideas can be used in percent problems.

EXAMPLE 6

If you scored 93% (to the nearest whole percent) of a maximum of 40 points on a math quiz, how many points did you score?

Solution

The ratio of your points p to 40 is equal to $\dfrac{93}{100}$. To find p, solve the proportion.

$$\frac{p}{40} = \frac{93}{100}$$

$$100p = 93 \cdot 40 \qquad\qquad \text{Cross multiplication}$$

$$100p = 3720 \qquad\qquad \text{Multiplying}$$

$$p = \frac{3720}{100} \qquad \text{Dividing both sides by 100}$$

$$p = 37.2 \qquad \text{Simplifying}$$

$$\approx 37 \text{ points} \qquad \blacksquare$$

Many consumer problems involve percent and can be solved using equations. Some helpful relationships are the following.

Relationships involving percent

Amount of discount = (rate of discount) × (original price)
Amount of simple interest
    = (rate of interest) × (principal) × (time)
Amount of increase = (% of increase) × (amount before increase)
Amount of decrease
    = (% of decrease) × (amount before decrease)

Note that in the second relationship, the unit of time must match the rate period. For example, if the rate is 6% per year, then the time must be in years.

EXAMPLE 7

The cost of a certain candy bar has increased by 300% since 1960. If it cost 10 cents in 1960, what does it cost now?

Solution

The problem involves an increase, so we need this relationship:

Amount of increase = (% of increase) × (amount before increase)

Amount of increase = 300% × (cost in 1960)

$$= 300\% \times 10 \text{ cents}$$

$$= 3.00 \times 10$$

$$= 30 \text{ cents}$$

Finally, to compute the present cost of the candy bar, we add the amount of the increase to the 1960 cost:

Present cost = 10 + 30 = 40 cents   ■

DISCUSSION QUESTIONS 3.6

1. Write the ratio of 7 to 12 in four ways. Write the ratio of 12 to 7 in four ways.

2. $\dfrac{2}{3} = \dfrac{4}{6}$ is a proportion. Make three other proportions using the numbers 2, 3, 4, and 6.

3. If your salary increased from \$20,000 to \$22,000, the amount of increase is \$2000. Is your rate of increase determined by finding what percent \$2000 is of \$20,000, or what percent it is of \$22,000? Why?

EXERCISES 3.6

In Exercises 1–10, write the ratio as a fraction in simplest terms.

1. $3:8$

2. $2\dfrac{1}{2}:10$

3. $1.8:1$

4. $16:6\dfrac{3}{4}$

5. $0.25:0.7$

6. $2500:350$

7. $8.34:3.15$

8. $5\dfrac{1}{2}:6\dfrac{1}{3}$

9. $3x:9x$

10. $\dfrac{-5.3x^2}{4.6x^2}$

In Exercises 11–20, write each ratio as a decimal. Round to hundredths.

11. $\dfrac{5}{2}$

12. $\dfrac{1.7}{5}$

13. $\dfrac{260}{15}$

14. $\dfrac{\frac{1}{3}}{\frac{4}{5}}$

15. $\dfrac{2.5}{7}$

16. $\dfrac{18.5}{12.5}$

17. $-8\dfrac{1}{2}:3.6$

18. $-5.36:2\dfrac{1}{5}$

19. $27y:(-3.8y)$

20. $\dfrac{18x^2}{2.9x^2}$

In Exercises 21–30, determine whether each pair of ratios is equal.

21. $\dfrac{9}{15},\ \dfrac{6}{10}$

22. $\dfrac{\frac{1}{2}}{\frac{1}{4}},\ \dfrac{\frac{3}{8}}{\frac{3}{4}}$

23. $\dfrac{8}{12.4},\ \dfrac{10}{15.5}$

24. $\dfrac{3\frac{1}{2}}{7\frac{2}{3}},\ \dfrac{\frac{3}{7}}{\frac{3}{4}}$

25. $\dfrac{14}{5},\ \dfrac{420}{150}$

26. $\dfrac{9.8}{15},\ \dfrac{2\frac{7}{10}}{3\frac{3}{4}}$

27. $\dfrac{1352}{840},\ \dfrac{0.75}{0.44}$

28. $\dfrac{0.9}{4.6},\ \dfrac{7.2}{36.8}$

29. $\dfrac{(-4)}{(-5)},\ \dfrac{524}{655}$

30. $\dfrac{9.7}{(-63)},\ \dfrac{(-53)}{346}$

In Exercises 31–42, solve each proportion for x.

31. $\dfrac{x}{18} = \dfrac{12}{8}$

32. $\dfrac{8}{18} = \dfrac{12}{x}$

33. $\dfrac{12}{x} = \dfrac{8}{18}$

34. $\dfrac{3.2}{5} = \dfrac{x}{8}$

35. $\dfrac{\dfrac{1}{2}}{x} = \dfrac{\dfrac{3}{4}}{4}$

36. $\dfrac{0.3}{0.04} = \dfrac{100}{x}$

37. $\dfrac{(x+1)}{4} = \dfrac{3.6}{2.7}$

38. $\dfrac{5}{(2-x)} = \dfrac{3.4}{5.6}$

39. $\dfrac{9}{8} = \dfrac{(2x-1)}{3.5}$

40. $\dfrac{a}{b} = \dfrac{2x}{a}$

41. $\dfrac{2}{3x} = \dfrac{(m+n)}{3}$

42. $\dfrac{c^2}{d^2} = \dfrac{5}{(x+3)}$

43. Suppose $ax = by$. Write four proportions with terms a, x, b, and y, assuming none of them is equal to zero.

44. In Example 4, suppose the 16-oz can costs 69 cents. Is it a better buy than the 10-oz can? If the 16-oz can costs 83 cents and the 10-oz can costs 49 cents, which is the better buy?

45. In a recent doubleheader with the White Sox, Wade Boggs got two hits in three times at bat in the first game, and one hit in four times at bat in the second game.
   a. What was Boggs' ratio of hits to times at bat in the first game? In the second game?
   b. What was his ratio of hits to times at bat for both games?
   c. Is your answer in part b the sum of the two fractions in part a? Why or why not?
   d. Convert the ratio you found in part b to a decimal correct to the nearest thousandth. This is Wade Boggs' batting average for the doubleheader.

46. The rectangle in Figure 3.19 is very nearly a *golden rectangle*, which was discovered by the ancient Greeks and appears frequently in art and architecture. What is the ratio of the width to the length of this rectangle? What is the ratio of the length to the sum of the length and the width?

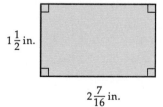

$1\frac{1}{2}$ in.

$2\frac{7}{16}$ in.

FIGURE 3.19

47. The two geometric figures in Figure 3.20 are similar. Find the lengths a, b, c, and d.

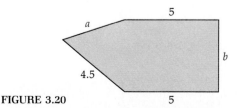
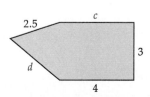

FIGURE 3.20

48. Suppose that $\frac{1}{4}$-oz lead fishing weights sell in packages of seven for 39 cents. At that rate, how much should a package of nine $\frac{1}{8}$-oz weights cost? Assume that the weight of the lead is the only factor affecting the cost.

49. The directions for Clear-aid, a liquid to clear the water in an aquarium, say to add 1 teaspoonful of Clear-aid for every 2 gal of aquarium water. If you have a 15-gal aquarium, how many teaspoonfuls should you add?

50. Feature-length 16-mm movies are made to project on a rectangular screen with a height-to-width ratio of 2:3. A television screen is square except for rounded corners. Use these facts to explain why you can't see the entire picture when some movies are shown on television.

51. After a vigorous antismoking campaign by the federal government, cigarette smoking is down by $23\frac{1}{2}\%$. If 258 adults per 1000 adults smoke now, how many smoked before the antismoking campaign?

52. The average salary for female professors was $33,000 in 1987. In 1990, it was $38,000. During the same three-year period, the average salary for male professors went from $36,000 to $40,000. What was the percent of increase for women? For men? If you were arguing for women's rights, would you emphasize the actual salaries or the percents of increase?

53. Last year you bought a motorbike, but you've forgotten the exact price. Although you just found the sales receipt, the price was smudged. However, the sales tax was $206. If your state has a 4% sales tax, what was the price of the motorbike?

Preparing for Section 3.7

54. Twice a number is 5 more than the number.
   a. If n represents the unknown number, write an equation showing this relationship.
   b. Solve the equation for n and check your answer.

55. The sum of two numbers is 45, and one number is twice the other.
   a. If x represents the smaller of the two numbers, what is the other number in terms of x?
   b. Write an equation in terms of x that shows the above relationship.
   c. Solve the equation for x and check your answer.

3.7 Problems and Applications

FOCUS

We discuss and use additional strategies for solving word problems with equations.

As we have seen in previous sections, equations can be used to solve problems that are presented verbally. The procedure for finding such solutions can be outlined as follows.

Solving word problems algebraically

1. Read the problem carefully. Draw a diagram or make a table if it will help you understand the problem.
2. Use a variable to represent one of the quantities to be found. Use the information in the problem to represent other unknown quantities in terms of this variable. (Later we will see that two variables can sometimes be used.)
3. Use the given conditions and numerical values, and perhaps a formula, to write an equation that describes the problem situation.
4. Solve the equation.
5. Check the solution against the conditions of the problem to be sure it is correct.

Step 3 is the most difficult one for most students. How do you translate the situation into an equation? There is no single answer to this question, since even slight changes in the wording of a problem can change the required equation considerably. One strategy is to recall an appropriate formula and use it as the basis for an equation. In earlier sections, we often based our equations on geometric formulas, such as those for perimeter, area, or circumference.

Another useful formula gives the distance traveled at rate r during a time t; we write it as

$$d = rt$$

It can also be used to find the rate or the travel time in its two other forms:

$$r = \frac{d}{t} \text{ or } t = \frac{d}{r}$$

EXAMPLE 1   A speeding train travels 420 mi in the same time it takes a car to travel 340 mi. The car's speed is 16 mph slower than that of the train. What is the car's speed?

Solution   In this problem, two trips are under consideration: the car's and the train's. If we let x represent the rate of the train, then $x - 16$ is the rate of the car. Do you see why? Then we can use the formula $t = \dfrac{d}{r}$ to write the times for the car and the train. A table will help us organize this information:

	Distance (d)	Rate (r)	Time $\left(t = \dfrac{d}{r} \right)$
Train	420	x	$\dfrac{420}{x}$
Car	340	$x - 16$	$\dfrac{340}{(x - 16)}$

We are told that the train makes the trip in the same time as the car. In other words, we must set the expressions for the time equal:

$$\frac{420}{x} = \frac{340}{(x - 16)}$$

Next we solve for x. Our first step is to cross multiply:

$420(x - 16) = 340x$	Cross multiplying
$420x - 6720 = 340x$	Removing parentheses
$80x - 6720 = 0$	Subtracting $340x$ from both sides
$80x = 6720$	Adding 6720 to both sides
$x = 84$	Dividing both sides by 80

The train averages 84 mph, so the car's rate is $84 - 16 = 68$ mph. We check by using this value in the original problem. If the train averages 84 mph, it will take $\dfrac{420}{84}$ or 5 h to travel 420 mi. If the car averages 68 mph, it will take $\dfrac{340}{68}$ or 5 h to travel 340 mi. Our answer checks.   ∎

Suppose you are trying to solve a word problem, and there is no formula that can be directly applied. A method for finding a useful equation, sometimes called the **guess-and-test method,** is illustrated in the next two examples.

EXAMPLE 2   A man had three equal piles of apples. He discovered eight rotten apples, threw them away, and divided the remaining apples into two equal piles.

   a. If the final two piles contained 17 apples each, how many apples were in each of the original three piles?

   b. If the final two piles contained 32 apples each, how many apples were in each of the original three piles?

Solution   **a.** We begin by guessing the number of apples in each of the original three piles. Let's guess 10 apples per pile and test to see whether we are right. If there are three piles of apples with 10 per pile, then

$$3 \cdot 10 = 30 \text{ apples originally}$$

Eight rotten apples are thrown away. Thus

$$3 \cdot 10 - 8 = 22 \text{ apples left}$$

These apples are divided equally into two piles. Hence

$$\frac{3 \cdot 10 - 8}{2} = 11 \text{ apples per pile}$$

Thus we see that if the original three piles contained 10 apples each, the resulting two piles would each contain 11 apples. Our guess is not correct (we are told in the problem that the final two piles contain 17, not 11, apples). However, we can use the test procedure to find the correct answer. To do this, we note that the number of apples in each of the original three piles is unknown. We'll call that number x. Now x has the same role as 10 in the guess-and-test procedure. We must perform the same operations on x that we performed on 10 to arrive at the number of apples in each of the remaining two piles. Let's follow the steps we used to test our earlier guess:

$$3 \cdot x \qquad \text{(total apples originally)}$$

Eight rotten apples are thrown away:

$$3x - 8 \qquad \text{(apples left)}$$

These apples are divided equally into two piles:

$$\frac{3x - 8}{2} \qquad \text{(apples per pile)}$$

But the number of apples in each of the remaining two piles is given to be 17. Thus we write our equation by setting the last expression above equal to 17. We then solve the equation:

$$\frac{3x - 8}{2} = 17$$

$3x - 8 = 34$ Multiplying both sides by 2

$3x = 42$ Adding 8 to both sides

$x = 14$ Dividing both sides by 3

There were 14 apples in each of the original three piles. The check is left as an exercise.

b. The only difference between part b and part a is that there are 32, instead of 17, apples in each of the remaining two piles. All the reasoning in part a applies here, too, with 32 replacing 17 in the equation. The solution is left as an exercise.   ∎

EXAMPLE 3

A flagpole is one-fourth as tall as the building on which it stands. The top of the pole is 65 meters (m) above the level of the ground. How tall is the flagpole? How tall is the building?

Solution

After reading the problem carefully, we note that a diagram may be helpful. Figure 3.21 shows a flagpole atop a building; the top of the flagpole is 65 m above the ground.

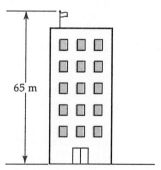

FIGURE 3.21

We are asked to find the height of the flagpole. Suppose for a moment that we do not see how to write an equation to solve the problem. Let's guess that the height is 12 m and test our guess.

Our guess: Flagpole is 12 m long.

To test the guess, we note that the building must be four times as tall as the flagpole, and the top of the flagpole must be 65 m above the ground. So we have

$$4 \cdot 12 = 48 \text{ m} \qquad \text{(height of building)}$$

from which we find

$$12 + 48 = 60 \text{ m} \qquad \text{(from ground to top of flagpole)}$$

Our guess, 12 m, is incorrect because $12 + 48$ is not 65, but 60. However, the guess-and-test process does give us a guide for writing the needed equation.

As in previous problems, we assign a variable to an unknown quantity. We will use h for the height of the flagpole:

$$h = \text{height of flagpole in meters}$$

Now h takes the role of our guess, 12, in the guess-and-test process. We multiplied $4 \cdot 12$ to get the height of the building, so we must multiply $4 \cdot h$:

$$4h = \text{height of building in meters}$$

Then in the guess-and-test process we added the two heights, and their sum was supposed to be the given height, 65 m. We do so again, and then solve:

$$h + 4h = 65$$
$$5h = 65 \qquad \text{Combining like terms}$$
$$h = 13 \qquad \text{Dividing both sides by 5}$$

The flagpole is 13 m long, and it follows that the height of the building is $4 \cdot 13$ or 52 m. The check is left as an exercise. ∎

Sometimes the ideas of ratio and proportion can be used with algebra to solve problems. For this purpose, it is helpful to recognize that if we are told that two unknown quantities are in the ratio of $\dfrac{a}{b}$, then we can write that ratio as $\dfrac{ax}{bx}$, where ax is one quantity and bx is the other and $x \neq 0$. We use this idea in the next example.

EXAMPLE 4

A livestock farmer owns cattle and hogs, in the ratio of 5 cattle for every 3 hogs. The combined total of his cattle and hogs is 576. Find the number of cattle and the number of hogs the farmer owns.

Solution

In the language of ratio and proportion, the ratio of the number of cattle to the number of hogs owned by the farmer is $\frac{5}{3}$. We thus can let $5x$ and $3x$ be the number of cattle and hogs, respectively, owned by the farmer:

$$5x = \text{number of cattle}$$
$$3x = \text{number of hogs}$$

We also know that the total number of animals is 576, and that forms the basis for our equation:

$$5x + 3x = 576$$

Next we solve the equation.

$$8x = 576 \qquad \text{Adding like terms}$$
$$x = 72 \qquad \text{Dividing both sides by 8}$$

We are not finished yet. The number of cattle is $5x = 5 \cdot 72 = 360$, and there are $3x = 3 \cdot 72 = 216$ hogs. The check is left as an exercise. ∎

One additional word of advice: When you attempt to solve a word problem, be flexible.

BE CAREFUL!

The strategies for solving word problems are "things to try." A particular strategy may be helpful for one problem, but useless for another. If one strategy does not seem to be getting you anywhere, try another. Perhaps it will help.

DISCUSSION QUESTIONS 3.7

1. What are the five steps in the procedures for solving word problems algebraically?

2. How can you tell when a formula might help in solving a particular word problem?

3. In what kinds of word problems might it be helpful to draw a diagram?

EXERCISES 3.7

1. Complete the check in part a of Example 2.

2. Solve part b of Example 2.

3. Complete the check in Example 3.

4. Complete the check in Example 4.

5. The ratio of two numbers is 6 to 5. Which of the following are possible values of the numbers? (Here k and x represent nonzero numbers.)
    a. 12 and 10   **b.** 210 and 175   **c.** -36 and -30
    d. 24 and 15   **e.** $6x$ and $5x$ **f.** $6k$ and $5k$

In Exercises 6–9, answer each part.

6. A farmer used 400 m of chicken wire (not counting waste) to build a fence around a rectangular lot. The width of the lot is 50 m.
    a. Draw a figure representing the farmer's lot. Use l for the length of the lot.
    b. Use the formula for the perimeter of a rectangle to write an equation representing this situation. Solve for l and check your answer.

7. A collection of nickels and dimes is worth $2.25. It consists of 6 more dimes than nickels. Use the variable n to represent the number of nickels.
    a. Write an expression for the number of dimes in terms of n.
    b. Write an expression in terms of n for the value in cents of the nickels.
    c. Write an expression in terms of n for the value in cents of the dimes.
    d. Write an expression in terms of n for the total value of the collection.
    e. Write and solve an equation formed by setting the expression in part d equal to 225.
    f. How many nickels and how many dimes are in the collection?
    g. Check your answer by substituting into the original problem.

8. A 50% acid solution is mixed with twice as much of a 10% acid solution to make a new solution. The new solution contains 7 gal of acid.
    a. Let x represent the number of gallons of 50% acid solution. Complete this table:

	Amount	Amount of Acid
50% solution	x	$0.50x$
10% solution	___	___

   b. Use the fact that the new solution contains 7 gal of acid and your answer in part a to write an equation that can be solved for x.

   c. Solve the equation for x.

   d. Find the total amount of the new solution.

   e. Find the percent of acid in the new solution.

 9. On a tour of the American West, a man drove for 8 h one day. On the second day, he was anxious to get to the Grand Canyon, so he drove for 10 h at an average rate of 5 mph more than his rate on the first day. He traveled 160 mi further on day 2 than on day 1.

   a. What was his average rate on day 1?

   b. What was his average rate on day 2?

   c. What was the total distance he traveled on both days combined?

In Exercises 10–30, use equations to solve the problems. Draw figures, rely on formulas, or use the guess-and-test method when appropriate. A calculator would be helpful for many of these exercises.

 10. If 4 is subtracted from a number and the result is doubled, the new number is 6. Find the number.

 11. A plane flies at an average rate of 450 mph from New York to San Francisco, a distance of 3000 mi. How long will it take for the plane to be 1125 mi from New York?

 12. A desk and chair together cost $352, and the desk costs three times as much as the chair. What was the cost of the desk?

 13. One day Jody completed a 420-mi trip. If she drove twice as far in the afternoon as she did in the morning, how far did she drive in the morning?

 14. The perimeter of a rectangle is 20 cm less than three times its length. The width of the rectangle is 10 cm. Find the perimeter.

 15. A student's average score on six math quizzes was 78. On five of the quizzes his scores were 63, 87, 80, 71, and 95. Find his score on the sixth quiz.

 16. A small town began the year with a treasury deficit. Before collecting taxes, it incurred debts equal to three times the deficit. During the first week of tax payments, the town collected $17,480, which just erased its total debt. What was the amount of the original deficit?

 17. On June 21 in Pittsburgh there is daylight for 6 h and 6 min longer than there is darkness. How long is the night? How long is the day?

 18. A builder built a new (rectangular) bedroom for his house that measured 32 ft by 24 ft. The building code required that the window space be at least one-twelfth of the floor space. What is the minimum number of 2-ft by 4-ft rectangular windows that the builder will need?

19. Four power plants supply a small city with 28 million kilowatts (kW) of electrical power. The first plant supplies twice as much as the second, and the third and fourth plants each supply 1 million kW less than the second. How much power comes from each plant?

20. Is it possible to make change for a $20 bill so that there are twice as many quarters as half dollars and half as many $1 bills as half dollars? If so, how? If not, show why not.

21. A woman sold her car for $6250. This was $122 more than 68% of the original price. What was the original price?

22. A child's piggy bank contains only nickels and dimes, which total $1.35. The number of dimes is 1 more than twice the number of nickels. How many dimes are in the bank?

23. A suit was priced at $450. That price is lowered by 20% and then raised by 10% of the new price. What is the resulting price of the suit?

24. Jeff paid an $8 debt with equal numbers of nickels, dimes, and quarters. How many of each were there?

25. A soccer team won 10 more games than it lost, and the ratio of its wins to its losses was 7 to 5. How many games did the team win?

26. A race driver is in second position 3 mi behind the leader. If the leading car's speed is 150 mph and the car in second place is capable of 165 mph, how many minutes would it take for the second-place car to overtake the leader?

27. A restaurant bought 80 glasses and 60 coffee mugs for a total cost of $120. If a mug costs twice as much as a glass, what is the cost of a glass?

28. An airplane that can fly at a speed of 200 mph in still air can fly 800 mi with the wind at the same time it takes to fly 640 mi against the wind. Find the speed of the wind.

29. The measures of the three angles of any triangle sum to 180°. One angle of a certain triangle is twice as large as the second. The third angle is 5° more than the larger of these. Find the measure of each angle.

30. In an isosceles triangle, two of the angles have equal measure. If the third angle measures 36° less than the sum of the two equal angles, find the measure of each angle.

Exercises 31–34 are applications from business and engineering that can be solved algebraically. You may need to have your instructor explain the meaning of some of the terms.

31. The depreciation on an automobile during the first year is 20% of the

initial cost. After that, the yearly depreciation is 10% of the initial cost. Suppose a car has depreciated $6000 in 5 years. What was the initial cost of the car? What will be the value of the car when it is 8 years old?

32. An engineer received a raise of $800 per year. Shortly afterward, however, her company became financially unstable, so she agreed to take a 4% pay cut in annual salary. She found that after the cut her yearly salary was $600 less than it had been before the raise. What was her salary before the raise?

33. A small plane leaves the deck of an aircraft carrier and travels north at a speed of 178 mph. The carrier is traveling north at a speed of 28 mph. If the plane's radio has a range of 600 mi, when will the pilot lose radio communication with the carrier?

34. The population of a town in 1980 was 10% greater than in 1970. The population in 1990 was 10% greater than 1980. Suppose the population in 1990 was 24,200.
   a. What was the population in 1970?
   b. What was the percent of increase in population from 1970 to 1990?

Preparing for Section 4.1

35. In 5^6, what is the exponent? What is the base?

36. Which of the following is equal to 4^3?
   a. $4 + 4 + 4$ **b.** $3 + 3 + 3 + 3$ **c.** $4 \cdot 4 \cdot 4$ **d.** $3 \cdot 3 \cdot 3 \cdot 3$

In Exercises 37–42, evaluate each expression for $x = 3$ and $y = -2$.

37. x^5 **38.** $7 + y^3$ **39.** $x^2 - y^2$ **40.** y^x

41. $3x^3 + 2y^4 - 15$ **42.** $5(y^2 - x^4 + 3)$

Chapter Summary

Important Ideas

Symmetric property of equality
If $a = b$, then $b = a$, where a and b are real numbers or expressions representing real numbers.

Addition property of equality
Any number or expression can be added to both sides of an equation without changing the relationship of the sides.

Subtraction property of equality
Any number or expression can be subtracted from both sides of an equation without changing the relationship of the sides.

Roots of a linear equation in one variable
The number of roots of a linear equation in one variable is either (1) zero, (2) exactly one, or (3) infinitely many (if the equation is an identity).

Multiplication property of equality
Both sides of an equation can be multiplied by the same number without changing the relationship of the sides.

Division property of equality
Both sides of an equation can be divided by the same nonzero number without changing the relationship of the sides.

Asymmetric property of inequality
If $a < b$, then $b > a$.   If $a \leq b$, then $b \geq a$.
If $a > b$, then $b < a$.   If $a \geq b$, then $b \leq a$.

Addition and subtraction properties of inequality
Any number or expression can be added to both sides of an inequality without changing the relationship between the sides.

   Any number or expression can be subtracted from both sides of an inequality without changing the relationship between the sides.

Multiplication property of inequality
Both sides of an inequality can be multiplied by the same *positive* number without changing the relationship of the sides.

$$\text{If } b > c \text{ and } a \text{ is positive, then } ab > ac.$$

If both sides of an inequality are multiplied by the same *negative* number, the relationship of the sides is reversed.

$$\text{If } b > c \text{ and } a \text{ is negative, then } ab < ac.$$

Division property of inequality
Both sides of an inequality can be divided by the same *positive* number without changing the relationship of the sides.

$$\text{If } b > c \text{ and } a \text{ is positive, then } b \div a > c \div a.$$

If both sides of an inequality are divided by the same *negative* number, the relationship of the sides is reversed.

$$\text{If } b > c \text{ and } a \text{ is negative, then } b \div a < c \div a.$$

Cross multiplication
If $\dfrac{a}{b} = \dfrac{c}{d}$, then $ad = bc$.

Chapter Review

Section 3.1

In Exercises 1–8, plot the points on a rectangular coordinate system.

1. $(-3, 6)$ 2. $(0, -2)$ 3. $(-7, -3)$ 4. $(5, -1)$

5. $(3, 2)$ 6. $(-5, 2)$ 7. $(4.4, -1.5)$ 8. $\left(2\frac{1}{3}, -4\frac{4}{5}\right)$

9. Could $(-4, -4)$, $(-1, 3)$, and $(0, 0)$ all satisfy the same linear equation? Explain.

In Exercises 10–15, graph the first equation. Use the graph to find the solution of the second equation. Check by substituting your graphical solution into the second equation for x.

10. $2x - 10 = y$; $2x - 10 = 4$

11. $2x - 10 = y$; $2x - 10 = -3$

12. $2x + y = 6$; $2x + 4 = 6$

13. $2x + y = 6$; $2x - 5 = 6$

14. $y = 10 - 4x$; $10 - 4x = 3.5$

15. $y = 10 - 4x$; $10 - 4x = -2\frac{1}{4}$

Section 3.2

In Exercises 16–26, solve each equation. Check by substitution.

16. $x + 6 = 13$ 17. $x - 3 = 7$ 18. $\frac{1}{3}y - \frac{1}{6} = \frac{1}{2}y$ 19. $\frac{3}{4}x - \frac{5}{8} = \frac{1}{4}$

20. $10 - 3x = 5$ 21. $4z - 10 = 8 + 3z$ 22. $8t + 6 = 2(7t + 9)$ 23. $4\frac{1}{5}(x - 5) = 3\frac{1}{2} + \frac{1}{5}x$

24. $24y - 19 = 8y - 23(6 - y)$ 25. $-4(2 - x) = 3(7 - 3x) + 5$ 26. $3(5x - 6) - 2(8 + 7x) = 24$

27. Write and solve an equation to find a number such that seven more than twice the number is 19.

28. A man received a bonus check. He then lost $25 betting on the lottery and paid a bill for $86.98. If the man had $581.76 left, write and solve an equation to find the amount of the bonus.

Section 3.3

29. Explain each step in the following solution:

$$8x - 3 = 37$$
$$8x = 40$$
$$x = 5$$

In Exercises 30–39, solve each equation. Check by substitution.

30. $3x - 9 = 3$

31. $-6y + 3 = 15$

32. $12 - 5x = 7x + 3$

33. $4(2x + 3) - 5 = 7(6 - 3x)$

34. $\frac{2}{7}t + \frac{3}{5} = \frac{1}{3}(3t + 5)$

35. $\frac{3}{4} + \frac{1}{2}w = \frac{1}{5}w - \frac{2}{5}$

36. $6(x - 3) - (8 - 3x) = 6 + 4(3x - 7)$

37. $9x = 5 - 2\{x - 8[2x - (4 - x)]\}$

38. $3[12 - 5(x + 2)] - 21 - 10x = -5x$

39. $8(x + 2) - 3(2 - 5x) = 6[2x - (7 - x)] + x$

40. Write and solve an equation to find a number such that if 7 is subtracted from five times the number, the result is -6.

41. The circumference of a circular window is 132 in. What is the diameter of the window?

Section 3.4

42. Write the inequality that results when both sides of $-5x < -25$ are multiplied by $-\frac{1}{5}$.

In Exercises 43–44, graph each inequality on a number line.

43. $x \geq 4.7$

44. $-6y > 36$

45. Explain each step in this solution.

$$8 + 4x \leq 5(3 - 2x) - 3$$
$$8 + 4x \leq 15 - 10x - 3$$
$$8 + 4x \leq 12 - 10x$$
$$8 + 14x \leq 12$$
$$14x \leq 4$$
$$x \leq \frac{4}{14}$$
$$x \leq \frac{2}{7}$$

In Exercises 46–51, solve each inequality.

46. $7r + 5 < 12$

47. $14 > 10 - 4x$

48. $2y + 21 \geq 12 - 16y$

49. $8 - 5x \geq 12 + 2x$

50. $1.4 - 3.2(5x - 4.6) < 4.7 + 2.5(14 - 6x)$

51. $36x + 45(12 - 24x) \geq 85 - 67(22x - 52)$

Section 3.5

52. **a.** Solve $I = prt$ for t.   **b.** Find t if $I = 240$, $p = 1000$, and $r = 0.08$.

53. **a.** Solve $Mg - T = Ma$ for a.   **b.** Find a when $M = 48$, $g = 32$, and $T = 96$.

54.   **a.** Solve $5x + 2y = 4$ for y.   **b.** Find y if $x = -3$.

55.   **a.** Solve $18 - 2y = 5x - 3$ for x.   **b.** Find x if $y = 2\dfrac{1}{3}$.

56.   **a.** Solve $ax - m^2t = \dfrac{m}{5}$ for x.   **b.** Find x if $a = 1$, $m = -3$, and $t = 1.7$.

57.   **a.** Solve $ax + by + c = 0$ for y.   **b.** Find y in terms of a, b, and c if $x = -2$.

58.   Use the $d = rt$ formula to solve: A bus drove from New Weston to Sharpsburg, a distance of 4 mi, at an average speed of 48 mph. How long, in minutes, did the trip take?

59.   Choose the appropriate measurement formula to solve: The area of a rectangular lot is 14,000 ft². If it is 100 ft wide, how long is the lot?

Section 3.6

60. $\dfrac{4}{5} = \dfrac{8}{10}$ is a proportion. Write three other true proportions that use the same four numbers.

61.   Write the ratio 2.5:1.5 as a common fraction.

62.   Determine whether $\dfrac{8}{18}$ and $\dfrac{20}{45}$ are equal ratios.

In Exercises 63–66, solve each proportion for x.

63. $\dfrac{3}{4} = \dfrac{2x}{30}$ **64.** $\dfrac{-6}{(2x - 3)} = \dfrac{-2}{x}$ **65.** $\dfrac{1.4x}{2.5(x + 1)} = \dfrac{2}{3}$ **66.** $\dfrac{(a + 2)}{2(x - 1)} = \dfrac{a}{x}$

67.   A recipe calls for 2 cans of orange juice concentrate for 24 servings. Find the amount of orange juice concentrate required for 72 servings.

68.   A dress originally priced at $60 was marked down by 20% for a sale. What is the sale price of the dress?

Section 3.7

69.   The circumference of a circle is 154 cm. Find the radius and area of the circle.

70.   A woman owns 50 shares of S-3 stock and 40 shares of S-4 stock. Today, the value of a share of S-3 stock is $6. If the total value of her investment is $540, what is the value of a share of S-4 stock?

71.   A man sold his motorcycle for $2400. This was $480 less than 80% of the price he paid for the motorcycle. How much did he pay for the motorcycle?

72. The ratio of mules to sheep owned by a farmer is 2 to 5. If he owns 133 animals altogether, how many mules and how many sheep does he own?

73. Two trains are traveling on parallel tracks with train B ahead of train A. Train A and train B are traveling at average rates of 70 mph and 60 mph, respectively. If it takes 0.6 h for train A to overtake train B, how many miles ahead was train B originally?

Chapter Test

In Problems 1–8, solve each equation or inequality.

1. $5 + y = 10$ **2.** $13 - y = -8$ **3.** $5m - 6 = 9m$ **4.** $10 - 2x > 4$

5. $\frac{1}{8}(3x - x) = \frac{9}{10}$ **6.** $3x - 1 = 5 - x$ **7.** $-3x + 2 \geq x - 5$ **8.** $2(r + 5) \geq 3r$

9. Graph $x - 2y = -5$. Use the graph to solve these equations.
   a. $5 - 2y = -5$ **b.** $x - 10 = -5$

10. Solve the equation $a = rs + tb$ for b.

11. Solve $3x + 4 = 2y + 7$ for x.

12. On a 420-mi trip, a woman drove three times as far in the afternoon as she did in the morning. How far did she drive in the afternoon?

13. A man wishes to enclose a rectangular section of lawn with a low fence to make a play area. The length of the section of lawn is 9 ft, and its area is 63 ft^2. How many feet of fencing will he need to make the play area?

14. A sum of $360 was shared by three people. The first person received three times as much as the second, and the third received twice as much as the second. How much did the second person receive?

15. In a certain week Moe earned 20% more than Larry, and Larry earned 40% more than Shemp. If Shemp earned $400, how much did Larry earn?

16. A woman withdrew $243 from a bank in $5 bills and $2 bills. She withdrew twice as many $2 bills as $5 bills. How many $5 bills did she withdraw?

17. A parking meter contains only nickels and dimes, for a total of $1.35. The number of dimes is one more than twice the number of nickels. How many dimes are in the parking meter?

4 *Exponents and Polynomials*

4.1 Basic Properties of Exponents

FOCUS

We review the definition of positive integer exponents, and discuss three properties of exponents. Then we use these properties to simplify algebraic expressions that contain exponents.

Recall that in Chapter 1 we used formulas to find the area of a square and the volume of a cube. For the square in Figure 4.1,

$$\text{Area} = s^2$$
$$= 4^2$$
$$= 4 \cdot 4$$
$$= 16$$

4

4

FIGURE 4.1

For the cube in Figure 4.2,

$$\text{Volume} = s^3$$
$$= 4^3$$
$$= 4 \cdot 4 \cdot 4$$
$$= 64$$

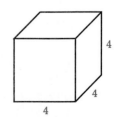

4

4

4

FIGURE 4.2

The expression 4^2 is read "the second power of 4" or "4 squared"; the 4 is the *base* and the 2 is the *exponent*. Notice that $4^2 = 4 \cdot 4$ is the product of two factors of 4.

Similarly, 4 is the base and 3 is the exponent of 4^3, which is read "the third power of 4" or "4 cubed." Note again that $4^3 = 4 \cdot 4 \cdot 4$ is the product of three factors of 4.

As you can see, we use exponents as a convenient way to write products of repeated factors. More precisely, we have the following definition of **positive integer exponents,** which was given in Chapter 1:

Positive integer exponents

For any real number a and any positive integer n,

$$a^n = \underbrace{a \cdot a \cdot \ldots \cdot a}_{n \text{ factors}}$$

The number a is called the **base** and the number n is called the **exponent.** The expression a^n is read "the *n*th **power of *a*.**"

It is understood that $a^1 = a$.

EXAMPLE 1

a. Find the value of x^4 when $x = 3$.

b. Find the value of x^4 when $x = -3$.

c. Find the value of $-x^4$ when $x = 3$.

Solution   We substitute the given value and then use the definition of exponents to evaluate each expression.

a. $\quad x^4 = 3^4$ $\qquad\qquad$ Substituting $x = 3$

$\quad = 3 \cdot 3 \cdot 3 \cdot 3$ $\qquad$ By the definition of exponents

$\quad = 81$

b. $\quad x^4 = (-3)^4$ $\qquad\qquad$ Substituting $x = -3$

$\quad = (-3) \cdot (-3) \cdot (-3) \cdot (-3)$ $\qquad$ By the definition of exponents

$\quad = 81$

c. $\quad -x^4 = -3^4$ $\qquad\qquad$ Substituting $x = 3$

$\quad = (-1) \cdot 3^4$ $\qquad$ Because $-a^n = (-1)a^n$

$\quad = (-1) \cdot 3 \cdot 3 \cdot 3 \cdot 3$ $\qquad$ By the definition of exponents

$\quad = -81$ ∎

Parts b and c of Example 1 illustrate a point that often causes confusion—and errors.

BE CAREFUL!

Be sure to distinguish between $(-a)^n$ and $-a^n$. For example, the parentheses in $(-3)^4$ indicate that the exponent 4 is applied to the base -3. However, the lack of parentheses in -3^4 indicates that the exponent 4 is applied only to the base 3, *not to* -3. It helps to think of $-a^n$ as $(-1)a^n$ or as the opposite of a^n.

EXAMPLE 2

Write each expression using exponents. Identify the base and the exponent.

a. $2 \cdot 2 \cdot 2 \cdot 2 \cdot 2$ b. $(-5) \cdot (-5) \cdot (-5)$

c. $-x \cdot x \cdot x \cdot x$ d. $-(-y) \cdot (-y)$

Solution   We apply the definition of exponents.

a. $2 \cdot 2 \cdot 2 \cdot 2 \cdot 2 = 2^5$ By the definition of exponents

     The base is 2 and the exponent is 5.

b. $(-5) \cdot (-5) \cdot (-5) = (-5)^3$ By the definition of exponents

     The base is -5 and the exponent is 3.

c. $-x \cdot x \cdot x \cdot x = -x^4$ By the definition of exponents

     The base is x and the exponent is 4.

d. $-(-y) \cdot (-y) = -(-y)^2$ By the definition of exponents

     The base is $-y$ and the exponent is 2.   ■

The $\boxed{y^x}$ key on your calculator is used to find powers. To find the values of 3^4, $(-3)^4$, and -3^4 on your calculator, follow these keying sequences:

Expression	Key Sequence	Display
3^4	3 $\boxed{y^x}$ 4 $\boxed{=}$	81
$(-3)^4$	3 $\boxed{+/-}$ $\boxed{y^x}$ 4 $\boxed{=}$	81
-3^4	3 $\boxed{y^x}$ 4 $\boxed{=}$ $\boxed{+/-}$	-81

The $\boxed{+/-}$ key produces the opposite of a displayed number. Unfortunately, some calculators cannot evaluate powers of negative numbers, like $(-3)^4$, with the $\boxed{y^x}$ key. Instead of displaying the correct result, the calculator displays the word "ERROR." The error occurs because the calculator does not use our definition of exponents.

We are now ready to turn to some properties of exponents that are of help in simplifying algebraic expressions. For example, how would you find the product $2^3 \cdot 2^5$? One way would be to proceed as follows:

$$2^3 \cdot 2^5 = (2 \cdot 2 \cdot 2) \cdot (2 \cdot 2 \cdot 2 \cdot 2 \cdot 2) \qquad \text{By the definition of exponents}$$

$$= 2 \cdot 2 \cdot 2 \cdot 2 \cdot 2 \cdot 2 \cdot 2 \cdot 2 \qquad \text{By the associative property}$$

$$= 2^8 \qquad \text{By the definition of exponents}$$

Notice that this result can be obtained by adding the exponents:

$$2^3 \cdot 2^5 = 2^{3+5} = 2^8$$

Similarly, when the base is a variable, we have

$$x^2 x^3 = (x \cdot x)(x \cdot x \cdot x)$$

$$= x \cdot x \cdot x \cdot x \cdot x$$

$$= x^5$$

and $$x^2 x^3 = x^{2+3} = x^5$$

These examples illustrate the *product property of exponents,* which tells us how to find the product of powers with the *same base:*

Product property of exponents

For any real number a and any positive integers m and n,

$$a^m \cdot a^n = a^{m+n}$$

EXAMPLE 3

Simplify each expression.

a. $\left(\dfrac{1}{2}\right)^2 \cdot \left(\dfrac{1}{2}\right)^4$ **b.** $(-3) \cdot (-3)^2$ **c.** $y^3 y^6$ **d.** $4z^3 \cdot 7z^4$

Solution

We apply the product property of exponents.

a. $\left(\dfrac{1}{2}\right)^2 \cdot \left(\dfrac{1}{2}\right)^4 = \left(\dfrac{1}{2}\right)^{2+4}$ By the product property of exponents

$= \left(\dfrac{1}{2}\right)^6$

$= \dfrac{1}{64}$

b.   Recall that $(-3) = (-3)^1$. Therefore

$(-3) \cdot (-3)^2 = (-3)^1 \cdot (-3)^2$

$= (-3)^{1+2}$ By the product property of exponents

$= (-3)^3$

$= -27$

c. $y^3 y^6 = y^{3+6}$ By the product property of exponents

$= y^9$

d.   We can use the commutative and associative properties to group the coefficients and to group the powers of z:

$4z^3 \cdot 7z^4 = (4 \cdot 7)(z^3 \cdot z^4)$ By the commutative and associative properties

$= 28z^{3+4}$ By the product property of exponents

$= 28z^7$

We can check this result by substituting a number for z, say $z = 1$, into each side of the last equation:

$$4z^3 \cdot 7z^4 = 4 \cdot 7 = 28 \quad \text{and} \quad 28z^7 = 28$$

Both sides of the equation have the same value, which suggests that $4z^3 \cdot 7z^4 = 28z^7$.   ∎

Our second property of exponents is concerned with powers of powers, and is illustrated in the following computation.

$(3^2)^4 = 3^2 \cdot 3^2 \cdot 3^2 \cdot 3^2$ By the definition of exponents

$= 3^{2+2+2+2}$ By the product property of exponents

$= 3^8$

This same result can be obtained by multiplying the exponents:

$$(3^2)^4 = 3^{2 \cdot 4} = 3^8$$

Similarly,

$$(w^3)^2 = w^3 \cdot w^3$$
$$= w^{3+3}$$
$$= w^6$$

and

$$(w^3)^2 = w^{3 \cdot 2} = w^6$$

These examples suggest the *power-to-a-power property of exponents:*

Power-to-a-power property of exponents

For any real number a and any positive integers m and n,

$$(a^m)^n = a^{m \cdot n}$$

EXAMPLE 4

Simplify each expression.

a. $(2^4)^3$ **b.** $(7^3)^3$ **c.** $(h^5)^8$ **d.** $(k^9)^2(k^6)^7$

Solution

We apply the power-to-a-power property of exponents.

a. $(2^4)^3 = 2^{4 \cdot 3} = 2^{12}$ By the power-to-a-power property of exponents

b. $(7^3)^3 = 7^{3 \cdot 3} = 7^9$ By the power-to-a-power property of exponents

c. $(h^5)^8 = h^{5 \cdot 8} = h^{40}$ By the power-to-a-power property of exponents

d.   We must use both properties of exponents to simplify $(k^9)^2(k^6)^7$:

$$(k^9)^2(k^6)^7 = k^{18}k^{42}$$ By the power-to-a-power property of exponents

$$= k^{18+42}$$ By the product property of exponents

$$= k^{60}$$

BE CAREFUL!

Don't confuse the power-to-a-power property of exponents with the product property of exponents. For example,

$$(h^5)^8 = h^{5 \cdot 8} = h^{40}$$

but

$$h^5 \cdot h^8 = h^{5+8} = h^{13}$$

In Figure 4.3, one square has sides of length s and the other square has sides of length $3s$. Since the sides of the larger square are three times the sides of the smaller square, is the area of the larger square three times the area of the smaller?

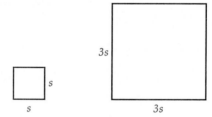

FIGURE 4.3

We can answer this question by finding the area of each square. The area of the smaller square is s^2, and the area of the larger square is $(3s)^2$. However, by the definition of exponents,

$$(3s)^2 = 3s \cdot 3s = 9s^2$$

so the area of the larger square is nine times the area of the smaller. By writing 9 as 3^2 in the last equation, we get

$$(3s)^2 = 3^2 s^2$$

which shows that the second power of the product $3s$ is equal to the product of each factor, 3 and s, to the second power. This example illustrates the *product-to-a-power property of exponents:*

Product-to-a-power property of exponents

For any real numbers a and b, and any positive integer n,

$$(ab)^n = a^n \cdot b^n$$

EXAMPLE 5

Simplify each expression.

a. $(5n)^3$ **b.** $3(2m)^4$ **c.** $(-4pq^3)^2$ **d.** $(3r^2)^4(rt^3)^5$

Solution

a. $(5n)^3 = 5^3 \cdot n^3$ By the product-to-a-power property of exponents

$= 125n^3$

b. $3(2m)^4 = 3 \cdot 2^4 \cdot m^4$ By the product-to-a-power property of exponents

$= 3 \cdot 16 \cdot m^4$

$= 48m^4$

c. $(-4pq^3)^2 = (-4)^2 p^2 (q^3)^2$ By the product-to-a-power property of exponents

$= 16p^2 q^6$ By the power-to-a-power property of exponents

d. $(3r^2)^4(rt^3)^5 = 3^4 (r^2)^4 r^5 (t^3)^5$ By the product-to-a-power property of exponents

$= 81r^8 r^5 t^{15}$ By the power-to-a-power property of exponents

$= 81r^{13} t^{15}$ By the product property of exponents ∎

BE CAREFUL!

A common mistake when simplifying a product to a power is forgetting to raise each factor to the given power. For example, $(5n)^3 \neq 5n^3$ because 5, which is a factor of the product $5n$, has not been raised to the third power.

DISCUSSION QUESTIONS 4.1

1. What is the meaning of the "nth power of a"?

2. Write the fifth power of x. What is the base? What is the exponent?

3. Explain the difference between $(-a)^n$ and $-a^n$.

4. In your own words, state the three properties of exponents given in this section.

EXERCISES 4.1

In Exercises 1–10, write the given expression using exponents. Identify the base and the exponent.

1. $5 \cdot 5 \cdot 5$ 2. $4 \cdot 4 \cdot 4 \cdot 4 \cdot 4$ 3. $-8 \cdot 8$ 4. $-6 \cdot 6 \cdot 6 \cdot 6$

5. $-(-6) \cdot (-6) \cdot (-6) \cdot (-6) \cdot (-6)$ 6. $-(-7) \cdot (-7) \cdot (-7)$ 7. $x \cdot x \cdot x \cdot x \cdot x \cdot x$

8. $y \cdot y \cdot y \cdot y \cdot y \cdot y \cdot y \cdot y$ 9. $-(-m) \cdot (-m)$ 10. $-(-n) \cdot (-n) \cdot (-n) \cdot (-n)$

In Exercises 11–26, evaluate each expression when $x = 2$ and $y = -3$.

11. x^3 12. x^5 13. y^5 14. y^4 15. $-x^4$ 16. $-x^2$

17. $-y^2$ 18. $-y^3$ 19. $7x^5$ 20. $4x^3$ 21. $-5y^3$ 22. $-6y^2$

23. $-8x^2y^4$ 24. $-7x^4y^2$ 25. $2x^3y^5$ 26. $10x^5y^3$

In Exercises 27–40, use the product property of exponents to simplify each expression.

27. $5^3 \cdot 5^2$ 28. $6^3 \cdot 6^4$ 29. $\left(\dfrac{3}{4}\right) \cdot \left(\dfrac{3}{4}\right)^3$ 30. $\left(\dfrac{3}{2}\right)^5 \cdot \left(\dfrac{3}{2}\right)^2$

31. $(-8)^5 \cdot (-8)^4$ 32. $(-9)^2 \cdot (-9)^4$ 33. $x^2 x^6$ 34. $y^7 y^3$

35. $(-w)^8 (-w)^6$ 36. $(-z)^6 (-z)^5$ 37. $h^2 h^4 h^3$ 38. $k^9 k^2 k^6$

39. $n^4 n^3 n^9 n^5$ 40. $m^7 m^6 m^2 m^4$

In Exercises 41–48, use the power-to-a-power property of exponents to simplify each expression.

41. $(4^3)^2$ 42. $(5^2)^4$ 43. $\left[\left(\dfrac{1}{4}\right)^3\right]^5$ 44. $\left[\left(\dfrac{2}{3}\right)^8\right]^4$

45. $(t^6)^9$ 46. $(s^2)^{10}$ 47. $(p^8)^2$ 48. $(q^{12})^5$

In Exercises 49–56, use the product-to-a-power property of exponents to simplify each expression.

49. $(3x)^4$ 50. $(6y)^2$ 51. $(-4z)^3$ 52. $(-7w)^4$

53. $(2pq)^5$ 54. $(6mn)^3$ 55. $\left(-\dfrac{3}{5}hk\right)^2$ 56. $\left(-\dfrac{1}{2}rs\right)^5$

In Exercises 57–72, simplify each expression.

57. $3q^2 \cdot 5q^7$ 58. $6p^3 \cdot 4p^6$ 59. $8u^4 \cdot (-4u^9)$ 60. $(-2v^5) \cdot 3v^8$

61. $(m^2)^5 (m^6)^3$ 62. $(n^4)^2 (n^5)^7$ 63. $(7x)^2 (2x)^4$ 64. $(3y)^2 (-5y)^4$

65. $-5(3p^4)^2$ 66. $-6(4q^2)^3$ 67. $4m(-6m^7)^3$ 68. $8n(-3n^2)^5$

69. $6z^5 (2w^2 z)^4$ 70. $5k^3 (2h^4 k^2)^6$ 71. $(-5rt^2)^3 (2r^3 t)^6$ 72. $(4u^5 v)^2 (-9uv^2)^3$

73. For the rectangle in Figure 4.4:
  a. Write an expression for the area in terms of x.
  b. Find the area when $x = 6$.

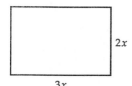

FIGURE 4.4 $3x$

74. For the square in Figure 4.5:
  a. Write an expression for the area in terms of y.
  b. Find the area when $y = 5$.

FIGURE 4.5 $2y$

75. For the rectangular solid in Figure 4.6:
  a. Write an expression for the total surface area in terms of w.
  b. Write an expression for the volume in terms of w.
  c. Find the total surface area and volume when $w = 4.5$.

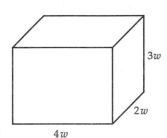

FIGURE 4.6 $4w$

76. For the cube in Figure 4.7:
  a. Write an expression for the total surface area in terms of z.
  b. Write an expression for the volume in terms of z.
  c. Find the total surface area and the volume when $z = 3.25$.

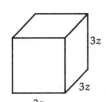

FIGURE 4.7 $3z$

Preparing for Section 4.2

In Exercises 77–83, evaluate each expression when $x = 6$ and $y = -2$.

77. $\dfrac{x}{y}$ **78.** $\dfrac{y}{x}$ **79.** $\dfrac{y}{4x}$ **80.** $\dfrac{x}{3y}$ **81.** $\left(\dfrac{y}{x}\right)^2$ **82.** $\left(\dfrac{x}{y}\right)^3$ **83.** $\dfrac{x^3}{y^3}$

4.2   Quotient Properties of Exponents

FOCUS

We extend our definition of a^n to include zero and negative integer exponents. Then we discuss two additional properties of exponents.

Consider what happens when we take decreasing integer powers of 10:

$$10^4 = 10,000$$

$$10^3 = 1,000$$

$$10^2 = 100$$

$$10^1 = 10$$

$$10^0 =$$

As the exponents on the left side of the equations decrease by 1, the values on the right side of the equations are divided by 10. For this pattern to continue, we should define $10^0 = 1$, since 10 divided by 10 is 1. That is exactly what we do. More generally, we have the following definition of the **zero exponent:**

Zero exponent

For any nonzero real number a,

$$a^0 = 1$$

Note that the expression 0^0 is not defined.

EXAMPLE 1

Simplify each expression. Assume $x \neq 0$.

a. $(-5)^0$ **b.** -5^0 **c.** $(3x)^0$ **d.** $3x^0$

Solution

a. The zero exponent applies to the base -5, so we have

$$(-5)^0 = 1 \qquad \text{By the definition of zero exponent}$$

b. Recall that $-5^0 = (-1) \cdot 5^0$. Therefore, the zero exponent applies only to the base 5:

$$-5^0 = (-1) \cdot 5^0$$

$$= (-1) \cdot 1 \qquad \text{By the definition of zero exponent}$$

$$= -1$$

c. Here the zero exponent applies to everything in the parentheses:

$$(3x)^0 = 1$$

d.   Recall that $3x^0 = 3 \cdot x^0$. Therefore, the zero exponent applies only to x:

$$3x^0 = 3 \cdot x^0$$
$$= 3 \cdot 1$$
$$= 3 \qquad \blacksquare$$

The three properties of exponents given in Section 4.1 also apply to the zero exponent because any expression that contains a zero exponent can be replaced by 1.

Negative exponents are produced when we continue the pattern for powers of 10. Decreasing the exponent by 1 on the left side of the equations, and dividing the right side by 10, we get

$$10^2 = 100$$
$$10^1 = 10$$
$$10^0 = 1$$
$$10^{-1} = \frac{1}{10} = \frac{1}{10^1}$$
$$10^{-2} = \frac{1}{100} = \frac{1}{10^2}$$
$$10^{-3} = \frac{1}{1000} = \frac{1}{10^3}$$
$$10^{-4} = \frac{1}{10,000} = \frac{1}{10^4}$$

and so on. If we compare the numbers on the left and right side of each line, we see another pattern: Each negative power of 10 is equal to the reciprocal of the corresponding positive power of 10. This observation leads us to the definition of **negative integer exponents:**

Negative integer exponents

For any nonzero real number a and any positive integer n,

$$a^{-n} = \frac{1}{a^n}$$

We can also define a^{-n} in terms of the reciprocal of a:

$$a^{-n} = \left(\frac{1}{a}\right)^n$$

To show that this is so, we can write

$$\left(\frac{1}{a}\right)^n = \underbrace{\frac{1}{a} \cdot \frac{1}{a} \cdot \ldots \cdot \frac{1}{a}}_{n \text{ factors}} = \frac{1}{a^n}$$

so the two definitions are equivalent. The alternative definition says that a^{-n} equals the nth power of the reciprocal of a. This definition is often helpful in simplifying expressions that contain negative exponents.

EXAMPLE 2

Simplify each expression.

a. $(-4)^{-2}$ **b.** -4^{-2} **c.** $\left(\frac{2}{3}\right)^{-1}$ **d.** $2y^{-3}$

Solution

a. The exponent -2 applies to the base -4, so we get

$$(-4)^{-2} = \frac{1}{(-4)^2} \qquad \text{By the definition of negative exponents}$$

$$= \frac{1}{16}$$

The alternative definition of negative exponents gives us the same result:

$$(-4)^{-2} = \left(\frac{1}{-4}\right)^2$$

$$= \frac{1}{16}$$

b. Here the exponent -2 applies only to the base 4:

$$-4^{-2} = (-1) \cdot 4^{-2}$$

$$= (-1) \cdot \frac{1}{4^2} \qquad \text{By the definition of negative exponents}$$

$$= (-1) \cdot \frac{1}{16}$$

$$= -\frac{1}{16}$$

c. $\left(\dfrac{2}{3}\right)^{-1} = \dfrac{1}{\left(\dfrac{2}{3}\right)^1}$ By the definition of negative exponents

$$= \dfrac{1}{\dfrac{2}{3}}$$

$$= 1 \cdot \dfrac{3}{2}$$

$$= \dfrac{3}{2}$$

By using the alternative definition of negative exponents, we get the same result more simply:

$$\left(\dfrac{2}{3}\right)^{-1} = \left(\dfrac{3}{2}\right)^1$$

$$= \dfrac{3}{2}$$

d. $2y^{-3} = 2 \cdot \dfrac{1}{y^3}$ By the definition of negative exponents

$$= \dfrac{2}{y^3}$$ ∎

The value of an expression that contains a negative exponent does not have to be a negative number. To see that this is so, compare parts a and b of Example 2. Remember that a negative exponent indicates only the reciprocal.

The definition of negative exponents allows us to write any expression that contains a negative exponent as an expression with a positive exponent. Hence the three properties of exponents given in Section 4.1 are also valid for negative exponents. We show this in the following example.

EXAMPLE 3

Simplify each expression. Write the result with positive exponents.

a. $w^{-2}\, w^5$ **b.** $(z^{-3})^4$ **c.** $(5q)^{-2}$ **d.** $(2p^{-1})^{-3}$

Solution   **a.** By the definition of negative exponents,

$$w^{-2}w^5 = \frac{1}{w^2} \cdot w^5$$

$$= \frac{w^5}{w^2} = \frac{w \cdot w \cdot w \cdot w \cdot w}{w \cdot w} \qquad \text{By the definition of exponents}$$

$$= \frac{w \cdot w \cdot w}{1} \qquad \qquad \text{Dividing numerator and denominator by } w \cdot w$$

$$= w^3$$

We get the same result with the product property of exponents:

$$w^{-2} \, w^5 = w^{(-2) + 5}$$

$$= w^3$$

b. By the definition of negative exponents,

$$(z^{-3})^4 = \left(\frac{1}{z^3}\right)^4$$

$$= \frac{1}{z^3} \cdot \frac{1}{z^3} \cdot \frac{1}{z^3} \cdot \frac{1}{z^3} \qquad \text{By the definition of exponents}$$

$$= \frac{1}{z^{3 + 3 + 3 + 3}} \qquad \qquad \text{By the product property of exponents}$$

$$= \frac{1}{z^{12}}$$

We get the same result with the power-to-a-power property of exponents:

$$(z^{-3})^4 = z^{(-3) \cdot 4} \qquad \text{By the power-to-a-power property of exponents}$$

$$= z^{-12}$$

$$= \frac{1}{z^{12}} \qquad \qquad \text{By the definition of negative exponents}$$

c. We first apply the definition of negative exponents:

$$(5q)^{-2} = \frac{1}{(5q)^2} \qquad \text{By the definition of negative exponents}$$

$$= \frac{1}{5^2 q^2} \qquad \qquad \text{By the product-to-a-power property of exponents}$$

$$= \frac{1}{25q^2}$$

We get the same result by first applying the product-to-a-power property of exponents:

$$(5q)^{-2} = 5^{-2}q^{-2} \qquad \text{By the product-to-a-power property of exponents}$$

$$= \frac{1}{5^2} \cdot \frac{1}{q^2} \qquad \text{By the definition of negative exponents}$$

$$= \frac{1}{25q^2}$$

d. $(2p^{-1})^{-3} = 2^{-3}(p^{-1})^{-3}$ By the product-to-a-power property of exponents

$$= \frac{1}{2^3} \cdot p^3 \qquad \text{By the definition of negative exponents and the power-to-a-power property of exponents}$$

$$= \frac{p^3}{8}$$

■

Another property of exponents is illustrated in the following example:

$$\left(\frac{a}{b}\right)^3 = \frac{a}{b} \cdot \frac{a}{b} \cdot \frac{a}{b} = \frac{a \cdot a \cdot a}{b \cdot b \cdot b} = \frac{a^3}{b^3}$$

The result $\left(\dfrac{a}{b}\right)^3 = \dfrac{a^3}{b^3}$ shows that the third power of the quotient $\dfrac{a}{b}$ is found by raising both the numerator and the denominator of the quotient to the third power. This suggests the *quotient-to-a-power property of exponents*:

Quotient-to-a-power property of exponents

For any real number a and nonzero real number b, and any integer n,

$$\left(\frac{a}{b}\right)^n = \frac{a^n}{b^n}$$

EXAMPLE 4

Simplify each expression. Write the result with positive exponents.

a. $\left(\dfrac{3}{4}\right)^2$ **b.** $\left(\dfrac{1}{3}\right)^{-3}$ **c.** $\left(\dfrac{t^4}{2}\right)^5$ **d.** $\left(\dfrac{m}{6n}\right)^{-2}$

Solution

a. $\left(\dfrac{3}{4}\right)^2 = \dfrac{3^2}{4^2}$ By the quotient-to-a-power property of exponents

$= \dfrac{9}{16}$

b. We give two equivalent solutions. First, using the quotient-to-a-power of exponents, we have

$\left(\dfrac{1}{3}\right)^{-3} = \dfrac{1^{-3}}{3^{-3}}$ By the quotient-to-a-power property of exponents

$= \dfrac{\dfrac{1}{1^3}}{\dfrac{1}{3^3}}$ By the definition of negative exponents

$= \dfrac{1}{1^3} \cdot \dfrac{3^3}{1}$

$= 3^3$

$= 27$

Second, using the alternative definition of negative exponents, we have

$\left(\dfrac{1}{3}\right)^{-3} = \left(\dfrac{3}{1}\right)^3$ By the alternative definition of negative exponents

$= 3^3$

$= 27$

c. $\left(\dfrac{t^4}{2}\right)^5 = \dfrac{(t^4)^5}{2^5}$ By the quotient-to-a-power property of exponents

$= \dfrac{t^{20}}{32}$ By the power-to-a-power property of exponents

d. Again, we give two equivalent solutions. Using the quotient-to-a-power property of exponents, we obtain

$\left(\dfrac{m}{6n}\right)^{-2} = \dfrac{m^{-2}}{(6n)^{-2}}$ By the quotient-to-a-power property of exponents

$= \dfrac{\dfrac{1}{m^2}}{\dfrac{1}{(6n)^2}}$ By the definition of negative exponents

$$= \frac{1}{m^2} \cdot \frac{(6n)^2}{1}$$

$$= \frac{(6n)^2}{m^2}$$

$$= \frac{36n^2}{m^2} \qquad \text{By the product-to-a-power property of exponents}$$

Or, using the alternative definition of negative exponents, we have

$$\left(\frac{m}{6n}\right)^{-2} = \left(\frac{6n}{m}\right)^2 \qquad \text{By the alternative definition of negative exponents}$$

$$= \frac{(6n)^2}{m^2} \qquad \text{By the quotient-to-a-power property of exponents}$$

$$= \frac{36n^2}{m^2} \qquad \text{By the product-to-a-power property of exponents} \qquad ■$$

Our last property of exponents tells how to simplify a quotient of powers with the *same base*. The fraction $\frac{5^6}{5^2}$ is such a quotient. To simplify it, we could write

$$\frac{5^6}{5^2} = \frac{5 \cdot 5 \cdot 5 \cdot 5 \cdot 5 \cdot 5}{5 \cdot 5} = \frac{5 \cdot 5 \cdot 5 \cdot 5}{1} = 5^4$$

This result can be obtained by subtracting the denominator exponent from the numerator exponent:

$$\frac{5^6}{5^2} = 5^{6-2} = 5^4$$

Similarly,

$$\frac{y^3}{y^5} = \frac{y \cdot y \cdot y}{y \cdot y \cdot y \cdot y \cdot y} = \frac{1}{y \cdot y} = \frac{1}{y^2}$$

and

$$\frac{y^3}{y^5} = y^{3-5} = y^{-2} = \frac{1}{y^2}$$

These examples illustrate the *quotient property of exponents:*

> **Quotient property of exponents**
>
> For any nonzero real number a and any integers m and n,
>
> $$\frac{a^m}{a^n} = a^{m-n}$$

EXAMPLE 5   Simplify each expression. Write the result with positive exponents.

a. $\dfrac{2^5}{2^9}$ **b.** $\dfrac{r^6}{r^{-3}}$ **c.** $\dfrac{4h^{-1}}{(3h)^2}$ **d.** $\left(\dfrac{7k^3x^5}{k^8x^{-4}}\right)^{-2}$

Solution   **a.** $\dfrac{2^5}{2^9} = 2^{5-9}$ By the quotient property of exponents

$= 2^{-4}$

$= \dfrac{1}{2^4}$ By the definition of negative exponents

$= \dfrac{1}{16}$

b. $\dfrac{r^6}{r^{-3}} = r^{6-(-3)}$ By the quotient property of exponents

$= r^{6+3}$

$= r^9$

c. $\dfrac{4h^{-1}}{(3h)^2} = \dfrac{4h^{-1}}{9h^2}$ By the product-to-a-power property of exponents

$= \dfrac{4}{9}h^{(-1)-2}$ By the quotient property of exponents

$= \dfrac{4}{9}h^{-3}$

$= \dfrac{4}{9h^3}$ By the definition of negative exponents

d.   There are several equivalent ways to simplify this expression. We start by simplifying inside the parentheses:

$\left(\dfrac{7k^3x^5}{k^8x^{-4}}\right)^{-2} = (7k^{3-8}x^{5-(-4)})^{-2}$ By the quotient property of exponents

$$= (7k^{-5}x^9)^{-2}$$

$$= 7^{-2}(k^{-5})^{-2}(x^9)^{-2} \qquad \text{By the product-to-a-power property of exponents}$$

$$= 7^{-2}k^{10}x^{-18} \qquad \text{By the power-to-a-power property of exponents}$$

$$= \frac{k^{10}}{49x^{18}} \qquad \text{By the definition of negative exponents} \qquad \blacksquare$$

DISCUSSION QUESTIONS 4.2

1. Does a^0 equal $(-a)^0$? Explain.

2. Does $(-a)^0$ equal $-a^0$? Explain.

3. Is a^{-n} a negative number when $a > 0$? When $a < 0$? Explain.

4. Does $a^n \cdot a^{-n}$ equal 0? Explain.

EXERCISES 4.2

In Exercises 1–20, evaluate the given expression.

1. 8^0
2. 5^0
3. $\left(\dfrac{4}{5}\right)^0$
4. $\left(-\dfrac{4}{3}\right)^0$
5. -9^0

6. -2^0
7. $(6x)^0$
8. $(-3y)^0$
9. 3^{-2}
10. 6^{-4}

11. $(-4)^{-3}$
12. $(-5)^{-2}$
13. -7^{-1}
14. -10^{-6}
15. $\left(\dfrac{1}{5}\right)^{-4}$

16. $\left(\dfrac{5}{6}\right)^{-3}$
17. $\left(-\dfrac{2}{3}\right)^{-5}$
18. $\left(-\dfrac{1}{7}\right)^{-1}$
19. 4.25^{-2}
20. 0.5^{-4}

In Exercises 21–28, simplify the given expression. Write the result with positive exponents.

21. $15k^{-5}$
22. $-12h^{-2}$
23. $-2m^{-8}$
24. $3n^{-9}$

25. $\dfrac{3}{9s^{-6}}$
26. $\dfrac{4t^{-8}}{2}$
27. $2u^{-1} - (2u)^{-1}$
28. $(-8v)^{-2} + 8y^{-2}$

In Exercises 29–36, use the quotient-to-a-power property to simplify the given expression. Write the result with positive exponents.

29. $\left(\dfrac{3}{8}\right)^2$
30. $\left(\dfrac{4}{9}\right)^3$
31. $\left(\dfrac{5}{2}\right)^{-3}$
32. $\left(\dfrac{2}{7}\right)^{-2}$

33. $\left(\dfrac{y}{2}\right)^4$
34. $\left(\dfrac{9}{x}\right)^2$
35. $\left(\dfrac{w}{z}\right)^{-5}$
36. $\left(\dfrac{q}{p}\right)^{-3}$

In Exercises 37–48, use the quotient property of exponents to simplify the given expression. Write the result with positive exponents.

37. $\dfrac{3^6}{3^4}$ **38.** $\dfrac{8^3}{8^5}$ **39.** $\dfrac{5^{-3}}{5^2}$ **40.** $\dfrac{10^3}{10^{-1}}$ **41.** $\dfrac{4^{-2}}{4^{-5}}$ **42.** $\dfrac{9^{-7}}{9^{-4}}$

43. $\dfrac{h^2}{h^7}$ **44.** $\dfrac{k^8}{k^3}$ **45.** $\dfrac{m^6}{m^{-3}}$ **46.** $\dfrac{n^{-1}}{n^4}$ **47.** $\dfrac{t^{-8}}{t^{-6}}$ **48.** $\dfrac{r^{-7}}{r^{-14}}$

In Exercises 49–66, simplify the given expression. Write the result with positive exponents.

49. $\dfrac{4v^{10}}{16v^5}$ **50.** $\dfrac{8u^{-3}}{6u^9}$ **51.** $\dfrac{-45y^9}{10y^{-7}}$ **52.** $\dfrac{16x^{-8}}{-24x^{-12}}$ **53.** $\left(\dfrac{z^{-6}z^6}{9}\right)^3$

54. $\left(\dfrac{2}{w^2w^{-8}}\right)^4$ **55.** $\dfrac{m^2m^{-3}}{m^4}$ **56.** $\dfrac{n^{-9}}{n^{-6}n^3}$ **57.** $\dfrac{(k^{-3})^2}{4k^{-6}}$ **58.** $\dfrac{-6h^{-5}}{(h^2)^{-3}}$

59. $\dfrac{(-5t^{-2})^3}{15t^8}$ **60.** $\dfrac{24s^{-7}}{(4s^9)^2}$ **61.** $\left(\dfrac{u^{-1}}{2u^3}\right)^{-4}$ **62.** $\left(\dfrac{6v^{-5}}{-4v^{-8}}\right)^2$ **63.** $\left(\dfrac{4x^7x^{-5}}{-12x^{-6}x^8}\right)^{-3}$

64. $\left(\dfrac{-10y^{-3}y^{-8}}{2y^{-6}y^4}\right)^0$ **65.** $\left[\dfrac{(3m^{-2}n)(4mn^{-3})}{(-6m^{-4}n^5)^2}\right]^{-1}$ **66.** $\left[\dfrac{(-2h^3k^{-1})^{-5}(7h^{-2}k^{-8})}{-16h^{-4}k^{-15}}\right]^2$

Preparing for Section 4.3

In Exercises 67–72, evaluate the given expression when $x = 10$.

67. $2.5x$ **68.** $5.8x^2$ **69.** $\dfrac{6.1}{x}$ **70.** $\dfrac{4.8}{x^2}$ **71.** $\dfrac{7.29}{x^3}$ **72.** $\dfrac{9.65}{x^4}$

4.3   Scientific Notation

FOCUS

We discuss how to write a number in scientific notation. Scientific notation is useful when computing with very large or very small numbers.

Astronomers often measure the vast distances of our solar system in terms of light years. One light year is the distance that light travels in one year. Since the speed of light is approximately 186,000 miles per second (mi/s), we can find the distance that light travels in one year by multiplying the number of seconds in one year by the speed of light:

Number of seconds in one year

> = 60 s/minute (min) · 60 min/hour (h) · 24 h/day · 365 days/year

> = 31,536,000 s/year

so in one year light travels

> 186,000 mi/s · 31,536,000 s/year = 5,865,696,000,000 mi/year

A light year is approximately 6 trillion miles!

You can use your calculator to find the number of miles in a light year by following this keying sequence:

Expression	*Key Sequence*	*Display*
186,000 · 31,536,000	186000 $\boxed{\times}$ 31536000 $\boxed{=}$	5.865696 12

Your calculator cannot display the result as 5,865,696,000,000 because this number has too many digits. Instead, the calculator displays 5,865,696,000,000 in *scientific notation*.

Scientific notation

A positive real number is in **scientific notation** when it is written in the form

$$a \cdot 10^n$$

where $1 \le a < 10$ and n is an integer.

This definition says that a positive number is in scientific notation when it is written as the product of a number between 1 and 10 and an integer power of 10. We can also use the form $a \times 10^n$ to express a number in scientific notation. In the calculator display

> 5.865696 12

the first part, 5.865696, is a number between 1 and 10; the second part of the display, 12, indicates the power of 10 that multiplies 5.865696. Therefore, in scientific notation,

$$5,865,696,000,000 = 5.865696 \cdot 10^{12}$$

To write a number in scientific notation, we must usually move the decimal point either to the left or to the right to produce a number between 1 and 10. Then, for each decimal place moved to the left the number must be multiplied by a factor of 10, and for each decimal place moved to the right the number must be multiplied by a factor of $\frac{1}{10}$. The factors of 10 and $\frac{1}{10}$ are needed to keep the value of the original number unchanged, because moving the decimal point one place to the left divides the original number by 10, whereas moving it one place to the right multiplies the original number by 10.

To convert one light year to scientific notation, we must move the decimal point 12 places to the left and multiply by 12 factors of 10:

$$5,865,696,000,000$$

$$= 5.\underbrace{865696000000}_{\text{12 places}} \cdot \underbrace{10 \cdot 10 \cdot 10 \cdot 10 \cdot 10 \cdot 10 \cdot 10 \cdot 10 \cdot 10 \cdot 10 \cdot 10 \cdot 10}_{\text{12 factors of 10}}$$

$$= 5.865696 \cdot 10^{12}$$

When the number is smaller than 1, we move the decimal point to the right:

$$0.000017 = \underbrace{00001.7}_{\text{5 places}} \cdot \underbrace{\frac{1}{10} \cdot \frac{1}{10} \cdot \frac{1}{10} \cdot \frac{1}{10} \cdot \frac{1}{10}}_{\text{5 factors of } \frac{1}{10}}$$

$$= 1.7 \cdot \frac{1}{10^5}$$

$$= 1.7 \cdot 10^{-5}$$

Note that the decimal point is moved only until there is one nonzero digit to the left of the decimal point. Furthermore, moving the decimal point to the left requires multiplying by a positive power of 10, whereas moving it to the right requires multiplying by a negative power of 10.

EXAMPLE 1   Write each number in scientific notation.

a.   47,900   **b.**   0.0000002816   **c.**   5,000,000   **d.**   0.0076

Solution   **a.**   We must move the decimal point 4 places to the left to produce a number between 1 and 10. Hence we must multiply by 10^4:

$$47,900 = 4.\underbrace{7900}_{\text{4 places}} \cdot 10^4$$

$$= 4.79 \cdot 10^4$$

b. We must move the decimal point 7 places to the right to produce a number between 1 and 10. So we need to multiply by 10^{-7}:

$$0.0000002816 = \underbrace{0000002}.816 \cdot 10^{-7}$$
$$7 \text{ places}$$

$$= 2.816 \cdot 10^{-7}$$

c. We must move the decimal point 6 places to the left and multiply by 10^6:

$$5,000,000 = 5.0 \cdot 10^6$$

d. We must move the decimal point 3 places to the right and multiply by 10^{-3}:

$$0.0076 = 7.6 \cdot 10^{-3} \qquad\blacksquare$$

The usual decimal notation, as in 0.0076, is called **standard notation.** We can easily convert from scientific notation to standard notation by performing the indicated multiplication.

EXAMPLE 2

Write each number in standard notation.

a. $7.98 \cdot 10^5$ **b.** $8.9 \cdot 10^{-3}$ **c.** $2.501 \cdot 10^8$ **d.** $3.66 \cdot 10^{-4}$

Solution

a. Recall that $10^5 = 10 \cdot 10 \cdot 10 \cdot 10 \cdot 10 = 100,000$. Therefore

$$7.98 \cdot 10^5 = 7.98 \cdot 100,000 = 798,000$$

Notice that multiplying 7.98 by 10^5 or 100,000 moved the decimal point 5 places to the right.

b. Recall that $10^{-3} = \dfrac{1}{10^3} = \dfrac{1}{1000}$. Therefore

$$8.9 \cdot 10^{-3} = 8.9 \cdot \frac{1}{1000}$$

$$= \frac{8.9}{1000}$$

$$= 0.0089$$

Notice that multiplying 8.9 by 10^{-3} is equivalent to moving the decimal point 3 places to the left.

c. Since we are multiplying 2.501 by 10^8 or 100,000,000, we move the decimal point 8 places to the right:

$$2.501 \cdot 10^8 = 250,100,000$$

d. Since we are multiplying 3.66 by 10^{-4}, which is equivalent to dividing by 10^4, we move the decimal point 4 places to the left:

$$3.66 \cdot 10^{-4} = 0.000366 \qquad ■$$

A number written in scientific notation can be entered into your calculator with the $\boxed{\text{EXP}}$ key. For example, to enter the numbers $5.865696 \cdot 10^{12}$ and $1.7 \cdot 10^{-5}$ into your calculator (one at a time), follow these keying sequences:

Expression	Key Sequence	Display
$5.865696 \cdot 10^{12}$	$5.865696 \ \boxed{\text{EXP}}\ 12$	5.865696 12
$1.7 \cdot 10^{-5}$	$1.7 \ \boxed{\text{EXP}}\ 5\ \boxed{+/-}$	1.7 −05

Scientific notation comes in handy when we are computing with very large or very small numbers. As an example, let us compute how long it would take a rocket going 18,000 mi/h to travel a distance of one light year. If we use $5.9 \cdot 10^{12}$ mi as an approximation for one light year and write the speed of the rocket in scientific notation as $1.8 \cdot 10^4$ mi/h, then we have

$$\text{Time} = \frac{\text{Distance}}{\text{Speed}}$$

$$= \frac{5.9 \cdot 10^{12}}{1.8 \cdot 10^4}$$

$$= \frac{5.9}{1.8} \cdot 10^{12-4} \qquad \text{By the quotient property of exponents}$$

$$\approx 3.3 \cdot 10^8 \text{ h}$$

It would take approximately 330,000,000 h, or slightly more than 37,671 years, for the rocket to travel a distance of one light year!

EXAMPLE 3

Perform each computation. Write the result in scientific notation.

a. $(4.3 \cdot 10^{11})(1.5 \cdot 10^{-4})$ **b.** $\dfrac{7.2 \cdot 10^6}{8 \cdot 10^{12}}$ **c.** $(8.03 \cdot 10^{-13})(4.7 \cdot 10^{-9})$

Solution

a. We can use the commutative and associative properties to write

$$(4.3 \cdot 10^{11})(1.5 \cdot 10^{-4})$$

$$= (4.3 \cdot 1.5)(10^{11} \cdot 10^{-4})$$

$$= 6.45 \cdot 10^7 \qquad \text{By the product property of exponents}$$

b. $\dfrac{7.2 \cdot 10^6}{8 \cdot 10^{12}} = \dfrac{7.2}{8} \cdot \dfrac{10^6}{10^{12}}$

$\qquad\qquad = \dfrac{7.2}{8} \cdot 10^{6-12}$ \qquad By the quotient property of exponents

$\qquad\qquad = 0.9 \cdot 10^{-6}$

In scientific notation, $0.9 = 9.0 \cdot 10^{-1}$, so

$\dfrac{7.2 \cdot 10^6}{8 \cdot 10^{12}} = 9.0 \cdot 10^{-1} \cdot 10^{-6}$

$\qquad\qquad = 9.0 \cdot 10^{-7}$ \qquad By the product property of exponents

c. We can use the commutative and associative properties to write

$(8.03 \cdot 10^{-13})(4.7 \cdot 10^{-9})$

$\qquad = (8.03 \cdot 4.7)(10^{-13} \cdot 10^{-9})$

$\qquad = 37.741 \cdot 10^{-22}$ \quad By the product property of exponents

In scientific notation, $37.741 = 3.7741 \cdot 10$, so

$(8.03 \cdot 10^{-13}) \cdot (4.7 \cdot 10^{-9})$

$\qquad = 3.7741 \cdot 10 \cdot 10^{-22}$

$\qquad = 3.7741 \cdot 10^{-21}$ \quad By the product property of exponents   ∎

DISCUSSION QUESTIONS 4.3

1. What is the definition of scientific notation?

2. In your own words, state how to write a number in scientific notation.

3. Why must a number be multiplied by 10 when the decimal point is moved one place to the right?

4. Why must a number be multiplied by $\dfrac{1}{10}$ when the decimal point is moved one place to the left?

5. When a number is written in scientific notation, does the power of 10 indicate whether the number is greater than or less than 1? Explain.

EXERCISES 4.3

In Exercises 1–10, write the given number in scientific notation.

1. 7000	**2.** 180,000	**3.** 0.009246	**4.** 0.000002	**5.** 240,000,000
6. 43,500	**7.** 0.0005	**8.** 0.000083	**9.** 0.000000038	**10.** 0.0000618

In Exercises 11–20, write the given number in standard notation.

11. $9.8 \cdot 10^4$　　**12.** $5.12 \cdot 10^3$　　**13.** $6 \cdot 10^{-5}$　　**14.** $8.4 \cdot 10^{-7}$　　**15.** $2.478 \cdot 10^9$

16. $3 \cdot 10^6$　　**17.** $7.8034 \cdot 10^0$　　**18.** $6.8027 \cdot 10^8$　　**19.** $1.004 \cdot 10^{-6}$　　**20.** $2.09 \cdot 10^{-10}$

In Exercises 21–40, perform the given computation. Write the result in scientific notation.

21. $(4 \cdot 10^6)(2 \cdot 10^9)$

22. $(2 \cdot 10^7)(3 \cdot 10^5)$

23. $\dfrac{8 \cdot 10^7}{4 \cdot 10^3}$

24. $\dfrac{9 \cdot 10^6}{3 \cdot 10^{10}}$

25. $(6.3 \cdot 10^{-2})(5 \cdot 10^8)$

26. $(8 \cdot 10^4)(7.5 \cdot 10^{-9})$

27. $\dfrac{4.2 \cdot 10^8}{3 \cdot 10^{-3}}$

28. $\dfrac{7.2 \cdot 10^{-5}}{4 \cdot 10^{11}}$

29. $(4.7 \cdot 10^4)(9.3 \cdot 10^{-12})$

30. $(5.3 \cdot 10^{-8})(3.9 \cdot 10^{14})$

31. $\dfrac{2.7 \cdot 10^{-5}}{5.2 \cdot 10^9}$

32. $\dfrac{8.4 \cdot 10^2}{9.8 \cdot 10^{-6}}$

33. $(1.03 \cdot 10^{-9})(6.4 \cdot 10^{-4})$

34. $(3.1 \cdot 10^{-10})(2.87 \cdot 10^{-2})$

35. $\dfrac{4.39 \cdot 10^{-2}}{3.8 \cdot 10^{-10}}$

36. $\dfrac{5.9 \cdot 10^{-15}}{1.78 \cdot 10^{-4}}$

37. $\dfrac{(4.8 \cdot 10^{-6})(8.4 \cdot 10^5)}{1.2 \cdot 10^7}$

38. $\dfrac{1.5 \cdot 10^{-3}}{(3.9 \cdot 10^{12})(4.5 \cdot 10^{-23})}$

39. $\dfrac{(5.1 \cdot 10^{-8})(4.9 \cdot 10^5)}{(9.6 \cdot 10^{12})(2.7 \cdot 10^{-9})}$

40. $\dfrac{(3.2 \cdot 10^{14})(7.4 \cdot 10^{-9})}{(6.8 \cdot 10^{-17})(5.9 \cdot 10^5)}$

In Exercises 41–48, express the result in scientific notation.

41. How far does light travel in one day? Assume that the speed of light is 186,000 mi/s.

42. Water consumption in the United States is approximately 1800 gallons (gal) per person per day. How many gallons of water are consumed in an average day? Assume that the population of the United States is 245,000,000 people.

43. A computer can perform an arithmetic computation in 0.000000025 s. How many of these computations can the computer perform in 24 h?

44. How many minutes does it take light to travel from San Francisco to Los Angeles? Assume that the distance from San Francisco to Los Angeles is 400 mi.

45. In a recent corporate merger, 425,000,000 shares of stock were purchased at a cost of $12.50 per share. What was the total cost of the stock purchased?

46. One dollar will purchase approximately 2400 Mexican pesos. The Mexican government recently obtained $72,000,000 from oil sales. How many pesos can these dollars purchase?

47. The area of all the land on the earth is approximately 150,000,000 square kilometers (km^2). How many square kilometers of land would each person receive if the land were distributed in equal-sized plots? Assume that the earth's population is 4,000,000,000 people.

48. One atom of silver has an area of approximately 0.0000000000000007 square centimeters (cm^2). How many atoms of silver would it take to cover the head of a pin of area 0.049 cm^2? Assume that the layer of silver on the pin head is one atom thick.

Preparing for Section 4.4

In Exercises 49–56, simplify the given expression.

49. $7x + 5x$ **50.** $8y - 2y$ **51.** $-(6w - 9w)$ **52.** $-(z + 3z)$

53. $13k^2 - 5k^2$ **54.** $6p^3 - 10p^3$ **55.** $-(3m^4 - 2m^4)$ **56.** $-5(q^6 + q^6 - 3q^6)$

4.4 Addition and Subtraction of Polynomials

FOCUS

We review some ideas that lead to the definition of polynomials. Then we discuss how to add and subtract polynomials.

The number 5376 is equal to 5 thousands plus 3 hundreds plus 7 tens plus 6 ones. Using powers of 10 to represent the ones, tens, hundreds, and thousands, we can write

$$5376 = 5 \cdot 10^3 + 3 \cdot 10^2 + 7 \cdot 10 + 6$$

Now consider the algebraic expression $5x^3 + 3x^2 + 7x + 6$. Notice that if we substitute $x = 10$ into this expression, the result is 5376. Algebraic expressions like $5x^3 + 3x^2 + 7x + 6$ are called **polynomials** and are often used to represent real numbers and to describe mathematical relationships.

Before we begin our study of polynomials, we need to review some ideas from Chapter 1. Recall that a **term** is either a number or the product of a number and one or more variables. The variables may be raised to positive integer powers. Some examples of terms are

$$-2, \ x^5, \ \frac{-5y^3}{8}, \ 4xy^2, \text{ and } -x^7yz^4$$

Note that an expression like $\dfrac{3}{x^2}$ is *not* a term because $\dfrac{3}{x^2} = 3x^{-2}$ (it is the product of a number and a variable raised to a *negative* integer power).

Recall from Chapter 1 that if a term contains one or more variables, then the **numerical coefficient** of the term, or simply the coefficient, is the number that multiplies the variable(s). For the examples above, we have

Term	Coefficient
x^5	1
$\dfrac{-5y^3}{8}$	$-\dfrac{5}{8}$
$4xy^2$	4
$-x^7yz^4$	-1

The **degree of a term** is either the exponent of the variable (if the term contains one variable) or the sum of the exponents of the variables (if the term contains more than one variable). A term that does not contain a variable has degree zero. For example,

Term	Degree
-2	0
x^5	5
$\dfrac{-5y^3}{8}$	3
$4xy^2$	3
$-x^7yz^4$	12

The polynomial $5x^3 + 3x^2 + 7x + 6$ is the sum of four terms. This observation leads us to the following definition:

Polynomial

A **polynomial** is either a single term or the sum of a finite number of terms.

Some other examples of polynomials are

$$-8z, \ x^5 - 2x^3, \ 3y^4 - 7y + 9, \text{ and } 6x^2y^2 + 4xy^2 - 5x + 8y - 11$$

We can think of the polynomial $x^5 - 2x^3$ as the sum of two terms because $x^5 - 2x^3 = x^5 + (-2)x^3$. In a similar way, the polynomials $3y^4 - 7y + 9$ and $6x^2y^2 + 4xy^2 - 5x + 8y - 11$ are sums of three terms and five terms, respectively.

The **degree of a polynomial** is the highest degree of the polynomial's terms. For the polynomials above, we have

Polynomial	Degree
$-8z$	1
$x^5 - 2x^3$	5
$3y^4 - 7y + 9$	4
$5x^3 + 3x^2 + 7x + 6$	3
$6x^2y^2 + 4xy^2 - 5x + 8y - 11$	4

When a polynomial contains only one variable, the degree of the polynomial is the degree of the term with the highest power of that variable. We usually write a polynomial in one variable in **descending powers** of the variable; that is, we arrange the terms so that the exponents of the variable (the degrees of the terms) decrease from left to right.

EXAMPLE 1

Write each polynomial in descending powers of the variable. Find the degree of the polynomial and the coefficient of each term.

a. $12 - 9z^6 + 3z$
b. $-7x + 18x^5 - 20x^4 + 6$
c. $4w^5 - 6w + w^8 - w^3 - 1$

Solution

a. We rearrange the terms of the polynomial so that the exponents of z decrease from left to right:

$$12 - 9z^6 + 3z = -9z^6 + 3z + 12$$

The degree of the polynomial is 6. The variable terms and their coefficients are

Term	Coefficient
$-9z^6$	-9
$3z$	3

b. Rearranging $-7x + 18x^5 - 20x^4 + 6$ in order of decreasing powers of x gives

$$18x^5 - 20x^4 - 7x + 6$$

The degree of the polynomial is 5. The variable terms and their coefficients are

Term	Coefficient
$18x^5$	18
$-20x^4$	-20
$-7x$	-7

c. $4w^5 - 6w + w^8 - w^3 - 1 = w^8 + 4w^5 - w^3 - 6w - 1$

The degree of the polynomial is 8. The variable terms and their coefficients are

Term	Coefficient
w^8	1
$4w^5$	4
$-w^3$	-1
$-6w$	-6

∎

Polynomials that contain one, two, or three terms are given the special names **monomial, binomial,** or **trinomial,** respectively:

Number of Terms	Examples	Name
One	-2, x^5, $\dfrac{-5y^3}{8}$, $4xy^2$	Monomial
Two	$x^5 - 2x^3$, $4z + 7$	Binomial
Three	$3y^4 - 7y + 9$, $-9z^6 + 3z + 12$	Trinomial

Recall from Chapter 1 that **like terms** are terms that contain the same variables raised to the same powers. Some examples of like terms are

$$8x^2 \text{ and } -5x^2$$

$$-10y^5, 4y^5, \text{ and } 3y^5$$

$$7w^3z \text{ and } 2w^3z$$

We use the distributive property to combine like terms. That is, we combine like terms by combining the coefficients of the terms. Here are some examples:

$$8x^2 - 5x^2 = (8 - 5)x^2 = 3x^2$$

$$-10y^5 + 4y^5 + 3y^5 = [(-10) + 4 + 3]y^5 = -3y^5$$

$$7w^3z + 2w^3z = (7 + 2)w^3z = 9w^3z$$

BE CAREFUL!

A common error occurs when the coefficients and the exponents of *unlike* terms are combined. For example, the terms $3x$ and $5x^2$ contain the same variable but are *not* like terms. Hence $3x + 5x^2 \neq 8x^3$. Remember, *only like terms can be combined.*

EXAMPLE 2

Combine like terms. Identify the result, where appropriate, as a monomial, binomial, or trinomial.

a. $4x^3 + 8x^2 - 9x^3$ **b.** $-3k^4 + 7k - k^8 - 9k + 5k^8$

c. $8p^6 - 13p^6 + 7p^6$ **d.** $5m^2n - 10m^3n^2 + 8mn^2 - 4m^2n + 3$

Solution

a. We write the polynomial in descending powers of x and then combine like terms:

$$4x^3 + 8x^2 - 9x^3 = 4x^3 - 9x^3 + 8x^2$$
$$= (4 - 9)x^3 + 8x^2 \qquad \text{By the distributive property}$$
$$= -5x^3 + 8x^2$$

The polynomial $-5x^3 + 8x^2$ is a binomial because it contains two terms.

b. We write the polynomial in descending powers of k and then combine like terms:

$$-3k^4 + 7k - k^8 - 9k + 5k^8$$
$$= -k^8 + 5k^8 - 3k^4 + 7k - 9k$$
$$= (-1 + 5)k^8 - 3k^4 + (7 - 9)k \qquad \text{By the distributive property}$$
$$= 4k^8 - 3k^4 - 2k$$

The polynomial $4k^8 - 3k^4 - 2k$ is a trinomial because it contains three terms.

c. $8p^6 - 13p^6 + 7p^6 = (8 - 13 + 7)p^6 = 2p^6$

The polynomial $2p^6$ is a monomial because it contains one term.

d. $5m^2n - 10m^3n^2 + 8mn^2 - 4m^2n + 3$
$$= -10m^3n^2 + 5m^2n - 4m^2n + 8mn^2 + 3$$
$$= -10m^3n^2 + (5 - 4)m^2n + 8mn^2 + 3$$
$$= -10m^3n^2 + m^2n + 8mn^2 + 3$$

∎

Adding Polynomials

We add two or more polynomials by combining like terms. Each term of each polynomial represents a real number. Hence, the terms can be combined in any order (by the commutative property) and grouped in any way (by the associative property). We illustrate the procedure in the next example.

EXAMPLE 3

Add the polynomials.

a. $(5x^3 + 3x^2 - 4) + (2x^3 - x^2 + 6)$
b. $(4y^5 - 8y^3 + 9y^2 + 5y) + (y^4 + 2y^3 - 7y)$

Solution

a.   We can use the commutative and associative properties to write

$$(5x^3 + 3x^2 - 4) + (2x^3 - x^2 + 6) = 5x^3 + 2x^3 + 3x^2 - x^2 - 4 + 6$$
$$= 7x^3 + 2x^2 + 2 \qquad \text{Combining like terms}$$

We check by substituting $x = 3$ into each side of the last equation:

$$(5x^3 + 3x^2 - 4) + (2x^3 - x^2 + 6) = 158 + 51 = 209$$

and $$7x^3 + 2x^2 + 2 = 189 + 18 + 2 = 209$$

Both sides of the equation have the same value, which suggests that

$$(5x^3 + 3x^2 - 4) + (2x^3 - x^2 + 6) = 7x^3 + 2x^2 + 2$$

b.   We can use the commutative and associative properties to write

$$(4y^5 - 8y^3 + 9y^2 + 5y) + (y^4 + 2y^3 - 7y)$$
$$= 4y^5 + y^4 - 8y^3 + 2y^3 + 9y^2 + 5y - 7y$$
$$= 4y^5 + y^4 - 6y^3 + 9y^2 - 2y \qquad \text{Combining like terms} \quad ∎$$

Polynomials can be added in a column format by placing like terms in the same column and then adding the columns. For example, the sum $(3z^4 - 7z^2 + 8z + 4) + (2z^3 + 3z^2 - 8)$ is found as follows:

$$
\begin{array}{l}
3z^4 \qquad\quad - 7z^2 + 8z + 4 \\
\underline{\qquad 2z^3 + 3z^2 \qquad\quad - 8} \qquad \text{Like terms in same column} \\
3z^4 + 2z^3 - 4z^2 + 8z - 4 \qquad \text{Columns added}
\end{array}
$$

Note that we left a space in each polynomial for each "missing" power of z. These spaces ensure that we will place like terms in the same column and add them correctly. In summary, we have the following simple rule:

> To add polynomials, combine like terms.

Subtracting Polynomials

The difference of two real numbers, $a - b$, is obtained by adding the opposite of b to a. In symbols, $a - b = a + (-b)$. The difference of two polynomials is obtained similarly by adding the opposite of the second polynomial to the first polynomial. The **opposite of a polynomial** is found by changing the sign of each of its terms. For example, the opposite of $2w^2 - 3w + 2$ is written $-(2w^2 - 3w + 2)$, but simplifying gives us

$$-(2w^2 - 3w + 2) = (-1)(2w^2 - 3w + 2)$$

$$= -2w^2 + 3w - 2 \qquad \text{By the distributive property}$$

which is what we would get by changing the sign of each term. Hence,

> To subtract two polynomials, change the sign of each term of the second polynomial and then add the polynomials.

EXAMPLE 4

Subtract the polynomials.

a. $(6w^2 + w - 3) - (2w^2 - 3w + 2)$
b. $(9z^4 + 7z^3 + z - 1) - (-5z^3 + 2z^2 + 3z)$
c. $(4p^3q^2 - 6p^2q^2 + 9pq^2) - (2p^3q^2 + 5pq^2)$

Solution

a. We change the sign of each term of $2w^2 - 3w + 2$ and then add:

$$(6w^2 + w - 3) - (2w^2 - 3w + 2) = (6w^2 + w - 3) + (-2w^2 + 3w - 2)$$

We can use the commutative and associative properties to write

$$(6w^2 + w - 3) - (2w^2 - 3w + 2) = 6w^2 - 2w^2 + w + 3w - 3 - 2$$

$$= 4w^2 + 4w - 5 \qquad \text{Combining like terms}$$

We check by substituting $w = 4$ into each side of the last equation:

$$(6w^2 + w - 3) - (2w^2 - 3w + 2) = 97 - 22 = 75$$

and $\qquad\qquad 4w^2 + 4w - 5 = 64 + 16 - 5 = 75$

Both sides of the equation have the same value, which suggests that $(6w^2 + w - 3) - (2w^2 - 3w + 2) = 4w^2 + 4w - 5$

b. Two polynomials can be subtracted in a column format by changing the sign of each term of the second polynomial, placing like terms in the same column and then adding the columns. For example, the difference $(9z^4 + 7z^3 + z - 1) - (-5z^3 + 2z^2 + 3z)$ is obtained as follows:

$$
\begin{array}{ll}
9z^4 + 7z^3 + z - 1 & \\
\underline{5z^3 - 2z^2 - 3z} & \text{Each sign changed} \\
9z^4 + 12z^3 - 2z^2 - 2z - 1 & \text{Columns added}
\end{array}
$$

c. We change the sign of each term of $2p^3q^2 + 5pq^2$ and then add:

$$(4p^3q^2 - 6p^2q^2 + 9pq^2) - (2p^3q^2 + 5pq^2)$$
$$= (4p^3q^2 - 6p^2q^2 + 9pq^2) + (-2p^3q^2 - 5pq^2)$$

We can use the commutative and associative properties to write

$$(4p^3q^2 - 6p^2q^2 + 9pq^2) - (2p^3q^2 + 5pq^2)$$
$$= 4p^3q^2 - 2p^3q^2 - 6p^2q^2 + 9pq^2 - 5pq^2$$
$$= 2p^3q^2 - 6p^2q^2 + 4pq^2 \qquad \text{Combining like terms} \qquad ∎$$

DISCUSSION QUESTIONS 4.4

1. What is a polynomial?

2. What is the degree of a polynomial?

3. Is the sum of two monomials always a binomial? Explain.

4. In your own words, state how to add polynomials.

5. In your own words, state how to subtract two polynomials.

6. Why is a space left for a missing power of the variable when adding or subtracting polynomials in one variable in the column format?

EXERCISES 4.4

In Exercises 1–12, combine like terms. Write the resulting polynomial in descending powers of the variable.

1. $2u^5 - u^5 + 4u$

2. $7v^3 + 2v^3 - 6v^2$

3. $6x^3 - 4x^5 - 2x^3$

4. $-5y^4 + 8y^7 + y^4$

5. $7w^2 + 9w^6 - 4w^3$

6. $2z^3 - 4z^8 + 3z^6$

7. $-6p^7 - 5p^3 + p^7 + 10p^4$

8. $8q^3 + 13q^9 - 10q^2 + 6q^3$

9. $4m^6 - 3m^2 + 2m^5 - 5m^4$

10. $5n^9 - n^4 - 6n^8 + 3n$

11. $3k^7 - k^9 + 4k^5 - 8k^7 + 2k^5$

12. $6k^8 + 2k^3 - k^{10} + 9k^3 - 3k$

In Exercises 13–26, combine like terms. Find the degree of the resulting polynomial and identify any monomials, binomials, or trinomials.

13. $6r^3 - 8r + 2r^3$

14. $5t^3 + 3t^4 - 9t^3$

15. $-2h^5 + 7h^5 - 6h^5$

16. $7k^4 + k^4 - 5k^4$

17. $3x^5 - 8x^2 - 6x^5 + 9x^2$

18. $-6y^7 + 2y^9 - y^7 - 5y^9$

19. $8p^4 - 5p^7 + 3p^4 - p^7$

20. $-2q^5 + 9q^2 + 6q^5 - 11q^2$

21. $-5m^{10} + 2m^8 + 7m^3 + 3m^{10} - 6m^8$

22. $4n^6 - 5n^9 - 10n^4 + 3n^9 - 7n^6$

23. $-9r^2s + 7s + 8r^2 + 5r^2s$

24. $2hk^3 - 8k^4 + h^5 - hk^3$

25. $6x^4y^2 - 2x^2y^5 + 10xy^3 - 12x^4y^2 - 5x^4$

26. $-7w^3z^8 + 13w^5z^6 - w^3 + 4w^3z^8 - w^6z^6$

In Exercises 27–36, use the column format to perform the indicated operation.

27. $(5h^3 - 4h) + (2h^3 + 7h)$

28. $(8k^2 + k) + (3k^2 + 4k)$

29. $(4x^2 + 3) - (7x^2 - 5)$

30. $(2y^6 - 7y) - (3y^6 - 10y)$

31. $(7m^3 - 3m^2 + 4m) + (8m^2 - 5m + 6)$

32. $(2n^7 + 8n^5 + 6n^3) + (4n^5 - 9n^4 - 5n^3)$

33. $(3p^5 - 6p^3 + p) - (8p^5 - 2p^4 + 4p)$

34. $(q^4 + 10q^2 - 6) - (-3q^2 + 8q - 9)$

35. $(-2t^6 + 10t^4 - 3t^2 + 9) + (4t^5 - 13t^4 - 10)$

36. $(6s^9 - 3s^6 + 9s^3 - s^2) + (-7s^8 + 5s^6 + 3s^2)$

In Exercises 37–50, perform the indicated operation.

37. $(3m + 4) + (2m - 7)$

38. $(6n - 5) + (4n + 9)$

39. $(5x^4 + 7x^3) - (3x^4 - 4x^3)$

40. $(y^2 - 6y) - (4y^2 - 5y)$

41. $(-2p^3 + 4p - 11) - (6p^3 + 9p^2 + 7)$

42. $(5q^4 + 9p^2 - 3q) - (6q^4 - q^3 + 4p)$

43. $(10r^8 + 4r^7 - 6r^5 + 3r^2) + (5r^8 - 8r^6 + 12r^5 - r^3 + 3r^2)$

44. $(4t^6 - 9t^3 + 6r^2 - 15) + (2r^5 - 5r^3 - 7r^2 + 8r - 1)$

45. $(3m^7 + 7m^6 - 4m^3 - 6m + 9) - (5m^6 + 8m^5 - 6m^3 - 2m)$

46. $(5n^{10} - 2n^8 + 7n^4 - 6n^2) - (8n^9 - 3n^7 + n^6 + 11n^4 - 4n^2)$

47. $(7p^2q^2 + 3pq + 5q) - (-3p^2q^2 - 8pq + 6)$

48. $(-4s^2t + 7st^2 + 5t) + (3s^2t + 2st^2 - 1)$

49. $(6h^3k + 14h^2k - 11hk) + (3h^2k + 7hk - 2hk^2)$

50. $(13x^2y^2 + 4xy + 16) + (9xy + 5x - 14)$

51. The total cost in thousands of dollars to manufacture x sailboats per week is given by the formula

$$\text{Total cost} = -x^2 + 8x + 2$$

What is the total cost to manufacture 4 sailboats per week?

52. A colony of bacteria was treated with a poison, and the number of survivors (in thousands) after t hours was found to be

$$\text{Number of survivors} = t^2 - 10t + 25$$

How many survivors were there after 3 h?

53. A market research firm estimates that the number of units a company can sell after spending x thousands of dollars on advertising is given by the formula

$$\text{Units sold} = -2x^3 + 100x^2 - 850x + 2100$$

How many units can be sold after spending $15,000 on advertising?

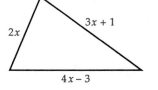

FIGURE 4.8

54. One hour after y milligrams of a particular drug are given to a person, the change in body temperature in degrees Fahrenheit is given by the formula

$$\text{Change in body temperature} = y^2 - 0.11y^3$$

What is the change in body temperature after taking 5 milligrams (mg) of the drug?

55. Find the perimeter of the triangle in Figure 4.8.

56. Find the perimeter of the rectangle in Figure 4.9.

57. **a.** Write the polynomial that represents the perimeter of each figure in Figure 4.10. Are these expressions monomials, binomials, or trinomials? Find the degree of each expression.

    b. Repeat part a for the area of each figure.

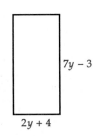

FIGURE 4.9

58. **a.** Write the polynomial that represents the sum of the perimeters of the four figures in Exercise 57. How many terms are in this polynomial? What is its degree?

    b. Repeat part a for the sum of the areas.

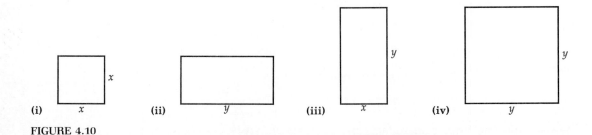

FIGURE 4.10

59. **a.** Write the polynomial that represents the total surface area of each figure in Figure 4.11. Are these expressions monomials, binomials, or trinomials? Find the degree of each expression.

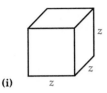 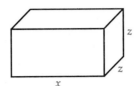 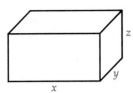

(i) z **(ii)** x **(iii)** x

FIGURE 4.11

b. Repeat part a for the volume of each figure.

60. **a.** Write the polynomial that represents the sum of the surface areas of the three figures in Exercise 59. How many terms are in this polynomial? What is its degree?
   b. Repeat part a for the sum of the volumes.

61. Refer to Exercises 57–60.
   a. What is the degree of a polynomial that represents a perimeter or a sum of perimeters?
   b. What is the degree of a polynomial that represents an area or a sum of areas?
   c. What is the degree of a polynomial that represents a volume or a sum of volumes?

Preparing for Section 4.5

In Exercises 62–69, simplify the given expression.

62. $5(4x + 3)$ **63.** $-2(8y - 6)$ **64.** $-3(5w^2 - w + 9)$ **65.** $4(-7z^3 + z - 2)$

66. $p^2(2p^4)$ **67.** $-q^3(8q^9)$ **68.** $(-5h^2)(10h^6)$ **69.** $(-2k^8)(-7k^5)$

4.5 **Multiplication of Polynomials**

FOCUS

We first use the distributive property to multiply one polynomial by another. Next we develop the FOIL method for multiplying any two binomials, and three formulas for special products of binomials.

The distributive property can be used to multiply a polynomial by a polynomial. It may be applied to any two polynomials, but we shall begin by multiplying a polynomial by a monomial. We can, for example, find the product $5x(x^3 + 3x^2)$ by multiplying each term of $x^3 + 3x^2$ by $5x$:

$$5x(x^3 + 3x^2) = 5x \cdot x^3 + 5x \cdot 3x^2 \qquad \text{By the distributive property}$$
$$= 5x^4 + 15x^3$$

EXAMPLE 1

Use the distributive property to find each product.

    a. $3y^2(2y^5 - 7y^3 + 8y)$ **b.** $-6z^3(z^5 - 5z^2 + 3z - 2)$

Solution

    a. We multiply each term of $2y^5 - 7y^3 + 8y$ by $3y^2$:

$$3y^2(2y^5 - 7y^3 + 8y)$$
$$= 3y^2 \cdot 2y^5 - 3y^2 \cdot 7y^3 + 3y^2 \cdot 8y \qquad \text{By the distributive property}$$
$$= 6y^7 - 21y^5 + 24y^3$$

    b. We multiply each term of $z^5 - 5z^2 + 3z - 2$ by $-6z^3$:

$$-6z^3(z^5 - 5z^2 + 3z - 2)$$
$$= (-6z^3) \cdot z^5 - (-6z^3) \cdot 5z^2 + (-6z^3) \cdot 3z - (-6z^3) \cdot 2$$
$$= -6z^8 + 30z^5 - 18z^4 + 12z^3 \qquad \blacksquare$$

The product of two multiterm polynomials can be found by repeatedly applying the distributive property. As an example, consider the product $(2x + 5)(x + 4)$. We can treat $(2x + 5)$ as a single real number and use the distributive property to obtain

$$(2x + 5)(x + 4) = (2x + 5) \cdot x + (2x + 5) \cdot 4$$

Now the distributive property can be used again to find the products $(2x + 5) \cdot x$ and $(2x + 5) \cdot 4$:

$$(2x + 5)(x + 4) = 2x \cdot x + 5 \cdot x + 2x \cdot 4 + 5 \cdot 4$$
$$= 2x^2 + 5x + 8x + 20$$
$$= 2x^2 + 13x + 20 \qquad \text{Combining like terms}$$

The product $(2x + 5)(x + 4)$ also represents the area of the rectangle in Figure 4.12. Notice that the area of this rectangle is equal to the sum of the areas of the four smaller rectangular regions.

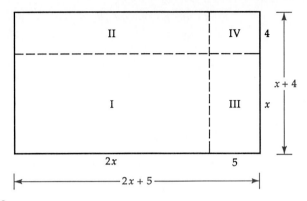

FIGURE 4.12

Area of large rectangle = Area I + Area II + Area III + Area IV

$$(2x + 5)(x + 4) = 2x \cdot x + 2x \cdot 4 + 5 \cdot x + 5 \cdot 4$$

$$= 2x^2 + 8x + 5x + 20$$

$$= 2x^2 + 13x + 20$$

The four products $2x \cdot x$, $2x \cdot 4$, $5 \cdot x$, and $5 \cdot 4$ show that *each* term of $x + 4$ was multiplied by *each* term of $2x + 5$ to obtain the total area. This suggests the following procedure:

To multiply two polynomials, multiply each term of one polynomial by each term of the other polynomial.

EXAMPLE 2

Multiply the polynomials.

a. $(y + 5)(7y - 2)$ **b.** $(2k - 3)(k^2 - 4)$ **c.** $(3n^2 + 2)(8n^3 - 5n + 6)$

Solution

a. To multiply each term of $7y - 2$ by each term of $y + 5$, we first multiply $7y - 2$ by y and then multiply $7y - 2$ by 5:

$$(y + 5)(7y - 2) = y \cdot 7y - y \cdot 2 + 5 \cdot 7y - 5 \cdot 2$$

$$= 7y^2 - 2y + 35y - 10$$

$$= 7y^2 + 33y - 10 \qquad \text{Combining like terms}$$

We check by substituting $y = 1$ into each side of the last equation:

$$(y + 5)(7y - 2) = 6 \cdot 5 = 30 \quad \text{and} \quad 7y^2 + 33y - 10 = 7 + 33 - 10 = 30$$

Both sides of the equation have the same value, which suggests that $(y + 5)(7y - 2) = 7y^2 + 33y - 10$.

b. We multiply each term of $k^2 - 4$ by $2k$ and by -3:

$$(2k - 3)(k^2 - 4) = 2k \cdot k^2 - 2k \cdot 4 - 3 \cdot k^2 - 3 \cdot (-4)$$
$$= 2k^3 - 8k - 3k^2 + 12$$
$$= 2k^3 - 3k^2 - 8k + 12$$

c. We multiply each term of $8n^3 - 5n + 6$ by each term of $3n^2 + 2$:

$$(3n^2 + 2)(8n^3 - 5n + 6)$$
$$= 3n^2 \cdot 8n^3 - 3n^2 \cdot 5n + 3n^2 \cdot 6 + 2 \cdot 8n^3 - 2 \cdot 5n + 2 \cdot 6$$
$$= 24n^5 - 15n^3 + 18n^2 + 16n^3 - 10n + 12$$
$$= 24n^5 + n^3 + 18n^2 - 10n + 12$$
■

Polynomials can be multiplied in a column format by a method that is similar to multiplication of whole numbers. For example, we can find the product $(5m + 3)(7m^2 - 6m + 4)$ in the column format as follows:

$$
\begin{array}{r}
7m^2 - 6m + 4 \\
5m + 3 \\
\hline
21m^2 - 18m + 12 \quad \leftarrow 3(7m^2 - 6m + 4) \\
35m^3 - 30m^2 + 20m \qquad \leftarrow 5m(7m^2 - 6m + 4) \\
\hline
35m^3 - 9m^2 + 2m + 12 \quad \text{Columns added}
\end{array}
$$

We multiplied each term of $7m^2 - 6m + 4$ first by 3 and then by $5m$, carefully placed like terms in the same columns, and then added the columns.

The FOIL Method

A convenient way to find the product of two *binomials* is called the **FOIL method.** FOIL stands for

Product of the + Product of the + Product of the + Product of the
first terms   outer terms   inner terms   last terms

Consider again the product $(2x + 5)(x + 4)$. The first terms are $2x$ and x, the outer terms are $2x$ and 4, the inner terms are 5 and x, and the last terms are 5 and 4:

Therefore, the product of the first terms is $2x \cdot x$, the product of the outer terms is $2x \cdot 4$, the product of the inner terms is $5 \cdot x$, and the product of the last terms is $5 \cdot 4$. By the FOIL method, we would write

$$
\begin{array}{cccc}
\text{F} & \text{O} & \text{I} & \text{L}
\end{array}
$$

$$(2x + 5)(x + 4) = 2x \cdot x + 2x \cdot 4 + 5 \cdot x + 5 \cdot 4$$
$$= 2x^2 + 8x + 5x + 20$$
$$= 2x^2 + 13x + 20$$

Notice again that in the FOIL method, each term of one binomial is multiplied by each term of the other.

EXAMPLE 3

Multiply by the FOIL method.

a. $(r - 8)(r + 5)$ b. $(2p + 7)(p - 1)$
c. $(5m - 2)(3m - 4)$ d. $(4t^2 + 9)(7t^2 + t)$

Solution

a.

$$
\begin{array}{cccc}
\text{F} & \text{O} & \text{I} & \text{L}
\end{array}
$$

$$(r - 8)(r + 5) = r \cdot r + 5 \cdot r - 8 \cdot r - 8 \cdot 5$$
$$= r^2 + 5r - 8r - 40$$
$$= r^2 - 3r - 40$$

b.

$$
\begin{array}{cccc}
\text{F} & \text{O} & \text{I} & \text{L}
\end{array}
$$

$$(2p + 7)(p - 1) = 2p \cdot p + 2p \cdot (-1) + 7 \cdot p + 7 \cdot (-1)$$
$$= 2p^2 - 2p + 7p - 7$$
$$= 2p^2 + 5p - 7$$

c.

$$
\begin{array}{cccc}
\text{F} & \text{O} & \text{I} & \text{L}
\end{array}
$$

$$(5m - 2)(3m - 4) = 15m^2 - 20m - 6m + 8$$
$$= 15m^2 - 26m + 8$$

d.

$$
\begin{array}{cccc}
\text{F} & \text{O} & \text{I} & \text{L}
\end{array}
$$

$$(4t^2 + 9)(7t^2 + t) = 28t^4 + 4t^3 + 63t^2 + 9t$$

■

Special Products

Consider the binomial $a + b$, whose square is $(a + b)^2$. Using the FOIL method, we find that

$$(a + b)^2 = (a + b)(a + b)$$
$$= a^2 + ab + ba + b^2$$
$$= a^2 + 2ab + b^2$$

This last equation says that the square of the sum of two terms is equal to the square of the first term, plus twice the product of the two terms, plus the square of the last term. Interestingly enough, $(a + b)^2 = a^2 + 2ab + b^2$ is also both the area of the large square in Figure 4.13 and the sum of the four smaller areas within the square.

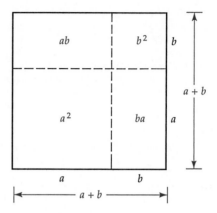

FIGURE 4.13

We can also use the FOIL method to find the square of the binomial $(a - b)$:

$$(a - b)^2 = (a - b)(a - b)$$
$$= a^2 - ab - ba + b^2$$
$$= a^2 - 2ab + b^2$$

This last equation says that the square of a difference of two terms is equal to the square of the first term, minus twice the product of the two terms, plus the square of the last term. To summarize:

Square of a binomial

$$(a + b)^2 = a^2 + 2ab + b^2$$
$$(a - b)^2 = a^2 - 2ab + b^2$$

Notice that the binomial $a + b$ is the *sum* of two terms, and the middle term of its square is $+2ab$, whereas the binomial $a - b$ is the *difference* of two terms and the middle term of its square is $-2ab$.

EXAMPLE 4

Find each square.

a. $(x + 5)^2$ **b.** $(3y + 7)^2$ **c.** $(w - 4)^2$ **d.** $(2k - 6m)^2$

Solution

a. We use $a = x$ and $b = 5$ in the formula for $(a + b)^2$:

$$(x + 5)^2 = x^2 + 2 \cdot x \cdot 5 + 5^2$$
$$= x^2 + 10x + 25$$

b. We use $a = 3y$ and $b = 7$ in the formula for $(a + b)^2$:

$$(3y + 7)^2 = (3y)^2 + 2 \cdot 3y \cdot 7 + 7^2$$
$$= 9y^2 + 42y + 49$$

c. We use $a = w$ and $b = 4$ in the formula for $(a - b)^2$:

$$(w - 4)^2 = w^2 - 2 \cdot w \cdot 4 + 4^2$$
$$= w^2 - 8w + 16$$

d. We use $a = 2k$ and $b = 6m$ in the formula for $(a - b)^2$:

$$(2k - 6m)^2 = (2k)^2 - 2 \cdot (2k) \cdot (6m) + (6m)^2$$
$$= 4k^2 - 24km + 36m^2$$

∎

We can also use the FOIL method to find a formula for the product of the sum and difference of two terms:

$$(a + b)(a - b) = a^2 - ab + ba - b^2$$
$$= a^2 - b^2$$

Notice that there is no middle term in the last equation because the opposite terms $-ab$ and $+ba$ cancel. The result, $a^2 - b^2$, is the difference of two squares.

Product of the sum and difference of two terms

$$(a + b)(a - b) = a^2 - b^2$$

> ## *BE CAREFUL!*
>
> Do not confuse the square of a binomial with the product of the sum and difference of its two terms. It is *incorrect* to write $(a + b)^2 = a^2 + b^2$ and $(a - b)^2 = a^2 - b^2$. Remember to include the middle term when you square a binomial.

EXAMPLE 5   Find each product.

    a. $(m + 6)(m - 6)$ **b.** $(2k - 1)(2k + 1)$

    c. $(11 + 4r)(11 - 4r)$ **d.** $(3s - 5t)(3s + 5t)$

Solution   **a.** We use $a = m$ and $b = 6$ in the formula for $(a + b)(a - b)$:

$$(m + 6)(m - 6) = m^2 - 6^2$$
$$= m^2 - 36$$

We check by substituting $m = 10$ into each side of the last equation:

$$(m + 6)(m - 6) = 16 \cdot 4 \quad \text{and} \quad m^2 - 36 = 100 - 36$$
$$= 64 \qquad\qquad\qquad = 64$$

Both sides of the equation have the same value, which suggests that $(m + 6)(m - 6) = m^2 - 36$.

b. We use $a = 2k$ and $b = 1$ in the formula for $(a + b)(a - b)$:

$$(2k - 1)(2k + 1) = (2k)^2 - 1^2$$
$$= 4k^2 - 1$$

c. We use $a = 11$ and $b = 4r$ in the formula for $(a + b)(a - b)$:

$$(11 + 4r)(11 - 4r) = 11^2 - (4r)^2$$
$$= 121 - 16r^2$$

d. We use $a = 3s$ and $b = 5t$ in the formula for $(a + b)(a - b)$:

$$(3s - 5t)(3s + 5t) = (3s)^2 - (5t)^2$$
$$= 9s^2 - 25t^2$$

DISCUSSION QUESTIONS 4.5

1. Write the area of the rectangle in Figure 4.14 as the product of two binomials. The areas of the four regions inside the rectangle are given.

2. In your own words, state how to multiply two polynomials.

3. What is the FOIL method?

4. Can the FOIL method be used to find the product $(y - 3)(y^2 + 2y + 1)$? Explain.

5. In your own words, state how to find the square of a binomial.

6. Write $n^2 - 16$ as the product of the sum and difference of two terms.

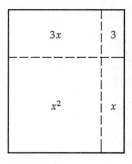

FIGURE 4.14

EXERCISES 4.5

In Exercises 1–10, use the distributive property to find the product.

1. $-5w(3w^2 - 4)$

2. $-2z(7z^3 + 8)$

3. $8p(p^2 - 2p + 3)$

4. $7q(3q^2 + q - 6)$

5. $6t^2(t^6 - 4t^3 - 2t)$

6. $5r^2(9r^4 - 7r + 4)$

7. $2x^2y(4x^2 + 7xy - y^2)$

8. $8pq^2(5p^3 - 10pq - 3p^4)$

9. $-9w^3z^2(6w^2z - 3w^2z^2 + 5wz^3)$

10. $-4r^2s^2(r^4s - 3r^2s^2 - 8rs^4)$

In Exercises 11–18, use the column format to find the product.

11. $(h + 3)(h^2 + 4h + 6)$

12. $(k + 4)(k^2 + 3k + 7)$

13. $(2x - 5)(x^2 - 6x + 3)$

14. $(3y - 7)(y^2 + 9y - 2)$

15. $(4w + 1)(3w^2 - 7w + 2)$

16. $(6z + 3)(9z^2 - 2z - 5)$

17. $(3p^2 - 4)(8p^3 - p^2 + 5p - 7)$

18. $(4q^2 - 3)(q^3 - 5q^2 - 9q + 2)$

In Exercises 19–38, use the FOIL method to find the product.

19. $(n + 1)(n + 2)$

20. $(m + 5)(m + 8)$

21. $(4 + h)(5 - h)$

22. $(k - 7)(k + 6)$

23. $(x - 8)(x - 6)$

24. $(y - 3)(y - 4)$

25. $(2w + 1)(w + 4)$

26. $(3z + 5)(z + 1)$

27. $(4p - 3)(p + 6)$

28. $(5q - 4)(q + 9)$

29. $(3m + 1)(2m - 5)$

30. $(7 - 4n)(8 + 3n)$

31. $(7r - 4)(9r - 2)$

32. $(2t + 6)(5t + 4)$

33. $(3h^2 + 7)(2h^2 + 5)$

34. $(2k^3 + 9)(5k^3 - 1)$

35. $(4x^2 - x)(2x + 11)$

36. $(5y - 2)(y^3 - 6y)$

37. $(8w - 3z)(2w + 9z)$

38. $(6p - 7q)(4p + 3q)$

In Exercises 39–76, find the product.

39. $(x + 8)^2$

40. $(y + 3)^2$

41. $(2 + p)(2 - p)$

42. $(q - 7)(q + 7)$

43. $(4m - 5)^2$ **44.** $(8n - 4)^2$ **45.** $(2h + 9)(2h - 9)$ **46.** $(6 + 5k)(6 - 5k)$

47. $(6x + 7y)^2$ **48.** $(4w + 3z)^2$ **49.** $(8p - 3q)^2$ **50.** $(10m - 7n)^2$

51. $3x(6x + 1)(x + 7)$ **52.** $6y(y - 4)(3y + 8)$ **53.** $-7w(4 - w)(9 + 2w)$

54. $-5z(2z - 6)(z - 1)$ **55.** $5p(p + 10)^2$ **56.** $3q(q + 2)^2$

57. $-4m(3m - 8)^2$ **58.** $-9n(5n - 7)^2$ **59.** $6r(5r + 4)(5r - 4)$

60. $7s(3 - 4s)(3 + 4s)$ **61.** $(4h + 8)(h + 3)(2h + 1)$ **62.** $(k - 4)(3k + 9)(2k + 5)$

63. $(x - 6)(7x + 2)(6x - 3)$ **64.** $(5y + 1)(y - 6)(3y - 7)$ **65.** $(w - 1)(3w + 2)^2$

66. $(z - 3)(2z + 5)^2$ **67.** $(r + 1)^3$ **68.** $(t + 2)^3$

69. $(h - 4)^3$ **70.** $(3 - k)^3$ **71.** $(3x + 7)(7x^2 - 2x + 1)$

72. $(4y - 6)(y^3 + 9y^2 - 2y - 5)$ **73.** $\left(3p - \dfrac{1}{2}q\right)\left(p + \dfrac{1}{3}q\right)$ **74.** $\left(\dfrac{3}{4}m + \dfrac{1}{2}n\right)^2$

75. $\left(\dfrac{2}{5}r - \dfrac{3}{8}t\right)^2$ **76.** $\left(4u + \dfrac{5}{8}v\right)\left(4u - \dfrac{5}{8}v\right)$

77. Find the area of the rectangle in Figure 4.15.

78. Find the area of the square in Figure 4.16.

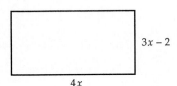

FIGURE 4.15

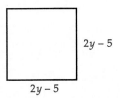

FIGURE 4.16

79. For the rectangular solid in Figure 4.17:
   a. Find the total surface area.
   b. Find the volume.

80. For the cube in Figure 4.18:
   a. Find the total surface area.
   b. Find the volume.

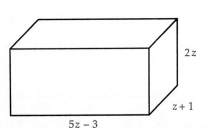

FIGURE 4.17

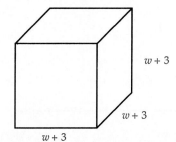

FIGURE 4.18

81. Find the area of the shaded region in Figure 4.19.

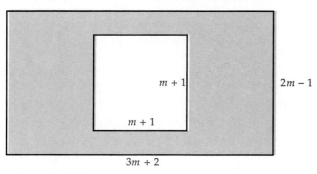

FIGURE 4.19

82. Find the area of the shaded region in Figure 4.20.

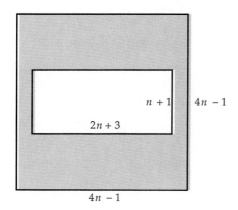

FIGURE 4.20

Preparing for Section 4.6

In Exercises 83–90, simplify the given expression.

83. $\dfrac{3x^2}{x}$

84. $\dfrac{7y}{y^3}$

85. $\dfrac{12w^5}{4w^9}$

86. $\dfrac{20z^7}{5z^3}$

87. $\dfrac{-10p^6}{6p^3}$

88. $\dfrac{-24q^8}{16q^5}$

89. $\dfrac{18m^4}{-30m^7}$

90. $\dfrac{20n^2}{-24n^9}$

4.6   Division of Polynomials

FOCUS

We first divide polynomials by a monomial. Then we develop a method to divide polynomials with two or more terms by a procedure similar to long division.

In Section 4.2 we used the quotient property of exponents to divide certain monomials. For example, we can use it to find

$$\frac{20x^3}{5x} = \frac{20}{5} \cdot \frac{x^3}{x} = 4x^2 \quad \text{and} \quad \frac{15x^2}{5x} = \frac{15}{5} \cdot \frac{x^2}{x} = 3x$$

Now consider the problem $20x^3 + 15x^2$ divided by $5x$, or $\dfrac{20x^3 + 15x^2}{5x}$. Let's see what happens if we divide each term of $20x^3 + 15x^2$ by $5x$:

$$\frac{20x^3 + 15x^2}{5x} = \frac{20x^3}{5x} + \frac{15x^2}{5x}$$

$$= 4x^2 + 3x$$

Is this result correct? Recall that multiplication is used to check division. We can check our result by multiplying $4x^2 + 3x$ by $5x$:

$$5x(4x^2 + 3x) = 20x^3 + 15x^2$$

which checks. This example illustrates the procedure for dividing a polynomial by a monomial.

To divide a polynomial by a monomial, divide each term of the polynomial by the monomial.

Notice that $\dfrac{20x^3 + 15x^2}{5x}$ is not defined when $x = 0$, because then the denominator would be zero, and division by zero is not defined. We will assume throughout the remainder of this chapter that denominators are not zero.

EXAMPLE 1

Divide and check the result by multiplication.

a. $\dfrac{12y^5 + 9y^3}{3y^2}$ b. $\dfrac{8k^{11} - 24k^9 + 12k^3}{4k^3}$ c. $\dfrac{36p^3 - 6p^2 + 18p - 12}{-6p}$

Solution

a.   We divide each term of $12y^5 + 9y^3$ by $3y^2$:

$$\frac{12y^5 + 9y^3}{3y^2} = \frac{12y^5}{3y^2} + \frac{9y^3}{3y^2}$$
$$= 4y^3 + 3y \qquad \text{By the quotient property of exponents}$$

Check: $3y^2(4y^3 + 3y) = 12y^5 + 9y^3$

b.   We divide each term of $8k^{11} - 24k^9 + 12k^3$ by $4k^3$:

$$\frac{8k^{11} - 24k^9 + 12k^3}{4k^3} = \frac{8k^{11}}{4k^3} - \frac{24k^9}{4k^3} + \frac{12k^3}{4k^3}$$
$$= 2k^8 - 6k^6 + 3$$

Check: $4k^3(2k^8 - 6k^6 + 3) = 8k^{11} - 24k^9 + 12k^3$

c.   We divide each term of $36p^3 - 6p^2 + 18p - 12$ by $-6p$:

$$\frac{36p^3 - 6p^2 + 18p - 12}{-6p} = \frac{36p^3}{-6p} - \frac{6p^2}{-6p} + \frac{18p}{-6p} - \frac{12}{-6p}$$
$$= -6p^2 + p - 3 + \frac{2}{p}$$

Check: $(-6p)\left(-6p^2 + p - 3 + \dfrac{2}{p}\right) = 36p^3 - 6p^2 + 18p - 12$ ∎

BE CAREFUL!

A common error when dividing a polynomial by a monomial is illustrated below:

$$\frac{x^3 + 2x}{x} = \frac{x^3 + 2\cancel{x}}{\cancel{x}} = x^3 + 2$$

The result is *incorrect*. Cancelling x from the numerator and the denominator is equivalent to dividing only the term $2x$ by x. Remember to divide each term of the polynomial by the monomial.

The procedure for dividing a polynomial by a polynomial with two or more terms is similar to long division with whole numbers. We begin by reviewing division of whole numbers. Pay careful attention to each step in the process, because similar steps are needed to divide polynomials.

Consider the problem 739 divided by 21, which is written as $21\overline{)739}$ in the long-division format. Recall that 739 is the *dividend* and 21 is the *divisor*:

Step 1. *Divide:*

$$\begin{array}{r} 3 \\ 21\overline{)739} \end{array} \leftarrow 73 \div 21 = 3 \text{ (plus a remainder)}$$

Step 2. *Multiply:*

$$\begin{array}{r} 3 \\ 21\overline{)739} \\ 63 \end{array} \leftarrow 3 \cdot 21 = 63$$

Step 3. *Subtract:*

$$\begin{array}{r} 3 \\ 21\overline{)739} \\ 63 \\ \overline{10} \end{array} \leftarrow 73 - 63 = 10$$

Step 4. *Bring down the next number in the dividend:*

$$\begin{array}{r} 3 \\ 21\overline{)739} \\ 63\downarrow \\ \overline{109} \end{array} \leftarrow \text{New dividend}$$

Step 5. *Repeat the first four steps with the new dividend:*
These four steps are now repeated for 109 divided by 21:

$$\begin{array}{r} \text{Divide:} \\ 35 \leftarrow 109 \div 21 = 5 \text{ (plus a remainder)} \\ 21\overline{)739} \\ 63 \\ \overline{109} \quad \text{Multiply:} \\ 105 \leftarrow 5 \cdot 21 = 105 \\ \overline{} \quad \text{Subtract:} \\ \leftarrow 109 - 105 = 4 \end{array}$$

There is no number remaining in the original dividend, so we are finished. The *quotient* is 35 and the *remainder* is 4. The answer is

$35 + \dfrac{4}{21}$ or $35\dfrac{4}{21}$. We check this result by multiplying the quotient by the divisor and then adding the remainder to obtain the dividend:

$$21 \cdot 35 + 4 = 735 + 4 = 739$$

Now consider the problem $8x^2 + 22x + 20$ divided by $2x + 3$, or $\dfrac{8x^2 + 22x + 20}{2x + 3}$. The dividend is $8x^2 + 22x + 20$, and the divisor is $2x + 3$. To begin the division, we write this problem in the long-division format. Then we follow the five-step procedure described above.

Step 1. *Divide* the first term of the dividend by the first term of the divisor:

$$
\begin{array}{r}
4x \qquad\qquad \leftarrow \dfrac{8x^2}{2x} = 4x \\[2pt]
2x + 3\overline{)8x^2 + 22x + 20}
\end{array}
$$

Step 2. *Multiply:*

$$
\begin{array}{r}
4x \qquad\qquad\qquad\quad \\
2x + 3\overline{)8x^2 + 22x + 20} \\
8x^2 + 12x \qquad \leftarrow 4x(2x + 3) = 8x^2 + 12x
\end{array}
$$

Step 3. *Subtract:*

$$
\begin{array}{r}
4x \qquad\qquad\qquad\qquad\qquad \\
2x + 3\overline{)8x^2 + 22x + 20} \\
8x^2 + 12x \qquad\qquad\qquad\qquad \\
\overline{10x} \qquad \leftarrow (8x^2 + 22x) - (8x^2 + 12x) = 10x
\end{array}
$$

(Recall that to subtract here, we change the sign of each term of $8x^2 + 12x$ and then add.)

Step 4. *Bring down the next term in the dividend:*

$$
\begin{array}{r}
4x \qquad\qquad\qquad \\
2x + 3\overline{)8x^2 + 22x + 20} \\
8x^2 + 12x \quad\downarrow \qquad\quad \\
\overline{10x + 20} \leftarrow \text{New dividend}
\end{array}
$$

Step 5. *Repeat the first four steps with the new dividend:*
We follow the same four steps for $10x + 20$ divided by $2x + 3$:

Divide:

$$\frac{10x}{2x} = 5$$

$$
\begin{array}{r}
4x + 5 \quad \leftarrow \\
\hline
2x + 3)\overline{8x^2 + 22x + 20} \\
8x^2 + 12x \\
\hline
10x + 20 \\
10x + 15 \\
\hline
5
\end{array}
$$

Multiply:

$5(2x + 3) = 10x + 15$

Subtract:

$(10x + 20) - (10x + 15) = 5$

There is no term remaining in the original dividend, so we are finished. The quotient is $4x + 5$ and the remainder is 5, so the answer may be written as $4x + 5 + \dfrac{5}{2x + 3}$. We can check this result by multiplying the quotient by the divisor and then adding the remainder to obtain the dividend:

$$(2x + 3)(4x + 5) + 5 = 8x^2 + 22x + 15 + 5 = 8x^2 + 22x + 20$$

You should note several things about using the five-step procedure for dividing one polynomial by another polynomial with two or more terms:

1. The polynomials must be written in descending powers of the variable.
2. The degree of the dividend must be *greater* than the degree of the divisor. Thus, $8x^2 + 22x + 20$ has degree 2 and $2x + 3$ has degree 1.
3. The quotient is written above the dividend so that like terms are aligned.
4. The procedure ends when the result of a subtraction is a polynomial with degree *less* than the degree of the divisor. In the example above, the polynomial 5 has degree 0, which is less than the degree of $2x + 3$.

EXAMPLE 2

Divide $2y^2 - 11y + 12$ by $y - 4$. Check the answer.

Solution

$$
\begin{array}{r}
2y - 3 \\
\hline
y - 4)\overline{2y^2 - 11y + 12} \\
2y^2 - 8y \\
\hline
-3y + 12 \\
-3y + 12 \\
\hline
0
\end{array}
$$

$\leftarrow 2y(y - 4) = 2y^2 - 8y$

$\leftarrow (2y^2 - 11y) - (2y^2 - 8y) = -3y$

$\leftarrow -3(y - 4) = -3y + 12$

$\leftarrow (-3y + 12) - (-3y + 12) = 0$

Check: $(y - 4)(2y - 3) + 0 = 2y^2 - 11y + 12$

∎

EXAMPLE 3

Divide $4z - 2z^2 + z^3 - 2$ by $z^2 + 1$. Check the answer.

Solution

We must write the dividend as $z^3 - 2z^2 + 4z - 2$ (in descending powers of z) before starting the five-step procedure. Also notice that the divisor $z^2 + 1$ does not have a z term. It is a good idea to *insert missing powers of either the dividend or divisor with 0 coefficients* to ensure that like terms are placed in the same column. Here we write the divisor as $z^2 + 0z + 1$ and then divide:

$$
\begin{array}{r}
z - 2 \\
z^2 + 0z + 1 \overline{)z^3 - 2z^2 + 4z - 2} \\
\underline{z^3 + 0z^2 + z} \qquad \leftarrow z(z^3 + 0z^2 + 1) \\
-2z^2 + 3z - 2 \\
\underline{-2z^2 - 0z - 2} \leftarrow (-2)(z^3 + 0z^2 + 1) \\
3z
\end{array}
$$

The degree of $3z$ is less than the degree of $z^2 + 1$, so we stop here. The quotient is $z - 2$ and the remainder is $3z$, so the answer is $z - 2 + \dfrac{3z}{z^2 + 1}$.

Check: $(z^2 + 1)(z - 2) + 3z = z^3 - 2z^2 + z - 2 + 3z$

$$= z^3 - 2z^2 + 4z - 2 \qquad \blacksquare$$

EXAMPLE 4

Divide $k^3 - 2$ by $k - 1$. Check the answer.

Solution

We write $k^3 - 2$ as $k^3 + 0k^2 + 0k - 2$ and then divide by $k - 1$.

$$
\begin{array}{r}
k^2 + k + 1 \\
k - 1 \overline{)k^3 + 0k^2 + 0k - 2} \\
\underline{k^3 - k^2} \\
k^2 + 0k \\
\underline{k^2 - k} \\
k - 2 \\
\underline{k - 1} \\
-1
\end{array}
$$

The quotient is $k^2 + k + 1$ and the remainder is -1, so the answer is

$$k^2 + k + 1 + \frac{-1}{k - 1} \quad \text{or} \quad k^2 + k + 1 - \frac{1}{k - 1}.$$

The check is left to you. $\qquad \blacksquare$

DISCUSSION QUESTIONS 4.6

1. In your own words, state how to divide a polynomial by a monomial.

2. How are the quotient and the remainder used to check the answer of a division problem?

3. For the problem $\dfrac{2w^3 + 6w - 5w^2 - 13}{w^2 + 3}$, what must be done to the dividend before division can begin? What must be done to the denominator?

4. Can the five-step procedure be used for the division problem $\dfrac{z^3 + 3z^2 - 7z + 4}{z^4 + 1}$? Explain.

5. When does the five-step procedure stop?

EXERCISES 4.6

In Exercises 1–4, divide the given polynomial by $4x$. Check the answer.

1. $8x^2 + 24x$ 2. $4x^3 - 12x^2 + 8x$ 3. $-20x^5 + 16x^3 - 28x^2$ 4. $36x^2 + 48x - 12$

In Exercises 5–8, divide the given polynomial by $-3y$. Check the answer.

5. $6y^3 - 15y$ 6. $9y^4 - 3y^2 + 21y$ 7. $18y^6 - 30y^3 - 42$ 8. $-21y^8 + 33y^4 - 9y^2$

In Exercises 9–12, divide the given polynomial by $2z^3$. Check the answer.

9. $10z^8 - 14z^4 + 6z^3$

10. $-12z^9 + 20z^7 - 4z^4$

11. $32z^7 - 40z^5 + 22z^3 - 18z^2$

12. $-38z^{10} - 52z^6 + 46z^3 + 64z$

In Exercises 13–16, divide the given polynomial by $-5kw^2$. Check the answer.

13. $10k^6w^3 - 25k^4w^2$

14. $30k^8w^9 - 20k^4w^5 + 45k^2w^3$

15. $-35k^5w^7 + 5k^2w^6 - 15kw^4$

16. $60k^7w^{10} + 40k^2w^6 - 55kw^2 + 100w$

In Exercises 17–50, perform the indicated division. Check the answer.

17. $\dfrac{h^2 + 7h + 10}{h + 2}$ 18. $\dfrac{m^2 + 7m + 12}{m + 4}$ 19. $\dfrac{q^2 + 5q + 8}{q + 3}$ 20. $\dfrac{r^2 + 9r + 4}{r + 2}$

21. $\dfrac{14 - 6h + h^2}{h - 7}$ 22. $\dfrac{30 - 13k + k^2}{k - 10}$ 23. $\dfrac{s^2 - 5s - 6}{s - 1}$ 24. $\dfrac{t^2 - 3t + 4}{t - 3}$

25. $\dfrac{6x^2 + 13x + 5}{2x + 1}$ 26. $\dfrac{10y^2 + 37y + 7}{5y + 1}$ 27. $\dfrac{9p^2 - 3p + 7}{3p + 2}$ 28. $\dfrac{8q^2 + 2q - 5}{2q + 3}$

29. $\dfrac{2u^2 + 13u + 21}{2u + 7}$

30. $\dfrac{10v^2 - 11v - 6}{5v + 2}$

31. $\dfrac{20k^2 - 3 + 7k}{4k - 1}$

32. $\dfrac{1 - 8h + 15h^2}{3h - 1}$

33. $\dfrac{12m^2 + 7m - 1}{4m - 3}$

34. $\dfrac{6n^2 - 5n + 2}{3n - 4}$

35. $\dfrac{w^3 + 4w^2 - 8}{w + 2}$

36. $\dfrac{3r^3 - 2r + 21}{r + 2}$

37. $\dfrac{z^3 - 8z - 3}{z - 3}$

38. $\dfrac{2t^3 + 7t^2 - 6}{2t + 3}$

39. $\dfrac{4p^3 + 3p^2 - 4p - 5}{p^2 - 1}$

40. $\dfrac{2q^3 - 5q^2 + 6q - 14}{q^2 + 3}$

41. $\dfrac{8m^3 - 1}{2m - 1}$

42. $\dfrac{27n^3 - 64}{3n - 4}$

43. $\dfrac{9h^4 + 12h^2 + 6}{3h^2 + 2}$

44. $\dfrac{36k^4 - 60k^2 + 29}{6k^2 - 5}$

45. $\dfrac{15x^4 + 3x^3 + 4x^2 + 4}{3x^2 - 1}$

46. $\dfrac{4y^4 + 6y^3 + 3y - 1}{2y^2 + 1}$

47. $\dfrac{r^4 - 5r^2 + 6}{r^2 - r - 2}$

48. $\dfrac{2t^3 + t^2 - 2t + 1}{t^2 + 2t + 2}$

49. $\dfrac{4 + 10u - 7u^2 + u^4}{u^2 + 2u - 4}$

50. $\dfrac{2v^4 + 3v^3 + 4v^2 + 9v - 5}{3 + v^2}$

51. The area of the rectangle in Figure 4.21 is $x^2 + 9x + 20$. Find the length of the rectangle.

52. The area of the parallelogram in Figure 4.22 is $8y^2 + 26y + 15$. Find the height of the parallelogram.

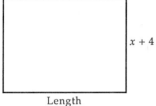

Length

FIGURE 4.21

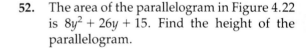

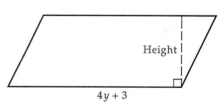

Height

$4y + 3$

FIGURE 4.22

53. The area of the triangle in Figure 4.23 is $6w^2 + w - 2$. Find its height.

54. The volume of the rectangular solid in Figure 4.24 is $z^3 + 6z^2 + 11z + 6$. Find its height.

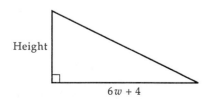

Height

$6w + 4$

FIGURE 4.23

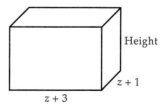

Height

$z + 1$

$z + 3$

FIGURE 4.24

Preparing for Section 5.1

In Exercises 55–58, find all positive integers that divide the given number evenly (with zero remainder).

55. 18   **56.** 24   **57.** 42   **58.** 105

In Exercises 59–62, find the product.

59. $6p(2p + 9)$ **60.** $4q(7q^2 - 5q + 1)$ **61.** $-3r^2(6r^3 - 4r - 10)$ **62.** $-5t^3(2t^4 - t^2 + 8t)$

Chapter Summary

Important Ideas

Exponent properties

Product property: $a^m \cdot a^n = a^{m+n}$

Power-to-a-power property: $(a^m)^n = a^{m \cdot n}$

Product-to-a-power property: $(ab)^n = a^n b^n$

Quotient-to-a-power property: $\left(\dfrac{a}{b}\right)^n = \dfrac{a^n}{b^n}$ $[b \neq 0]$

Quotient property: $\dfrac{a^m}{a^n} = a^{m-n}$ $[a \neq 0]$

Scientific notation
A real number is in *scientific notation* when it is written in the form $a \cdot 10^n$, where $1 \leq a < 10$ and n is an integer.

FOIL method
To multiply two binomials by the FOIL method, find the sum $F + O + I + L$. F is the product of the first terms, O is the product of the outer terms, I is the product of the inner terms, and L is the product of the last terms.

Square of a binomial

$$(a + b)^2 = a^2 + 2ab + b^2$$
$$(a - b)^2 = a^2 - 2ab + b^2$$

Product of the sum and difference of two terms

$$(a + b)(a - b) = a^2 - b^2$$

Chapter Review

Section 4.1

In Exercises 1–4, write the given expression using exponents and then evaluate it when $x = 2$.

1. $x \cdot x \cdot x \cdot x$

2. $-x \cdot x \cdot x \cdot x \cdot x$

3. $(-x) \cdot (-x) \cdot (-x)$

4. $-(-x) \cdot (-x)$

In Exercises 5–14, simplify the given expression.

5. $y^5 y^2$

6. $(w^4)^3$

7. $(9z)^2$

8. $p^5(2p)^3$

9. $(4t^7)^3$

10. $-4q^6(q^2)^7$

11. $(-3r)^4(r^3)^2$

12. $7u(6u^5v^4)^2$

13. $9mn^2(-5m^6n)^3$

14. $(-3h^4k^8)^3(4h^9k^7)^2$

Section 4.2

In Exercises 15–20, evaluate the given expression.

15. -5^0

16. $(-7)^0$

17. 8^{-1}

18. $(-2)^{-3}$

19. $\dfrac{5}{10^{-4}}$

20. $\left(\dfrac{2}{3}\right)^{-2}$

In Exercises 21–28, simplify the given expression. Write the result with positive exponents.

21. $\dfrac{h^7}{h^2}$

22. $\dfrac{k^2}{k^{-3}}$

23. $\dfrac{9t^{-8}}{-12t^{-4}}$

24. $\dfrac{(-2w)^5}{16w^0}$

25. $\dfrac{15z^{-7}z^2}{10z^{-3}}$

26. $\left(\dfrac{10q^{-4}}{-8q^3}\right)^{-3}$

27. $\left(\dfrac{u^{-4}v^6}{9u^3v^{-5}}\right)^{-2}$

28. $\left[\dfrac{(-2r^4)^2s^{-6}}{-6r^{-8}s^{-1}}\right]^3$

Section 4.3

In Exercises 29–32, write the given number in scientific notation.

29. 0.0000038 **30.** $905,000,000$ **31.** $40,570,000,000,000$ **32.** 0.000000000000000001

In Exercises 33–36, perform the indicated computation. Write the result in scientific notation.

33. $(7 \cdot 10^4)(2.4 \cdot 10^3)$

34. $\dfrac{3.5 \cdot 10^{-6}}{5 \cdot 10^6}$

35. $\dfrac{(3 \cdot 10^{-7})(6.02 \cdot 10^{12})}{4.8 \cdot 10^{-18}}$

36. $\dfrac{(2.9 \cdot 10^8)(8 \cdot 10^{-14})}{(6 \cdot 10^{-10})(7.1 \cdot 10^{17})}$

In Exercises 37–38, write the given number in standard notation.

37. $8.3 \cdot 10^4$ **38.** $2.08 \cdot 10^{-9}$

In Exercises 39–40, express the result in scientific notation.

39.   The nearest star to the earth (other than the sun) is approximately 25,000,000,000,000 mi away. How long would it take light to travel from this star to the earth? Assume the speed of light is 186,000 mi/s.

40.   A certain computer can do 4,500,000 computations per second. How many computations can the computer do in a year? Assume 365 days in a year.

Section 4.4

In Exercises 41–44:

a. Combine like terms and write the resulting polynomial in descending powers of the variable.
b. Determine the degree of the polynomial and identify any monomials, binomials, or trinomials.

41. $3x^5 + 6x^4 - 4x^5$ **42.** $4y^2 - 9y^3 + 5y^2 - 3y + 7y^3$

43. $-8w^2 + 11w^2 - w^2$ **44.** $5p + 6p^2 - 3p^4 - 12p + 7$

In Exercises 45–50, perform the indicated operation.

45. $(4z^2 - 3z + 5) + (z^2 - 2z - 6)$ **46.** $(-2q^4 + q^3 + 3q) - (5q^4 - 4q^3 + 7q)$

47. $(4m^7 + 6m^5 - 3m^4 - 9m^2 + m) - (m^6 - 8m^5 + 10m^4 - 2m + 1)$

48. $(3n^8 + n^5 - 7n^3 - 9n^2 + n) + (4n^7 - 2n^6 + 10n^3 + 6n^2)$

49. $(8u^3v - 2uv^2 + 6v^4) + (4u^3v + uv^2 - 9v^4)$

50. $(s^4r^3 - 5s^2r^2 + 8r) - (3s^4r^3 + 6rs - 10r - 9s) + (3s^2r^2 - 13s)$

Section 4.5

In Exercises 51–61, find the product.

51. $7h^3(9h^5 - 3h^2 + 5)$ **52.** $-4k^5(2k^3 - 8k^2 + k - 10)$

53. $(3r + 5)(r + 8)$ **54.** $(6s + 1)(4s - 5)$

55. $(x - 2y)(3x + y)$ **56.** $(w + 12)^2$

57. $(2z - 7)^2$ **58.** $(7s + 10r)(7s - 10r)$

59. $(8 - 4w)(3 + 2w)(5 - w)$ **60.** $(q + 8)(5q^2 - q + 4)$

61. $(9u^2 - 6u - 3)(2u^5 + 7u^3 - 8u)$

62. Find the volume of the cube in Figure 4.25.

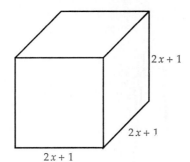

FIGURE 4.25

Section 4.6

In Exercises 63–70, perform the indicated division. Check the answer.

63. $\dfrac{16n^5 - 12n^3 + 20n^2}{2n}$

64. $\dfrac{24m^6 - 15m^5 + 30m^3 - 18m^2}{-3m^2}$

65. $\dfrac{2h^2 + 7h - 15}{h + 5}$

66. $\dfrac{6k^2 - 11k - 10}{2 + 3k}$

67. $\dfrac{2 - 7x + 4x^2}{4x - 3}$

68. $\dfrac{8y^3 - 4y + 4}{2y - 3}$

69. $\dfrac{3q^4 + 2q^3 - 11q^2 - 2q + 5}{q^2 - 2}$

70. $\dfrac{9z^3 - 6z + 1}{3z^2 - 3z + 1}$

71. The area of the rectangle in Figure 4.26 is $15z^2 + z - 2$. Find the width of the rectangle.

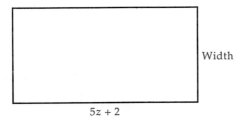

FIGURE 4.26

Chapter 4 Test

In Problems 1–4, evaluate the given expression.

1. -7^2

2. $\left(\dfrac{5}{8}\right)^3$

3. 9^{-3}

4. $-\dfrac{4}{(-2)^{-5}}$

In Problems 5–8, simplify the given expression. Write the result with positive exponents.

5. $-2x^7 \cdot 3x^{-5}$ **6.** $12y^3(4y)^{-2}$ **7.** $\left(\dfrac{p^{-2}p^5}{2p^{-1}}\right)^3$ **8.** $-8q^{-10}\left(\dfrac{q^{-5}}{4q^0}\right)^{-3}$

In Problems 9–10, write the given number in scientific notation.

9. 0.000284   **10.** 9,201,000,000,000

11. Compute $\dfrac{(9 \cdot 10^6)(5 \cdot 10^{-4})}{1.5 \cdot 10^{-7}}$ and express the result in scientific notation.

In Problems 12–20, perform the indicated operation.

12. $(4m^2 - 2m - 5) + (-7m^2 + 8m - 1)$ **13.** $5n^2(n^3 + n - 1) - 4n^3(n^2 + 2n + 1)$

14. $(3k - 1)(6k^3 - 4k^2 - k)$ **15.** $(2 - 7t)(5 + 3t)$ **16.** $(9 - 2r)^2$ **17.** $(s^2 - 3)(s^2 + 3)$

18. $\dfrac{24x^5 - 8x^3 + 16x}{8x}$ **19.** $\dfrac{6y^2 + 7y - 20}{3y - 4}$ **20.** $\dfrac{4 - 8w + 4w^2}{2w + 1}$

21. The distance around the equator of the earth is approximately 25,000 mi. How many times must you travel around the equator to go a distance of one light year? Write the result in scientific notation.

22. Find the area of the shaded region in Figure 4.27.

23. For the rectangular solid in Figure 4.28:
    a. Find the total surface area.
    b. Find the volume.

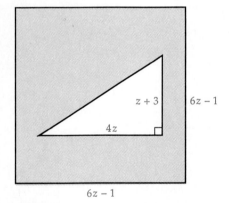

FIGURE 4.27

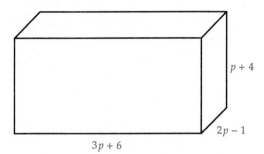

FIGURE 4.28

5 *Factoring Polynomials*

5.1 Greatest Common Factor and Grouping

FOCUS

We discuss some concepts involved in factoring. Then we use two methods to factor polynomials—finding the greatest common factor and grouping terms.

In this chapter we explore the process of factoring. **Factoring** an expression means writing it as a product of expressions. The expressions that form the product are called **factors.** The process of factoring is important in simplifying algebraic expressions and solving certain types of equations.

You can think of factoring as reversing the process of multiplication. For example, we can factor 18 by writing it as the product of 3 and 6:

$$\text{Product} \rightarrow \underset{\text{Multiplying}}{\overset{\text{Factoring}}{18 = 3 \cdot 6}} \leftarrow \text{Factors}$$

The integers 3 and 6 are *factors* of 18. We can find all integer factors of 18 by listing all the ways that 18 can be written as the product of two integers:

$$18 = 1 \cdot 18 \qquad 18 = (-1)(-18)$$
$$18 = 2 \cdot 9 \qquad 18 = (-2)(-9)$$
$$18 = 3 \cdot 6 \qquad 18 = (-3)(-6)$$

This shows that the integers 1, −1, 2, −2, 3, −3, 6, −6, 9, −9, 18, and −18 are the factors of 18. Each of these factors divides 18 evenly (with zero

221

remainder). For this reason the factors of 18 are also called the *divisors* of 18.

Another way to factor 18 is

$$18 = 2 \cdot 3 \cdot 3$$

The factors 2 and 3 are prime numbers. A **prime number** is a positive integer greater than 1 whose only factors are itself and 1. The first fifteen prime numbers are

$$2, 3, 5, 7, 11, 13, 17, 19, 23, 29, 31, 37, 41, 43, 47$$

An important result in mathematics states that *every integer greater than 1 is either a prime number or a product of prime numbers*. When a number is written as a product of prime numbers, it is said to be in **prime factored form**. Hence, the prime factored form of 18 is $2 \cdot 3 \cdot 3 = 2 \cdot 3^2$.

To write a number in prime factored form, we must find its prime factors. As an aid in finding them, we can use simple tests to determine whether a number has 2, 3, or 5 as a prime factor:

Divisibility tests

A number is *divisible by 2* if its last digit is 0, 2, 4, 6, or 8 (the number is even).
A number is *divisible by 3* if the sum of its digits is divisible by 3. (For example, 126 is divisible by 3 because $1 + 2 + 6 = 9$ is divisible by 3; in fact, $126 = 3 \cdot 42$.)
A number is *divisible by 5* if its last digit is 0 or 5.

The next example illustrates how to use these divisibility tests to find the prime factored form of a number.

EXAMPLE 1   Write each number in prime factored form.

a. 85   **b.** 114   **c.** 108   **d.** 126

Solution   **a.** Since 85 is divisible by 5, we can write

$$85 = 5 \cdot 17 \qquad \text{Dividing 85 by 5}$$

Recall that 17 is a prime number. Hence, the prime factored form of 85 is $5 \cdot 17$.

b. Since 114 is divisible by 2, we can write

$$114 = 2 \cdot 57 \qquad \text{Dividing 114 by 2}$$

Now 57 is divisible by 3 because the sum of its digits is 12, and that is divisible by 3. Therefore

$$114 = 2 \cdot 3 \cdot 19 \qquad \text{Dividing 57 by 3}$$

Recall that 19 is a prime number. Hence, the prime factored form of 114 is $2 \cdot 3 \cdot 19$.

c.
$$
\begin{aligned}
108 &= 2 \cdot 54 && \text{Dividing 108 by 2} \\
&= 2 \cdot 2 \cdot 27 && \text{Dividing 54 by 2} \\
&= 2 \cdot 2 \cdot 3 \cdot 9 && \text{Dividing 27 by 3} \\
&= 2 \cdot 2 \cdot 3 \cdot 3 \cdot 3 && \text{Dividing 9 by 3} \\
&= 2^2 \cdot 3^3
\end{aligned}
$$

The prime factored form of 108 is $2^2 \cdot 3^3$.

d.
$$
\begin{aligned}
126 &= 2 \cdot 63 && \text{Dividing 126 by 2} \\
&= 2 \cdot 3 \cdot 21 && \text{Dividing 63 by 3} \\
&= 2 \cdot 3 \cdot 3 \cdot 7 && \text{Dividing 21 by 3} \\
&= 2 \cdot 3^2 \cdot 7
\end{aligned}
$$

The prime factored form of 126 is $2 \cdot 3^2 \cdot 7$. ■

Greatest Common Factor

The largest number that is a factor of two or more integers is called the **greatest common factor** of the integers. For example, the greatest common factor of 4 and 6 is 2 because 2 is the largest factor of both 4 and 6. When the greatest common factor cannot be found by inspection, we can find it from the prime factored form of the numbers.

To find the greatest common factor of two or more integers

1. Write each integer in prime factored form.
2. Find the lowest power of each prime number common to *all* the factorizations.
3. Multiply the powers, if any, found in step 2. If there are none, then the greatest common factor is 1.

EXAMPLE 2

Find the greatest common factor of:

a.   18 and 24   b.   85 and 114
c.   108 and 126   d.   280, 420, and 700

Solution

a.   We begin by writing 18 and 24 in prime factored form:

$$18 = 2 \cdot 3^2$$

$$24 = 2^3 \cdot 3$$

Powers of 2 and 3 are common to both factorizations. The lowest power of 2 that appears in both is $2^1 = 2$, and the lowest power of 3 is $3^1 = 3$. Therefore, the greatest common factor is $2 \cdot 3 = 6$.

b. $85 = 5 \cdot 17$

$114 = 2 \cdot 3 \cdot 19$

There are no prime numbers common to both factorizations, so the greatest common factor is 1.

c. $108 = 2^2 \cdot 3^3$

$126 = 2 \cdot 3^2 \cdot 7$

Powers of 2 and 3 are common to both factorizations (7 is not a factor of 108). The lowest power of 2 is 2, and the lowest power of 3 is 3^2, so the greatest common factor is $2 \cdot 3^2 = 18$.

d. $280 = 2^3 \cdot 5 \cdot 7$

$420 = 2^2 \cdot 3 \cdot 5 \cdot 7$

$700 = 2^2 \cdot 5^2 \cdot 7$

Powers of 2, 5, and 7 are common to all factorizations (3 is not a factor of 280 or 700). The lowest power of 2 is 2^2, the lowest power of 5 is 5, and the lowest power of 7 is 7. Therefore, the greatest common factor is $2^2 \cdot 5 \cdot 7 = 140$. ∎

We can find the greatest common factor of two or more terms in much the same way. For example, to find the greatest common factor of the terms $15x^3$ and $6x^2$, we first factor the two terms:

$$15x^3 = 3 \cdot 5 \cdot x^3 \quad \text{and} \quad 6x^2 = 2 \cdot 3 \cdot x^2$$

Powers of 3 and x are common to both factorizations. The lowest power of 3 is 3, and the lowest power of x is x^2. Therefore, the greatest common factor is $3x^2$.

Once we know the greatest common factor of the terms $15x^3$ and $6x^2$, we can use it to factor the polynomial $15x^3 + 6x^2$. We begin by writing each term as a product, with the greatest common factor as one of the factors:

$$15x^3 + 6x^2 = 3x^2 \cdot 5x + 3x^2 \cdot 2$$

Next we apply the distributive property to the right side of the equation to "factor out" $3x^2$:

$$15x^3 + 6x^2 = 3x^2(5x + 2)$$

The expression $3x^2(5x + 2)$ is the factored form of $15x^3 + 6x^2$. This process is called *factoring out the greatest common factor*. We can check that the factored form of $15x^3 + 6x^2$ is $3x^2(5x + 2)$ by multiplication:

$$3x^2(5x + 2) = 3x^2 \cdot 5x + 3x^2 \cdot 2$$
$$= 15x^3 + 6x^2$$

To factor out the greatest common factor of a polynomial

1. Find the greatest common factor of the terms of the polynomial.
2. Use the distributive property to factor the greatest common factor out of each term of the polynomial.
3. Check the result by multiplication.

EXAMPLE 3

Factor out the greatest common factor of each polynomial.

a. $4x + 6$
b. $5y^3 + 3y^2 + y$
c. $6z^4 - 12z^3 + 30z^2$
d. $60p^3q^8 - 45p^7q^2$

Solution

a. Since $4x = 2^2 \cdot x$ and $6 = 2 \cdot 3$, the greatest common factor of the terms $4x$ and 6 is 2. We write each term as a product using 2 as one of the factors, and then factor out 2 using the distributive property:

$$4x + 6 = 2 \cdot 2x + 2 \cdot 3$$
$$= 2(2x + 3) \qquad \text{By the distributive property}$$

We check this result by multiplication: $2(2x + 3) = 4x + 6$.

b. Since $5y^3 = 5 \cdot y^3$, $3y^2 = 3 \cdot y^2$, and $y = 1 \cdot y$, the greatest common factor of these three terms is y. We write each term as a product with y as one of the factors, and then use the distributive property to factor out y:

$$5y^3 + 3y^2 + y = y \cdot 5y^2 + y \cdot 3y + y \cdot 1$$
$$= y(5y^2 + 3y + 1) \qquad \text{By the distributive property}$$

Check this result by multiplication.

c. The greatest common factor of the terms $6z^4$, $-12z^3$, and $30z^2$ is $6z^2$. We factor out $6z^2$:

$$6z^4 - 12z^3 + 30z^2 = 6z^2 \cdot z^2 - 6z^2 \cdot 2z + 6z^2 \cdot 5$$
$$= 6z^2(z^2 - 2z + 5)$$

Check this result by multiplication.

d. The greatest common factor of the terms $60p^3q^8$ and $-45p^7q^2$ is $15p^3q^2$. We factor out $15p^3q^2$:

$$60p^3q^8 - 45p^7q^2 = 15p^3q^2(4q^6 - 3p^4)$$

Check this result by multiplication. ∎

BE CAREFUL!

Make sure you don't omit any terms when factoring out the greatest common factor. It is *incorrect* to factor out y from the polynomial $5y^3 + 3y^2 + y$ as follows:

$$5y^3 + 3y^2 + y = y(5y^2 + 3y)$$

The correct result is

$$5y^3 + 3y^2 + y = y(5y^2 + 3y + 1)$$

It helps to write the term y in the polynomial as $y \cdot 1$ before factoring out y. Remember to write each term of a polynomial as a product with the greatest common factor as one of the factors before factoring out.

Factoring by Grouping

When a polynomial cannot be factored by factoring out the greatest common factor, sometimes we can factor it by *grouping* terms that have a common factor. For example, we cannot factor the polynomial $2x^3 + 3x^2 + 4x + 6$ by factoring out the greatest common factor because the greatest common factor of all four terms is 1. However, the first two terms (the first group) have a common factor of x^2, and the last two terms (the second group) have a common factor of 2. By factoring out x^2 from the first two terms and 2 from the last two terms, we can write

$$2x^3 + 3x^2 + 4x + 6 = x^2(2x + 3) + 2(2x + 3)$$

Now we can see that $2x + 3$ is a common factor of $x^2(2x + 3)$ and $2(2x + 3)$. Hence, we can now factor out $2x + 3$ to obtain

$$x^2(2x + 3) + 2(2x + 3) = (2x + 3)(x^2 + 2)$$

The expression $(2x + 3)(x^2 + 2)$ is the factored form of $2x^3 + 3x^2 + 4x + 6$. We check by multiplication using the FOIL method: $(2x + 3)(x^2 + 2) = 2x^3 + 3x^2 + 4x + 6$.

Another way to factor $2x^3 + 3x^2 + 4x + 6$ is to group the terms $2x^3$ and $4x$, which have a common factor of $2x$, and to group the terms $3x^2$ and 6, which have a common factor of 3:

$$2x^3 + 3x^2 + 4x + 6 = 2x^3 + 4x + 3x^2 + 6$$
$$= 2x(x^2 + 2) + 3(x^2 + 2) \qquad \text{Factoring out}$$

Now we can factor out $x^2 + 2$ to obtain the same result as before:

$$2x(x^2 + 2) + 3(x^2 + 2) = (x^2 + 2)(2x + 3)$$

The key idea in factoring by grouping is to produce a common factor *after* (1) the terms of the given polynomial have been grouped, and (2) a common factor has been factored out of each group. In the example above, we found the common factor $2x + 3$ in the terms $x^2(2x + 3)$ and $2(2x + 3)$ after we factored out x^2 from the first two terms and 2 from the last two terms of the original polynomial.

EXAMPLE 4

Factor each polynomial by grouping.

a. $k^3 + 3k^2 + 4k + 12$ **b.** $3h^2 + h - 6h - 2$
c. $6w^2 - 18wz + wz - 3z^2$

Solution

a. We can factor out k^2 from the first two terms and 4 from the last two terms to obtain

$$k^3 + 3k^2 + 4k + 12 = k^2(k + 3) + 4(k + 3)$$

Now we can factor out $k + 3$:

$$k^2(k + 3) + 4(k + 3) = (k + 3)(k^2 + 4)$$

The factored form of $k^3 + 3k^2 + 4k + 12$ is $(k + 3)(k^2 + 4)$. We check this result by multiplication: $(k + 3)(k^2 + 4) = k^3 + 3k^2 + 4k + 12$.

b. We can factor out h from the first two terms and -2 from the last two terms:

$$3h^2 + h - 6h - 2 = h(3h + 1) - 2(3h + 1)$$

Now we can factor out $3h + 1$:

$$h(3h + 1) - 2(3h + 1) = (3h + 1)(h - 2)$$

The factored form of $3h^2 + h - 6h - 2$ is $(3h + 1)(h - 2)$. Check this result by multiplication.

c. We can factor out $6w$ from the first two terms and z from the last two terms:

$$6w^2 - 18wz + wz - 3z^2 = 6w(w - 3z) + z(w - 3z)$$

Now we can factor out $w - 3z$:

$$6w(w - 3z) + z(w - 3z) = (w - 3z)(6w + z)$$

The factored form of $6w^2 - 18wz + wz - 3z^2$ is $(w - 3z)(6w + z)$. Check this result by multiplication. ∎

DISCUSSION QUESTIONS 5.1

1. Why is a factor of a number also called a divisor?

2. What is a prime number?

3. What is the prime factored form of a number?

4. In your own words, state how to find the greatest common factor of two or more integers.

5. In your own words, state how to factor out the greatest common factor of a polynomial.

EXERCISES 5.1

In Exercises 1–8, write the given number in prime factored form.

1. 36   **2.** 28   **3.** 63   **4.** 84   **5.** 105   **6.** 189   **7.** 315   **8.** 498

In Exercises 9–14, find the greatest common factor of the given numbers.

9. 20 and 24   **10.** 32 and 56   **11.** 60 and 72

12. 75 and 90   **13.** 220 and 441   **14.** 225 and 308

In Exercises 15–44, factor out the greatest common factor of the given polynomial.

15. $2u - 8$ **16.** $3v - 18$ **17.** $21z - 14$ **18.** $24w - 32$

19. $3p^2 + p$ **20.** $8q^4 + q^2$ **21.** $18r^9 + 2r^4$ **22.** $12t^5 + 4t^3$

23. $20h^3 - 25h^4$ **24.** $14k^2 - 35k^6$ **25.** $2x^2 + 16x + 28$

26. $3y^2 - 24y + 30$ **27.** $3z^3 - 4z^2 + 6z$ **28.** $2w^5 + 3w^4 - 5w^2$

29. $32v^3 + 16v^4 - 24v^5$ **30.** $10u^5 - 12u^7 + 18u^9$ **31.** $14p^4 + 21p^3 - 7p^2$

32. $18q^5 - 27q^4 + 9q^3$ **33.** $33m^2 - 121n^2$ **34.** $48r^3 - 36t^3$

35. $16x^2y^3 + 80xy^2$ **36.** $35w^3z^2 + 14wz^3$ **37.** $36p^2q - 72pq^2 + 81pq$

38. $24m^3n^2 - 40m^2n^3 + 56m^2n^2$ **39.** $49h^4k^6 - 21h^5k^4 - 35h^6k^2$ **40.** $15r^7s^4 + 65r^5s^6 + 20r^3s^8$

41. $3x(y - 1) + 7(y - 1)$ **42.** $4z(w + 3) - 11(w + 3)$

43. $p^3(q^2 - 7) - 3p(q^2 - 7)$ **44.** $5m^2(2n^3 + 9) + 4m(2n^3 + 9)$

In Exercises 45–56, factor the given polynomial by grouping.

45. $3y^3 + 4y^2 + 9y + 12$ **46.** $2x^3 + 8x^2 + 5x + 20$ **47.** $10z^3 - 15z^2 + 4z - 6$

48. $2w^3 - 7w^2 + 10w - 35$ **49.** $18p^2 + 12p - 15p - 10$ **50.** $12q^2 + 8q - 75q - 50$

51. $7r^2 - 14r - 6r + 12$ **52.** $15s^2 - 45s - 16s + 48$ **53.** $8x^2 - 4xy + 6xy - 3y^2$

54. $9w^2 - 6wz - 12wz + 8z^2$ **55.** $5p^2 + pq - 50pq - 10q^2$ **56.** $6h^2 + hk - 72hk - 12k^2$

57. The total surface area of a rectangular solid with a square base of side x and height h is $4xh + 2x^2$. Write this expression in factored form.

58. The total surface area of a right circular cylinder with radius r and height h is $2\pi rh + 2\pi r^2$. Write this expression in factored form.

Preparing for Section 5.2

In Exercises 59–64, find the product.

59. $(x + 3)(x + 7)$ **60.** $(y - 2)(y + 5)$ **61.** $(z + 1)(z - 9)$

62. $(w - 6)(w - 4)$ **63.** $(p - 6)(p + 2)$ **64.** $(q + 9)(q + 7)$

5.2   Trinomials

FOCUS

We see how to factor trinomials of the form $x^2 + bx + c$ as the product of two binomials.

Recall that the FOIL method is used to multiply two binomials. For example,

$$\overset{\text{F}\quad\text{O}\quad\text{I}\quad\text{L}}{(x + 3)(x + 4) = x^2 + 4x + 3x + 12}$$

$$= x^2 + 7x + 12$$

Now suppose we wanted to factor $x^2 + 7x + 12$. The last equation shows that the factored form of the trinomial $x^2 + 7x + 12$ is $(x + 3)(x + 4)$. In this section, we see how to find that factored form by reversing the FOIL method.

If we did not know that $x^2 + 7x + 12 = (x + 3)(x + 4)$, how could we factor $x^2 + 7x + 12$ into the product of two binomials? First of all, we must realize that to factor $x^2 + 7x + 12$ we must find integers p and q such that

$$x^2 + 7x + 12 = (x + p)(x + q)$$

The first term in each binomial factor must be x because their product, by the FOIL method, is x^2.

To find the integers p and q, we multiply the two binomials by the FOIL method:

$$(x + p)(x + q) = x^2 + qx + px + pq$$

$$= x^2 + (p + q)x + pq$$

This must be equal to $x^2 + 7x + 12$, so we have

$$x^2 + 7x + 12 = x^2 + (p + q)x + pq$$

By comparing the terms on the two sides of this last equation, we see that the sum $p + q$ must equal 7 (because $p + q$ and 7 are the coefficients of x). Similarly, the product pq must equal 12:

$$\overset{\displaystyle 12 = pq}{x^2 + 7x + 12 = x^2 + (p + q)x + pq}$$
$$7 = p + q$$

We now must look for a pair of integers whose product is 12 and whose sum is 7:

Product	Sum
$1 \cdot 12 = 12$	$1 + 12 = 13$
$(-1)(-12) = 12$	$(-1) + (-12) = -13$
$2 \cdot 6 = 12$	$2 + 6 = 8$
$(-2)(-6) = 12$	$(-2) + (-6) = -8$
$3 \cdot 4 = 12$	$3 + 4 = 7$
$(-3)(-4) = 12$	$(-3) + (-4) = -7$

The required integers are 3 and 4 because $3 \cdot 4 = 12$ and $3 + 4 = 7$. Therefore

$$x^2 + 7x + 12 = (x + 3)(x + 4)$$

This result can be checked by multiplication.

This method of factoring trinomials works when the coefficient of the squared term is 1, that is, when the trinomial can be expressed in the form $x^2 + bx + c$, where b and c are real numbers. As you become proficient in factoring trinomials, you will be able to do many of the steps in this method mentally.

EXAMPLE 1

Factor each trinomial.

a. $w^2 + 5w + 6$ **b.** $y^2 - 9y + 14$
c. $z^2 + 3z - 10$ **d.** $p^2 - 2pq - 15q^2$

Solution

a. To factor $w^2 + 5w + 6$, we must find two integers whose product is 6 and whose sum is 5:

Product	Sum
$1 \cdot 6 = 6$	$1 + 6 = 7$
$(-1)(-6) = 6$	$(-1) + (-6) = -7$
$2 \cdot 3 = 6$	$2 + 3 = 5$
$(-2)(-3) = 6$	$(-2) + (-3) = -5$

The required integers are 2 and 3 because $2 \cdot 3 = 6$ and $2 + 3 = 5$. Therefore

$$w^2 + 5w + 6 = (w + 2)(w + 3)$$

We check this result by multiplication:

$$(w + 2)(w + 3) = w^2 + 3w + 2w + 6$$
$$= w^2 + 5w + 6$$

b. To factor $y^2 - 9y + 14$, we must find two integers whose product is 14 and whose sum is -9:

Product	Sum
$1 \cdot 14 = 14$	$1 + 14 = 15$
$(-1)(-14) = 14$	$(-1) + (-14) = -15$
$2 \cdot 7 = 14$	$2 + 7 = 9$
$(-2)(-7) = 14$	$(-2) + (-7) = -9$

The required integers are -2 and -7 because $(-2) \cdot (-7) = 14$ and $(-2) + (-7) = -9$. Therefore

$$y^2 - 9y + 14 = (y - 2)(y - 7)$$

Check this result by multiplication.

c. To factor $z^2 + 3z - 10$, we must find two integers whose product is -10 and whose sum is 3:

Product	Sum
$1 \cdot (-10) = -10$	$1 + (-10) = -9$
$(-1) \cdot 10 = -10$	$(-1) + 10 = 9$
$2 \cdot (-5) = -10$	$2 + (-5) = -3$
$(-2) \cdot 5 = -10$	$(-2) + 5 = 3$

The required integers are -2 and 5 because $(-2) \cdot 5 = -10$ and $(-2) + 5 = 3$. Therefore

$$z^2 + 3z - 10 = (z - 2)(z + 5)$$

Check this result by multiplication.

d. To factor $p^2 - 2pq - 15q^2$, we must find two *expressions* whose product is $-15q^2$ and whose sum is $-2q$. Both expressions must contain q because their sum contains q. Hence we try the following:

Product	Sum
$q \cdot (-15q) = -15q^2$	$q + (-15q) = -14q$
$(-q) \cdot 15q = -15q^2$	$(-q) + 15q = 14q$
$3q \cdot (-5q) = -15q^2$	$3q + (-5q) = -2q$
$(-3q) \cdot 5q = -15q^2$	$(-3q) + 5q = 2q$

The required expressions are $3q$ and $-5q$ because $3q \cdot (-5q) = -15q^2$ and $3q + (-5q) = -2q$. Therefore

$$p^2 - 2qp - 15q^2 = (p + 3q)(p - 5q)$$

Check this result by multiplication.   ■

Some trinomials cannot be factored with integer factors. For example, consider the trinomial $x^2 + 2x + 2$. To factor $x^2 + 2x + 2$, we must find two integers whose product is 2 and whose sum is 2. However, there is no such pair of integers, as you can see by trying the two possibilities:

Product	Sum
$1 \cdot 2 = 2$	$1 + 2 = 3$
$(-1)(-2) = 2$	$(-1) + (-2) = -3$

Therefore the trinomial $x^2 + 2x + 2$ cannot be factored with integer factors. A polynomial that cannot be factored with integer factors is called a **prime polynomial.**

Is the polynomial $6y^4 + 30y^3 + 36y^2$ prime? Our method of factoring trinomials does not apply to $6y^4 + 30y^3 + 36y^2$ because it is not of the form $x^2 + bx + c$. However, the trinomial can be factored! To do so, we first factor out $6y^2$, the greatest common factor of the terms of $6y^4 + 30y^3 + 36y^2$:

$$6y^4 + 30y^3 + 36y^2 = 6y^2(y^2 + 5y + 6)$$

Now we try to factor the trinomial $y^2 + 5y + 6$. We must find two integers whose product is 6 and whose sum is 5. The required integers are 2 and 3, so we have

$$y^2 + 5y + 6 = (y + 2)(y + 3)$$

We can now write the complete factored form of $6y^4 + 30y^3 + 36y^2$:

$$6y^4 + 30y^3 + 36y^2 = 6y^2(y + 2)(y + 3)$$

This result can be checked by multiplication.

The example above illustrates the following important rule:

> The first step in factoring a polynomial is to look for the greatest common factor of its terms. If the greatest common factor is not 1, then begin by factoring it out.

DISCUSSION QUESTIONS 5.2

1. If $x^2 - 5x - 14 = (x + p)(x + q)$, what is the value of pq? What is the value of $p + q$?

2. What is a prime polynomial?

3. Is $y^2 - 10y + 20$ a prime polynomial? Explain.

EXERCISES 5.2

In Exercises 1–20, factor the given trinomial. If a trinomial cannot be factored, then identify it as a prime polynomial.

1. $w^2 + 8w + 15$ 2. $z^2 + 11z + 10$ 3. $u^2 + 8u - 9$ 4. $v^2 + 12v + 32$

5. $m^2 - 20m + 48$ 6. $n^2 - 30n + 60$ 7. $r^2 - 23r + 24$ 8. $s^2 - 5s - 24$

9. $z^2 + 45z + 120$ 10. $w^2 + 30w + 160$ 11. $h^2 - 7hk + 12k^2$ 12. $p^2 - 9pq + 20q^2$

13. $x^2 - 5xy - 36y^2$ 14. $w^2 - 2wz - 8z^2$ 15. $s^2 - 13st + 12t^2$ 16. $u^2 - 3uv - 28v^2$

17. $w^2 + 5wz + 15z^2$ 18. $x^2 - 8xy + 24y^2$ 19. $p^2 - 16pq + 64q^2$ 20. $h^2 - 14hk + 49k^2$

In Exercises 21–46, completely factor the given polynomial.

21. $2q^2 + 12q + 16$ 22. $3p^2 - 21p + 30$ 23. $r^6 + r^5 - 12r^4$

24. $t^8 - 3t^7 + 4t^6$ 25. $x^4 + 2x^3 + x^2$ 26. $y^7 - 2y^6 + y^5$

27. $z^3 + 6z^2 - 27z$ 28. $w^6 - 5w^5 - 14w^4$ 29. $4u^2 - 48u + 128$

30. $7v^2 + 70v - 63$ 31. $2t^5 + 4t^4 - 30t^3$ 32. $5r^4 - 60r^3 + 55r^2$

33. $5h^8 + 45h^7 + 100h^6$ 34. $6k^5 - 48k^4 - 120k^3$ 35. $9t^7 - 18t^6 + 9t^5$

36. $4s^9 + 8s^8 + 4s^7$ 37. $x^2y + 6xy^2 - 18y^3$ 38. $w^3 - 18w^2z + 32wz^2$

39. $6p^2q^3 - 48p^2q^2 + 72p^2q$ 40. $8m^4n^3 + 56m^3n^3 - 96m^2n^3$

41. $3u^3v + 15u^2v^2 - 96uv^3$ 42. $10x^3y + 90x^2y^2 - 360xy^3$

43. $(w + z)p^2 + 17(w + z)p + 72(w + z)$ 44. $(x - y)q^2 - 4(x - y)q - 96(x - y)$

45. $r^4 + 5r^2 + 6$ 46. $t^6 + 8t^3 + 7$

Preparing for Section 5.3

In Exercises 47–52, find the product.

47. $(2x + 1)(x + 3)$ 48. $(y - 5)(3y + 7)$ 49. $(w + 9)(4w - 5)$

50. $(2z - 9)(z - 8)$ 51. $(3p - 7)(5p + 6)$ 52. $(4q + 1)(6q + 3)$

5.3   Other Trinomials

FOCUS

We factor trinomials of the form $ax^2 + bx + c$, where $a \neq 1$. We use two approaches, one based on the FOIL method and the other on grouping.

When the coefficient of the squared term of a trinomial is not 1, and the terms do not have a common factor, we cannot use the methods of Section 5.2. However, we can reverse the FOIL method to factor such trinomials. We begin by noting that, for example, to factor $3x^2 + 7x + 2$ as the product of two binomials, we must find integers m, n, p, and q such that

$$3x^2 + 7x + 2 = (mx + p)(nx + q)$$

By using the FOIL method to multiply the binomials on the right side of the equation, we find that

$$3x^2 + 7x + 2 = mnx^2 + (mq + np)x + pq$$

with $3 = mn$ and $2 = pq$.

Comparing the terms on the two sides of this last equation, we see that $mn = 3$ because mn and 3 are both the coefficient of x^2. Also, $pq = 2$ because pq and 2 are both the constant term. But in addition, $(mq + np) = 7$ because the coefficients of x must be equal. Hence, our task now is to find factors of 3 (m and n) and factors of 2 (p and q) that produce an outer product and an inner product whose sum is the middle term $7x$. We prefer that the first term in each binomial be positive, so we consider only the positive factors of 3, which are 3 and 1. For factors of 2, we try 2 and 1, -2 and -1, 1 and 2, and -1 and -2:

$$(mx + p)(nx + q)(mq + np)x$$

Possible Factorization	Middle Term
$(3x + 2)(x + 1)$	$5x$
$(3x - 2)(x - 1)$	$-5x$
$(3x + 1)(x + 2)$	$7x$
$(3x - 1)(x - 2)$	$-7x$

The third pair of factors produces the correct middle term, $7x$. Therefore, we may immediately write

$$3x^2 + 7x + 2 = (3x + 1)(x + 2)$$

We check this result by multiplication:

$$(3x + 1)(x + 2) = 3x^2 + 6x + x + 2$$
$$= 3x^2 + 7x + 2$$

As you become proficient in factoring trinomials, you will learn to eliminate many of the possible factorizations simply by examining them. In the example above, we could have eliminated the factorizations $(3x - 2)(x - 1)$

and $(3x - 1)(x - 2)$ because each produces a negative middle term when multiplied out. This suggests the following principle:

> If all the terms of a trinomial are positive, then all the terms of its binomial factors (when completely factored) are positive.

EXAMPLE 1

Factor the trinomial $4x^2 - 9x + 2$.

Solution

To factor $4x^2 - 9x + 2$, we must find two binomial factors such that the product of the first terms is $4x^2$, the product of the last terms is 2, and the sum of the inner and outer products is $-9x$. We work with the positive factors of 4 and the factors of 2:

Possible Factorization	Middle Term
$(4x + 2)(x + 1)$	$6x$
$(4x - 2)(x - 1)$	$-6x$
$(4x + 1)(x + 2)$	$9x$
$(4x - 1)(x - 2)$	$-9x$
$(2x + 1)(2x + 2)$	$6x$
$(2x - 1)(2x - 2)$	$-6x$

The fourth pair of factors produces the correct middle term; therefore

$$4x^2 - 9x + 2 = (4x - 1)(x - 2)$$

We check this result by multiplication:

$$(4x - 1)(x - 2) = 4x^2 - 8x - x + 2$$
$$= 4x^2 - 9x + 2 \qquad \blacksquare$$

In the "possible factorization" $(4x - 2)(x - 1)$ of Example 1, the binomial $4x - 2$ has a common factor of 2. By factoring out 2, we can write

$$(4x - 2)(x - 1) = 2(2x - 1)(x - 1)$$

This indicates that the original trinomial must have the same common factor of 2. But our original trinomial, $4x^2 - 9x + 2$, has no common factor except 1; hence, the factorization $(4x - 2)(x - 1)$ cannot be correct. A similar argument shows that the factorizations $(4x + 2)(x + 1)$, $(2x + 1)(2x + 2)$, and $(2x - 1)(2x - 2)$ cannot be correct. This, then, is another way to eliminate factorizations:

> If a trinomial has no common factor (except 1), then none of its binomial factors has a common factor (except 1).

EXAMPLE 2

Solution

Factor the trinomial $6y^2 + 17y - 3$.

To factor $6y^2 + 17y - 3$, we must find two binomial factors such that the product of the first terms is $6y^2$, the product of the last terms is -3, and the sum of the inner and outer products is $17y$. We work with the positive factors of 6 and the factors of -3. We can eliminate the factorization $(3y - 3) \cdot (2y + 1)$ because the binomial $3y - 3$ has a common factor of 3, but $6y^2 + 17y - 3$ has no common factor except 1. Similarly, we can eliminate the factorization $(3y + 3)(2y - 1)$. Hence, we try

Possible Factorization	Middle Term
$(6y - 1)(y + 3)$	$17y$
$(6y + 1)(y - 3)$	$-17y$
$(3y - 1)(2y + 3)$	$7y$
$(3y + 1)(2y - 3)$	$-7y$

The first pair of factors produces the correct middle term; therefore

$$6y^2 + 17y - 3 = (6y - 1)(y + 3)$$

Check this result by multiplication. ∎

A second way to factor trinomials of the form $ax^2 + bx + c$ $(a \neq 1)$ involves grouping. We illustrate the method by again factoring the trinomial $3x^2 + 7x + 2$. To begin, we look for two integers whose product is $3 \cdot 2 = 6$ and whose sum is 7:

Sum must be 7

$$3x^2 + 7x + 2$$

Product must be $3 \cdot 2 = 6$

The required integers are 1 and 6. We can now use these integers to write the middle term, $7x$, as $7x = 1x + 6x$. This is what we would do if we were multiplying the factors using the FOIL method:

$$3x^2 + 7x + 2 = 3x^2 + 1x + 6x + 2$$

Finally, we factor the right side of the equation by grouping:

$$3x^2 + 7x + 2 = \underbrace{3x^2 + 1x} + \underbrace{6x + 2}$$

$$= x(3x + 1) + 2(3x + 1)$$

$$= (3x + 1)(x + 2)$$

If the middle term had been written as $7x = 6x + 1x$, then we would have

$$3x^2 + 7x + 2 = \underbrace{3x^2 + 6x} + \underbrace{1x + 2}$$

$$= 3x(x + 2) + 1(x + 2)$$

$$= (x + 2)(3x + 1)$$

Both results are correct, and the same as we found before.

EXAMPLE 3

Solution

Factor the trinomial $16z^4 - 28z^3 - 30z^2$.

Notice that $16z^4 - 28z^3 - 30z^2$ has a greatest common factor of $2z^2$. Our first step is to factor it out:

$$16z^4 - 28z^3 - 30z^2 = 2z^2(8z^2 - 14z - 15)$$

We will use the grouping method to factor $8z^2 - 14z - 15$. We must find two integers whose product is $8 \cdot (-15) = -120$ and whose sum is -14. Considering the factors of -120, we find that the required integers are 6 and -20 because $6 \cdot (-20) = -120$ and $6 + (-20) = -14$. We now write the middle term of the trinomial, $-14z$, as $-14z = 6z - 20z$ and factor by grouping:

$$8z^2 - 14z - 15 = 8z^2 + 6z - 20z - 15$$

$$= 2z(4z + 3) - 5(4z + 3)$$

$$= (4z + 3)(2z - 5)$$

Therefore, $16z^4 - 28z^3 - 30z^2 = 2z^2(4z + 3)(2z - 5)$.

Check this result by multiplication.   ∎

EXAMPLE 4

Solution

Factor the trinomial $6m^2 + 19mn + 10n^2$.

We will use the grouping method to factor $6m^2 + 19mn + 10n^2$. We must find two integers whose product is $6 \cdot 10 = 60$ and whose sum is 19. By considering the factors of 60, we find that the required integers are 4 and 15 because $4 \cdot 15 = 60$ and $4 + 15 = 19$. We now write the middle term of the trinomial as $19mn = 4mn + 15mn$ and factor by grouping:

$$6m^2 + 19mn + 10n^2 = 6m^2 + 4mn + 15mn + 10n^2$$
$$= 2m(3m + 2n) + 5n(3m + 2n)$$
$$= (3m + 2n)(2m + 5n)$$

Check this result by multiplication. ■

We can summarize the procedure for factoring a trinomial by grouping as follows:

To factor a trinomial of the form $ax^2 + bx + c$ by grouping

1. Find two integers, u and v, such that $uv = ac$ and $u + v = b$.
2. Replace the term bx with the sum $ux + vx$, and then factor by grouping.
3. Check the result by multiplication.

DISCUSSION QUESTIONS 5.3

1. If $6x^2 + 13x - 5 = (mx + p)(nx + q)$, what is the value of mn? What is the value of pq? What is the value of $mq + np$?

2. If a trinomial has a positive first term, a negative middle term, and a positive last term, then the last terms of its binomial factors must have the same sign. Are the last terms both positive or both negative? Explain.

3. A trinomial has a positive first term and a negative last term. Are the last terms of its binomial factors negative? Explain.

4. Is $3y + 9$ a binomial factor of $3y^2 + 19y + 18$? Explain your answer without factoring $3y^2 + 19y + 18$.

EXERCISES 5.3

In Exercises 1–18, factor the given trinomial. If a trinomial cannot be factored, then identify it as a prime polynomial.

1. $2w^2 + 5w + 2$
2. $2z^2 + 7z + 3$
3. $7n^2 - 16n + 4$
4. $5r^2 - 16r + 3$

5. $6x^2 + 10x - 3$
6. $4y^2 + 13y - 5$
7. $4h^2 + 17h - 15$
8. $2k^2 + 19k - 10$

9. $10z^2 + 19z + 6$
10. $6w^2 + 19w + 10$
11. $36r^2 - 5r - 20$
12. $20t^2 - 48t - 5$

13. $14m^2 + 11m - 15$ **14.** $6n^2 + 5n - 8$ **15.** $10x^2 - 21xy + 9y^2$

16. $15w^2 - 17wz + 4z^2$ **17.** $14m^2 + 29mn - 15n^2$ **18.** $12r^2 + 7rs - 12s^2$

In Exercises 19–40, completely factor the given polynomial.

19. $x^4 + 2x^3 + x^2$ **20.** $y^7 - 2y^6 + y^5$ **21.** $w^5 - 5w^4 - 24w^3$ **22.** $z^4 - 6z^3 + 27z^2$

23. $4m^2 - 36m + 56$ **24.** $2n^2 + 2n - 40$ **25.** $8h^2 - 16h + 40$ **26.** $7k^2 + 21k + 28$

27. $5r^3 + 70r^2 + 65r$ **28.** $2t^6 + 4t^5 - 48t^4$ **29.** $4q^2 + 2q - 6$ **30.** $6p^2 - 51p + 63$

31. $9y^4 + 6y^3 + y^2$ **32.** $4x^6 + 4x^5 + x^4$ **33.** $6k^3 - 14k^2 - 40k$ **34.** $10h^4 - 23h^3 + 12h^2$

35. $10s^6 - 6s^5 - 4s^4$ **36.** $6r^5 + 15r^4 + 9r^3$ **37.** $8pq^3 + 14pq^2 + 3pq$

38. $18m^5n^2 + 3m^4n^2 - 6m^3n^2$ **39.** $32r^4s^7 - 56r^3s^8 + 12r^2s^9$ **40.** $30k^7h^3 + 35k^6h^4 + 10k^5h^5$

Preparing for Section 5.4

In Exercises 41–48, find the product.

41. $(x + 5)^2$ **42.** $(y - 7)^2$ **43.** $(z + 6)(z - 6)$ **44.** $(4w + 3)^2$

45. $(7p - 2)^2$ **46.** $(3q + 1)(3q - 1)$ **47.** $(2m + 9n)^2$ **48.** $(5q - 11p)^2$

5.4   Special Polynomials

FOCUS

We discuss and apply formulas for factoring perfect-square trinomials and a difference of two squares.

Recall from Chapter 4 the two special formulas for squaring a binomial:

$$(a + b)^2 = a^2 + 2ab + b^2$$
$$(a - b)^2 = a^2 - 2ab + b^2$$

By reversing these formulas we obtain formulas for factoring the trinomials on the right. These trinomials are called **perfect-square trinomials.**

Factoring perfect-square trinomials

$$a^2 + 2ab + b^2 = (a + b)^2$$
$$a^2 - 2ab + b^2 = (a - b)^2$$

You can recognize a perfect-square trinomial by its terms: Its first and last terms are perfect squares, and its middle term is twice the product of the terms that are squared. These are also the terms in the squared binomial.

EXAMPLE 1

Factor each polynomial.

a. $x^2 + 8x + 16$ **b.** $4y^2 - 12y + 9$

c. $12z^3 + 60z^2 + 75z$ **d.** $25h^2 - 80hk + 64k^2$

Solution

a. In the trinomial $x^2 + 8x + 16$, the first and last terms, x^2 and 16, are perfect squares (since $16 = 4^2$). We suspect that

$$x^2 + 8x + 16 = (x + 4)^2$$

To check, we find twice the product of the terms x and 4, the two terms in the squared binomial:

$$2 \cdot x \cdot 4 = 8x$$

Since $8x$ is also the middle term of the trinomial, $x^2 + 8x + 16$ is a perfect-square trinomial and

$$x^2 + 8x + 16 = (x + 4)^2$$

b. The first and last terms of $4y^2 - 12y + 9$ are perfect squares because $4y^2 = (2y)^2$ and $9 = 3^2$. Since the middle term is negative, we suspect that

$$4y^2 - 12y + 9 = (2y - 3)^2$$

To check, we find twice the product of the terms $2y$ and -3, the two terms in the squared binomial:

$$2 \cdot 2y \cdot (-3) = -12y$$

Since $-12y$ is also the middle term of the trinomial, $4y^2 - 12y + 9$ is a perfect-square trinomial and

$$4y^2 - 12y + 9 = (2y - 3)^2$$

c. Notice that $12z^3 + 60z^2 + 75z$ has a greatest common factor of $3z$. Our first step is to factor it out:

$$12z^3 + 60z^2 + 75z = 3z(4z^2 + 20z + 25)$$

But $4z^2 + 20z + 25$ is a perfect-square trinomial because $4z^2 = (2z)^2$, $25 = 5^2$, and $2 \cdot 2z \cdot 5 = 20z$. Therefore $4z^2 + 20z + 25 = (2z + 5)^2$, and the complete factored form is

$$12z^3 + 60z^2 + 75z = 3z(2z + 5)^2$$

d. $25h^2 - 80hk + 64k^2$ looks like a perfect-square trinomial because $25h^2 = (5h)^2$ and $64k^2 = (8k)^2$. Because the middle term is negative, we compute $2 \cdot 5h \cdot (-8k) = -80hk$ which is the middle term of the trinomial. Therefore, $25h^2 - 80hk + 64k^3 = (5h - 8k)^2$. ∎

BE CAREFUL!

A common error when factoring trinomials is to assume that any trinomial whose first and last terms are perfect squares must be a perfect-square trinomial. Consider the trinomials

$$x^2 + 29x + 100 \quad \text{and} \quad x^2 + 20x + 100$$

The first and last terms of both are perfect squares, but only $x^2 + 20x + 100$ is a perfect-square trinomial. Nevertheless, $x^2 + 29x + 100$ can be factored: $x^2 + 29x + 100 = (x + 4)(x + 25)$.

Another special formula from Chapter 4 produced the difference of two squares:

$$(a + b)(a - b) = a^2 - b^2$$

By reversing this formula we obtain a formula for factoring the **difference of two squares**:

Factoring the difference of two squares

$$a^2 - b^2 = (a + b)(a - b)$$

EXAMPLE 2

Factor each polynomial.

a. $p^2 - 36$ **b.** $18q^3 - 2q$ **c.** $81m^2 - 121n^2$ **d.** $t^4 - 16$

Solution

a. We recognize that p^2 and 36 are both squares. Hence we can factor $p^2 - 36$ by setting $a = p$ and $b = 6$ in the formula for $a^2 - b^2$:

$$p^2 - 36 = p^2 - 6^2 = (p + 6)(p - 6)$$

b. We note that $18q^3 - 2q$ has a greatest common factor of $2q$. Our first step, then, is to factor out $2q$:

$$18q^3 - 2q = 2q(9q^2 - 1)$$

But $9q^2 - 1$ is the difference of two squares:

$$9q^2 - 1 = (3q)^2 - 1^2 = (3q + 1)(3q - 1)$$

Therefore the complete factored form of $18q^3 - 2q$ is

$$18q^3 - 2q = 2q(3q + 1)(3q - 1)$$

c. Since $81m^2 = (9m)^2$ and $121n^2 = (11n)^2$, we have

$$81m^2 - 121n^2 = (9m)^2 - (11n)^2 = (9m + 11n)(9m - 11n)$$

d. We can factor $t^4 - 16$ by setting $a = t^2$ and $b = 4$ in the formula for $a^2 - b^2$:

$$t^4 - 16 = (t^2)^2 - 4^2 = (t^2 + 4)(t^2 - 4)$$

Notice that $t^2 - 4$ can be factored further as the difference of two squares, but $t^2 + 4$ is a prime polynomial and cannot be factored. Therefore the complete factored form of $t^4 - 16$ is

$$t^4 - 16 = (t^2 + 4)(t + 2)(t - 2) \qquad \blacksquare$$

BE CAREFUL!

Do not try to apply our factoring formulas to a *sum of two squares*, such as $t^2 + 4$. To do so would be *incorrect*:

$$t^2 + 4 \neq (t + 2)^2 \text{ because } (t + 2)^2 = t^2 + 4t + 4$$
$$t^2 + 4 \neq (t - 2)^2 \text{ because } (t - 2)^2 = t^2 - 4t + 4$$
$$t^2 + 4 \neq (t + 2)(t - 2) \text{ because } (t + 2)(t - 2) = t^2 - 4$$

In general, the sum of two squares, $a^2 + b^2$, is a prime polynomial and cannot be factored.

We conclude this section with some guidelines for factoring polynomials. These guidelines may help you to recognize which factoring technique is appropriate for a given polynomial.

Factoring guidelines

1. If the greatest common factor of a polynomial is not 1, then factor it out.
2. If a polynomial has two terms, is it the difference of two squares? If so, then factor it using the formula

$$a^2 - b^2 = (a + b)(a - b)$$

3. If a polynomial has three terms, is it:
   a. A perfect-square trinomial? If so, then factor it using the formulas

$$a^2 + 2ab + b^2 = (a + b)^2$$
$$a^2 - 2ab + b^2 = (a - b)^2$$

   b. A trinomial that can be factored by the methods of Section 5.2 or 5.3?
4. If a polynomial has four terms, then try factoring by grouping.
5. Continue factoring until all factors are prime polynomials.

DISCUSSION QUESTIONS 5.4

1. Squaring a number always produces a nonnegative number. How can the middle term of a perfect-square trinomial be negative even though the trinomial is produced by squaring a binomial?

2. What is the factored form of the difference of two squares?

3. In your own words, state the guidelines for factoring a polynomial.

EXERCISES 5.4

In Exercises 1–58, completely factor the given polynomial. If a polynomial cannot be factored, then identify it as a prime polynomial.

1. $w^2 + 8w + 16$
2. $z^2 + 6z + 9$
3. $m^2 + 16m + 64$
4. $n^2 + 10n + 25$
5. $h^2 + 2h + 4$
6. $k^2 - k + 1$
7. $u^2 - 22u + 121$
8. $v^2 + 24v + 144$
9. $9r^2 - 12r + 4$
10. $49s^2 - 14s + 1$
11. $49y^2 + 14y + 1$
12. $9x^2 + 12x + 4$
13. $81m^2 + 18m + 1$
14. $64n^2 - 80n + 25$
15. $18h^2 + 48h + 32$
16. $12k^2 - 36k + 27$
17. $50m^4 + 40m^3 + 8m^2$
18. $12n^4 + 72n^3 + 108n^2$
19. $16w^2z^2 + 80wz + 100$

20. $36u^2v^2 - 24uv + 4$ **21.** $4t^2 - 28tr + 49r^2$ **22.** $16x^2 - 40xy + 25y^2$

23. $16p^2 + 40pq + 25q^2$ **24.** $4w^2 + 28wz + 49z^2$ **25.** $12m^2n - 36mn^2 + 27n^3$

26. $24h^3k^2 - 24h^4k + 6h^5$ **27.** $24s^6t^3 + 24s^5t^4 + 6s^4t^5$ **28.** $12x^3y^3 + 36x^2y^4 + 27xy^5$

29. $w^2 - 9$ **30.** $z^2 - 16$ **31.** $p^2 - 49$ **32.** $q^2 - 100$

33. $k^2 - 225$ **34.** $h^2 - 169$ **35.** $25z^2 - 1$ **36.** $49w^2 - 16$

37. $16x^2 - 64$ **38.** $9y^2 - 81$ **39.** $5v^3 - 180v$ **40.** $25u^4 - 49u^2$

41. $r^2 + 4t^2$ **42.** $100h^2 + k^2$ **43.** $36p^2 - 225q^2$ **44.** $64h^2 - 121k^2$

45. $4t^2 - 196r^2$ **46.** $36m^2 - 25n^2$ **47.** $24x^2y - 54y$ **48.** $50w^2z^3 - 72z^3$

49. $32r^4s - 8r^2s^3$ **50.** $16uv^4 - 36u^3v^2$ **51.** $6p^3q - 150pq^3$ **52.** $32h^5k^2 - 128h^3k^4$

53. $n^4 - 144$ **54.** $m^4 - 169$ **55.** $16y^4 - 1$ **56.** $9x^4 - 9$

57. $3r^4 - 243$ **58.** $5s^4 - 80$

Preparing for Section 5.5

In Exercises 59–66, solve the given equation.

59. $n + 5 = 0$ **60.** $m - 11 = 0$ **61.** $2p - 3 = 0$ **62.** $5q + 1 = 0$

63. $8u + 16 = 0$ **64.** $6v - 24 = 0$ **65.** $7k - 12 = 0$ **66.** $12h + 5 = 0$

5.5 Quadratic Equations and Applications

FOCUS

We apply the factoring methods of the preceding sections to solve quadratic equations and applications involving quadratic equations.

A baseball player, celebrating an important victory, throws a ball vertically upward with an initial velocity of 96 feet per second (ft/s) (about 65 mph). When will the ball return to the ground?

To solve this problem, we must use a formula from physics. If an object is projected vertically upward with an initial velocity of v feet per second, then (ignoring air resistance) its height h in feet after t seconds is given by the formula

$$h = vt - 16t^2$$

For our baseball problem, $v = 96$ ft/s, so the height of the ball after t seconds is

$$h = 96t - 16t^2$$

Since $h = 0$ ft when the ball returns to the ground, we must find a value of t such that

$$0 = 96t - 16t^2$$

How can we find this value of t? One strategy is to substitute values for t in the formula $h = 96t - 16t^2$ until one is found that makes $h = 0$. Integer values of t and the corresponding values of h are displayed in the following table:

Time t	Height $h = 96t - 16t^2$
0	$h = 96 \cdot 0 - 16 \cdot 0 = 0$
1	$h = 96 \cdot 1 - 16 \cdot 1 = 80$
2	$h = 96 \cdot 2 - 16 \cdot 4 = 128$
3	$h = 96 \cdot 3 - 16 \cdot 9 = 144$
4	$h = 96 \cdot 4 - 16 \cdot 16 = 128$
5	$h = 96 \cdot 5 - 16 \cdot 25 = 80$
6	$h = 96 \cdot 6 - 16 \cdot 36 = 0$

Notice that $h = 0$ ft when $t = 0$ s. That's the situation just before the player has thrown the ball upward. But also $h = 0$ ft when $t = 6$ s, so it will take 6 s for the ball to return to the ground.

One way to understand the solution is to make a graph of the table. In Figure 5.1 we have plotted the ordered pairs (t, h) from the table and connected the points with a smooth curve. The graph shows that the ball reaches its highest point after 3 s, and then begins to fall. It hits the ground at $t = 6$ s.

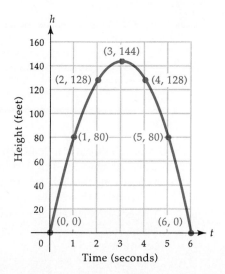

FIGURE 5.1

Solution by Factoring

The equation $0 = 96t - 16t^2$ can also be solved by factoring, using the **zero product property** of real numbers:

Zero product property

If a and b are real numbers such that $ab = 0$, then either $a = 0$ or $b = 0$ or both.

The zero product property says that if the product of two numbers is zero, then at least one of the numbers must be zero.

   To solve the baseball problem by factoring, we first rewrite the equation $0 = 96t - 16t^2$ as $16t^2 - 96t = 0$. We now factor the left side of the equation:

$$16t^2 - 96t = 0$$
$$16t(t - 6) = 0$$

Since the product of the factors $16t$ and $t - 6$ is zero, the zero product property tells us that at least one of these factors must be zero. That is,

$$16t = 0 \quad \text{or} \quad t - 6 = 0$$

We solve both these equations to obtain the possible solutions:

$$t = 0 \quad \text{or} \quad t = 6$$

Each solution must be checked by substitution into the original equation, $16t^2 - 96t = 0$. Both values check, so both values of t are solutions of the equation. However, only the solution $t = 6$ s is appropriate for the baseball problem.

   Equations that contain a squared term and no terms of higher degree, like $16t^2 - 96t = 0$, are called *quadratic equations*:

Quadratic equation

An equation that can be expressed in the form

$$ax^2 + bx + c = 0$$

where a, b, and c are real numbers and $a \neq 0$, is a **quadratic equation.**

A quadratic equation is said to be in **standard form** when it is in the form $ax^2 + bx + c = 0$, with powers of the variable in descending order on one side, and zero on the other side. The quadratic equation $16t^2 - 96t = 0$ is in standard form.

The solution of the baseball problem suggests the following procedure for solving a quadratic equation by factoring:

To solve a quadratic equation by factoring

1. Write the equation in standard form.
2. Factor it completely.
3. Set each factor equal to zero, and solve the resulting equations.
4. Check each solution in the original equation.

EXAMPLE 1

Solve each equation.

a. $x^2 - 7x + 12 = 0$ b. $2y^2 - 2 = 3y$
c. $3z^2 = 12z$ d. $4z^3 + 12z^2 + 8z = 0$

Solution

a. The quadratic equation $x^2 - 7x + 12 = 0$ is already in standard form. We factor the left side of the equation to obtain

$$(x - 3)(x - 4) = 0$$

Because the product of the factors $x - 3$ and $x - 4$ is zero, at least one of the factors must be zero. Therefore, we write

$$x - 3 = 0 \quad \text{or} \quad x - 4 = 0$$

We solve these equations to obtain

$$x = 3 \quad \text{or} \quad x = 4$$

We check each value by substitution into the original equation $x^2 - 7x + 12 = 0$.

If $x = 3$, then: $x^2 - 7x + 12 = 0$

$$9 - 21 + 12 = 0$$

$$0 = 0 \qquad \text{True}$$

If $x = 4$, then: $x^2 - 7x + 12 = 0$

$$16 - 28 + 12 = 0$$

$$0 = 0 \qquad \text{True}$$

Both $x = 3$ and $x = 4$ are solutions.

b. The quadratic equation $2y^2 - 2 = 3y$ is not in standard form because the term $3y$ is on the right side of the equation. We subtract $3y$ from both sides of the equation to obtain the standard form:

$$2y^2 - 3y - 2 = 0$$

We factor the left side of the equation:

$$(2y + 1)(y - 2) = 0$$

Then we apply the zero product property and solve:

$$2y + 1 = 0 \qquad \text{or} \qquad y - 2 = 0$$
$$2y = -1 \qquad\qquad\qquad y = 2$$
$$y = -\frac{1}{2}$$

Each of these values can be checked by substitution into the original equation $2y^2 - 2 = 3y$. Both are solutions.

c. The quadratic equation $3z^2 = 12z$ is not in standard form because the term $12z$ is on the right side of the equation. We subtract $12z$ from both sides of the equation to obtain the standard form:

$$3z^2 - 12z = 0$$

in which the constant term is zero. We factor the left side of the equation:

$$3z(z - 4) = 0$$

and apply the zero product property and solve:

$$3z = 0 \qquad \text{or} \qquad z - 4 = 0$$
$$z = 0 \qquad\qquad\qquad z = 4$$

You should check that each value is a solution by substitution into the original equation, $3z^2 = 12z$.

d. The equation $4z^3 + 12z^2 + 8z = 0$ is not a quadratic equation because it contains the third-degree term $4z^3$. Nevertheless, this equation can be solved by factoring. We begin by factoring out $4z$, the greatest common factor of the terms of $4z^3 + 12z^2 + 8z$. This gives us

$$4z(z^2 + 3z + 2) = 0$$

The trinomial $z^2 + 3z + 2$ can be factored as $(z + 1)(z + 2)$, so we have

$$4z(z + 1)(z + 2) = 0$$

This tells us that the product of the factors $4z$, $z + 1$, and $z + 2$ is zero. By an extension of the zero product property, we know that this product can be zero only if at least one of the factors is zero. Therefore we set each factor equal to zero and solve:

$$4z = 0 \qquad \text{or} \qquad z + 1 = 0 \qquad \text{or} \qquad z + 2 = 0$$
$$z = 0 \qquad\qquad\qquad z = -1 \qquad\qquad\qquad z = -2$$

Again, check that these values are solutions by substitution into the original equation, $4z^3 + 12z^2 + 8z = 0$. ■

BE CAREFUL!

Do not solve a quadratic equation by dividing both sides by the variable, because then you may lose one solution:

$$3z^2 = 12z \qquad \text{Original equation}$$
$$\frac{3z^2}{3z} = \frac{12z}{3z} \qquad \text{Dividing both sides by } 3z$$
$$z = 4$$

But the result is only a single solution. *The solution $z = 0$ has been lost* because, when we divide by $3z$, we assume that $z \neq 0$ (division by zero is undefined). Remember to put a quadratic equation in standard form before solving.

Applications Involving Quadratic Equations

Quadratic equations frequently arise in applications. One example is the baseball problem at the beginning of this section. Another application is given in the following example.

EXAMPLE 2

The length of a rectangular carpet is 1 foot (ft) more than twice its width. The area of the carpet is 55 ft^2. Find the length and the width of the carpet.

Solution

Let w represent the width of the carpet. Since the length is 1 ft more than twice the width, the length is $2w + 1$. It helps to make a sketch, so we draw a rectangle to represent the carpet, and label its length and width, as in Figure 5.2.

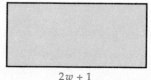

FIGURE 5.2

Recall that the area of a rectangle equals the length times the width. Since the area of the carpet is 55 ft^2, we can write the following equation:

$$\text{Area} = \text{Length} \cdot \text{Width}$$
$$55 = (2w + 1)w$$

We can simplify the right side of the equation to obtain

$$55 = 2w^2 + w$$

This is a quadratic equation that we can write in standard form (in this case $0 = ax^2 + bx + c$) by subtracting 55 from both sides:

$$0 = 2w^2 + w - 55$$

To solve it, we factor and apply the zero product property:

$$0 = (2w + 11)(w - 5)$$
$$0 = 2w + 11 \quad \text{or} \quad 0 = w - 5$$
$$-11 = 2w \qquad\qquad 5 = w$$
$$-\frac{11}{2} = w$$

We ignore the solution $w = -\dfrac{11}{2}$ because the width cannot be negative.

Therefore the width of the carpet is 5 ft and the length of the carpet is $2w + 1 = 10 + 1 = 11$ ft. We check by computing the area of the carpet: $A = lw = 11 \cdot 5 = 55$ ft^2. ∎

BE CAREFUL!

To apply the zero product property, we must have a *zero product*. It is *incorrect* to try to solve the equation

$$55 = (2w + 1)w$$

by writing

$$55 = 2w + 1 \quad \text{or} \quad 55 = w$$
$$54 = 2w$$
$$27 = w$$

Neither $w = 27$ nor $w = 55$ checks in the original equation $55 = (2w + 1)w$, because the product of the factors $2w + 1$ and w is not zero.

We close this section by noting that many quadratic equations cannot be solved by factoring. For example, we cannot solve the quadratic equation $x^2 + 2x - 2 = 0$ by factoring because $x^2 + 2x - 2$ is a prime polynomial. However, there are other methods we can use for solving such an equation. In fact, in Chapter 10 you will learn a method for solving *any* quadratic equation.

DISCUSSION QUESTIONS 5.5

1. What is the zero product property?

2. Is $2x(x^2 - 1) = 0$ a quadratic equation? Explain.

3. Is $2y(y - 1) = 4$ a quadratic equation? If so, is it in standard form? Explain your answers.

4. Do the equations $4z^2 = 10z$ and $2z = 5$ have the same solutions? Explain.

EXERCISES 5.5

In Exercises 1–12, solve the given equation by applying the zero product property.

1. $x(x + 3) = 0$

2. $y(y - 5) = 0$

3. $(w + 6)(w - 4) = 0$

4. $(z - 8)(z - 11) = 0$

5. $(2p - 1)(p - 5) = 0$

6. $(3q + 2)(q - 7) = 0$

7. $(5h - 8)(2h + 3) = 0$

8. $(6k + 1)(3k + 10) = 0$

9. $r(r + 4)(r + 7) = 0$

10. $t(t - 8)(t + 3) = 0$

11. $3m(m - 4)(2m - 5) = 0$

12. $4n(3n + 7)(n - 1) = 0$

In Exercises 13–48, solve the given equation.

13. $z^2 + 5z + 6 = 0$

14. $w^2 + 6w + 8 = 0$

15. $p^2 - 2p - 24 = 0$

16. $q^2 - q - 12 = 0$

17. $6m^2 - m - 2 = 0$

18. $3n^2 + 5n - 2 = 0$

19. $3r^2 + 7r - 20 = 0$

20. $2t^2 - 15t + 18 = 0$

21. $q^2 + 16 = -8q$

22. $p^2 + 64 = 16p$

23. $2y^2 - 10 = y$

24. $2x^2 - 20 = 3x$

25. $25x^2 = 20x - 4$

26. $9y^2 + 20y = -4$

27. $16p^2 - 25 = 0$

28. $9q^2 - 36 = 0$

29. $n^2 - 8n = 0$

30. $m^2 + 5m = 0$

31. $4h^2 = 48h$

32. $2k^2 = -16k$

33. $5t^2 = -100t$

34. $6s^2 = 36s$

35. $k(k + 1) = 4$

36. $h(h + 2) = 5$

37. $x(2x - 3) = 20$

38. $y(3y - 5) = 12$

39. $3q(2q - 5) = 10 - 4q$

40. $14p - 21 = p(2p - 3)$ **41.** $(r - 3)(r + 1) = 12$ **42.** $(s - 2)(s - 3) = 12$

43. $(h + 2)^2 = 100 - h$ **44.** $k = (k - 6)^2 + 4$ **45.** $s^3 + 22s^2 + 121s = 0$

46. $t^3 - 18t^2 + 81t = 0$ **47.** $y^3 - 2y^2 - 8y = 0$ **48.** $x^3 - 8x^2 + 15x = 0$

49. One number is three more than another. Their product is 28. What are the two numbers?

50. One number is four less than another. Their product is 45. What are the two numbers?

51. One number is four times another. Their product is 144. What are the two numbers?

52. One number is three times another. Their product is 75. What are the two numbers?

53. A toy rocket is shot vertically upward with an initial velocity of 96 ft/s.
   a. When will the rocket be 144 ft above the ground?
   b. When will the rocket hit the ground?

54. A bullet is shot vertically upward with an initial velocity of 1200 ft/s.
   a. When will the bullet be 10,400 ft above the ground?
   b. When will the bullet hit the ground?

55. The cost C in dollars to manufacture x hundred videotapes is given by the formula $C = 1400 + 300x - 100x^2$. If the cost is $1600, how many videotapes can be manufactured?

56. The cost C in dollars to manufacture y thousand pens is given by the formula $C = 49.25 + 1.5y - y^2$. If the cost is $49.75, how many pens can be manufactured?

57. The number of calculators n that an electronics store sells per week is related to the price per calculator p by the equation $n = 2200 - 100p$. At what price should the store sell the calculators if they want a weekly revenue of $11,200? (Remember: Revenue $= np$.)

58. The number of pairs of earrings n that a jewelry store sells per day is related to the price per pair by the equation $n = 260 - 20p$. At what price should the store sell the earrings if it wants a daily revenue of $800?

59. The area of a rectangle is 28 square centimeters (cm^2). The length of the rectangle is 1 cm less than twice its width. Find the length and the width of the rectangle.

60. The area of a rectangle is 30 square inches (in.^{2}). The length of the rectangle is 1 in. more than three times its width. Find the length and the width of the rectangle.

61. The base of a triangle is twice its height. The area of the triangle is 100 in.^{2}. Find the base and the height of the triangle.

62. The height of a triangle is 5 cm more than its base. The area of the triangle is 12 cm^2. Find the base and the height of the triangle.

63. A 12-ft by 15-ft rug is placed in the center of a rectangular living room, leaving uncovered a strip of floor of uniform width around each edge. The area of the living room is 378 ft^2. How wide is the strip? (Hint: See Figure 5.3.)

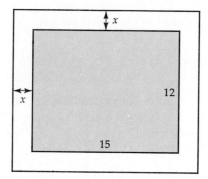

FIGURE 5.3

64. A gardener must mow a 30-ft by 40-ft rectangular lawn. She decides to mow a uniform strip along each edge of the lawn and then take a break. If she mows half the area of the lawn before taking her break, how wide is the strip?

65. The area of the entire figure in Figure 5.4, including the shaded triangle, is 70. Find the area of the shaded triangle.

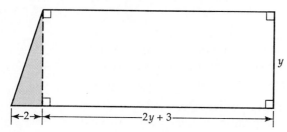

FIGURE 5.4

Preparing for Section 6.1

In Exercises 66–69, evaluate the given expression when $x = -3$ and $y = 2$.

66. $\dfrac{6x^2}{y}$

67. $\dfrac{xy}{12}$

68. $\dfrac{4xy^2}{54}$

69. $\dfrac{-24x^3}{36y^2}$

In Exercises 70–73, simplify the given expression.

70. $\dfrac{x^2}{x^6}$ **71.** $\dfrac{18y^5}{30y^2}$ **72.** $\dfrac{12w^2z^3}{8wz^4}$ **73.** $\dfrac{42p^6q^2}{63p^4q^7}$

Chapter Summary

Important Ideas

Factoring perfect-square trinomials

$$a^2 + 2ab + b^2 = (a + b)^2$$
$$a^2 - 2ab + b^2 = (a - b)^2$$

Factoring the difference of two squares

$$a^2 - b^2 = (a + b)(a - b)$$

Zero product property
If a and b are real numbers such that $ab = 0$, then $a = 0$ or $b = 0$ or both.

To solve a quadratic equation by factoring
a. Write the equation in standard form.
b. Factor it completely.
c. Set each factor equal to zero, and solve the resulting equations.
d. Check each solution in the original equation.

Chapter Review

Section 5.1

In Exercises 1–2, write the given number in prime factored form.

1. 234   **2.** 532

In Exercises 3–4, find the greatest common factor of each set of integers.

3. 76, 190   **4.** 220, 260, 340

In Exercises 5–10, factor the greatest common factor out of the given polynomial.

5. $14x^2 + 35x$ **6.** $y^3 - y^2$ **7.** $9z^2 + 3z + 6$

8. $10u^5 - 5u^3 + 25u$ **9.** $36p^4 - 27p^3 + 18p^2$ **10.** $w^2z^6 + 3w^2z^4 - 6w^2z^2$

In Exercises 11–14, factor the given polynomial by grouping.

11. $v^3 + 7v^2 + v + 7$ **12.** $3st - 8s + 9t - 24$

13. $6r^2 + 8r - 15r - 20$ **14.** $2n^2 + 10n + 7n + 35$

Section 5.2

In Exercises 15–22, completely factor the given trinomial.

15. $s^2 + 14s + 13$ **16.** $x^2 - 11x + 30$ **17.** $k^2 + 3k - 40$ **18.** $r^2 - 6r - 72$

19. $h^2 + 5h - 84$ **20.** $s^2 - 13s - 48$ **21.** $3w^2 - 45w - 162$ **22.** $u^5 + 10u^4 - 56u^3$

Section 5.3

In Exercises 23–30, completely factor the given trinomial.

23. $2x^2 - 17x + 30$ **24.** $5w^2 + 21w + 4$ **25.** $4y^2 + y - 3$ **26.** $3z^2 - 24z + 20$

27. $6k^2 - 17k - 45$ **28.** $12t^2 + 25t + 7$ **29.** $21s^4 + 44s^3 + 20s^2$ **30.** $18r^7 - 66r^6 - 24r^5$

Section 5.4

In Exercises 31–42, completely factor the given polynomial.

31. $t^2 - 18t + 81$ **32.** $s^2 - 14s + 49$ **33.** $4k^2 + 28k + 49$

34. $32h^2 + 42h + 10$ **35.** $36x^2 + 48xy + 16y^2$ **36.** $25p^2q^2 + 20pq + 4$

37. $27m^3n^2 - 72m^2n + 48m$ **38.** $75w^3z - 90w^2z^2 + 27wz^3$ **39.** $100 - u^2$

40. $v^2 - \dfrac{1}{16}$ **41.** $48s^2t - 108t^3$ **42.** $q^4 - 625$

Section 5.5

In Exercises 43–54, solve the given equation.

43. $(n + 3)(n - 2) = 0$ **44.** $h^2 - 15h + 54 = 0$ **45.** $m^2 + 2m = 48$

46. $12 - x - x^2 = 0$ **47.** $4k^2 + 13k + 3 = 0$ **48.** $3r^2 + 16r = 35$

49. $25s^2 + 81 = 90s$ **50.** $5t(3t + 2) = 21t + 14$ **51.** $16q^2 - 256 = 0$

52. $8p^2 = 72p$ **53.** $100x^3 = x$ **54.** $28y^3 + 84y^2 + 63y = 0$

55. The product of two consecutive integers is 132. What are the integers?

56. The square of a negative number is 18 more than 3 times the negative number. Find the number.

57. A ball is thrown vertically upward with an initial velocity of 48 ft/s.
a. When will the ball be 32 ft above the ground?
b. When will the ball hit the ground?

58. The revenue R in dollars made by an airline when x seats are unsold on a certain airplane is given by the formula $R = x^2 + 30x + 139{,}000$. If the revenue is \$140,000, how many seats are unsold?

59. The length of a rectangle is 5 cm less than three times its width. The area of the rectangle is 78 cm^2. Find the length and the width of the rectangle.

60. The base of a triangle is 4 in. more than twice its height. The area of the triangle is 35 in.^{2}. Find the base and the height of the triangle.

61. An 8-in. by 12-in. photograph is surrounded by a frame of uniform width. The area of the photograph and the frame is 140 in.^{2}. Find the width of the frame.

62. A rectangular swimming pool has a length of 60 ft and a width of 40 ft. The pool is surrounded by a concrete path of uniform width. If the area of the concrete path is 180 ft^2, what is its width?

Chapter 5 Test

1. Write 5544 in prime factored form.

2. Find the greatest common factor of 2100 and 5580.

In Problems 3–14, completely factor the given polynomial.

3. $6x^2 - 14x$ **4.** $w^2 - 9w - 52$ **5.** $2z^2 + 14z + 20$

6. $8y^6 - 20y^4 + 4y^2$ **7.** $9s^2 + 42s + 49$ **8.** $14s^3 + 94s^2 - 28s$

9. $6p^2 + 7pq - 5q^2$ **10.** $h^4k^3 - 4h^3k^4 + 4h^2k^5$ **11.** $12m^2 - 24mn - 36mn + 96n^2$

12. $9r^4t^2 + 64r^2t^4$ **13.** $81y^2 - 16$ **14.** $27u^2v^2 - 363$

In Problems 15–23, solve the given equation.

15. $x^2 - 14x + 24 = 0$ **16.** $3y^2 + 11y - 4 = 0$ **17.** $2w^2 + 2 = 5w$

18. $16k^2 = 16k - 4$ **19.** $25z^2 = 100z$ **20.** $(p + 4)(p - 6) = 56$

21. $9(5q + 3) = 6q - 10q^2$ **22.** $h^3 - 49h = 0$ **23.** $42n^3 + 114n^2 - 36n = 0$

24. The sum of the squares of two consecutive integers is 145. Find the two integers.

25. The base of a parallelogram is 1 in. less than three times its height. The area of the parallelogram is 70 in.^{2}. Find the base and the height of the parallelogram.

6 *Rational Expressions*

6.1 Simplifying Rational Expressions

FOCUS

We discuss the definition of rational expressions. Then we simplify rational expressions by factoring.

Political pollsters found that the percentage of voters that recognize a certain candidate's name x months after the campaign started is given by

$$\text{Percentage} = \frac{0.2x^2 + 7.2x + 8}{x^2 + 36}$$

The expression $\dfrac{0.2x^2 + 7.2x + 8}{x^2 + 36}$ represents the division of $0.2x^2 + 7.2x + 8$ by $x^2 + 36$. It is, in fact, an algebraic fraction with the polynomial $0.2x^2 + 7.2x + 8$ as its numerator and the polynomial $x^2 + 36$ as its denominator, and hence is an example of a rational expression.

Rational expression

A **rational expression** is an expression of the form

$$\frac{A}{B}$$

where A and B are polynomials and $B \neq 0$.

A polynomial P can be considered to be a rational expression with a denominator of one: $P = \dfrac{P}{1}$.

In evaluating rational expressions, we must be careful of zero denominators.

EXAMPLE 1

Evaluate $\dfrac{x^2 + 5x + 4}{x + 2}$ for each of the following values of x:

a. $x = 2$ **b.** $x = 0$ **c.** $x = -2$

Solution

We substitute the given value and then evaluate.

a. When $x = 2$:

$$\frac{x^2 + 5x + 4}{x + 2} = \frac{4 + 10 + 4}{2 + 2}$$

$$= \frac{18}{4}$$

$$= \frac{9}{2} \text{ or } 4.5$$

b. When $x = 0$:

$$\frac{x^2 + 5x + 4}{x + 2} = \frac{0 + 0 + 4}{0 + 2}$$

$$= \frac{4}{2}$$

$$= 2$$

c. When $x = -2$:

$$\frac{x^2 + 5x + 4}{x + 2} = \frac{4 - 10 + 4}{(-2) + 2}$$

$$= \frac{-2}{0}$$

Since the denominator of a fraction cannot be zero, the rational expression $\dfrac{x^2 + 5x + 4}{x + 2}$ is *undefined* when $x = -2$.

■

As illustrated in the solution of part c of Example 1, a rational expression is undefined when a numerical value of the variable produces a zero denominator. When working with rational expressions, you should determine all values of the variable that make the denominator zero.

EXAMPLE 2   Find all values of the variable for which each rational expression is undefined.

a. $\dfrac{x-3}{9}$ b. $\dfrac{3y^2 - 4y + 5}{y - 6}$ c. $\dfrac{2z + 7}{z^2 - 4z + 3}$ d. $\dfrac{8w^2}{w^2 + 1}$

Solution

a. Since the denominator of $\dfrac{x-3}{9}$ is 9 for every value of x, there are no values of x that make $\dfrac{x-3}{9}$ undefined. We also say that $\dfrac{x-3}{9}$ is defined for every value of x.

b. The rational expression $\dfrac{3y^2 - 4y + 5}{y - 6}$ is undefined when the denominator, $y - 6$, is equal to zero. We solve the equation $y - 6 = 0$ to obtain $y = 6$, meaning that the rational expression $\dfrac{3y^2 - 4y + 5}{y - 6}$ is undefined when $y = 6$. We also say that $\dfrac{3y^2 - 4y + 5}{y - 6}$ is defined for all $y \neq 6$.

c. The rational expression $\dfrac{2z + 7}{z^2 - 4z + 3}$ is undefined when $z^2 - 4z + 3 = 0$. Solving this equation gives

$$z^2 - 4z + 3 = 0$$
$$(z - 1)(z - 3) = 0$$
$$z - 1 = 0 \quad \text{or} \quad z - 3 = 0$$
$$z = 1 \qquad\qquad z = 3$$

Therefore $\dfrac{2z + 7}{z^2 - 4z + 3}$ is undefined when $z = 1$ or when $z = 3$.

d. Since $w^2 + 1$ is a positive number for every value of w, there are no values of w for which the rational expression $\dfrac{8w^2}{w^2 + 1}$ is undefined.   ■

The fractional form of rational expressions suggests that the properties of fractions apply to rational expressions. Recall that you can simplify fractions by dividing the numerator and the denominator by the same nonzero number. Similarly, for rational expressions we have the following property:

Fundamental principle for rational expressions

If $\dfrac{A}{B}$ is a rational expression with $B \neq 0$, and if K is a nonzero expression, then

$$\frac{A \cdot K}{B \cdot K} = \frac{A}{B}$$

In other words, a common factor of the numerator and denominator can be divided out of a rational expression. For example, to simplify $\dfrac{6x^2}{9x}$ we can first write the numerator and the denominator in factored form:

$$\frac{6x^2}{9x} = \frac{2 \cdot 3 \cdot x \cdot x}{3 \cdot 3 \cdot x}$$

Both 3 and x are common factors of the numerator and the denominator. By the fundamental principle we can divide out these factors:

$$\frac{6x^2}{9x} = \frac{2 \cdot \cancel{3} \cdot \cancel{x} \cdot x}{3 \cdot \cancel{3} \cdot \cancel{x}} \qquad \text{Dividing out 3 and } x$$

$$= \frac{2x}{3}$$

We can check this result by substituting a number, say $x = 1$, into each side of the last equation:

$$\frac{6x^2}{9x} = \frac{6}{9} = \frac{2}{3} \qquad \text{and} \qquad \frac{2x}{3} = \frac{2}{3}$$

Both sides of the equation have the same value, which suggests that $\dfrac{6x^2}{9x} = \dfrac{2x}{3}$.

In the example above we can divide out x from the numerator and the denominator because we have assumed that $x \neq 0$ (if $x = 0$, then $\dfrac{6x^2}{9x}$ is undefined). We will assume from now on that the only numerical values permitted for the variable are those for which the rational expression is defined.

A rational expression is **reduced to lowest terms** when its numerator and denominator contain no common factors. The procedure for doing so is as follows:

To reduce a rational expression to lowest terms

1. Completely factor the numerator and the denominator of the rational expression.
2. Divide out any common factors of the numerator and the denominator.

EXAMPLE 3

Reduce each rational expression to lowest terms.

a. $\dfrac{4p - 20}{5p - 25}$ **b.** $\dfrac{3r + 6}{r^2 - 4}$ **c.** $\dfrac{q^2 + 2q - 3}{q^2 - 5q + 4}$ **d.** $\dfrac{t - 3}{3 - t}$

Solution

a. We reduce $\dfrac{4p - 20}{5p - 25}$ to lowest terms by factoring the numerator and the denominator and then dividing out any common factors:

$$\frac{4p - 20}{5p - 25} = \frac{4(p - 5)}{5(p - 5)} = \frac{4}{5} \qquad \text{Dividing out } p - 5$$

b. We factor the numerator and the denominator and then divide out any common factors:

$$\frac{3r + 6}{r^2 - 4} = \frac{3(r + 2)}{(r + 2)(r - 2)} = \frac{3}{r - 2} \qquad \text{Dividing out } r + 2$$

c. $\dfrac{q^2 + 2q - 3}{q^2 - 5q + 4} = \dfrac{(q - 1)(q + 3)}{(q - 1)(q - 4)} = \dfrac{q + 3}{q - 4} \qquad \text{Dividing out } q - 1$

d. Here, $t - 3$ and $3 - t$ are *opposites* because $3 - t = -t + 3 = (-1)(t - 3)$. Therefore

$$\frac{t - 3}{3 - t} = \frac{1(t - 3)}{(-1)(t - 3)} = \frac{1}{-1} = -1 \qquad \text{Dividing out } t - 3$$

Remember, a rational expression is equal to -1 when the numerator and the denominator are opposites.   ∎

> ### BE CAREFUL!
>
> When reducing a rational expression to lowest terms, do not "cancel" as in the following *incorrect* example:
>
> $$\frac{x}{x^2 + x} = \frac{\cancel{x}}{x^2 + \cancel{x}} = \frac{1}{x^2 + 1}$$
>
> Only common *factors* can be cancelled. Remember to first factor the numerator and the denominator, and then divide out any common factors.

DISCUSSION QUESTIONS 6.1

1. Is a polynomial a rational expression? Explain.

2. When is a rational expression undefined? How do we find the values for which it is undefined?

3. Is the rational expression $\dfrac{y - x}{x - y}$ reduced to lowest terms? Explain.

EXERCISES 6.1

In Exercises 1–8, evaluate the given rational expression when:
a. $x = 2$.   **b.** $x = -2$.

1. $\dfrac{x - 1}{x}$

2. $\dfrac{x + 2}{2x}$

3. $\dfrac{x^2}{2x + 1}$

4. $\dfrac{4x}{x - 2}$

5. $\dfrac{x + 1}{x^2 - 4}$

6. $\dfrac{x^2 - 3}{3x - 6}$

7. $\dfrac{3x - 2}{x^2 + 4x - 1}$

8. $\dfrac{2x + 3}{x^2 + 3x + 5}$

In Exercises 9–16, find all values of the variable for which the given rational expression is undefined.

9. $\dfrac{2}{x - 1}$

10. $\dfrac{4}{y + 2}$

11. $\dfrac{3z}{6z + 9}$

12. $\dfrac{5w}{4w - 10}$

13. $\dfrac{3p - 4}{p^2 - 6p + 8}$

14. $\dfrac{q + 6}{q^2 - 8q + 12}$

15. $\dfrac{2r^2}{r^2 + 4}$

16. $\dfrac{3t^2}{t^2 + 9}$

In Exercises 17–50, reduce the given rational expression to lowest terms.

17. $\dfrac{6h^2}{2h^3}$

18. $\dfrac{3k^5}{9k^2}$

19. $\dfrac{8x^7}{12x^4}$

20. $\dfrac{18y}{12y^3}$

21. $\dfrac{28p^6q^3}{35p^4q}$

22. $\dfrac{54r^8s^2}{36r^2s^4}$

23. $\dfrac{4y-8}{y-2}$

24. $\dfrac{6x+18}{x+3}$

25. $\dfrac{q-4}{3q-12}$

26. $\dfrac{p-5}{6p-30}$

27. $\dfrac{3z+12}{7z+28}$

28. $\dfrac{5w+10}{4w+8}$

29. $\dfrac{6k^2}{3k^2+18k}$

30. $\dfrac{10h^3}{4h^4-6h^2}$

31. $\dfrac{r^2-2r}{r^3-r^2}$

32. $\dfrac{s^3+2s}{s^2+4s}$

33. $\dfrac{m-5}{5-m}$

34. $\dfrac{6-n}{n-6}$

35. $\dfrac{p^2-25}{5-p}$

36. $\dfrac{6-q}{q^2-36}$

37. $\dfrac{1+w^2}{w^2-1}$

38. $\dfrac{z^2-4}{4+z^2}$

39. $\dfrac{h-3}{h^2-7h+12}$

40. $\dfrac{k+2}{k^2+2k-15}$

41. $\dfrac{y^2-4}{y^2+8y+12}$

42. $\dfrac{x^2-9}{x^2+3x-18}$

43. $\dfrac{q^2-6q+9}{3q^2-9q}$

44. $\dfrac{4p^2+8p}{p^2+4p+4}$

45. $\dfrac{z^2+4z+3}{z^2+8z+15}$

46. $\dfrac{w^2+3w-10}{w^2-8w-20}$

47. $\dfrac{4m^2-m-3}{3m^2+m-4}$

48. $\dfrac{4n^2-7n-2}{5n^2-9n-2}$

49. $\dfrac{m^2+3mn+m+3n}{3n^2+6mn}$

50. $\dfrac{p^2q+2q-p^2t-2t}{pq-pt}$

51. A communications company found that the total number of subscribers (in thousands) to a cable television system t months after installation of the system is given by the formula

$$\text{Total number of subscribers} = \frac{250t}{t+5}$$

   a. How many subscribers are there 3 months after installation of the system? after 6 months? after 1 year?
   b. After how many months will there be more than 195,000 subscribers?

52. Medical researchers have found that the concentration in milligrams per cubic centimeter of a particular drug in a person's bloodstream is given by the formula

$$\text{Concentration} = \frac{0.2t}{t^2+4t+4}$$

   where t is the number of hours after the drug has been taken orally.
   a. What is the concentration of the drug after 2 hours (h)? after 4 h? after 8 h?
   b. After how many hours will the concentration be less than 0.0014 milligrams per cubic centimeter (mg/cm³)?

Preparing for Section 6.2

Recall the rules for multiplying and dividing fractions:

$$\frac{a}{b}\cdot\frac{c}{d}=\frac{a\cdot c}{b\cdot d} \quad \text{and} \quad \frac{a}{b}\div\frac{c}{d}=\frac{a}{b}\cdot\frac{d}{c}$$

In Exercises 53–58, perform the indicated operation. Write the result in lowest terms.

53. $\dfrac{1}{3} \cdot \dfrac{4}{5}$ **54.** $\dfrac{3}{7} \cdot \dfrac{2}{5}$ **55.** $\dfrac{9}{10} \cdot \dfrac{5}{6}$ **56.** $\dfrac{15}{21} \cdot \dfrac{14}{20}$ **57.** $\dfrac{3}{4} \div \dfrac{2}{3}$ **58.** $\dfrac{5}{8} \div \dfrac{4}{15}$

6.2 Multiplication and Division of Rational Expressions

FOCUS

We discuss how to multiply and divide rational expressions. The procedures are similar to those for multiplying and dividing fractions.

To multiply two fractions, we multiply the numerators and multiply the denominators. For example,

$$\frac{8}{15} \cdot \frac{3}{4} = \frac{8 \cdot 3}{15 \cdot 4} = \frac{24}{60}$$

The result, $\dfrac{24}{60}$, is not in lowest terms. We can reduce it to lowest terms by dividing out the common factors of 24 and 60:

$$\frac{24}{60} = \frac{\not{2} \cdot \not{2} \cdot 2 \cdot \not{3}}{\not{2} \cdot \not{2} \cdot \not{3} \cdot 5} = \frac{2}{5}$$

The procedure for multiplying rational expressions is essentially the same as that for multiplying fractions:

To multiply rational expressions

1. Completely factor the numerator and the denominator of each rational expression.
2. Multiply the numerators and multiply the denominators.
3. Reduce the result of step 2 to lowest terms.

EXAMPLE 1   Multiply the rational expressions.

a. $\dfrac{2x}{3x + 9} \cdot \dfrac{5x + 15}{6x^2}$ **b.** $\dfrac{y^2 + 5y}{2y + 4} \cdot \dfrac{7y + 14}{y^2 - 25}$ **c.** $\dfrac{w^2 + 4w - 5}{w^2 - 4w + 3} \cdot \dfrac{4w^2 - 12w}{w^2 + 8w + 15}$

Solution | We follow the procedure outlined above.

a. $\dfrac{2x}{3x + 9} \cdot \dfrac{5x + 15}{6x^2} = \dfrac{2 \cdot x}{3 \cdot (x + 3)} \cdot \dfrac{5 \cdot (x + 3)}{2 \cdot 3 \cdot x \cdot x}$ Factoring completely

$$= \dfrac{2 \cdot x \cdot 5 \cdot (x + 3)}{3 \cdot (x + 3) \cdot 2 \cdot 3 \cdot x \cdot x} \quad \text{Multiplying}$$

$$= \dfrac{\cancel{2} \cdot \cancel{x} \cdot 5 \cdot \cancel{(x + 3)}}{3 \cdot \cancel{(x + 3)} \cdot \cancel{2} \cdot 3 \cdot \cancel{x} \cdot x} \quad \begin{array}{l}\text{Dividing out} \\ \text{common factors}\end{array}$$

$$= \dfrac{5}{9x}$$

We check by substituting $x = 1$ into each side of the last equation:

$$\dfrac{2x}{3x + 9} \cdot \dfrac{5x + 15}{6x^2} = \dfrac{2}{3 + 9} \cdot \dfrac{5 + 15}{6} \text{ and } \dfrac{5}{9x} = \dfrac{5}{9}$$

$$= \dfrac{2}{12} \cdot \dfrac{20}{6}$$

$$= \dfrac{40}{72}$$

$$= \dfrac{\cancel{8} \cdot 5}{\cancel{8} \cdot 9}$$

$$= \dfrac{5}{9}$$

Both sides of the equation have the same value, which suggests that

$$\dfrac{2x}{3x + 9} \cdot \dfrac{5x + 15}{6x^2} = \dfrac{5}{9x}.$$

b. $\dfrac{y^2 + 5y}{2y + 4} \cdot \dfrac{7y + 14}{y^2 - 25} = \dfrac{y(y + 5)}{2(y + 2)} \cdot \dfrac{7(y + 2)}{(y + 5)(y - 5)}$ Factoring completely

$$= \dfrac{y(y + 5) \cdot 7(y + 2)}{2(y + 2) \cdot (y + 5)(y - 5)} \quad \text{Multiplying}$$

$$= \dfrac{y\cancel{(y + 5)} \cdot 7\cancel{(y + 2)}}{2\cancel{(y + 2)} \cdot \cancel{(y + 5)}(y - 5)} \quad \begin{array}{l}\text{Dividing out} \\ \text{common factors}\end{array}$$

$$= \dfrac{7y}{2(y - 5)}$$

You should check the result.

c. $\dfrac{w^2 + 4w - 5}{w^2 - 4w + 3} \cdot \dfrac{4w^2 - 12w}{w^2 + 8w + 15}$

$$= \dfrac{(w - 1)(w + 5)}{(w - 1)(w - 3)} \cdot \dfrac{4w(w - 3)}{(w + 3)(w + 5)} \qquad \text{Factoring completely}$$

$$= \dfrac{(w - 1)(w + 5) \cdot 4w(w - 3)}{(w - 1)(w - 3) \cdot (w + 3)(w + 5)} \qquad \text{Multiplying}$$

$$= \dfrac{\cancel{(w - 1)}\cancel{(w + 5)} \cdot 4w\cancel{(w - 3)}}{\cancel{(w - 1)}\cancel{(w - 3)} \cdot (w + 3)\cancel{(w + 5)}} \qquad \text{Dividing out common factors}$$

$$= \dfrac{4w}{w + 3}$$

You should check the result. ∎

As you might suspect, the procedure for dividing rational expressions is similar to that for dividing fractions. For example, consider $\dfrac{3}{4} \div \dfrac{5}{6}$. Recall that the fraction $\dfrac{3}{4}$ is called the *dividend* and the fraction $\dfrac{5}{6}$ is the *divisor*. To divide the two fractions, we multiply the dividend by the reciprocal of the divisor. Since the reciprocal of $\dfrac{5}{6}$ is $\dfrac{6}{5}$, we have

$$\dfrac{3}{4} \div \dfrac{5}{6} = \dfrac{3}{4} \cdot \dfrac{6}{5} \qquad \text{Multiplying by the reciprocal}$$

$$= \dfrac{3 \cdot 6}{4 \cdot 5}$$

$$= \dfrac{18}{20}$$

We can reduce $\dfrac{18}{20}$ to lowest terms by dividing out the common factors of 18 and 20:

$$\dfrac{18}{20} = \dfrac{\cancel{2} \cdot 9}{\cancel{2} \cdot 10} = \dfrac{9}{10}$$

To divide two rational expressions

1. Multiply the dividend by the reciprocal of the divisor.
2. Reduce the result of step 1 to lowest terms.

EXAMPLE 2

Divide the rational expressions.

a. $\dfrac{3p + 15}{10p} \div \dfrac{p + 5}{5p^2}$ **b.** $\dfrac{4z + 8}{z^3 + z^2} \div \dfrac{6z + 12}{z}$ **c.** $\dfrac{q^2 + 5q + 6}{q^2 - 4} \div \dfrac{q - 1}{q^2 - 2q}$

Solution

We follow the procedure outlined above.

a. The reciprocal of $\dfrac{p + 5}{5p^2}$ is $\dfrac{5p^2}{p + 5}$. Therefore

$$\frac{3p + 15}{10p} \div \frac{p + 5}{5p^2} = \frac{3p + 15}{10p} \cdot \frac{5p^2}{p + 5} \qquad \text{Multiplying by the reciprocal}$$

Since we must multiply the rational expressions on the right side of the equation, we first completely factor each numerator and denominator:

$$\frac{3p + 15}{10p} \div \frac{p + 5}{5p^2} = \frac{3(p + 5)}{2 \cdot 5 \cdot p} \cdot \frac{5 \cdot p \cdot p}{(p + 5)} \qquad \text{Factoring completely}$$

$$= \frac{3(p + 5) \cdot 5 \cdot p \cdot p}{2 \cdot 5 \cdot p \cdot (p + 5)} \qquad \text{Multiplying}$$

$$= \frac{3\cancel{(p + 5)} \cdot \cancel{5} \cdot \cancel{p} \cdot p}{2 \cdot \cancel{5} \cdot \cancel{p} \cdot \cancel{(p + 5)}} \qquad \text{Dividing out common factors}$$

$$= \frac{3p}{2}$$

We check by substituting $p = 1$ into each side of the last equation:

$$\frac{3p + 15}{10p} \div \frac{p + 5}{5p^2} = \frac{18}{10} \div \frac{6}{5} \quad \text{and} \quad \frac{3p}{2} = \frac{3}{2}$$

$$= \frac{18}{10} \cdot \frac{5}{6}$$

$$= \frac{90}{60}$$

$$= \frac{3}{2}$$

Both sides of the equation have the same value, which suggests that

$$\frac{3p + 15}{10p} \div \frac{p + 5}{5p^2} = \frac{3p}{2}.$$

b. Since the reciprocal of $\dfrac{6z + 12}{z}$ is $\dfrac{z}{6z + 12}$, we can write

$$\frac{4z + 8}{z^3 + z^2} \div \frac{6z + 12}{z} = \frac{4z + 8}{z^3 + z^2} \cdot \frac{z}{6z + 12} \qquad \text{Multiplying by the reciprocal}$$

$$= \frac{4(z + 2)}{z^2(z + 1)} \cdot \frac{z}{6(z + 2)} \qquad \text{Factoring}$$

$$= \frac{2 \cdot 2 \cdot (z + 2)}{z \cdot z \cdot (z + 1)} \cdot \frac{z}{2 \cdot 3(z + 2)} \qquad \text{Factoring completely}$$

$$= \frac{2 \cdot 2 \cdot (z + 2) \cdot z}{z \cdot z \cdot (z + 1) \cdot 2 \cdot 3 \cdot (z + 2)} \qquad \text{Multiplying}$$

$$= \frac{\cancel{2} \cdot 2 \cdot \cancel{(z + 2)} \cdot \cancel{z}}{\cancel{z} \cdot z \cdot (z + 1) \cdot \cancel{2} \cdot 3 \cdot \cancel{(z + 2)}} \qquad \text{Dividing out common factors}$$

$$= \frac{2}{3z(z + 1)}$$

You should check the result.

c. The reciprocal of $\dfrac{q - 1}{q^2 - 2q}$ is $\dfrac{q^2 - 2q}{q - 1}$, so we have

$$\frac{q^2 + 5q + 6}{q^2 - 4} \div \frac{q - 1}{q^2 - 2q} = \frac{q^2 + 5q + 6}{q^2 - 4} \cdot \frac{q^2 - 2q}{q - 1} \qquad \text{Multiplying by the reciprocal}$$

$$= \frac{(q + 2)(q + 3)}{(q + 2)(q - 2)} \cdot \frac{q(q - 2)}{(q - 1)} \qquad \text{Factoring completely}$$

$$= \frac{\cancel{(q + 2)}(q + 3) \cdot q\cancel{(q - 2)}}{\cancel{(q + 2)}\cancel{(q - 2)} \cdot (q - 1)} \qquad \text{Multiplying}$$

$$= \frac{q(q + 3)}{q - 1} \qquad \text{Dividing out common factors}$$

You should check the result. ∎

The following summary may be of help to you in multiplying and dividing rational expressions:

Multiplication and division of rational expressions

If $\dfrac{A}{B}$ and $\dfrac{C}{D}$ are rational expressions, then

$$\frac{A}{B} \cdot \frac{C}{D} = \frac{A \cdot C}{B \cdot D} \qquad B \neq 0,\, D \neq 0$$

and

$$\frac{A}{B} \div \frac{C}{D} = \frac{A}{B} \cdot \frac{D}{C} \qquad B \neq 0,\, C \neq 0,\, D \neq 0$$

DISCUSSION QUESTIONS 6.2

1. In your own words, explain how to multiply rational expressions.

2. Consider the problem $\dfrac{4}{x-2} \div \dfrac{2x}{x^2-4}$. What is the dividend? What is the divisor?

3. Rational expressions are defined when the denominator is not zero. However, why do we assume that the *numerator* of the dividend is not zero when we divide rational expressions?

EXERCISES 6.2

In Exercises 1–8, multiply the given rational expressions. Write the result in lowest terms.

1. $\dfrac{1}{10x} \cdot \dfrac{5x^2}{3}$

2. $\dfrac{7y}{2} \cdot \dfrac{6}{y^3}$

3. $\dfrac{-24}{k^3} \cdot \dfrac{k^4}{8k}$

4. $\dfrac{h^5}{-9h} \cdot \dfrac{27h^7}{h^8}$

5. $\dfrac{(2p)^2}{14} \cdot \dfrac{p^7}{12p^6}$

6. $\dfrac{24}{45q^4} \cdot \dfrac{30q^9}{(3q)^2}$

7. $\dfrac{18r^5}{27r^7} \cdot \dfrac{(9r^3)^2}{r^8}$

8. $\dfrac{16t^4}{(2t^4)^3} \cdot \dfrac{t^2}{6t^5}$

In Exercises 9–16, divide the given rational expressions. Write the result in lowest terms.

9. $\dfrac{4}{u^2} \div \dfrac{u^2}{4}$

10. $\dfrac{-5v^3}{v^8} \div \dfrac{15v^4}{20v^2}$

11. $\dfrac{3w^2}{4w^5} \div \dfrac{15w^3}{-8w^4}$

12. $\dfrac{20z^6}{6z} \div \dfrac{4z^5}{9z^2}$

13. $\dfrac{12x^8}{28x^3} \div \dfrac{(3x)^2}{14x^7}$

14. $\dfrac{(6y^4)^2}{15y^3} \div \dfrac{9y^8}{40y^{12}}$

15. $\dfrac{63r^6}{(4r^2)^3} \div \dfrac{27r^4}{8r^8}$

16. $\dfrac{2s^5}{100s^{10}} \div \dfrac{s^9}{(5s^2)^3}$

In Exercises 17–42, perform the indicated operation. Write the result in lowest terms.

17. $\dfrac{7m+14}{10} \cdot \dfrac{20m}{3m+6}$

18. $\dfrac{4n}{6n-18} \cdot \dfrac{9n-27}{8}$

19. $\dfrac{3y-3}{8y} \div \dfrac{y^2-y}{40y^2}$

20. $\dfrac{6x^2}{2x-8} \div \dfrac{54x^4}{9x^2-36x}$

21. $(4h-2) \cdot \dfrac{h}{2h-1}$

22. $\dfrac{3k^2}{6k^2-10k} \cdot (3k-5)$

23. $\dfrac{n^2-10n}{5} \div (n^2-100)$

24. $(m^2+3m+2) \div \dfrac{8m+16}{12}$

25. $\dfrac{2r+10}{3r-12} \cdot \dfrac{5r-20}{4r+40}$

26. $\dfrac{9t-27}{12t-24} \cdot \dfrac{6t+12}{15t-45}$

27. $\dfrac{x^2-5x+6}{2x+4} \div \dfrac{2x-6}{3x+6}$

28. $\dfrac{4y+6}{y^2-3y-4} \div \dfrac{6y+9}{2y-8}$

29. $\dfrac{q^2-1}{q^2+4q} \cdot \dfrac{q^2+8q+16}{q^2-q-2}$

30. $\dfrac{3p-15}{p^2-10p+25} \cdot \dfrac{p^2-25}{p^2+4p-5}$

31. $\dfrac{h^2+3h-10}{h^2-25} \cdot \dfrac{h^2-2h-15}{4h^2-8h}$

32. $\dfrac{k^2+9k+14}{k^2+8k+7} \cdot \dfrac{k^2-9k}{k^2-7k-18}$

33. $\dfrac{r^2+7r+10}{r^2+6r+5} \div \dfrac{r^2-4r-12}{r^2-6r}$

34. $\dfrac{s^2+6s-7}{2s^3+4s^2} \div \dfrac{s^2-49}{s^2-5s-14}$

35. $\dfrac{x^2-x-2}{3x^2-x-4} \div \dfrac{2x^2+9x+10}{6x^2+7x-20}$

36. $\dfrac{y^2-2y-3}{4y^2+9y+5} \div \dfrac{2y^2+3y+9}{8y^2-2y-15}$

37. $\dfrac{h^3-7h^2k}{h^2-49k^2} \cdot \dfrac{h^2+7hk}{h^2+14hk+49k^2}$

38. $\dfrac{w^2-wz-2z^2}{4w^2+4wz+z^2} \cdot \dfrac{2w^2+wz}{w^2-z^2}$

39. $\dfrac{2m^2-m+8m-4}{m^2+2m-8} \cdot \dfrac{6m-4}{6m^2-3m+4m-2}$

40. $\dfrac{p^2-6p-16}{4p^2-3p-8p+6} \cdot \dfrac{8p^2-6p+4p-3}{3p^2-24p}$

41. $\dfrac{4x^2-y^2}{x^2+5xy+4y^2} \div \dfrac{y^2-4x^2}{2x^2+3xy+y^2}$

42. $\dfrac{w^2+5wz+6z^2}{w^2-9z^2} \div \dfrac{3w-z}{3z^2-4zw+w^2}$

Preparing for Section 6.3

Recall the rules for adding and subtracting fractions with the same denominator:

$$\frac{a}{b}+\frac{c}{b}=\frac{a+c}{b} \quad \text{and} \quad \frac{a}{b}-\frac{c}{b}=\frac{a-c}{b}$$

In Exercises 43–46, perform the indicated operation. Write the result in lowest terms.

43. $\dfrac{2}{5}+\dfrac{1}{5}$

44. $\dfrac{2}{7}+\dfrac{3}{7}$

45. $\dfrac{7}{12}-\dfrac{2}{12}$

46. $\dfrac{5}{8}-\dfrac{2}{8}$

Recall the rules for adding and subtracting fractions with different denominators:

$$\frac{a}{b}+\frac{c}{d}=\frac{ad+bc}{bd} \quad \text{and} \quad \frac{a}{b}-\frac{c}{d}=\frac{ad-bc}{bd}$$

In Exercises 47–50, perform the indicated operation. Write the result in lowest terms.

47. $\dfrac{1}{3}+\dfrac{3}{4}$

48. $\dfrac{5}{6}+\dfrac{7}{8}$

49. $\dfrac{4}{9}-\dfrac{6}{18}$

50. $\dfrac{7}{8}-\dfrac{4}{12}$

6.3   Addition and Subtraction of Rational Expressions

FOCUS

We first add and subtract rational expressions with the same denominator. Then we use the least common denominator to add or subtract any rational expressions.

To add or subtract fractions having the same denominator, as in

$$\frac{3}{5} + \frac{1}{5} = \frac{3+1}{5} = \frac{4}{5} \quad \text{and} \quad \frac{3}{5} - \frac{1}{5} = \frac{3-1}{5} = \frac{2}{5},$$

we add or subtract the numerators and put the result over the common denominator. This same procedure is used to add or subtract rational expressions having the *same denominator*:

> **To add or subtract rational expressions with the same denominator**
>
> 1. Add or subtract the numerators of the rational expressions and place the result over the common denominator.
> 2. Reduce the result of step 1 to lowest terms.

EXAMPLE 1   Perform the indicated operation. Write the result in lowest terms.

a. $\dfrac{2}{x+5} + \dfrac{3}{x+5}$ b. $\dfrac{9y}{3y+1} + \dfrac{3}{3y+1}$ c. $\dfrac{2w^2}{w^2-1} - \dfrac{2w}{w^2-1}$

Solution   a.   We add the numerators and put the result over the common denominator:

$$\frac{2}{x+5} + \frac{3}{x+5} = \frac{2+3}{x+5}$$

$$= \frac{5}{x+5}$$

The rational expression $\dfrac{5}{x+5}$ is in lowest terms because the numerator and denominator have no common factors that can be divided out.

b. We add the numerators and put the result over the common denominator:

$$\frac{9y}{3y + 1} + \frac{3}{3y + 1} = \frac{9y + 3}{3y + 1}$$

Now we factor the numerator and denominator of $\dfrac{9y + 3}{3y + 1}$, and divide out common factors:

$$\frac{9y}{3y + 1} + \frac{3}{3y + 1} = \frac{3(3y + 1)}{(3y + 1)} \qquad \text{Factoring}$$

$$= 3 \qquad \text{Dividing out } 3y + 1$$

c. $\dfrac{2w^2}{w^2 - 1} - \dfrac{2w}{w^2 - 1} = \dfrac{2w^2 - 2w}{w^2 - 1} \qquad \text{Subtracting}$

$$= \frac{2w(w - 1)}{(w + 1)(w - 1)} \qquad \text{Factoring}$$

$$= \frac{2w}{w + 1} \qquad \text{Dividing out } w - 1 \qquad \blacksquare$$

Least Common Denominator

How do we add or subtract rational expressions with different denominators? Before we answer this question, let us review a method for adding fractions with different denominators. For example, consider the problem $\dfrac{3}{10} + \dfrac{5}{12}$. Recall that we can add these fractions if we first find the least common denominator (LCD). The least common denominator of $\dfrac{3}{10}$ and $\dfrac{5}{12}$ is the smallest number that is evenly divisible by 10 and 12. We can find it by writing each denominator in prime factored form:

$$10 = 2 \cdot 5 \quad \text{and} \quad 12 = 2^2 \cdot 3$$

The **least common denominator** is found by multiplying together the highest powers of all the factors. The highest power of 2 is 2^2, the highest power of 3 is $3^1 = 3$, and the highest power of 5 is $5^1 = 5$; therefore

$$\text{Least common denominator} = 2^2 \cdot 3 \cdot 5 = 60$$

We check that 60 is the least common denominator by noting that it is the smallest number evenly divisible by 10 and 12.

We can find the least common denominator of rational expressions similarly:

To find the least common denominator of two or more rational expressions

1. Completely factor each denominator. Write numerical factors in prime factored form.
2. Multiply together the highest powers of all the factors in the denominators.

EXAMPLE 2

Find the least common denominator of each set of rational expressions.

a. $\dfrac{3}{p}$ and $\dfrac{2}{p-1}$ b. $\dfrac{1}{8q}$ and $\dfrac{q}{q^2+5q}$

c. $\dfrac{3}{4m^2}$ and $\dfrac{5}{6m}$ d. $\dfrac{n}{n^2-16}$ and $\dfrac{n-1}{n^2+8n+16}$

Solution

a. The denominators of $\dfrac{3}{p}$ and $\dfrac{2}{p-1}$ are already completely factored. Therefore

$$\text{LCD} = p(p-1)$$

b. We first completely factor the denominators of $\dfrac{1}{8q}$ and $\dfrac{q}{q^2+5q}$:

$$8q = 2^3 \cdot q \quad \text{and} \quad q^2 + 5q = q \cdot (q+5)$$

The factors are powers of 2, q, and $q + 5$. The highest power of 2 is 2^3, the highest power of q is q, and the highest power of $q + 5$ is $q + 5$. Therefore

$$\text{LCD} = 2^3 \cdot q \cdot (q+5) = 8q(q+5)$$

c. We first completely factor the denominators of $\dfrac{3}{4m^2}$ and $\dfrac{5}{6m}$:

$$4m^2 = 2^2 \cdot m^2 \quad \text{and} \quad 6m = 2 \cdot 3 \cdot m$$

The highest power of 2 is 2^2, the highest power of 3 is 3, and the highest power of m is m^2. Therefore

$$\text{LCD} = 2^2 \cdot 3 \cdot m^2 = 12m^2$$

d. We first completely factor the denominators of

$$\frac{n}{n^2 - 16} \quad \text{and} \quad \frac{n - 1}{n^2 + 8n + 16}$$

Then $n^2 - 16 = (n + 4)(n - 4)$ and $n^2 + 8n + 16 = (n + 4)^2$

The highest factor of $n + 4$ is $(n + 4)^2$, and the highest factor of $n - 4$ is $n - 4$. There are no other factors. Hence,

$$LCD = (n + 4)^2(n - 4)$$ ∎

Now let's return to the problem $\frac{3}{10} + \frac{5}{12}$ and see how the LCD can be used to solve it. Recall that the two fractions have a least common denominator of 60. The key step is to write each fraction as an equivalent fraction having the least common denominator as its denominator. We note that the denominator 10 must be multiplied by 6, and the denominator 12 must be multiplied by 5, to produce the least common denominator 60. Hence we write

$$\frac{3}{10} = \frac{3}{10} \cdot \frac{6}{6} = \frac{18}{60} \quad \text{and} \quad \frac{5}{12} = \frac{5}{12} \cdot \frac{5}{5} = \frac{25}{60}$$

We can now add the equivalent fractions because they have the same denominator:

$$\frac{3}{10} + \frac{5}{12} = \frac{18}{60} + \frac{25}{60} = \frac{43}{60}$$

The process described above can also be used to subtract two fractions with different denominators. It suggests a procedure for adding or subtracting rational expressions with *different denominators*:

To add or subtract rational expressions with different denominators

1. Find the least common denominator of the rational expressions.
2. Write each rational expression as an equivalent rational expression having the least common denominator as its denominator.
3. Add or subtract the equivalent rational expressions.
4. Reduce the result of step 3 to lowest terms.

EXAMPLE 3

Add the rational expressions.

a. $\dfrac{2}{3r} + \dfrac{3}{4r}$

b. $\dfrac{4}{t+2} + \dfrac{8}{t^2+2t}$

c. $\dfrac{1}{h^2+4h-5} + \dfrac{h}{h^2-25}$

d. $\dfrac{7}{k-1} + \dfrac{6}{1-k}$

Solution

a. We begin by finding the least common denominator of $\dfrac{2}{3r}$ and $\dfrac{3}{4r}$:

$$3r = 3 \cdot r \quad \text{and} \quad 4r = 2^2 \cdot r$$

$$\text{LCD} = 2^2 \cdot 3 \cdot r = 12r$$

Next we write each rational expression as an equivalent rational expression having the denominator $12r$. The denominator $3r$ must be multiplied by 4, and the denominator $4r$ must be multiplied by 3. Hence we multiply by $\dfrac{4}{4}$ and $\dfrac{3}{3}$, respectively:

$$\dfrac{2}{3r} = \dfrac{2}{3r} \cdot \dfrac{4}{4} \quad \text{and} \quad \dfrac{3}{4r} = \dfrac{3}{4r} \cdot \dfrac{3}{3}$$

$$= \dfrac{8}{12r} \qquad\qquad\qquad = \dfrac{9}{12r}$$

We now add the equivalent rational expressions:

$$\dfrac{2}{3r} + \dfrac{3}{4r} = \dfrac{8}{12r} + \dfrac{9}{12r}$$

$$= \dfrac{8+9}{12r}$$

$$= \dfrac{17}{12r}$$

The rational expression $\dfrac{17}{12r}$ is in lowest terms.

b. We first find the least common denominator of $\dfrac{4}{t+2}$ and $\dfrac{8}{t^2+2t}$:

$$t+2 \text{ is completely factored} \quad \text{and} \quad t^2+2t = t(t+2)$$

$$\text{LCD} = t(t+2)$$

Next we write each rational expression as an equivalent rational expression having the denominator $t(t + 2)$:

$$\frac{4}{t + 2} = \frac{4}{t + 2} \cdot \frac{t}{t} \quad \text{and} \quad \frac{8}{t^2 + 2t} = \frac{8}{t(t + 2)}$$

$$= \frac{4t}{t(t + 2)}$$

We now add the equivalent rational expressions and reduce to lowest terms:

$$\frac{4}{t + 2} + \frac{8}{t^2 + 2t} = \frac{4t}{t(t + 2)} + \frac{8}{t(t + 2)}$$

$$= \frac{4t + 8}{t(t + 2)} \qquad \text{Adding}$$

$$= \frac{4(t + 2)}{t(t + 2)} \qquad \text{Factoring}$$

$$= \frac{4}{t} \qquad \text{Dividing out } t + 2$$

c. We start by finding the least common denominator of $\dfrac{1}{h^2 + 4h - 5}$ and $\dfrac{h}{h^2 - 25}$:

$$h^2 + 4h - 5 = (h - 1)(h + 5) \quad \text{and} \quad h^2 - 25 = (h + 5)(h - 5)$$

$$\text{LCD} = (h - 1)(h + 5)(h - 5)$$

Next we write each rational expression as an equivalent rational expression having the denominator $(h - 1)(h + 5)(h - 5)$:

$$\frac{1}{h^2 + 4h - 5} = \frac{1}{(h - 1)(h + 5)} \cdot \frac{(h - 5)}{(h - 5)}$$

$$= \frac{(h - 5)}{(h - 1)(h + 5)(h - 5)}$$

and

$$\frac{h}{h^2 - 25} = \frac{h}{(h + 5)(h - 5)} \cdot \frac{(h - 1)}{(h - 1)}$$

$$= \frac{h(h - 1)}{(h - 1)(h + 5)(h - 5)}$$

We now add the equivalent rational expressions:

$$\frac{1}{h^2 + 4h - 5} + \frac{h}{h^2 - 25}$$

$$= \frac{(h - 5)}{(h - 1)(h + 5)(h - 5)} + \frac{h(h - 1)}{(h - 1)(h + 5)(h - 5)}$$

$$= \frac{(h - 5) + h(h - 1)}{(h - 1)(h + 5)(h - 5)} \qquad \text{Adding}$$

$$= \frac{h - 5 + h^2 - h}{(h - 1)(h + 5)(h - 5)} \qquad \text{By the distributive property}$$

$$= \frac{h^2 - 5}{(h - 1)(h + 5)(h - 5)} \qquad \text{Simplifying}$$

The rational expression $\dfrac{h^2 - 5}{(h - 1)(h + 5)(h - 5)}$ is in lowest terms.

d. The denominators of $\dfrac{7}{k - 1}$ and $\dfrac{6}{1 - k}$ are opposites. Since $k - 1 = (-1)(1 - k)$, we can obtain the common denominator $k - 1$ by multiplying the numerator and the denominator of $\dfrac{6}{1 - k}$ by -1:

$$\frac{7}{k - 1} + \frac{6}{1 - k} = \frac{7}{k - 1} + \frac{6}{1 - k} \cdot \frac{(-1)}{(-1)}$$

$$= \frac{7}{k - 1} + \frac{-6}{k - 1}$$

Now we can add the rational expressions:

$$\frac{7}{k - 1} + \frac{6}{1 - k} = \frac{7 + (-6)}{k - 1}$$

$$= \frac{1}{k - 1}$$

The rational expression $\dfrac{1}{k - 1}$ is in lowest terms.   ∎

EXAMPLE 4   Subtract the rational expressions.

a. $\dfrac{9}{4x} - \dfrac{3}{2x^2}$ b. $\dfrac{4y}{y^2 - 1} - \dfrac{2}{y + 1}$ c. $\dfrac{w}{w^2 + 2w + 1} - \dfrac{1}{w^2 - w - 2}$

Solution a. We first find the least common denominator of $\dfrac{9}{4x}$ and $\dfrac{3}{2x^2}$:

$$4x = 2^2 \cdot x \quad \text{and} \quad 2x^2 = 2 \cdot x^2$$
$$\text{LCD} = 2^2 \cdot x^2 = 4x^2$$

Next we must write each rational expression as an equivalent rational expression having the denominator $4x^2$. The denominator $4x$ must be multiplied by x and the denominator $2x^2$ must be multiplied by 2. Hence we write

$$\frac{9}{4x} = \frac{9}{4x} \cdot \frac{x}{x} \quad \text{and} \quad \frac{3}{2x^2} = \frac{3}{2x^2} \cdot \frac{2}{2}$$

$$= \frac{9x}{4x^2} \qquad\qquad = \frac{6}{4x^2}$$

We now subtract the equivalent rational expressions:

$$\frac{9}{4x} - \frac{3}{2x^2} = \frac{9x}{4x^2} - \frac{6}{4x^2}$$

$$= \frac{9x - 6}{4x^2} \qquad \text{Subtracting}$$

$$= \frac{3(3x - 2)}{4x^2} \qquad \text{Factoring}$$

The rational expression $\dfrac{3(3x - 2)}{4x^2}$ is in lowest terms.

b. We begin by finding the least common denominator of $\dfrac{4y}{y^2 - 1}$ and $\dfrac{2}{y + 1}$:

$$y^2 - 1 = (y + 1)(y - 1) \quad \text{and} \quad y + 1 \text{ is completely factored}$$
$$\text{LCD} = (y + 1)(y - 1)$$

Next we write each rational expression as an equivalent rational expression having the denominator $(y + 1)(y - 1)$:

$$\frac{4y}{y^2 - 1} = \frac{4y}{(y + 1)(y - 1)} \quad \text{and} \quad \frac{2}{y + 1} = \frac{2}{(y + 1)} \cdot \frac{(y - 1)}{(y - 1)}$$

$$= \frac{2(y - 1)}{(y + 1)(y - 1)}$$

We now subtract the equivalent rational expressions and reduce to lowest terms:

$$\frac{4y}{y^2 - 1} - \frac{2}{y + 1} = \frac{4y}{(y + 1)(y - 1)} - \frac{2(y - 1)}{(y + 1)(y - 1)}$$

$$= \frac{4y - 2(y - 1)}{(y + 1)(y - 1)} \qquad \text{Subtracting}$$

$$= \frac{2y + 2}{(y + 1)(y - 1)} \qquad \text{Simplifying}$$

$$= \frac{2(\cancel{y + 1})}{(\cancel{y + 1})(y - 1)} \qquad \text{Factoring}$$

$$= \frac{2}{y - 1} \qquad \text{Dividing out } y + 1$$

c. We start by finding the least common denominator of $\dfrac{w}{w^2 + 2w + 1}$ and $\dfrac{1}{w^2 - w - 2}$:

$$w^2 + 2w + 1 = (w + 1)^2 \quad \text{and} \quad w^2 - w - 2 = (w + 1)(w - 2)$$

$$\text{LCD} = (w + 1)^2(w - 2)$$

Next we write each rational expression as an equivalent rational expression having the denominator $(w + 1)^2(w - 2)$:

$$\frac{w}{w^2 + 2w + 1} = \frac{w}{(w + 1)^2} \cdot \frac{(w - 2)}{(w - 2)}$$

$$= \frac{w(w - 2)}{(w + 1)^2(w - 2)}$$

and

$$\frac{1}{w^2 - w - 2} = \frac{1}{(w + 1)(w - 2)} \cdot \frac{(w + 1)}{(w + 1)}$$

$$= \frac{(w + 1)}{(w + 1)^2(w - 2)}$$

We now subtract the equivalent rational expressions. We keep parentheses around $w + 1$ to avoid subtraction errors:

$$\frac{w}{w^2 + 2w + 1} - \frac{1}{w^2 - w - 2} = \frac{w(w - 2)}{(w + 1)^2(w - 2)} - \frac{(w + 1)}{(w + 1)^2(w - 2)}$$

$$= \frac{w(w - 2) - (w + 1)}{(w + 1)^2(w - 2)} \qquad \text{Subtracting}$$

$$= \frac{w^2 - 2w - w - 1}{(w + 1)^2(w - 2)} \qquad \text{By the distributive property}$$

$$= \frac{w^2 - 3w - 1}{(w + 1)^2(w - 2)} \qquad \text{Simplifying}$$

The rational expression $\dfrac{w^2 - 3w - 1}{(w + 1)^2(w - 2)}$ is in lowest terms. ∎

DISCUSSION QUESTIONS 6.3

1. In your own words, state how to add rational expressions with the same denominator.

2. Are the rational expressions $\dfrac{y}{y - 3}$ and $\dfrac{y^2 - y}{y^2 - 4y + 3}$ equivalent? Explain.

3. Is the following simplification correct? Explain.

$$\frac{w^2}{w + 1} - \frac{2w - 3}{w + 1} = \frac{w^2 - 2w - 3}{w + 1} = \frac{(w + 1)(w - 3)}{(w + 1)} = w - 3$$

EXERCISES 6.3

In Exercises 1–16, perform the indicated operation. Write the result in lowest terms.

1. $\dfrac{2}{x} + \dfrac{1}{x}$

2. $\dfrac{3}{y} + \dfrac{4}{y}$

3. $\dfrac{10}{w^2} - \dfrac{6}{w^2}$

4. $\dfrac{8}{z^3} - \dfrac{12}{z^3}$

5. $\dfrac{1}{8p} + \dfrac{3}{8p}$

6. $\dfrac{7}{9q} + \dfrac{11}{9q}$

7. $\dfrac{2}{h + 2} + \dfrac{h}{h + 2}$

8. $\dfrac{k}{3 + k} + \dfrac{3}{3 + k}$

9. $\dfrac{4}{m - 4} - \dfrac{m}{m - 4}$

10. $\dfrac{n}{5 - n} - \dfrac{5}{5 - n}$

11. $\dfrac{y^2}{y + 1} + \dfrac{y}{y + 1}$

12. $\dfrac{x^2}{x + 4} + \dfrac{4x}{x + 4}$

13. $\dfrac{2p^2}{p - 3} - \dfrac{18}{p - 3}$

14. $\dfrac{3q^2}{q - 1} - \dfrac{3}{q - 1}$

15. $\dfrac{z^2 + 10z}{z + 5} + \dfrac{25}{z + 5}$

16. $\dfrac{w^2 - 12w}{w - 6} + \dfrac{36}{w - 6}$

In Exercises 17–32, find the least common denominator of the given set of rational expressions.

17. $\dfrac{3}{7n}$ and $\dfrac{4}{5n}$

18. $\dfrac{1}{3n}$ and $\dfrac{9}{10n}$

19. $\dfrac{8}{10s^2}$ and $\dfrac{3}{6s^3}$

20. $\dfrac{10}{12r}$ and $\dfrac{4}{8r^3}$ **21.** $\dfrac{2}{6x}$ and $\dfrac{6}{4x-8}$ **22.** $\dfrac{2}{10y}$ and $\dfrac{3}{15y+30}$

23. $\dfrac{5}{w+2}$ and $\dfrac{7}{w-2}$ **24.** $\dfrac{6}{z+1}$ and $\dfrac{8}{z-1}$ **25.** $\dfrac{8}{q^2-81}$ and $\dfrac{4}{4q+36}$

26. $\dfrac{3}{p^2-49}$ and $\dfrac{7}{3p+21}$ **27.** $\dfrac{6}{k^2+6k}$ and $\dfrac{12}{k^2+2k-24}$ **28.** $\dfrac{15}{h^2+h-12}$ and $\dfrac{6}{h^2-3h}$

29. $\dfrac{36}{m^2-16m+64}$ and $\dfrac{24}{m^2-2m-48}$ **30.** $\dfrac{7}{n^2+9n+14}$ and $\dfrac{21}{n^2+14n+49}$

31. $\dfrac{2}{u-3},\ \dfrac{u}{2u-8},$ and $\dfrac{u+1}{u^2-7u+12}$ **32.** $\dfrac{4}{3v+15},\ \dfrac{2v}{v-2},$ and $\dfrac{v^2}{2v^2+6v-20}$

In Exercises 33–60, perform the indicated operation. Write the result in lowest terms.

33. $\dfrac{2}{3x}+\dfrac{1}{4x}$ **34.** $\dfrac{2}{5y}+\dfrac{3}{4y}$ **35.** $\dfrac{4}{w}-\dfrac{7}{2w^2}$ **36.** $\dfrac{10}{3z^2}-\dfrac{2}{z^3}$ **37.** $\dfrac{1}{6p^2}+\dfrac{3}{8p^3}$

38. $\dfrac{2}{6q^3}+\dfrac{1}{9q^5}$ **39.** $\dfrac{6}{uv^2}+\dfrac{v+2}{uv}$ **40.** $\dfrac{b+2}{4a^2b}-\dfrac{a-5}{3ab^2}$ **41.** $\dfrac{3}{r+1}-\dfrac{2}{r}$ **42.** $\dfrac{4}{s-2}-\dfrac{3}{s}$

43. $\dfrac{2}{2-m}-\dfrac{3}{m-2}$ **44.** $\dfrac{6}{n-7}-\dfrac{4}{7-n}$ **45.** $\dfrac{4}{3y+12}+\dfrac{2}{y+4}$ **46.** $\dfrac{7}{5x-10}+\dfrac{2}{x-2}$

47. $\dfrac{1}{2h}-\dfrac{3}{h^2+3h}$ **48.** $\dfrac{1}{3k}-\dfrac{2}{k^2-5k}$ **49.** $\dfrac{3z+35}{z^2-25}+\dfrac{2}{z+5}$ **50.** $\dfrac{5w-27}{w^2-9}+\dfrac{2}{w-3}$

51. $\dfrac{1}{2m-12}+\dfrac{4}{m^2-4m-12}$ **52.** $\dfrac{1}{4n+28}+\dfrac{2}{n^2+6n-7}$ **53.** $\dfrac{2r}{r^2+7r+12}-\dfrac{r}{r^2+5r+6}$

54. $\dfrac{s}{s^2+3s-4}-\dfrac{6s}{s^2-2s-24}$ **55.** $\dfrac{3}{x^2+6x+9}+\dfrac{2}{x^2+x-6}$ **56.** $\dfrac{3}{y^2-y-6}+\dfrac{4}{y^2-6y+9}$

57. $\dfrac{2}{u+3}+\dfrac{1}{u-3}-\dfrac{6}{u^2-9}$ **58.** $\dfrac{3}{v+1}+\dfrac{3}{v-1}-\dfrac{6}{v^2-1}$

59. $\dfrac{10}{q^2-2q-24}+\dfrac{2q}{q-6}+\dfrac{1}{q+4}$ **60.** $\dfrac{2p}{p+7}-\dfrac{12}{p^2+2p-35}+\dfrac{1}{p-5}$

Preparing for Section 6.4

In Exercises 61–66, find the value of the given expression by simplifying the numerator and the denominator and then dividing.

61. $\dfrac{1+\dfrac{1}{2}}{1-\dfrac{1}{2}}$ **62.** $\dfrac{2-\dfrac{3}{4}}{2+\dfrac{1}{4}}$ **63.** $\dfrac{\dfrac{1}{3}+\dfrac{2}{3}}{\dfrac{3}{4}+\dfrac{1}{4}}$ **64.** $\dfrac{\dfrac{5}{6}-\dfrac{3}{6}}{\dfrac{7}{8}-\dfrac{5}{8}}$ **65.** $\dfrac{\dfrac{4}{3}-\dfrac{1}{6}}{\dfrac{2}{4}-\dfrac{3}{8}}$ **66.** $\dfrac{\dfrac{4}{5}+\dfrac{3}{10}}{\dfrac{5}{6}+\dfrac{1}{8}}$

6.4 Complex Fractions

FOCUS

We discuss complex fractions and use two methods to simplify them.

The resistance of an electric circuit affects the strength of the current flowing in the circuit. According to the laws of physics, the total resistance of a circuit in which a 75-ohm (Ω) resistor is connected in parallel with a resistor of R ohms is $\dfrac{1}{\dfrac{1}{75} + \dfrac{1}{R}}$. The algebraic expression $\dfrac{1}{\dfrac{1}{75} + \dfrac{1}{R}}$ is an example of a **complex fraction,** a fraction that contains fractions or rational expressions in its numerator or denominator or both. The complex fraction $\dfrac{1}{\dfrac{1}{75} + \dfrac{1}{R}}$ contains both the numerical fraction $\dfrac{1}{75}$ and the rational expression $\dfrac{1}{R}$ in its denominator. Its numerator is 1.

There are two methods for simplifying complex fractions. In the first method we simplify the numerator and denominator of the complex fraction and then divide the numerator by the denominator. In the second method, the least common denominator of all the fractions and rational expressions in the complex fraction is used to simplify it. We illustrate these methods in the next example.

EXAMPLE 1

Simplify the complex fraction $\dfrac{2 + \dfrac{1}{x}}{\dfrac{1}{3} + \dfrac{1}{4}}$.

Solution

Method 1 (Simplify and Divide)

We begin by obtaining a single rational expression for the numerator and one for the denominator of the complex fraction. For the numerator, we must find the sum $2 + \dfrac{1}{x}$. The LCD is x, so we write

$$2 + \frac{1}{x} = 2 \cdot \frac{x}{x} + \frac{1}{x}$$

$$= \frac{2x}{x} + \frac{1}{x}$$

$$= \frac{2x + 1}{x}$$

For the denominator, we must find the sum $\frac{1}{3} + \frac{1}{4}$. The LCD is 12, so we have

$$\frac{1}{3} + \frac{1}{4} = \frac{4}{12} + \frac{3}{12}$$

$$= \frac{7}{12}$$

Therefore, the original complex fraction can be written as

$$\frac{2 + \frac{1}{x}}{\frac{1}{3} + \frac{1}{4}} = \frac{\frac{2x + 1}{x}}{\frac{7}{12}}$$

Now, since a fraction represents division, we can divide the numerator of the new complex fraction by the denominator:

$$\frac{\frac{2x + 1}{x}}{\frac{7}{12}} = \frac{2x + 1}{x} \div \frac{7}{12}$$

$$= \frac{2x + 1}{x} \cdot \frac{12}{7} \qquad \text{Multiplying by the reciprocal of } \frac{7}{12}$$

$$= \frac{12(2x + 1)}{7x} \qquad \text{Multiplying}$$

Method 2 (Use the LCD)

Our first step is to find the least common denominator of all the fractions and rational expressions in the complex fraction $\dfrac{2 + \frac{1}{x}}{\frac{1}{3} + \frac{1}{4}}$. The LCD of $\frac{1}{x}, \frac{1}{3}$,

and $\frac{1}{4}$ is $12x$. We now multiply the numerator and the denominator of the complex fraction by this LCD:

$$\dfrac{2 + \dfrac{1}{x}}{\dfrac{1}{3} + \dfrac{1}{4}} = \dfrac{12x\left(2 + \dfrac{1}{x}\right)}{12x\left(\dfrac{1}{3} + \dfrac{1}{4}\right)}$$

$$= \dfrac{12x \cdot 2 + 12x \cdot \dfrac{1}{x}}{12x \cdot \dfrac{1}{3} + 12x \cdot \dfrac{1}{4}} \qquad \text{By the distributive property}$$

$$= \dfrac{24x + 12}{4x + 3x} \qquad \text{Simplifying}$$

$$= \dfrac{12(2x + 1)}{7x} \qquad \text{Factoring} \qquad \blacksquare$$

EXAMPLE 2

Simplify the complex fraction $\dfrac{\dfrac{z + 1}{z - 1}}{\dfrac{z + 1}{2z}}$.

Solution

We shall use Method 1 here because the numerator and the denominator of the complex fraction are already single rational expressions. We need only divide the numerator of the complex fraction by its denominator.

$$\dfrac{\dfrac{z + 1}{z - 1}}{\dfrac{z + 1}{2z}} = \dfrac{z + 1}{z - 1} \div \dfrac{z + 1}{2z}$$

$$= \dfrac{z + 1}{z - 1} \cdot \dfrac{2z}{z + 1} \qquad \text{Multiplying by the reciprocal of } \dfrac{z + 1}{2z}$$

$$= \dfrac{2z(z + 1)}{(z - 1)(z + 1)} \qquad \text{Multiplying}$$

$$= \dfrac{2z}{z - 1} \qquad \text{Dividing out } z + 1 \qquad \blacksquare$$

EXAMPLE 3

Simplify the complex fraction $\dfrac{\dfrac{1}{16} - \dfrac{1}{y^2}}{\dfrac{1}{4} + \dfrac{1}{y}}$.

Solution   We use Method 2 here because the least common denominator of the fractions and rational expressions in the complex fraction is relatively easy to determine. The LCD of $\frac{1}{4}$, $\frac{1}{16}$, $\frac{1}{y}$, and $\frac{1}{y^2}$ is $16y^2$. We now multiply the numerator and the denominator of the complex fraction by $16y^2$:

$$\frac{\dfrac{1}{16} - \dfrac{1}{y^2}}{\dfrac{1}{4} + \dfrac{1}{y}} = \frac{16y^2\left(\dfrac{1}{16} - \dfrac{1}{y^2}\right)}{16y^2\left(\dfrac{1}{4} + \dfrac{1}{y}\right)}$$

$$= \frac{16y^2 \cdot \dfrac{1}{16} - 16y^2 \cdot \dfrac{1}{y^2}}{16y^2 \cdot \dfrac{1}{4} + 16y^2 \cdot \dfrac{1}{y}} \qquad \text{By the distributive property}$$

$$= \frac{y^2 - 16}{4y^2 + 16y} \qquad \text{Simplifying}$$

$$= \frac{(y + 4)(y - 4)}{4y(y + 4)} \qquad \text{Factoring}$$

$$= \frac{y - 4}{4y} \qquad \text{Dividing out } y + 4 \qquad ■$$

We conclude by summarizing the two methods for simplifying a complex fraction:

To simplify a complex fraction

Method 1 (Simplify and Divide): Obtain a single rational expression for the numerator and a single rational expression for the denominator of the complex fraction. Then divide the numerator by the denominator.

Method 2 (Use the LCD): Multiply the numerator and the denominator of the complex fraction by the least common denominator of all the fractions and rational expressions in the complex fraction. Then simplify the result.

DISCUSSION QUESTIONS 6.4

1. What is a complex fraction?

2. How is division of rational expressions used to simplify complex fractions?

3. In Method 2, we multiply the numerator and the denominator of a complex fraction by the least common denominator of all the fractions and rational expressions in the complex fraction. How does that help simplify the complex fraction?

EXERCISES 6.4

In Exercises 1–36, simplify the given complex fraction.

1. $\dfrac{\dfrac{3}{x}}{\dfrac{6}{x}}$

2. $\dfrac{\dfrac{8}{y^2}}{\dfrac{4}{y^2}}$

3. $\dfrac{\dfrac{w^2}{8z}}{\dfrac{w^3}{10z^2}}$

4. $\dfrac{\dfrac{p^2}{2q^5}}{\dfrac{p^4}{6q}}$

5. $\dfrac{\dfrac{h+1}{h}}{\dfrac{h-1}{h}}$

6. $\dfrac{\dfrac{k}{k-3}}{\dfrac{k}{k+3}}$

7. $\dfrac{\dfrac{r}{r^2-4}}{\dfrac{4}{r+2}}$

8. $\dfrac{\dfrac{t^2}{t^2-25}}{\dfrac{5}{t-5}}$

9. $\dfrac{1+\dfrac{5}{y}}{\dfrac{y+5}{3}}$

10. $\dfrac{\dfrac{x+3}{4}}{\dfrac{3}{x}+1}$

11. $\dfrac{\dfrac{w^2-1}{2}}{\dfrac{1}{w}-1}$

12. $\dfrac{1-\dfrac{2}{z}}{\dfrac{4-z^2}{6}}$

13. $\dfrac{3+\dfrac{1}{m}}{3-\dfrac{1}{m}}$

14. $\dfrac{4-\dfrac{2}{n}}{4+\dfrac{2}{n}}$

15. $\dfrac{\dfrac{1}{h}-\dfrac{1}{k}}{\dfrac{1}{hk}}$

16. $\dfrac{\dfrac{1}{xy}}{\dfrac{1}{x}+\dfrac{1}{y}}$

17. $\dfrac{k+\dfrac{1}{9}}{k^2-\dfrac{1}{81}}$

18. $\dfrac{h^2-\dfrac{1}{64}}{h-\dfrac{1}{8}}$

19. $\dfrac{1+\dfrac{2}{m}}{\dfrac{1}{4}-\dfrac{1}{m^2}}$

20. $\dfrac{1-\dfrac{3}{n}}{\dfrac{1}{9}-\dfrac{1}{n^2}}$

21. $\dfrac{\dfrac{1}{25}-\dfrac{1}{t^2}}{\dfrac{1}{t}-\dfrac{1}{5}}$

22. $\dfrac{\dfrac{1}{7}-\dfrac{1}{r}}{\dfrac{1}{r^2}-\dfrac{1}{49}}$

23. $\dfrac{\dfrac{1}{x}+\dfrac{1}{2}}{\dfrac{1}{x}-\dfrac{1}{2}}$

24. $\dfrac{\dfrac{1}{y}-\dfrac{1}{3}}{\dfrac{1}{y}+\dfrac{1}{3}}$

25. $\dfrac{2v-\dfrac{2}{2v+1}}{2-\dfrac{2v}{2v+1}}$

26. $\dfrac{3-\dfrac{6u-6}{3u-1}}{3u-\dfrac{12}{3u-1}}$

27. $\dfrac{6y-3-\dfrac{24}{2y-3}}{2y-\dfrac{4}{2y-3}}$

28. $\dfrac{4x-\dfrac{24}{2x+1}}{5x+15-\dfrac{90}{2x+1}}$

29. $\dfrac{n+3-\dfrac{10}{n-6}}{2n+4-\dfrac{40}{n-6}}$

30. $\dfrac{3m-21+\dfrac{15}{m-1}}{m-3+\dfrac{1}{m-1}}$

31. $\dfrac{\dfrac{1}{p-1}-\dfrac{3}{p+3}}{\dfrac{3}{p+3}+\dfrac{2}{p-2}}$

32. $\dfrac{\dfrac{1}{q+1}+\dfrac{2}{q-2}}{\dfrac{4}{q+4}-\dfrac{1}{q+1}}$

33. $\dfrac{1-\dfrac{2}{k}}{1-\dfrac{4}{k}+\dfrac{4}{k^2}}$

34. $\dfrac{1+\dfrac{3}{h}}{1+\dfrac{6}{h}+\dfrac{9}{h^2}}$

35. $\dfrac{\dfrac{1}{m}-\dfrac{1}{n}}{\dfrac{1}{n^2}-\dfrac{1}{m^2}}$

36. $\dfrac{\dfrac{1}{r^2}-\dfrac{1}{s^2}}{\dfrac{s^2}{r^2}-\dfrac{r^2}{s^2}}$

Preparing for Section 6.5

In Exercises 37–42, solve the given equation.

37. $2x + 5 = 3x$

38. $4 - 7y = 3y + 2$

39. $6(w + 4) - 1 = 11w + 3$

40. $5 - 2(z + 1) = 4(z - 3) + 10$

41. $q^2 + 10q + 16 = 0$

42. $p^2 - 12p + 35 = 0$

6.5　Solving Equations Containing Rational Expressions

FOCUS

We solve equations containing rational expressions by use of the least common denominator of the rational expressions.

In Sections 6.3 and 6.4 you learned how to use the least common denominator to add or subtract rational expressions and to simplify complex fractions. In this section we will use the least common denominator to solve equations that contain fractions or rational expressions. Our strategy is to **clear the equation of fractions** by multiplying both sides of the equation by the LCD of all the fractions and rational expressions in the equation. The result is an equation *without* fractions or rational expressions. We illustrate this procedure in the following example.

EXAMPLE 1　Solve the equation $\dfrac{k}{2} = 5 + \dfrac{k}{3}$.

Solution　The LCD of $\dfrac{k}{2}$ and $\dfrac{k}{3}$ is 6. We multiply both sides of the equation by 6:

$$6\left(\frac{k}{2}\right) = 6\left(5 + \frac{k}{3}\right)$$

$$3k = 6 \cdot 5 + 6 \cdot \frac{k}{3} \qquad \text{By the distributive property}$$

$$3k = 30 + 2k \qquad \text{Simplifying}$$

The original equation has now been "cleared of fractions." This happens because the denominator of each fraction in the equation is a factor (divisor) of the least common denominator. We can easily solve the resulting equation by subtracting $2k$ from both sides:

$$3k = 30 + 2k$$

$$k = 30 \qquad \text{Subtracting } 2k \text{ from both sides}$$

We check by substituting $k = 30$ into each side of the original equation:

$$\frac{k}{2} = \frac{30}{2} \quad \text{and} \quad 5 + \frac{k}{3} = 5 + \frac{30}{3}$$

$$= 15 \qquad\qquad = 5 + 10$$

$$= 15$$

Both sides of the equation have the same value, so $k = 30$ is the solution. ∎

EXAMPLE 2

Solve the equation $\dfrac{7}{2p} - \dfrac{3}{5} = \dfrac{1}{10}$.

Solution

The LCD of $\dfrac{7}{2p}, \dfrac{3}{5},$ and $\dfrac{1}{10}$ is $10p$. We multiply both sides of the equation by $10p$ and solve:

$$10p\left(\frac{7}{2p} - \frac{3}{5}\right) = 10p\left(\frac{1}{10}\right)$$

$$10p \cdot \frac{7}{2p} - 10p \cdot \frac{3}{5} = p \qquad \text{By the distributive property}$$

$$35 - 6p = p \qquad \text{Simplifying}$$

$$-7p = -35 \qquad \text{Subtracting 35 and } p \text{ from both sides}$$

$$p = 5 \qquad \text{Dividing both sides by } -7$$

We check by substituting $p = 5$ into the left side of the original equation:

$$\frac{7}{2p} - \frac{3}{5} = \frac{7}{10} - \frac{3}{5}$$

$$= \frac{7}{10} - \frac{6}{10}$$

$$= \frac{1}{10}$$

This is the same as the right side, so $p = 5$ is the solution. ∎

EXAMPLE 3

Solve the equation $\dfrac{1}{2} + \dfrac{3}{2x-2} = \dfrac{2}{x-1}$.

Solution

Since $2x - 2 = 2(x - 1)$, the least common denominator of $\dfrac{1}{2}$, $\dfrac{3}{2x-2}$, and $\dfrac{2}{x-1}$ is $2(x - 1)$. We multiply both sides of the equation by $2(x - 1)$ and solve:

$$2(x-1)\left(\frac{1}{2} + \frac{3}{2x-2}\right) = 2(x-1)\left(\frac{2}{x-1}\right)$$

$$2(x-1)\cdot\frac{1}{2} + 2(x-1)\cdot\frac{3}{2(x-1)} = 2(x-1)\cdot\frac{2}{(x-1)} \quad \text{By the distributive property}$$

$$(x-1) + 3 = 4 \qquad\qquad \text{Simplifying}$$

$$x + 2 = 4$$

$$x = 2$$

We check by substituting $x = 2$ into each side of the original equation:

$$\frac{1}{2} + \frac{3}{2x-2} = \frac{1}{2} + \frac{3}{4-2} \quad \text{and} \quad \frac{2}{x-1} = \frac{2}{2-1}$$

$$= \frac{1}{2} + \frac{3}{2} \qquad\qquad\qquad = \frac{2}{1}$$

$$= \frac{4}{2} \qquad\qquad\qquad\qquad = 2$$

$$= 2$$

Both sides of the equation have the same value, so $x = 2$ is the solution. ∎

EXAMPLE 4

Solve the equation $\dfrac{8}{y^2-2y-3} - \dfrac{2}{y-3} = \dfrac{y}{y+1}$.

Solution

Since $y^2 - 2y - 3 = (y+1)(y-3)$, the least common denominator of $\dfrac{8}{y^2-2y-3}$, $\dfrac{2}{y-3}$, and $\dfrac{y}{y+1}$ is $(y+1)(y-3)$. We multiply both sides of the equation by $(y+1)(y-3)$:

$$(y+1)(y-3)\left(\frac{8}{y^2-2y-3} - \frac{2}{y-3}\right) = (y+1)(y-3)\left(\frac{y}{y+1}\right)$$

$$(y + 1)(y - 3) \cdot \frac{8}{(y + 1)(y - 3)} - (y + 1)(y - 3) \cdot \frac{2}{(y - 3)}$$

$$= (y + 1)(y - 3) \cdot \frac{y}{(y + 1)} \qquad \text{By the distributive property}$$

$$8 - 2(y + 1) = y(y - 3) \qquad \text{Simplifying}$$

$$8 - 2y - 2 = y^2 - 3y \qquad \text{By the distributive property}$$

The last equation is a quadratic equation. We put the equation in standard form and solve by factoring:

$$y^2 - y - 6 = 0$$

$$(y + 2)(y - 3) = 0 \qquad \text{Factoring}$$

$$y + 2 = 0 \quad \text{or} \quad y - 3 = 0 \qquad \text{By the zero product property}$$

$$y = -2 \qquad\qquad y = 3$$

We check each value by substitution into each side of the original equation.

If $y = -2$, then

$$\frac{8}{y^2 - 2y - 3} - \frac{2}{y - 3} = \frac{8}{4 + 4 - 3} - \frac{2}{(-2) - 3}$$

$$= \frac{8}{5} - \frac{2}{-5}$$

$$= \frac{8}{5} + \frac{2}{5}$$

$$= \frac{10}{5}$$

$$= 2$$

and

$$\frac{y}{y + 1} = \frac{-2}{(-2) + 1}$$

$$= \frac{-2}{-1}$$

$$= 2$$

Both sides of the equation have the same value, so $y = -2$ is a solution.

If $y = 3$, then

$$\frac{8}{y^2 - 2y - 3} - \frac{2}{y - 3} = \frac{8}{9 - 6 - 3} - \frac{2}{3 - 3}$$

$$= \frac{8}{0} - \frac{2}{0}$$

Since denominators cannot be zero, we conclude that $y = 3$ is *not* a solution of the original equation.   ■

BE CAREFUL!

Remember to *check each solution in the original equation.* Any solution that produces a zero denominator must be discarded.

EXAMPLE 5   Solve the equation $\dfrac{w + 3}{2w} = \dfrac{3}{w^2} + \dfrac{1}{2}$.

Solution   The LCD of $\dfrac{w + 3}{2w}$, $\dfrac{3}{w^2}$, and $\dfrac{1}{2}$ is $2w^2$. We multiply both sides of the equation by $2w^2$ and solve:

$$2w^2\left(\frac{w + 3}{2w}\right) = 2w^2\left(\frac{3}{w^2} + \frac{1}{2}\right)$$

$$2w^2 \cdot \frac{(w + 3)}{2w} = 2w^2 \cdot \frac{3}{w^2} + 2w^2 \cdot \frac{1}{2} \qquad \text{By the distributive property}$$

Note that we have put parentheses around $w + 3$. This precaution helps to reduce multiplication errors. Continuing, we obtain

$$w(w + 3) = 6 + w^2 \qquad \text{Simplifying}$$

$$w^2 + 3w = 6 + w^2 \qquad \text{By the distributive property}$$

$$3w = 6 \qquad \text{Subtracting } w^2 \text{ from both sides}$$

$$w = 2$$

We check by substituting $w = 2$ into each side of the original equation:

$$\frac{w+3}{2w} = \frac{2+3}{4} \quad \text{and} \quad \frac{3}{w^2} + \frac{1}{2} = \frac{3}{4} + \frac{1}{2}$$

$$= \frac{5}{4} \qquad\qquad = \frac{3}{4} + \frac{2}{4}$$

$$= \frac{5}{4}$$

Both sides of the equation have the same value, so $w = 2$ is a solution. ∎

We can summarize the procedure for solving equations that contain fractions or rational expressions as follows:

To solve equations containing rational expressions

1. Multiply both sides of the equation by the LCD of all the fractions and rational expressions in the equation.
2. Solve the resulting equation.
3. Check each solution in the *original* equation.

DISCUSSION QUESTIONS 6.5

1. When an equation contains rational expressions, why do we multiply it by the least common denominator of those expressions?

2. For what values of x is the equation $\dfrac{x}{x^2 - 5x - 6} - \dfrac{2}{x - 6} = \dfrac{4}{x + 1}$ not defined? What does your answer imply about possible solutions of the equation?

3. In your own words, state how to solve equations containing rational expressions.

EXERCISES 6.5

In Exercises 1–34, solve the given equation. Check the result.

1. $\dfrac{17}{3} + \dfrac{5u}{3} = \dfrac{2}{3}$

2. $\dfrac{2v}{5} - \dfrac{4}{5} = \dfrac{2}{5}$

3. $\dfrac{x}{3} - \dfrac{x}{4} = 1$

4. $\dfrac{y}{3} + \dfrac{y}{2} = 5$

5. $\dfrac{2}{w} + 1 = \dfrac{1}{5}$

6. $\dfrac{4}{z} - 1 = \dfrac{1}{3}$

7. $\dfrac{1}{2v} + \dfrac{8}{5} = \dfrac{3}{v}$

8. $\dfrac{1}{6u} = \dfrac{1}{4} - \dfrac{1}{3u}$

9. $\dfrac{2m - 5}{m} = \dfrac{3}{2}$

10. $\dfrac{1}{n} = \dfrac{3}{4n - 1}$

11. $\dfrac{r - 2}{r - 4} = \dfrac{r + 1}{r + 3}$

12. $\dfrac{s + 2}{s - 3} = \dfrac{s + 1}{s + 6}$

13. $\dfrac{8}{y+6} + \dfrac{8}{y-6} = 1$ **14.** $\dfrac{3}{x+4} + \dfrac{3}{x-4} = 1$ **15.** $\dfrac{9}{2p-4} - \dfrac{3}{2} = \dfrac{p}{p-2}$

16. $\dfrac{14}{5q+5} + \dfrac{q}{q+1} = -\dfrac{4}{5}$ **17.** $\dfrac{12}{h^2-9} + \dfrac{3}{h+3} = \dfrac{2}{h-3}$ **18.** $\dfrac{8}{k^2-4} = \dfrac{1}{k-2} - \dfrac{3}{k+2}$

19. $\dfrac{2}{r^2-25} = \dfrac{1}{r^2-5r}$ **20.** $\dfrac{2}{s^2-16} = \dfrac{3}{s^2+4s}$ **21.** $2 + \dfrac{m}{m+4} = \dfrac{3m}{m-4}$

22. $\dfrac{2n}{n-3} = \dfrac{n}{n+3} + 1$ **23.** $\dfrac{3}{s+2} + \dfrac{s-1}{s+5} = \dfrac{5s+20}{6s+24}$ **24.** $\dfrac{t-1}{t} + \dfrac{t-3}{t-2} = \dfrac{22t-110}{3t^2-15t}$

25. $\dfrac{2}{x-2} - \dfrac{3}{x+3} = \dfrac{12}{x^2+x-6}$ **26.** $\dfrac{4}{y+4} - \dfrac{3}{y-3} = \dfrac{1}{y^2+y-12}$

27. $\dfrac{3}{z^2-6z+8} + \dfrac{4z}{z-4} = \dfrac{3z}{z-2}$ **28.** $\dfrac{2w}{w+2} + \dfrac{3}{w^2+5w+6} = \dfrac{w}{w+3}$

29. $\dfrac{q}{q+4} = \dfrac{3q+7}{q^2+7q+12} + \dfrac{2}{q+3}$ **30.** $\dfrac{4}{2p^2-3p-2} = \dfrac{p}{p-2} + \dfrac{4}{2p+1}$

31. $\dfrac{1}{k^2+2k+1} + \dfrac{2}{k+1} + 1 = 0$ **32.** $1 + \dfrac{6}{h-2} + \dfrac{5}{h^2-4h+4} = 0$

33. $\dfrac{5-m}{m^2-m-6} - \dfrac{3}{m^2+2m} + \dfrac{1}{m} = 0$ **34.** $\dfrac{7-n}{n^2-n-12} + \dfrac{1}{n} = \dfrac{5}{n^2+3n}$

Preparing for Section 6.6

35. A plane can fly 600 miles per hour (mph) in still air. What is its speed against a headwind of x mph?

36. A boat can go 12 mph in still water. What is its speed going with a river current of y mph?

37. Mr. J can read a magazine in 3 h. How much of the magazine can he read in 1 h?

38. Ms. D can paint a portrait in 5 h. How much of the portrait can she paint in 1 h?

39. If $y = kx$, find the value of k when $x = 2$ and $y = 12$.

40. If $y = kx^2$, find the value of k when $x = 6$ and $y = 4$.

41. If $y = \dfrac{k}{x}$, find the value of k when $x = 12$ and $y = \dfrac{2}{3}$.

42. If $y = \dfrac{k}{x^2}$, find the value of k when $x = 0.5$ and $y = 9$.

6.6 Applications of Rational Expressions

FOCUS

We solve some applications of rational expressions. Then we discuss the concepts of direct and inverse variation.

Certain types of problems involving distance, rate (speed), and time lead to equations containing rational expressions.

EXAMPLE 1

A jet airliner travels 1000 miles (mi) with the wind in the same time it takes to fly 800 mi against the wind. If the wind is blowing 60 mph, what is the cruising speed of the airliner in still air?

Solution

This example requires several ideas. First we need a formula we have used before:

$$\text{Distance} = \text{Speed} \cdot \text{Time}$$

$$d = rt$$

Next we must use the fact that the speed of an airliner with or against the wind is affected by wind speed as follows:

Speed with the wind = Cruising speed in still air + Wind speed

Speed against the wind = Cruising speed in still air − Wind speed

We are concerned with two trips—one with the wind and the other against the wind—and we need to use or to find the distance, speed, and time for each. A table will help us to organize the information. We let x = the cruising speed of the airliner in still air. Then we have, so far,

	Distance d	Speed r	Time t
With the wind	1000	$x + 60$	
Against the wind	800	$x - 60$	

Now, by solving the formula $d = rt$ for t, we obtain $t = \dfrac{d}{r}$. Substituting values for d and r from each line of the table gives the time of flight for that line: with the wind, we have $t = \dfrac{1000}{x + 60}$ hours, and against the wind $t = \dfrac{800}{x - 60}$ hours. Hence, the completed table is:

	Distance	Speed	Time
	d	r	$t = \dfrac{d}{r}$
With the wind	1000	$x + 60$	$\dfrac{1000}{x + 60}$
Against the wind	800	$x - 60$	$\dfrac{800}{x - 60}$

Our next step is to write an equation. We use the fact that the airliner's flying time with the wind is the same as its flying time against the wind:

$$\text{Time with wind} = \text{Time against wind}$$

$$\frac{1000}{x + 60} = \frac{800}{x - 60}$$

To solve this equation, we must find the least common denominator of $\dfrac{1000}{x + 60}$ and $\dfrac{800}{x - 60}$, which is $(x + 60)(x - 60)$. We now multiply both sides of the equation by $(x + 60)(x - 60)$ and solve:

$$(x + 60)(x - 60)\left(\frac{1000}{x + 60}\right) = (x + 60)(x - 60)\left(\frac{800}{x - 60}\right)$$

$1000(x - 60) = 800(x + 60)$	Simplifying
$1000x - 60{,}000 = 800x + 48{,}000$	By the distributive property
$200x = 108{,}000$	Adding 60,000 and subtracting $800x$ from both sides
$x = 540$	Dividing both sides by 200

The cruising speed of the airliner in still air is 540 mph. We check this solution by showing that the time of flight with the wind is the same as the time of flight against the wind:

$$\text{Time with wind} = \frac{1000}{x + 60}$$

$$= \frac{1000}{540 + 60}$$

$$= \frac{1000}{600}$$

$$= 1\frac{2}{3}\,\text{h}$$

and   Time against wind $= \dfrac{800}{x - 60}$

$$= \dfrac{800}{540 - 60}$$

$$= \dfrac{800}{480}$$

$$= 1\dfrac{2}{3} \text{ h}$$

The two times are the same, so $x = 540$ mph is the cruising speed of the airliner in still air.   ■

Rational expressions also occur in problems involving the amount of time needed to complete a job.

EXAMPLE 2   A mountain cabin has two water pumps to fill its reservoir. One pump can fill the reservoir in 2 h, and the other pump can fill the reservoir in 5 h. How long does it take to fill the reservoir when both pumps are working together?

Solution   It does not take the two pumps working together 7 h to fill the reservoir, since the slower pump can do the job in 5 h working alone. In fact, it must take less than 2 h for the pumps to fill the reservoir working together. The key to solving the problem is to determine the rate of work of each pump. The **rate of work** is the fractional part of the job that can be done in 1 h. Since the first pump can fill the entire reservoir in 2 h, this pump can fill $\dfrac{1}{2}$ the reservoir in 1 h. Similarly, the second pump can fill $\dfrac{1}{5}$ of the reservoir in 1 h.

We now let $x =$ the number of hours for the pumps to fill the reservoir working together. Then the rate of work of the two pumps working together must be $\dfrac{1}{x}$. We summarize this information in the following table:

	Hours to Fill Reservoir	Part of Reservoir Filled in 1 h
First pump	2	$\dfrac{1}{2}$
Second pump	5	$\dfrac{1}{5}$
Together	x	$\dfrac{1}{x}$

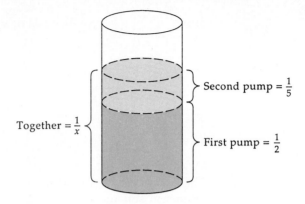

FIGURE 6.1

It may help to picture the water level in the reservoir after 1 h, as we have done in Figure 6.1.

We can now write an equation for the amount of work done by the pumps *in 1 h:*

$$\underset{\text{first pump}}{\text{Part filled by}} \quad + \quad \underset{\text{second pump}}{\text{Part filled by}} \quad = \quad \underset{\text{working together}}{\text{Part filled}}$$

$$\frac{1}{2} \quad + \quad \frac{1}{5} \quad = \quad \frac{1}{x}$$

This equation says that the amount of the reservoir filled by the first pump in 1 h plus the amount of the reservoir filled by the second pump in 1 h must equal the amount of the reservoir filled by both pumps working together in 1 h. Now, the least common denominator of $\frac{1}{2}$, $\frac{1}{5}$, and $\frac{1}{x}$ is $10x$. We multiply both sides of the equation by $10x$ and solve:

$$10x\left(\frac{1}{2} + \frac{1}{5}\right) = 10x\left(\frac{1}{x}\right)$$

$$10x \cdot \frac{1}{2} + 10x \cdot \frac{1}{5} = 10 \qquad \text{By the distributive property}$$

$$5x + 2x = 10 \qquad \text{Simplifying}$$

$$7x = 10$$

$$x = \frac{10}{7} \text{ or } 1\frac{3}{7}$$

It takes the pumps $1\frac{3}{7}$ h (approximately 1.4 h) to fill the reservoir working together. You should check this solution. ∎

Variation

Many real-world situations involve relationships between two quantities, such that a change in the first is accompanied by a change in the second. For example, a change in a person's weight may produce a change in the person's blood pressure. Some of the relationships between two quantities can be described by mathematical formulas. The simplest are those that show direct variation.

Direct variation

We say that a variable y **varies directly as** a variable x if

$$y = kx$$

where k is a constant called the **constant of variation**.

Sometimes the phrase "y is **directly proportional to** x" is used to describe direct variation, and the constant k is called the **constant of proportionality**.

EXAMPLE 3   The pressure of water at a certain point varies directly as the depth of the point below the surface of the water. At a depth of 100 meters (m), the pressure is approximately 10.5 atmospheres. What is the pressure at a depth of 250 m?

Solution   We let p = pressure and d = depth. Since p varies directly as d, there is a constant k such that

$$p = kd$$

We can use the given information to find k, the constant of variation. By substituting $p = 10.5$ and $d = 100$ into the equation, we obtain

$$10.5 = k \cdot 100$$

$$\frac{10.5}{100} = k$$

$$0.105 = k$$

Therefore, the relationship between pressure and depth is given by the equation

$$p = 0.105d$$

To find the pressure at a depth of 250 m, we substitute $d = 250$ into the equation above:

$$p = 0.105 \cdot 250$$

$$p = 26.25$$

There is a pressure of 26.25 atmospheres at a depth of 250 m. ∎

It may help to think of direct variation as the situation in which one quantity increases when the other quantity increases and vice versa. For example, as the depth below the surface increases, the water pressure increases.

EXAMPLE 4

When the brakes of a car are applied, the length of the skid marks made by its tires varies directly as the square of the speed of the car (before the brakes are applied). If skid marks 45 feet (ft) long are produced by a car that was going 30 mph, find the length of the skid marks produced by a car that was going 60 mph.

Solution

We let l = length of the car's skid marks and s = speed of the car. Since l varies directly as the *square* of s, there is a constant k such that

$$l = ks^2$$

As before, we can find the constant k by substituting the given information into the equation. We substitute $l = 45$ and $s = 30$ to obtain

$$45 = k \cdot 30^2$$

$$45 = k \cdot 900$$

$$\frac{45}{900} = k$$

$$0.05 = k$$

Therefore, the relationship between the length of the car's skid marks and the speed of the car before the brakes are applied is given by the equation

$$l = 0.05s^2$$

To find the length of the car's skid marks when it was going 60 mph, we substitute $s = 60$ into the equation above:

$$l = 0.05 \cdot 60^2$$

$$= 0.05 \cdot 3600$$

$$= 180$$

The car's skid marks are 180 ft long if it is going 60 mph. ∎

When two quantities are related so that they change in opposite directions, we have inverse variation.

Inverse variation

A variable y **varies inversely as** a variable x if

$$y = \frac{k}{x}$$

where k is a constant called the **constant of variation.**

Sometimes the phrase "y is **inversely proportional to** x" is used to describe inverse variation, and the constant k is called the **constant of proportionality.**

EXAMPLE 5   The current flowing in an electric circuit varies inversely as the resistance of the circuit. With a resistance of 8 ohms (Ω), the current in a circuit is approximately 12 amperes (A). What is the current when the resistance is 24 Ω?

Solution   We let c = current and r = resistance. Since c varies inversely as r, there is a constant k such that

$$c = \frac{k}{r}$$

To find the constant k, we substitute $c = 12$ and $r = 8$ into the equation and solve for k:

$$12 = \frac{k}{8}$$
$$8 \cdot 12 = k$$
$$96 = k$$

Therefore, the relationship between current and resistance is given by the equation

$$c = \frac{96}{r}$$

To find the current when the resistance is 24 Ω, we substitute $r = 24$ into the equation above:

$$c = \frac{96}{24}$$
$$= 4$$

The current is 4 A when the resistance is 24 Ω.   ∎

It may help to think of inverse variation as the situation in which one quantity increases as the other quantity decreases and vice versa. For example, as the resistance increases, the current decreases.

EXAMPLE 6

The weight of an object on or above the earth's surface varies inversely as the square of the distance of the object from the center of the earth. If an object weighs 200 pounds (lb) on the earth's surface, find its weight 1000 mi above the earth's surface. (Assume that the radius of the earth is 4000 mi.)

Solution

We let w = weight and d = distance from the center of the earth. Since w varies inversely as the *square* of d, there is a constant k such that

$$w = \frac{k}{d^2}$$

Since the object weighs 200 lb on the surface of the earth (4000 mi from the center), we can find k by substituting $w = 200$ and $d = 4000$ into the equation:

$$200 = \frac{k}{4000^2}$$

$$200 = \frac{k}{16,000,000}$$

$$16,000,000 \cdot 200 = k$$

$$3,200,000,000 = k$$

$$3.2 \cdot 10^9 = k$$

Therefore, the relationship between weight and distance from the center of the earth is given by the equation

$$w = \frac{3.2 \cdot 10^9}{d^2}$$

To find the weight of the object 1000 mi above the surface of the earth (5000 mi from the center), we substitute $d = 5000$ into the equation above:

$$w = \frac{3.2 \cdot 10^9}{5000^2}$$

Since $5000^2 = 25,000,000 = 2.5 \cdot 10^7$, we can write

$$w = \frac{3.2 \cdot 10^9}{2.5 \cdot 10^7}$$

$$= 1.28 \cdot 10^2$$

$$= 128$$

The object weighs 128 lb when it is 1000 mi above the surface of the earth. ∎

DISCUSSION QUESTIONS 6.6

1. What is the rate of work?

2. What is direct variation?

3. What is inverse variation?

EXERCISES 6.6

1. Vince can jog 4 mph faster than Raul. If Vince can jog 10 mi in the same time it takes Raul to jog 6 mi, how fast can Raul jog?

2. Sarah can ride a bike 10 mph faster than Joe. If Sarah can ride 75 mi in the same time it takes Joe to ride 60 mi, how fast can Sarah ride her bike?

3. A boat is traveling on a river that has a current of 4 mph. If the boat can travel 90 mi with the current in the same time it takes to travel 60 mi against the current, what is the speed of the boat in still water?

4. A helicopter can fly 125 mi against the wind in the same time it takes to fly 175 mi with the wind. If the wind is blowing at 25 mph, what is the speed of the helicopter in still air?

5. John can paint a room in 6 h working alone, and Antonio can paint the room in 8 h working alone. How long will it take John and Antonio to paint the room working together?

6. Karyl can mow a lawn in 4 h working alone, and Bill can mow the lawn in 6 h working alone. How long will it take Karyl and Bill to mow the lawn together?

7. Sean can type twice as fast as Judy. They can type a report in 6 h working together. How long would it take Sean to type the report working alone?

8. A mainframe computer can do a job three times faster than a personal computer. The two computers can do the job in 5 minutes (min) working together. How long would it take the mainframe computer to do the job working alone?

9. Maria and Bill can tune their car in 3 h working together. Bill can tune the car in 7 h working alone. How long would it take Maria to tune the car working alone?

10. Dave and Wayne can build a fence in 16 h working together. Dave can

build the fence in 30 h working alone. How long would it take Wayne to build the fence working alone?

11. A faucet can fill a sink in 9 min. The drain can empty the sink in 12 min. If the faucet is turned on and the drain is accidently left open, how long will it take to fill the sink?

12. A gasoline storage tank can be filled by an inlet pipe in 10 h. The outlet pipe can empty the tank in 15 h. If the inlet pipe and the outlet pipe are left open, how long will it be before the tank overflows?

In Exercises 13–18, write an equation that describes the given relationship.

13. D varies directly as p.

14. A is directly proportional to r^2

15. N is inversely proportional to d.

16. H varies inversely as y^2.

17. V varies directly as z^3.

18. S varies inversely as x^3.

19. If y varies directly as x, and $y = 12$ when $x = 3$, find y when $x = 20$.

20. If w is directly proportional to z, and $w = 80$ when $z = 10$, find w when $z = 4$.

21. If h is inversely proportional to l, and $h = 15$ when $l = 6$, find h when $l = 8$.

22. If p varies inversely as q, and $p = 21$ when $q = 3$, find p when $q = 9$.

23. If m varies directly as n^2, and $m = 50$ when $n = 5$, find m when $n = 2$.

24. If s is directly proportional to t^2, and $s = 36$ when $t = 2$, find s when $t = 3$.

25. Oxygen is required to burn carbon during combustion. The weight of oxygen needed varies directly as the weight of carbon. If 32 kilograms (kg) of oxygen are needed to burn 12 kg of carbon, how many kilograms of oxygen are needed to burn 50 kg of carbon?

26. The distance a spring stretches is directly proportional to the force applied. If a 30-lb force stretches the spring 18 inches (in.), how far will the spring stretch when a force of 12 lb is applied?

27. The exposure time needed to photograph an object varies directly as the square of the distance from the object to the light source. If the exposure time is 0.04 second (s) with a light 4 ft from the object, what is the exposure time with the light 10 ft from the object?

28. The pressure exerted by a gas on the container in which it is held varies inversely as the volume of the container. If the pressure is 40 lb/in.^{2} on a container whose volume is 125 in.^{3}, find the pressure on a container whose volume is 300 in.^{3}.

29. The force needed to raise an object with a crowbar is inversely proportional to the length of the crowbar. If it takes 45 lb to lift an object with a 3-ft crowbar, how much force will it take to lift the object with a 2-ft crowbar?

30. The rate of vibration of a string under constant tension varies inversely as the length of the string. If a string 48 in. long vibrates 256 times per second, what is the length of a string that vibrates 576 times per second?

31. The intensity of illumination varies inversely as the square of the distance from the source. Six feet from the light source the intensity is 10 footcandles. What is the intensity 4 ft from the light source?

32. The magnitude of the force that acts between two electrical charges is inversely proportional to the distance between the charges. A force of $8.4 \cdot 10^6$ newtons (N) acts on the charges when they are 0.1 m apart. What is the force when the charges are 0.25 m apart?

Preparing for Section 7.1

In Exercises 33–36, evaluate the given expression when $x = -1$ and $y = 3$.

33. $2x + y$ **34.** $3x + 2y$ **35.** $3y - 4x$ **36.** $2y - 5x$

In Exercises 37–40, find the value of y when $x = 2$.

37. $y = 3x + 8$ **38.** $y = 5 - 7x$ **39.** $x - 2y = 6$ **40.** $6x + 4y = 1$

Chapter Summary

Important Ideas

Rational expression

A rational expression is an expression of the form $\dfrac{A}{B}$, where A and B are polynomials and $B \neq 0$.

Fundamental principle for rational expressions

If $\dfrac{A}{B}$ is a rational expression with $B \neq 0$, and K is a nonzero expression, then $\dfrac{A \cdot K}{B \cdot K} = \dfrac{A}{B}$.

Complex fraction

A complex fraction is a fraction that contains fractions or rational expressions in its numerator or denominator or both.

Direct variation

We say that a variable y varies directly as a variable x if $y = kx$, where k is a constant called the constant of variation.

Inverse variation

We say that a variable y varies inversely as a variable x if $y = \dfrac{k}{x}$, where k is a constant called the constant of variation.

Chapter Review

Section 6.1

In Exercises 1–2, evaluate the given rational expression when $t = 2$.

1. $\dfrac{4t}{t+2}$ **2.** $\dfrac{t-4}{t^2-2t-8}$

In Exercises 3–6, find all values of the variable for which the given rational expression is undefined.

3. $\dfrac{3}{2v}$ **4.** $\dfrac{8}{3x-24}$ **5.** $\dfrac{10y}{y^2+4y-60}$ **6.** $\dfrac{u-3}{u^2+9}$

In Exercises 7–12, reduce the given rational expression to lowest terms.

7. $\dfrac{48w^2}{-18w^4}$ **8.** $\dfrac{3z-27}{2z^2-18z}$ **9.** $\dfrac{1-s^2}{s^2-3s+2}$

10. $\dfrac{5p^2-35p}{p^2-11p+28}$ **11.** $\dfrac{q^3-9q}{q^2-6q+9}$ **12.** $\dfrac{4r^2-8r+3}{6r^2-7r-3}$

Section 6.2

In Exercises 13–20, perform the indicated operation. Write the result in lowest terms.

13. $\dfrac{x^2}{2x+12} \cdot \dfrac{6}{x}$ **14.** $\dfrac{3y}{4y-2} \cdot \dfrac{2y-1}{6}$ **15.** $\dfrac{4k-20}{8k^3} \cdot \dfrac{12k}{k^2-25}$ **16.** $\dfrac{s^2-16}{3s-12} \cdot \dfrac{s^3}{s^2+4s}$

17. $\dfrac{8u^2}{u^2+5u+4} \cdot \dfrac{u^2+2u+1}{4u^3}$ **18.** $\dfrac{27v}{v^2+5v+6} \div \dfrac{3v^2}{v^2+v-2}$

19. $\dfrac{3n-9}{n^2+11n+24} \div \dfrac{18-6n}{n^2+10n+16}$ **20.** $\dfrac{m+6}{m^2-12m+36} \div \dfrac{m^2+3m-18}{m^2-9m+18}$

Section 6.3

In Exercises 21–26, find the least common denominator of the given pair of rational expressions.

21. $\dfrac{7}{54v}, \dfrac{13}{36v}$

22. $\dfrac{5}{12u^3}, \dfrac{11}{18u^7}$

23. $\dfrac{q}{4q^2 + 24q}, \dfrac{3}{8q^2}$

24. $\dfrac{15}{6p - 30}, \dfrac{p - 5}{p^2 + 4p - 45}$

25. $\dfrac{w}{w^2 - 4}, \dfrac{w + 2}{w^2 - 4w + 4}$

26. $\dfrac{5z}{3z^2 - z - 2}, \dfrac{z + 2}{2z^2 - z - 1}$

In Exercises 27–36, perform the indicated operation.

27. $\dfrac{3}{2h} + \dfrac{5}{2h}$

28. $\dfrac{m}{2m - 8} - \dfrac{4}{2m - 8}$

29. $\dfrac{3k}{k^2 - 9} - \dfrac{9}{k^2 - 9}$

30. $\dfrac{6}{6 - n} + \dfrac{n}{n - 6}$

31. $\dfrac{3r}{4r^2 - 1} + \dfrac{1}{2r + 1}$

32. $\dfrac{7}{6t - 42} - \dfrac{5}{t^2 - 10t + 21}$

33. $\dfrac{3}{v^2 + 8v} - \dfrac{1}{v^2 + 4v - 32}$

34. $\dfrac{2x + 3}{3x^2 + 2x - 8} + \dfrac{3x + 4}{2x^2 + x - 6}$

35. $\dfrac{1}{q + 4} - \dfrac{1}{2q - 8} + \dfrac{8}{q^2 - 16}$

36. $\dfrac{4}{y + 3} + \dfrac{y}{y - 6} - \dfrac{1}{y^2 - 3y - 18}$

Section 6.4

In Exercises 37–46, simplify the given complex fraction.

37. $\dfrac{w}{3 + \dfrac{1}{w}}$

38. $\dfrac{\dfrac{2}{y}}{y - \dfrac{1}{y}}$

39. $\dfrac{\dfrac{9}{h^3}}{\dfrac{12}{h}}$

40. $\dfrac{\dfrac{s^2}{s + 4}}{\dfrac{4s}{2s + 8}}$

41. $\dfrac{3 + \dfrac{1}{k}}{2 - \dfrac{3}{k}}$

42. $\dfrac{\dfrac{q^2 - 4}{6}}{\dfrac{2}{q} - 1}$

43. $\dfrac{2 + \dfrac{3}{p}}{\dfrac{2p + 3}{p^2}}$

44. $\dfrac{\dfrac{1}{n} - \dfrac{1}{3}}{\dfrac{1}{3} - \dfrac{1}{n}}$

45. $\dfrac{\dfrac{1}{100} - \dfrac{1}{m^2}}{\dfrac{1}{10} + \dfrac{1}{m}}$

46. $\dfrac{\dfrac{2}{u} + \dfrac{1}{u + 1}}{\dfrac{3}{u + 1} - \dfrac{1}{u}}$

Section 6.5

In Exercises 47–56, solve the given equation.

47. $\dfrac{12 + 3k}{8} = 15$

48. $\dfrac{4}{w} - \dfrac{1}{w} = 6$

49. $\dfrac{2}{3h} - \dfrac{9}{h} = 25$

50. $\dfrac{4}{5x + 3} = \dfrac{2}{3x}$

51. $\dfrac{5}{3z} + \dfrac{5}{4z} = \dfrac{7}{12}$

52. $\dfrac{1}{t} + \dfrac{1}{t - 1} = \dfrac{5}{t - 1}$

53. $1 - \dfrac{4}{q} - \dfrac{5}{q^2} = 0$

54. $\dfrac{6}{p + 8} = 1 - \dfrac{6}{p - 8}$

55. $\dfrac{m^2}{3m + 6} + 1 = \dfrac{2}{m + 2}$

56. $\dfrac{2n}{n - 4} - \dfrac{n}{n + 6} = \dfrac{80}{n^2 + 2n - 24}$

Section 6.6

57. Barbara can run 3 mph faster than Barry. If Barbara can run 10 mi in the same time it takes Barry to run 8 mi, how fast can Barbara run?

58. A small plane cruises at 180 mph in still air. If the plane can fly 300 mi with the wind in the same time it takes to fly 240 mi against the wind, what is the speed of the wind?

59. Ian can assemble a bicycle in 3 h and Mark can assemble the bicycle in 5 h. How long will it take Ian and Mark to assemble the bicycle working together?

60. Myrene can plant a garden twice as fast as Ann. They can plant the garden in 4 h working together. How long would it take Myrene to plant the garden working alone?

In Exercises 61–64, write an equation that describes the given relationship.

61. P varies directly as s.

62. V is directly proportional to r^2.

63. W is inversely proportional to l.

64. K varies inversely as h^2.

65. If y varies directly as x, and $y = 12$ when $x = 4$, find y when $x = 10$.

66. If w is inversely proportional to z, and $w = 0.25$ when $z = 2$, find w when $z = 0.5$.

67. If q is directly proportional to p^2, and $q = 13.5$ when $p = 3$, find q when $p = 1.5$.

68. If s varies inversely as t^2, and $s = 4$ when $t = 5$, find s when $t = 15$.

69. The heat loss per hour through a glass window of a house is inversely proportional to the thickness of the window. If 6000 calories of heat are lost per hour through a window 0.25 cm thick, what is the heat loss per hour through a window 0.6 cm thick?

70. The amount of oil used by a ship varies directly as the distance traveled. If the ship uses 500 barrels of oil to go 2000 mi, how much oil is used to go 4500 mi?

Chapter 6 Test

In Problems 1–2, find all values of the variable for which the given rational expression is undefined.

1. $\dfrac{2y}{y^2 - 4y}$

2. $\dfrac{6z + 3}{z^2 - 7z - 30}$

In Problems 3–4, reduce the given rational expression to lowest terms.

3. $\dfrac{8p^3}{4p^2 - 12p}$

4. $\dfrac{6q - 12}{3q^2 + 12q - 36}$

In Problems 5–6, find the least common denominator of each pair of rational expressions.

5. $\dfrac{5}{24m^3},\ \dfrac{3}{40m^7}$

6. $\dfrac{6n}{n^2 + n - 12},\ \dfrac{12}{n^2 - 9}$

In Problems 7–15, perform the indicated operation.

7. $\dfrac{8h}{18h^4} \cdot \dfrac{27h^5}{12h^2}$

8. $\dfrac{2k + 6}{4k^3} \cdot \dfrac{10k^2}{5k + 15}$

9. $\dfrac{3r - 4}{r^2 + 8r} \div \dfrac{8 - 6r}{r^2 + 5r - 24}$

10. $\dfrac{t^2 - 4}{t^2 + 4t + 3} \div \dfrac{t^2 - 7t - 18}{t^2 - 6t - 27}$

11. $\dfrac{2x}{x^2 - 3x} - \dfrac{6}{x^2 - 3x}$

12. $\dfrac{2}{3q} + \dfrac{4q}{3q + 24}$

13. $\dfrac{p}{p^2 - 1} + \dfrac{2}{p^2 - 3p + 2}$

14. $\dfrac{8 - 2s}{s - 4} - \dfrac{s^2}{4 - s}$

15. $\dfrac{6}{r^2 + 4r - 5} - \dfrac{5}{3r + 15}$

In Problems 16–17, simplify the given complex fraction.

16. $\dfrac{9 - \dfrac{1}{h^2}}{\dfrac{3h + 1}{h}}$

17. $\dfrac{1 - \dfrac{8}{k}}{\dfrac{1}{k^2} - \dfrac{1}{64}}$

In Problems 18–21, solve the given equation.

18. $\dfrac{1}{6m} + \dfrac{1}{9m} = \dfrac{5}{18}$

19. $\dfrac{3}{n + 3} - \dfrac{1}{2n} = \dfrac{2}{n}$

20. $\dfrac{1}{z + 4} + \dfrac{z + 6}{z^2 + 2z - 8} = 0$

21. $\dfrac{w}{w + 2} - \dfrac{3}{3w - 1} = \dfrac{6w - 9}{3w^2 + 5w - 2}$

22. A river is flowing at 6 mph. If it takes a boat as long to go 36 mi with the current as to go 27 mi against the current, what is the speed of the boat in still water?

23. A new copier can complete a certain job in 2 h. An old copier can do the job in 3 h. How long would it take both copiers to do the job working together?

24. If u varies inversely as v, and $u = 24$ when $v = 3$, find u when $v = 8$.

25. The value of a diamond is directly proportional to the square of its weight. If a diamond weighing 3 carats is valued at $8400, what is the value of a diamond weighing 4.5 carats?

7 Linear Equations and Inequalities in Two Variables

7.1 Solutions of Linear Equations

FOCUS

We discuss linear equations in two variables and find their solutions.

Recall from Chapter 1 that the cruise control feature of a car is used to keep the car moving at a constant speed on the open highway. For example, if the cruise control is set to 55 miles per hour (mph), then each hour the car travels 55 miles (mi). One way to describe the mathematical relationship between time and distance when the cruise control is set to 55 mph is by making a table:

Time (hours)	Distance (miles)
1	55
2	110
3	165
4	220
5	275

The two numbers in each row of the table form an **ordered pair.** For example, the ordered pair (1, 55) indicates that in 1 hour (h) the car travels 55 mi. The other ordered pairs in the table are (2, 110), (3, 165), (4, 220), and (5, 275). The pairs are said to be *ordered* because the first number in each ordered pair is always a time, and the second number is always the corresponding distance.

Another way to describe the relationship between time and distance when the cruise control is set to 55 mph is to use the equation $y = 55x$, where y is the number of miles traveled in x hours. The equation $y = 55x$ is an example of a linear equation in two variables. To see that it describes the same relationship, we can check that each ordered pair (x, y) in the table satisfies the equation. To do so, we substitute the x and y values of each ordered pair into the equation:

(x, y)	$y = 55x$
(1, 55)	$55 = 55 \cdot 1$
(2, 110)	$110 = 55 \cdot 2$
(3, 165)	$165 = 55 \cdot 3$
(4, 220)	$220 = 55 \cdot 4$
(5, 275)	$275 = 55 \cdot 5$

Note that each ordered pair makes the equation $y = 55x$ a true statement. Therefore, each ordered pair is a solution of the linear equation $y = 55x$.

The example above provides the basis for the following definition:

Linear equation in two variables

A **linear equation in two variables** is an equation that can be expressed in the form

$$ax + by = c$$

where a, b, and c are real numbers and a and b are not both zero.

The equation $y = 55x$ is a linear equation in two variables because it can be expressed in the form $55x - y = 0$. (Of course, other variables besides x and y can be used to write linear equations in two variables.) A linear equation in two variables indicates that there is a *linear relationship* between the variables.

A solution of a linear equation is an ordered pair that makes the equation a true statement.

Solution of a linear equation

A **solution of a linear equation** in two variables is an ordered pair of numbers (x, y) that satisfies the equation $ax + by = c$.

When the variables are x and y, the first number in the ordered pair is always associated with the variable x and is called the **x coordinate.** The second number in the ordered pair is always associated with the variable y and is called the **y coordinate.**

EXAMPLE 1

Determine whether each ordered pair is a solution of the equation $3x + 2y = 8$.

a. (2, 1)   **b.** (−4, 10)   **c.** (3, 5)   **d.** (6, −5)

Solution

a. To determine whether the ordered pair (2, 1) is a solution of $3x + 2y = 8$, we substitute $x = 2$ and $y = 1$ into the equation $3x + 2y = 8$:

$$3x + 2y = 8$$
$$3 \cdot 2 + 2 \cdot 1 = 8 \qquad \text{Substituting } x = 2 \text{ and } y = 1$$
$$6 + 2 = 8$$
$$8 = 8 \qquad \text{True}$$

Since the ordered pair (2, 1) makes the equation $3x + 2y = 8$ a true statement, (2, 1) is a solution of $3x + 2y = 8$.

b. We substitute $x = -4$ and $y = 10$ into the equation $3x + 2y = 8$:

$$3x + 2y = 8$$
$$3 \cdot (-4) + 2 \cdot 10 = 8 \qquad \text{Substituting } x = -4 \text{ and } y = 10$$
$$(-12) + 20 = 8$$
$$8 = 8 \qquad \text{True}$$

Therefore, (−4, 10) is a solution of $3x + 2y = 8$.

c. We substitute $x = 3$ and $y = 5$ into the equation $3x + 2y = 8$:

$$3x + 2y = 8$$
$$3 \cdot 3 + 2 \cdot 5 = 8$$
$$9 + 10 = 8$$
$$19 = 8 \qquad \text{False}$$

Since the ordered pair (3, 5) does *not* make the equation $3x + 2y = 8$ a true statement, (3, 5) is *not* a solution of $3x + 2y = 8$.

d. We substitute $x = 6$ and $y = -5$ into the equation $3x + 2y = 8$:

$$3x + 2y = 8$$
$$3 \cdot 6 + 2 \cdot (-5) = 8$$
$$18 + (-10) = 8$$
$$8 = 8 \qquad \text{True}$$

Therefore, (6, −5) is a solution of $3x + 2y = 8$.   ∎

BE CAREFUL!

Do not incorrectly interchange the x coordinate and the y coordinate in an ordered pair. For example, the ordered pair (2, 1) is not the same as the ordered pair (1, 2). The ordered pair (2, 1) means that $x = 2$ and $y = 1$, whereas the ordered pair (1, 2) means that $x = 1$ and $y = 2$.

The cruise control example and Example 1 illustrate that a linear equation in two variables has more than one solution. In general, a linear equation in two variables has an infinite number of solutions. To find a particular solution, we substitute a number for one of the variables and then solve the equation for the other variable. We demonstrate this procedure in the next example.

EXAMPLE 2

Complete each ordered pair to obtain a solution of $x - 3y = 9$.

a. (0,)   **b.** (, 5)   **c.** (−6,)   **d.** (, −2)

Solution

a. We complete the ordered pair (0,) by substituting $x = 0$ into the equation $x - 3y = 9$ and then solving for y:

$$x - 3y = 9$$
$$0 - 3y = 9 \qquad \text{Substituting } x = 0$$

Note that this substitution gives a linear equation in one variable, which you know how to solve:

$$-3y = 9$$
$$y = -3 \qquad \text{Dividing both sides by } -3$$

The ordered pair (0, −3) is a solution of the equation $x - 3y = 9$.

b. We complete the ordered pair (, 5) by substituting $y = 5$ into the equation $x - 3y = 9$ and then solving for x:

$$x - 3y = 9$$
$$x - 3 \cdot 5 = 9 \qquad \text{Substituting } y = 5$$
$$x - 15 = 9$$
$$x = 24 \qquad \text{Adding 15 to both sides}$$

The ordered pair (24, 5) is a solution of the equation $x - 3y = 9$.

c.   We substitute $x = -6$ into the equation $x - 3y = 9$ and then solve for y:

$$x - 3y = 9$$
$$(-6) - 3y = 9$$
$$-3y = 15$$
$$y = -5$$

The ordered pair $(-6, -5)$ is a solution of the equation $x - 3y = 9$.

d.   We substitute $y = -2$ into the equation $x - 3y = 9$ and then solve for x:

$$x - 3y = 9$$
$$x - 3 \cdot (-2) = 9$$
$$x + 6 = 9$$
$$x = 3$$

The ordered pair $(3, -2)$ is a solution of the equation $x - 3y = 9$.   ∎

The ordered-pair solutions of a linear equation in two variables are often listed in a **table of values.** For the four solutions of the equation $x - 3y = 9$ of Example 2, we have the following table of x and y values:

x	y
0	−3
24	5
−6	−5
3	−2

A table of values is useful in graphing linear equations, as you will see in the next section. To make one, we assign values to one of the variables (usually x), and then complete the ordered pairs as in Example 2.

DISCUSSION QUESTIONS 7.1

1.   What is a linear equation in two variables?

2.   Is $ax = by$ a linear equation in two variables? Explain.

3.   What is a solution of a linear equation in two variables?

4.   Is (x, y) different than (y, x) for all values of x and y? Explain.

EXERCISES 7.1

In Exercises 1–15, determine whether the ordered pair is a solution of the
given equation.

1. (4, 1); $x + 2y = 6$

2. (3, 7); $4x - y = 5$

3. (−2, 4); $y = 3x + 10$

4. (1, −2); $x = 5y + 11$

5. (5, −3); $3x - 2y = 9$

6. (−4, 2); $5x + 3y = -15$

7. $\left(\dfrac{1}{2}, 3\right)$; $2x + y = 4$

8. $\left(2, \dfrac{1}{3}\right)$; $x + 3y = 3$

9. (−8, −1); $x = 8y$

10. (−1, −5); $y = 5x$

11. (6, 2); $2x + 3y = -6$

12. (2, 3); $3x - 4y = 18$

13. (2, 0.8); $3x - 5y = 2$

14. (0.5, 1); $4x - y = 1$

15. (3, −2); $y = -2$

In Exercises 16–28, complete the ordered pair to make it a solution of the
given equation.

16. (, 3); $x + y = 4$

17. (, 3); $x = 5y - 1$

18. (5,); $y = 2x + 6$

19. (−5,); $3x + 4y = 1$

20. (, −1); $6x - 8y = 10$

21. (, 0.8); $2x - 5y = -4$

22. (0.4,); $10x + 5y = 9$

23. (, −4); $y = -10x$

24. (−2,); $x = -6y$

25. (1,); $4x - y = -7$

26. (, 5); $x + 3y = 1$

27. (, 9); $x = 6$

28. (4,); $y = 7$

In Exercises 29–34 complete the table of values for the given equation.

29. $x + 6y = 3$

x	y
0	
9	
	2
	5

30. $8x - y = 12$

x	y
	0
	4
3	
7	

31. $6x - 4y = 24$

x	y
	0
2	
	−6
−8	

32. $3x + 9y = 18$

x	y
0	
	1
−9	
	−2

33. $y = 3x$

x	y
	6
	−2
5	
−4	

34. $x = 2y$

x	y
4	
−1	
	3
	−5

In Exercises 35–44: **a.** Give three ordered pairs (x,y) that satisfy the given conditions.

b. Write an equation that describes the linear relationship between x and y.

35. The y coordinate is 2 less than the x coordinate.

36. The x coordinate is 7 more than the y coordinate.

37. The sum of the x coordinate and the y coordinate is 6.

38. The sum of the x coordinate and the y coordinate is -9.

39. The y coordinate is 12 times the x coordinate.

40. The x coordinate is 8 times the y coordinate.

41. The x coordinate is 5 more than 12 times the y coordinate.

42. The y coordinate is 6 less than 4 times the x coordinate.

43. The y coordinate is -11.

44. The x coordinate is 10.

45. The perimeter of a rectangle is 20 inches (in.). Write an equation that describes the relationship between the length l and the width w of the rectangle. Is it a linear relationship? Explain.

46. Write an equation that describes the relationship between the perimeter of a square and the length of a side of the square. (Let P = perimeter and s = length of a side.) Is it a linear relationship? Explain.

Preparing for Section 7.2

In Exercises 47–50, find y when $x = 0$.

47. $x + y = 7$ 48. $x - 3y = 8$ 49. $5x - 2y = 11$ 50. $4x + 7y = 16$

In Exercises 51–54, solve the given equation.

51. $3x - 13 = 2$ 52. $17 - 6y = 1$ 53. $14 + 5y = -8$ 54. $9x + 4 = -3$

7.2   Graphs of Linear Equations

FOCUS

We review the graphing of linear equations and discuss the special properties of lines through the origin, vertical lines, and horizontal lines.

Recall from Chapter 3 that ordered pairs (x, y) are graphed on a **rectangular coordinate system** that is constructed of two perpendicular number lines. The horizontal number line is called the **x axis,** and the vertical number line is called the **y axis.** The point of intersection of the axes is called the **origin.** To graph an ordered pair, say (3, 4), we start at the origin. Since the x coordinate is 3, we go 3 units to the right on the x axis. Since the y coordinate is 4, we then go up 4 units on a line parallel to the y axis. (See Figure 7.1.) The process of locating the point associated with the ordered pair (3, 4) is called *plotting* the point (3, 4).

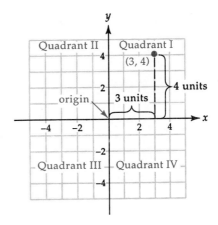

FIGURE 7.1

The rectangular coordinate system divides the plane into four regions, called **quadrants.** As shown in Figure 7.1, the quadrants are numbered counterclockwise, with the axes forming the boundary lines between the quadrants; the axes themselves are not part of any quadrant. The point (3, 4) is located in the first quadrant because both the x coordinate and the y coordinate are positive.

EXAMPLE 1

Plot each ordered pair and determine the quadrant in which it lies.

a. $A(-1, 3)$ **b.** $B(4, -2)$ **c.** $C(-3, -1)$ **d.** $D(2, 0)$

Solution

The ordered pairs are plotted in Figure 7.2.

a. To plot the point $A(-1, 3)$, we start at the origin. Since the x coordinate is -1, we move 1 unit to the left on the x axis. Since the y coordinate is 3, we then go up 3 units on a line parallel to the y axis. The point $A(-1, 3)$ is in quadrant II.

b. The point $B(4, -2)$ is in quadrant IV.

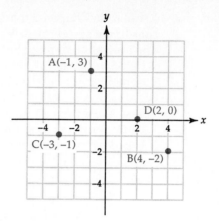

FIGURE 7.2

  c.   The point $C(-3, -1)$ is in quadrant III.

  d.   The point $D(2, 0)$ is on the positive x axis and therefore on the boundary between quadrants I and IV.   ■

    To graph a linear equation in two variables, we plot three points that are ordered-pair solutions of the equation. We find each ordered pair by assigning a real number to one of the variables in the equation and then solving for the other. For instance, to find an ordered-pair solution of the linear equation $x + y = 2$, we can substitute $x = 0$ into the equation and solve for y:

$$x + y = 2$$

$$0 + y = 2 \qquad \text{Substituting } x = 0$$

$$y = 2$$

Therefore, the ordered pair $(0, 2)$ is a solution of the equation. Similarly, the ordered pairs $(-2, 4)$ and $(3, -1)$ are also solutions. These points are plotted in Figure 7.3.

    By drawing the straight line through these points, as shown in Figure 7.4, we obtain the graph of the linear equation $x + y = 2$. This graph is a "picture" or visual representation of all the solutions of $x + y = 2$. That is, every ordered pair that satisfies the linear equation $x + y = 2$ lies on this straight line. Conversely, every point on the straight line has x and y coordinates that satisfy the linear equation $x + y = 2$.

    The example above shows that the **graph of a linear equation** in two variables consists of all ordered pairs that satisfy the equation. When these points are plotted, the result is a straight line. Since a straight line is determined by two distinct points, we can graph a linear equation by finding two distinct ordered-pair solutions of the equation. However, we always plot a third point as a check.

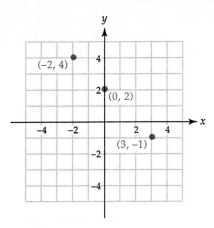

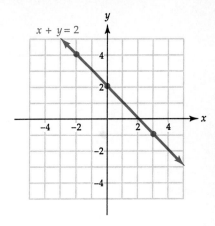

FIGURE 7.3   **FIGURE 7.4**

EXAMPLE 2

Graph the linear equation $x - 3y = 9$.

Solution

We want three distinct ordered-pair solutions to graph the equation $x - 3y = 9$. To simplify the computations, we will first let $x = 0$ in the equation and solve for y. Then we will let $y = 0$ and solve for x. Finally, we will let $x = 3$ and solve for y. The results are listed in the following table:

x	y
0	−3
9	0
3	−2

These points are plotted in Figure 7.5, along with the straight line passing through them. The line is the graph of the linear equation $x - 3y = 9$.

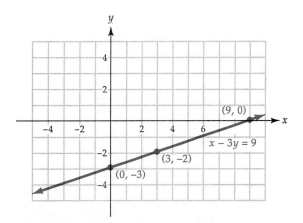

FIGURE 7.5   ∎

The graph of the linear equation $x - 3y = 9$ in Figure 7.5 intersects the x axis at the point $(9, 0)$, and it intersects the y axis at the point $(0, -3)$. The numbers 9 and -3 are called the x intercept and the y intercept, respectively, of the graph.

Intercepts of a graph of an equation in two variables

The x **intercept** of a graph is the x coordinate of the point where the graph intersects the x axis. Similarly, the y **intercept** of a graph is the y coordinate of the point where the graph intersects the y axis.

As we saw in Example 2, the intercepts are especially useful for graphing linear equations in two variables. They are found as follows:

1. To find the x intercept, let $y = 0$ in the given equation and solve for x.
2. To find the y intercept, let $x = 0$ in the given equation and solve for y.

EXAMPLE 3

Find the intercepts of the graph of $8x + 6y = -24$ and use them to graph the equation.

Solution

To find the x intercept, we substitute $y = 0$ into the equation and then solve for x:

$$8x + 6y = -24$$
$$8x + 6 \cdot 0 = -24 \qquad \text{Substituting } y = 0$$
$$8x = -24$$
$$x = -3 \qquad \text{Dividing both sides by 8}$$

The x intercept is -3, so the graph intersects the x axis at the point $(-3, 0)$. To find the y intercept, we substitute $x = 0$ into the equation and then solve for y:

$$8x + 6y = -24$$
$$8 \cdot 0 + 6y = -24 \qquad \text{Substituting } x = 0$$
$$6y = -24$$
$$y = -4 \qquad \text{Dividing both sides by 6}$$

The y intercept is -4, so the graph intersects the y axis at the point $(0, -4)$. As a check, we find a third point by letting x or y equal some other number, say $y = 2$; then $x = -4\frac{1}{2}$ or -4.5. Our third point is $(-4.5, 2)$. The coordinates of the three points are thus

x	y
-3	0
0	-4
-4.5	2

These points and the line connecting them are plotted in Figure 7.6. The line is the graph of the linear equation $8x + 6y = -24$.

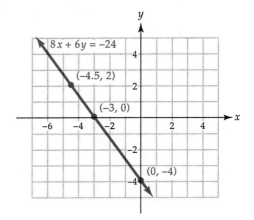

FIGURE 7.6

EXAMPLE 4   Graph the linear equation $y = 4x$.

Solution   The x intercept of the graph of $y = 4x$ is 0, because if $y = 0$, then $x = 0$. However, the y intercept of the graph is also 0, because if $x = 0$, then $y = 0$. Thus, the graph of $y = 4x$ intersects *both* the x axis and the y axis at the origin $(0, 0)$.

Therefore we still need at least a second point in order to graph the equation $y = 4x$. If we let $x = 1$, then $y = 4$, so a second point is $(1, 4)$. We also find a third point as a check: If $x = -2$, then $y = -8$, so the third point is $(-2, -8)$. The coordinates of the three points are

x	y
0	0
1	4
-2	-8

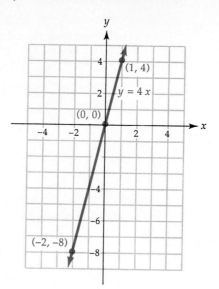

FIGURE 7.7

The points are plotted in Figure 7.7. The line passing through them is the graph of the linear equation $y = 4x$. ∎

The results of Example 4 suggest that *the graph of the linear equation $y = ax$ passes through the origin (0, 0).* Both the x intercept and the y intercept of the graph are zero.

We have seen in the examples above that the graph of a linear equation in two variables has an x intercept and a y intercept. However, as the next examples show, some linear equations have only one intercept.

EXAMPLE 5

Graph the given linear equation.

a. $x = 2$ **b.** $y = -1$

Solution

a. The linear equation $x = 2$ is equivalent to the equation in two variables $x + 0y = 2$. Hence the equation $x = 2$ indicates that for every value of y, the x value is 2. Ordered-pair solutions of the equation are thus of the form $(2, y)$, where y is any real number. The coordinates of three such ordered pairs are

x	y
2	1
2	0
2	-3

These points are plotted in Figure 7.8. When we draw the line through them, we find that *the graph of the linear equation $x = 2$ is a vertical*

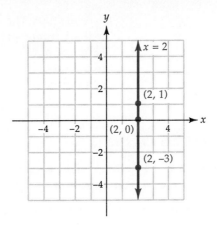

FIGURE 7.8

line. Note that this graph has an x intercept of 2 but does not have a y intercept.

b. The linear equation $y = -1$ indicates that for any value of x, the y value is -1. Therefore, ordered-pair solutions of the equation $y = -1$ are of the form $(x, -1)$, where x is any real number. The coordinates of three such ordered pairs are

x	y
4	-1
0	-1
-2	-1

These points are plotted in Figure 7.9; they show that the graph of the linear equation $y = -1$ is a horizontal line. Note that this graph has a y intercept of -1 but does not have an x intercept.

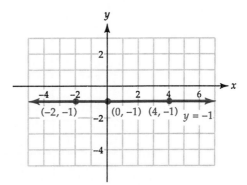

FIGURE 7.9

The results of Example 5 suggest that:

1. *The graph of the linear equation $x = c$ is a vertical line that passes through*

the point (c, 0). The x intercept of the graph is c. The graph does not have a y intercept.

2. *The graph of the linear equation y = c is a horizontal line that passes through the point (0, c). The y intercept of the graph is c. The graph does not have an x intercept.*

DISCUSSION QUESTIONS 7.2

1. If $x > 0$ and $y > 0$, in what quadrant does each point lie?
   a. $(x, -y)$ **b.** $(-x, y)$ **c.** $(-x, -y)$ **d.** (x, y)

2. If $x > 0$ and $y > 0$, state how to plot the point $(-x, y)$.

3. What is the graph of a linear equation in two variables?

4. Does the graph of the equation $x + 4y = 0$ pass through the origin? Explain.

EXERCISES 7.2

In Exercises 1–12, plot the given point and state the quadrant in which it lies.

1. $(-3, 4)$ **2.** $(6, -1)$ **3.** $(1, -2)$ **4.** $(-5, 3)$ **5.** $(7, 2)$ **6.** $(5, 9)$

7. $(-4, -9)$ **8.** $(-10, -4)$ **9.** $(2, 0)$ **10.** $(0, 6)$ **11.** $(0, -1)$ **12.** $(-5, 0)$

In Exercises 13–20, write the ordered pair that corresponds to the given point as plotted in Figure 7.10.

13. *A*   **14.** *B*   **15.** *C*   **16.** *D*   **17.** *E*   **18.** *F*   **19.** *G*   **20.** *H*

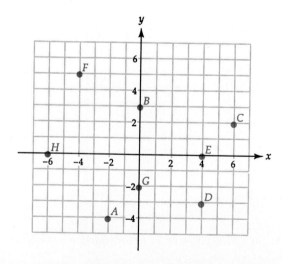

FIGURE 7.10

In Exercises 21–28, complete the table of values for the given equation. Graph the equation.

21.

x	y
0	
1	
-2	

$y = 3x + 1$

22.

x	y
0	
2	
-1	

$y = 2x - 5$

23.

x	y
2	
	0
-4	

$x + 2y = 6$

24.

x	y
3	
	0
-6	

$x + 3y = 9$

25.

x	y
	0
8	
	-1

$x = -4y$

26.

x	y
0	
	10
-3	

$y = 5x$

27.

x	y
2	
9	
-1	

$y = 3$

28.

x	y
	5
	-1
	-3

$x = -7$

In Exercises 29–50, graph the given equation and find its intercepts.

29. $y = 2x - 8$

30. $y = 6x + 12$

31. $x - 3y = 9$

32. $x - 4y = 8$

33. $x - y = -2$

34. $x + y = 5$

35. $2x + 3y = 18$

36. $3x - 2y = -6$

37. $4x - 6y = 24$

38. $4x + 3y = 12$

39. $x = 3y$

40. $y = -4x$

41. $y = -9x$

42. $x = 2y$

43. $2x - 5y = 15$

44. $3x + 2y = 21$

45. $5x + 4y = 3$

46. $7x - 10y = 4$

47. $x = 4$

48. $y = 6$

49. $y + 8 = 0$

50. $x - 3 = 0$

51. A rental car company charges $30 per day and $0.20 per mile to rent a mid-sized car. A salesperson rents the car for one day.
   a. What does it cost the salesperson to drive 125 mi?
   b. Write an equation that gives the salesperson's cost y to drive x miles.
   c. Graph the equation of part b.
   d. If it cost the salesperson $59.80 to rent the car, how many miles did she drive?

52. A salesperson is paid a salary of $250 per week plus a commission of 15% of his gross sales.
   a. What is the salesperson's weekly pay when gross sales are $1000?

b. Write an equation that gives the salesperson's weekly pay y when his gross sales are x dollars.

c. Graph the equation of part b.

d. If the salesperson's pay for one week is $681, what were his gross sales?

Preparing for Section 7.3

In Exercises 53–56, compute the value of each expression.

53. $\dfrac{6-1}{4-2}$ **54.** $\dfrac{8-2}{7-4}$ **55.** $\dfrac{1-6}{2-4}$ **56.** $\dfrac{2-8}{4-7}$

In Exercises 57–60, find m.

57. $m = \dfrac{3-(-6)}{(-7)-5}$ **58.** $m = \dfrac{(-8)-4}{7-(-9)}$ **59.** $m = \dfrac{(-10)-5}{(-4)-(-14)}$ **60.** $m = \dfrac{(-16)-(-12)}{(-7)-13}$

7.3   The Slope of a Line

FOCUS

We discuss the concept of slope as a measure of the steepness of a line and as a rate of change. The slope formula is developed and applied to graphing linear equations.

We usually think of slope as a measure of "steepness," as in the slope of a hill or the slope of a pitched roof. But we can also measure the steepness of a line. To do so, we use the notion that there is a horizontal change and a vertical change when we move from one point to another on a line. Consider the two points $A(x_1, y_1)$ and $B(x_2, y_2)$, both on the line in Figure 7.11.

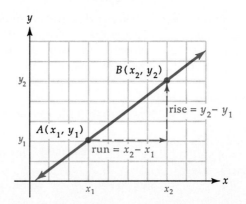

FIGURE 7.11

(The x coordinates x_1 and x_2 are read "x-sub-one" and "x-sub-two," respectively. The number 1 in x_1 and the number 2 in x_2 are called **subscripts.** Subscript notation allows us to distinguish between the coordinates of the two points while indicating which are x coordinates and which are y coordinates.) Note that we move through a vertical change, called the **rise,** and a horizontal change, called the **run,** as we move from A to B along the line.

As we move from $A(x_1, y_1)$ to $B(x_2, y_2)$, the y coordinate changes from y_1 to y_2. Therefore the rise is $y_2 - y_1$, the difference between the y coordinates. Similarly, the run is $x_2 - x_1$, the difference between the x coordinates. The slope of the line, usually designated by the letter m, is defined as the ratio of the rise to the run:

Slope of a line

The **slope** m of the line that passes through the points $A(x_1, y_1)$ and $B(x_2, y_2)$ is

$$m = \frac{\text{Rise}}{\text{Run}} = \frac{y_2 - y_1}{x_2 - x_1}$$

The slope m, as defined, is a measure of the steepness of the line. The slope of the line is the same no matter which two points on the line are chosen to compute m.

EXAMPLE 1   Find the slope of the line that passes through the points $(1, 4)$ and $(3, 7)$.

Solution   We let $(x_1, y_1) = (1, 4)$ and $(x_2, y_2) = (3, 7)$ and use the definition of the slope of a line:

$$m = \frac{\text{Rise}}{\text{Run}} = \frac{y_2 - y_1}{x_2 - x_1}$$

$$= \frac{7 - 4}{3 - 1}$$

$$= \frac{3}{2}$$

The slope of the line is $\frac{3}{2}$ or 1.5, and the line is graphed in Figure 7.12. The slope can also be found by letting $(x_1, y_1) = (3, 7)$ and $(x_2, y_2) = (1, 4)$. This gives

$$m = \frac{y_2 - y_1}{x_2 - x_1}$$

$$= \frac{4 - 7}{1 - 3}$$

$$= \frac{-3}{-2}$$

$$= \frac{3}{2}$$

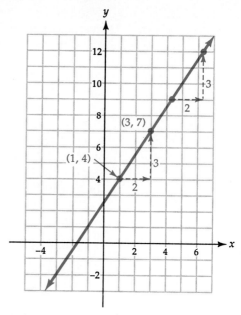

FIGURE 7.12

BE CAREFUL!

It makes no difference which point is called (x_1, y_1) and which is called (x_2, y_2) when computing the slope of a line. However, you must subtract the x and y coordinates of one point from the corresponding coordinates of the *other* point. Remember that the y difference must be in the numerator and the x difference in the denominator.

We can see in Figure 7.12 that every run of 2 units produces a rise of 3 units. Therefore, the slope $\frac{3}{2}$ indicates that every 2-unit increase in x produces a 3-unit increase in y. Since $m = \dfrac{\text{Rise}}{\text{Run}} = \dfrac{3}{2} = \dfrac{1.5}{1}$, the slope also indicates that y increases by 1.5 units for each unit increase in x. In general, we have the following important result:

> If m is the slope of a line, then y changes by m units for each unit increase in x.

The slope of a line is the **rate of change** in y, with respect to x, for a unit increase in x. That is, the slope tells how much y changes for each unit increase in x.

EXAMPLE 2

Find the slope of the line that passes through the points $(-2, 5)$ and $(3, -3)$.

Solution

We let $(x_1, y_1) = (-2, 5)$ and $(x_2, y_2) = (3, -3)$ and compute the slope of the line:

$$m = \frac{y_2 - y_1}{x_2 - x_1}$$

$$= \frac{(-3) - 5}{3 - (-2)}$$

$$= \frac{-8}{5}$$

$$= -\frac{8}{5}$$

The slope of the line is $-\frac{8}{5}$, which indicates that every run of 5 units produces a rise of -8 units (that is, a *fall* of 8 units). Since $m = -\frac{8}{5} = -1.6$, y changes by -1.6 units (y *decreases* by 1.6 units) for each unit increase in x. Figure 7.13 shows the line and the slope.

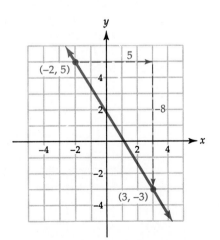

FIGURE 7.13

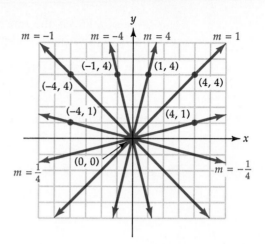

FIGURE 7.14

Examples 1 and 2 show that a line with positive slope rises from left to right, and a line with negative slope falls from left to right. Furthermore, the steeper the line, the greater its slope (in absolute value). These ideas are illustrated in Figure 7.14.

The slope of a line can be used to graph the line when one point on the line is known.

EXAMPLE 3   Graph the line that passes through the point $(-3, -1)$ and has slope $\dfrac{1}{4}$.

Solution   Recall that to graph the line we must find two points on the line. The slope $\dfrac{1}{4}$ indicates that every run of 4 units produces a rise of 1 unit. To graphically find a second point on the line, we would start at the given point $(-3, -1)$ in Figure 7.15a, and move 4 units to the right (because the run is 4) and 1 unit up (because the rise is 1). This would put us at the point $(1, 0)$. We would then draw the line that passes through the points $(-3, -1)$ and $(1, 0)$ as in Figure 7.15b.   ■

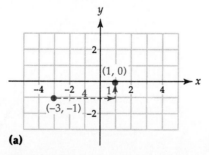

(a)

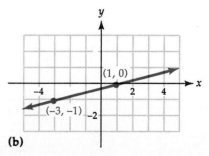

(b)

FIGURE 7.15

EXAMPLE 4

Find the slope of the line that passes through the points $(-1, 4)$ and $(2, 4)$.

Solution

As shown in Figure 7.16, the line that passes through the points $(-1, 4)$ and $(2, 4)$ is horizontal. We let $(x_1, y_1) = (-1, 4)$ and $(x_2, y_2) = (2, 4)$ and compute the slope of the line:

$$m = \frac{y_2 - y_1}{x_2 - x_1}$$

$$= \frac{4 - 4}{2 - (-1)}$$

$$= \frac{0}{3}$$

$$= 0$$

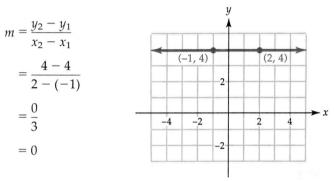

FIGURE 7.16

The slope of the line is 0 because the rise is 0. That is, there is no change in the y coordinates between the given points. In general, the slope of any horizontal line is 0 because every point on a horizontal line has the same y coordinate.   ■

EXAMPLE 5

Find the slope of the line that passes through the points $(-1, -5)$ and $(-1, 3)$.

Solution

As shown in Figure 7.17, the line through the points $(-1, -5)$ and $(-1, 3)$ is vertical. We let $(x_1, y_1) = (-1, -5)$ and $(x_2, y_2) = (-1, 3)$ and compute the slope of the line:

$$m = \frac{y_2 - y_1}{x_2 - x_1}$$

$$= \frac{3 - (-5)}{(-1) - (-1)}$$

$$= \frac{8}{0}$$

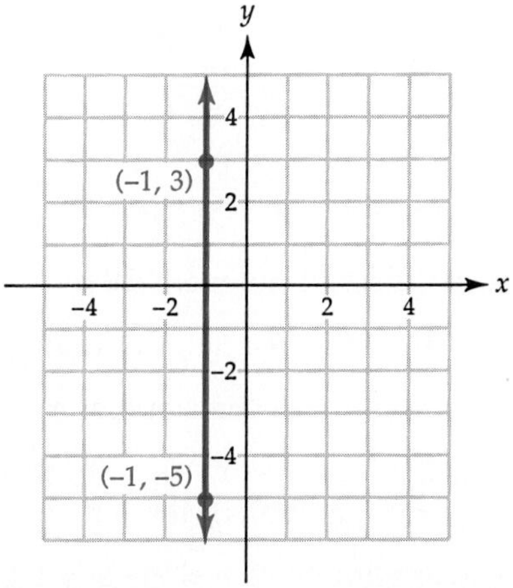

FIGURE 7.17

The slope of this line is *undefined* because the run, which is the denominator in the slope formula, is 0. That is, there is no change in the x coordinates between the given points. The slope of every vertical line is undefined because every point on a vertical line has the same x coordinate. ■

These facts are summarized in Figure 7.18.

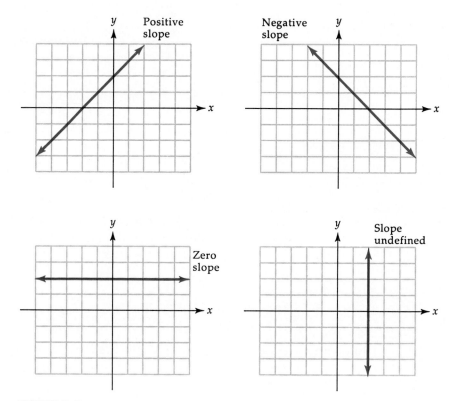

FIGURE 7.18

DISCUSSION QUESTIONS 7.3

1. What is subscript notation? Why is it useful?

2. How is the slope of a line computed?

3. In what sense does the slope of a line indicate a rate of change?

4. Two lines have the same slope but different y intercepts. What can you say about the graphs of the lines?

5. Does every line have a slope and a y intercept? Explain.

EXERCISES 7.3

In Exercises 1–20, find the slope of the line that passes through the given points.

1. (2, 5), (4, 9)

2. (6, 1), (3, 7)

3. (3, 4), (5, 2)

4. (5, 3), (0, 13)

5. (−3, 1), (2, −4)

6. (4, 8), (−4, −2)

7. (−4, −2), (6, 11)

8. (7, −3), (−6, 2)

9. (6, −7), (−2, −9)

10. (−1, 10), (−3, 6)

11. (−5, 8), (0, −3)

12. (4, 7), (−2, 0)

13. (−1, 2), (3, 2)

14. (2, −7), (−5, −7)

15. (6, −1), (6, −9)

16. (−3, 5), (−3, 4)

17. $\left(\frac{3}{8}, 2\right), \left(\frac{5}{8}, -6\right)$

18. $\left(-1, \frac{5}{3}\right), \left(7, \frac{1}{3}\right)$

19. $\left(\frac{2}{3}, -\frac{8}{3}\right), \left(-\frac{1}{3}, \frac{4}{3}\right)$

20. $\left(\frac{7}{4}, -\frac{3}{4}\right), \left(-\frac{1}{4}, -\frac{9}{4}\right)$

In Exercises 21–28, find the slope of the given line.

21.

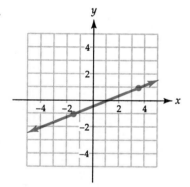

FIGURE 7.19

22.

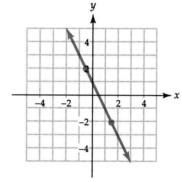

FIGURE 7.20

23.

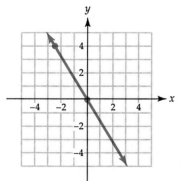

FIGURE 7.21

24.

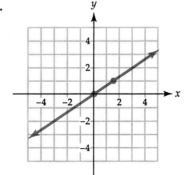

FIGURE 7.22

25.

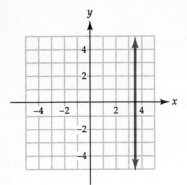

FIGURE 7.23

26.

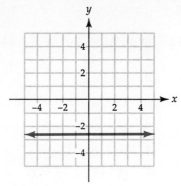

FIGURE 7.24

27.

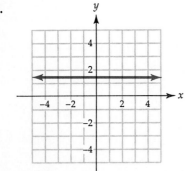

FIGURE 7.25

28.

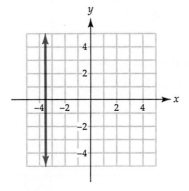

FIGURE 7.26

In Exercises 29–46, graph the line that passes through the given point and has the given slope.

29. $(0, 0)$, $m = 4$

30. $(0, 0)$, $m = -2$

31. $(0, 0)$, $m = -5$

32. $(0, 0)$, $m = 7$

33. x intercept -2, $m = 3$

34. y intercept 9, $m = -1$

35. y intercept 6, $m = -2$

36. x intercept -8, $m = 6$

37. $(-2, 5)$, $m = -\dfrac{3}{4}$

38. $(4, -7)$, $m = \dfrac{2}{7}$

39. $(3, -1)$, $m = \dfrac{3}{2}$

40. $(-2, 1)$, $m = -\dfrac{7}{3}$

41. $(0, 2)$, $m = -1.6$

42. $(1, 0)$, $m = 1.25$

43. $(-4, 0)$, $m = 0$

44. $(0, -3)$, $m = 0$

45. $(-1, -3)$, m is undefined

46. $(-5, -1)$, m is undefined

47. The recommended heart rate for a healthy person when exercising is given by the equation $y = 172 - 0.75x$, where y is the heart rate and x is the age of the person in years.

   a. Graph the equation $y = 172 - 0.75x$ and find the slope of the line.

   b. If a person's recommended heart rate is 142 today, what heart rate would be recommended 3 years from today?

48. The speed of an object thrown upward with an initial velocity of 80 ft/s is given by the equation $y = 80 - 32x$, where y is the speed of the object in feet per second after x seconds.

   a. Graph the equation $y = 80 - 32x$ and find the slope of the line.

   b. If the speed of an object thrown upward is currently 76 ft/s, what will be its speed 2 seconds from now?

Preparing for Section 7.4

In Exercises 49–54, solve the given equation for y.

49. $x + y = 10$

50. $x - 2y = 4$

51. $x = -0.5y$

52. $x + \dfrac{7}{8}y = 0$

53. $\dfrac{y - 9}{x + 1} = 6$

54. $\dfrac{y + 6}{x - 15} = -2$

7.4   The Slope-Intercept Form

FOCUS

We develop the slope-intercept form of a linear equation. Then we use it to find the equation of a line.

A linear relationship exists between temperatures measured in degrees Celsius (°C) and temperatures measured in degrees Fahrenheit (°F). This means that a linear equation in two variables can be used to convert degrees Celsius to degrees Fahrenheit and vice versa. You may already know how to convert some familiar temperatures. For example, at sea level water freezes at 0°C, which is equivalent to 32°F, and it boils at 100°C, which is equivalent to 212°F. We can use this information to find a linear equation that describes the relationship between degrees Celsius and degrees Fahrenheit, and then use it to convert other temperatures.

   If we let x = temperature measured in °C and y = temperature measured in °F, then the freezing point of water can be represented by the ordered pair (0, 32). Similarly, the boiling point of water can be represented by the ordered pair (100, 212). We plot these ordered pairs and draw the line through the points in Figure 7.27. We can compute the slope of the line by using the points (0, 32) and (100, 212) in the slope formula:

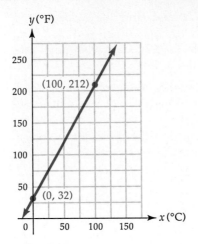

FIGURE 7.27   FIGURE 7.28

$$m = \frac{212 - 32}{100 - 0}$$

$$= \frac{180}{100}$$

$$= \frac{9}{5} \text{ or } 1.8$$

The slope 1.8 indicates that there is a 1.8° increase in the Fahrenheit temperature for each 1° increase in the Celsius temperature. Note that the y intercept of the line is 32, since the graph crosses the y axis at the point (0, 32).

To find the equation of the line, we first select any other point (x, y) on the line (see Figure 7.28). Next we substitute the slope, the coordinates of one of the given points, and the general point (x, y) into the slope formula. We use $m = 1.8$, $(x_1, y_1) = (100, 212)$, and $(x_2, y_2) = (x, y)$. [*Note:* We could also use $(x_1, y_1) = (0, 32)$ and get the same result.] This gives us

$$\frac{y_2 - y_1}{x_2 - x_1} = m$$

$$\frac{y - 212}{x - 100} = 1.8$$

We now solve this equation for y:

$$y - 212 = 1.8(x - 100) \qquad \text{Multiplying both sides by } x - 100$$

$$y - 212 = 1.8x - 180 \qquad \text{By the distributive property}$$

$$y = 1.8x + 32 \qquad \text{Adding 212 to both sides}$$

Note the form of the equation of the line: The slope of the line, 1.8, is the coefficient of x, and the y intercept, 32, is the constant term. We say that the linear equation $y = 1.8x + 32$ is in **slope-intercept form.** In general,

---

Slope-intercept form of the equation of a line

The equation of the line with slope m and y intercept b is

$$y = mx + b$$

We can express any linear equation in two variables in slope-intercept form by solving the equation for y. When the equation is written in slope-intercept form, the coefficient of x is the slope and the constant term is the y intercept.

EXAMPLE 1

Find the slope and the y intercept of the line whose equation is $5x + 2y = 14$.

Solution

We can express the equation $5x + 2y = 14$ in slope-intercept form by solving for y:

$$5x + 2y = 14$$
$$2y = -5x + 14 \qquad \text{Subtracting } 5x \text{ from both sides}$$
$$y = -\frac{5}{2}x + 7 \qquad \text{Dividing both sides by 2}$$

When the equation is in this form, the coefficient of x is the slope and the constant term is the y intercept. Therefore, the slope is $-\dfrac{5}{2}$, or -2.5, and the y intercept is 7. ∎

As shown in the following examples, the slope-intercept form of a line can be used to find an equation of the line.

EXAMPLE 2

Find an equation of the line that passes through the point $(0, -2)$ and has slope 3. Graph the equation.

Solution

The y intercept is -2, since the line passes through the point $(0, -2)$. The slope is given as 3. We can write the equation of the line in slope-intercept form by substituting $m = 3$ and $b = -2$ as follows:

$$y = mx + b$$
$$= 3x - 2$$

The equation of the line is $y = 3x - 2$. To find another point on the line, we start at the given point $(0, -2)$, as shown in Figure 7.29. The line has slope $3 = \dfrac{3}{1}$, so we move 1 unit to the right and then 3 units up. This produces the point $(1, 1)$. We draw the line through the points $(0, -2)$ and $(1, 1)$, as shown. (Another way to find a second point on the line is to substitute a number for x in the equation $y = 3x - 2$ and compute the value of y.)

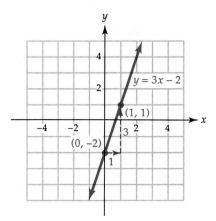

FIGURE 7.29

EXAMPLE 3

Find an equation of the line that passes through the point $(-1, 6)$ and has slope -2. Graph the equation.

Solution

Since the slope of the line is -2, we can substitute $m = -2$ into the slope-intercept form. We get

$$y = mx + b$$
$$y = -2x + b$$

We must now determine b, the y intercept. We know that the line passes through the point $(-1, 6)$, so $(-1, 6)$ is a solution of the equation $y = -2x + b$. Hence, we can substitute $x = -1$ and $y = 6$ into the equation $y = -2x + b$ and solve for b:

$$y = -2x + b$$
$$6 = -2(-1) + b$$
$$6 = 2 + b$$
$$4 = b$$

The y intercept is 4, and an equation of the line is $y = -2x + 4$. You should check that the point $(-1, 6)$ satisfies this equation. The graph of $y = -2x + 4$ is shown in Figure 7.30.

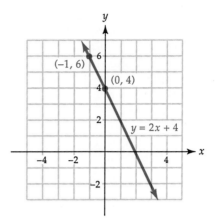

FIGURE 7.30   ■

EXAMPLE 4

Find an equation of the line that passes through the points $(-1, 1)$ and $(1, -2)$. Graph the equation.

Solution

We must find the slope and the y intercept of the line to use the slope-intercept form. We can find the slope of the line by using the points $(-1, 1)$ and $(1, -2)$ in the slope formula:

$$m = \frac{(-2) - 1}{1 - (-1)}$$

$$= \frac{-3}{2}$$

$$= -\frac{3}{2} \text{ or } -1.5$$

We can find the y intercept as in Example 3. We substitute $m = -1.5$ into the slope-intercept form to obtain

$$y = mx + b$$

$$y = -1.5x + b$$

Since the line passes through the point $(-1, 1)$, we substitute $x = -1$ and $y = 1$ into the equation $y = -1.5x + b$ and solve for b:

$$y = -1.5x + b$$
$$1 = -1.5(-1) + b$$
$$1 = 1.5 + b$$
$$-0.5 = b$$

[This result could also be obtained by using the point $(1, -2)$ and substituting $x = 1$ and $y = -2$ into the equation $y = -1.5x + b$.] The y intercept is -0.5, and an equation of the line is $y = -1.5x - 0.5$. You should check that the points $(-1, 1)$ and $(1, -2)$ satisfy this equation. The graph of $y = -1.5x - 0.5$ is shown in Figure 7.31.

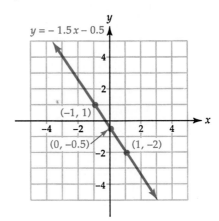

FIGURE 7.31

DISCUSSION QUESTIONS 7.4

1. Paper burns at approximately 451°F. At what temperature, in degrees Celsius, does paper burn?
2. If the temperature of a room decreases by 5°C, what is the decrease in degrees Fahrenheit?
3. What is the slope-intercept form of a horizontal line?
4. What is the slope-intercept form of a vertical line? Explain.

EXERCISES 7.4

In Exercises 1–12, find the slope and the y intercept of the line which has the given equation.

1. $y = 5x - 3$
2. $y = 4x - 7$
3. $x - 3 = 0$
4. $y + 5 = 0$

5. $3x + 2y = 10$
6. $4x - 3y = 9$
7. $4x - 6y = 0$
8. $2x + 5y = 0$

9. $\dfrac{x}{3} - 2y = 6$
10. $9x + \dfrac{y}{3} = 5$
11. $\dfrac{3}{4}x + \dfrac{2}{3}y = 1$
12. $\dfrac{3}{5}x - \dfrac{5}{6}y = -2$

In Exercises 13–22, write an equation of the line that passes through the given point and has the given slope.

13.   (0, 4), $m = 7$ **14.**   (0, −1), $m = 4$ **15.**   (0, 0), $m = \dfrac{2}{3}$ **16.**   (0, 0), $m = \dfrac{1}{5}$

17.   (−2, 5), $m = -4$ **18.**   (−3, −2), $m = -7$ **19.**   (2, 3), $m = 0$ **20.**   (4, −3), $m = 0$

21.   (6, −4), m is undefined   **22.**   (−1, −5), m is undefined

In Exercises 23–34, write an equation of the line that passes through the given points.

23.   (−3, 6), (1, 2)   **24.**   (2, −3), (5, −6)   **25.**   (−1, 9), (−5, −6)   **26.**   (3, −6), (−1, −9)

27.   (5, −3), (5, 8)   **28.**   (9, 5), (−2, 5)   **29.**   (7, −1), (−4, −1)   **30.**   (−2, 3), (−2, 0)

31. $\left(\dfrac{1}{5}, -3\right), \left(\dfrac{4}{5}, 6\right)$ **32.** $\left(7, -\dfrac{3}{4}\right), \left(-5, \dfrac{1}{4}\right)$ **33.** $\left(-1, \dfrac{7}{8}\right), \left(\dfrac{5}{8}, -3\right)$ **34.** $\left(-\dfrac{5}{6}, 4\right), \left(-7, -\dfrac{1}{6}\right)$

35.   Find an equation of the line with x intercept 3 and y intercept 5.

36.   Find an equation of the line with x intercept −4 and y intercept 1.

37.   Find an equation of the line with x intercept 2 and y intercept −3.

38.   Find an equation of the line with x intercept −2 and y intercept −7.

39.   A linear relationship exists between the speed of sound at sea level [measured in meters per second (m/s)] and the temperature (measured in degrees Celsius). Consider the information in the table below, where x = temperature and y = speed of sound:

x, °C	y, m/s
20	335
100	395

   a. Write an equation that describes the linear relationship between the speed of sound and temperature.
   b. What is the speed of sound at 0°C?
   c. At what temperature would sound have no velocity?

40.   The length of a stretched wire when carrying a load of 2 pounds (lb) is 2.08 feet (ft); when carrying a load of 20 lb, it is 2.36 ft. Let x = weight of a load, y = length of the wire, and assume that a linear relationship exists between the length of the wire and the weight of the load.
   a. Write an equation that describes the linear relationship between the length of the wire and the weight.
   b. What is the length of the unstretched wire (carrying no weight)?
   c. Does the equation in part a apply to all weights? Explain.

Preparing for Section 7.5

In Exercises 41–48, decide whether the given inequality is a true statement when $x = 3$ and $y = 2$.

41. $x + y > 4$ **42.** $x - y \leq 1$ **43.** $6x - y \leq 3$ **44.** $x + 7y > 17$

45. $3x + 5y < 15$ **46.** $2x - 9y \geq -7$ **47.** $2x - 7y \geq -8$ **48.** $6x - 3y < 12$

7.5   Graphs of Linear Inequalities

FOCUS

We discuss linear inequalities in two variables and graph their solutions.

A contractor must move at least 200 tons of earth before construction can begin on an office building. There are two trucks available to move the earth. The smaller truck has a capacity of 5 tons, and the larger truck has a capacity of 8 tons. How many truckloads are needed to move the earth?

To answer this question, we let $x =$ the number of loads of earth moved by the smaller truck and $y =$ the number of loads moved by the larger truck. Since the smaller truck moves 5 tons of earth with each load, it moves a total of $5x$ tons of earth. Similarly, the larger truck moves a total of $8y$ tons of earth. Therefore the number of tons of earth moved by the two trucks is $5x + 8y$, and the contractor's situation can now be described by the following linear inequality in two variables:

$$5x + 8y \geq 200$$

The linear inequality says that 200 or more tons of earth are moved by the two trucks.

If we replace the equals sign in a linear equation with one of the inequality symbols $<$, $>$, $\leq$, or $\geq$, we get a **linear inequality in two variables.** A **solution of a linear inequality** in two variables is an ordered pair of numbers (x, y) that makes the inequality a true statement. For example, the ordered pair (20, 15) is a solution of the linear inequality $5x + 8y \geq 200$. To show that it is, we substitute $x = 20$ and $y = 15$ into the inequality:

$$5x + 8y \geq 200$$
$$5 \cdot 20 + 8 \cdot 15 \geq 200 \qquad \text{Substituting } x = 20 \text{ and } y = 15$$
$$100 + 120 \geq 200$$
$$220 \geq 200 \qquad \text{True}$$

You can check that the ordered pair $(-5, 35)$ is also a solution of the inequality. In general, a linear inequality has an infinite number of solutions. In the contractor's situation described above, the solutions of $5x + 8y \geq 200$ are restricted to ordered pairs (x, y) where $x \geq 0$ and $y \geq 0$ because the number of loads carried by either truck cannot be negative.

The **graph of a linear inequality** in two variables is the region of the coordinate plane that contains all ordered-pair solutions of the inequality. The boundary of the region is a line; it is determined by replacing the inequality symbol with an equals sign and then graphing the resulting linear equation.

EXAMPLE 1

Graph the linear inequality $5x + 8y \geq 200$.

Solution

The graph of $5x + 8y \geq 200$ consists of all ordered pairs (x, y) that satisfy either the inequality $5x + 8y > 200$ or the equation $5x + 8y = 200$. We begin the solution by graphing the boundary line $5x + 8y = 200$. This is done in Figure 7.32, where we have used the x and y intercepts, $(40, 0)$ and $(0, 25)$, respectively, to draw the line.

The boundary line separates the plane into two regions, called **half-planes,** one above the line and one below the line. The graph of $5x + 8y \geq 200$ consists of one of these half-planes *and* the boundary line. The boundary line is included in the graph of $5x + 8y \geq 200$ because the line contains all ordered pairs (x, y) that satisfy the equation $5x + 8y = 200$. We have drawn the line *solid* to indicate that it belongs to the graph of $5x + 8y \geq 200$.

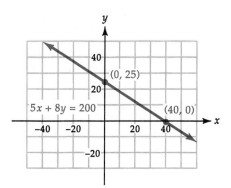

FIGURE 7.32

To determine which half-plane is included in the graph of $5x + 8y \geq 200$, we choose a **test point** that does *not* lie on the boundary line. Although we could choose any point not on the boundary line, we choose the origin $(0, 0)$ to make the computations easy. We now substitute $x = 0$ and $y = 0$ into the inequality to determine whether the resulting statement is true or false:

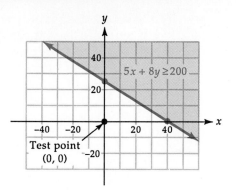

FIGURE 7.33

$$5x + 8y \geq 200$$
$$5 \cdot 0 + 8 \cdot 0 \geq 200$$
$$0 \geq 200 \quad \text{False}$$

Since the last statement is false, indicating that (0, 0) is *not* a solution of $5x + 8y \geq 200$, the half-plane that does *not* contain the test point (0, 0) must be part of the graph of $5x + 8y \geq 200$. As shown in Figure 7.33, the graph of $5x + 8y \geq 200$ thus consists of the upper half-plane and the solid boundary line (shown in color). We note that the solutions of $5x + 8y \geq 200$ for the contractor's situation would be restricted to the points on the graph that are also in quadrant I, where $x \geq 0$ and $y \geq 0$.   ■

EXAMPLE 2   Graph the linear inequality $y < 6x + 12$.

Solution   We begin by graphing the boundary line $y = 6x + 12$, as in Figure 7.34. We have used the x and y intercepts, $(-2, 0)$ and $(0, 12)$, respectively, to draw the line. The points on the boundary line, which satisfy the equation $y = 6x + 12$, are not solutions of the inequality $y < 6x + 12$ because the inequal-

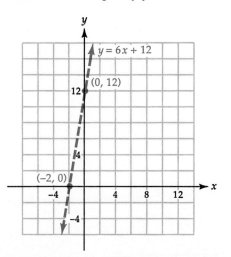

FIGURE 7.34

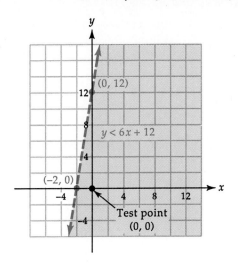

FIGURE 7.35

ity symbol is not $\leq$. We show this in the figure by drawing the boundary as a *broken* line.

A convenient test point is again the origin (0, 0). It gives us

$$y < 6x + 12$$
$$0 < 6 \cdot 0 + 12$$
$$0 < 12 \qquad \text{True}$$

Since the last statement is true, the ordered pair (0, 0) is a solution of the inequality $y < 6x + 12$. Therefore the graph of $y < 6x + 12$ is the shaded half-plane that contains the point (0, 0), as shown in Figure 7.35   ∎

EXAMPLE 3

Solution

Graph the linear inequality $y > -3x$.

We begin by graphing the boundary line $y = -3x$. The boundary line, drawn in Figure 7.36, is not included in the graph of $y > -3x$ because the

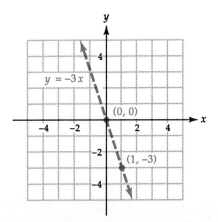

FIGURE 7.36

inequality symbol is not ≥. Notice that we cannot choose (0, 0) as the test point in this problem because (0, 0) lies on the boundary line. Therefore we choose another test point, say (1, 0). This gives us

$$y > -3x$$
$$0 > -3 \qquad \text{True}$$

Since the last statement is true, the ordered pair (1, 0) is a solution of $y > -3x$. Therefore the graph of $y > -3x$ consists of the half-plane that contains the point (1, 0), as shown shaded in Figure 7.37.

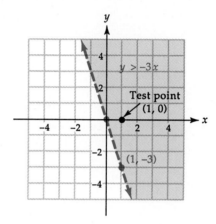

FIGURE 7.37 ■

We can summarize the procedure for graphing linear inequalities as follows:

To graph a linear inequality

1. Graph the boundary line whose equation is found by replacing the inequality symbol with an equals sign. Draw a solid boundary line if the inequality symbol is ≤ or ≥. Draw a broken boundary line if the inequality symbol is < or >.
2. Choose a test point not on the boundary line, and substitute the coordinates of the test point into the original inequality. If the result is a true statement, then shade the half-plane that contains the test point. If the result is a false statement, then shade the half-plane that does not contain the test point.

EXAMPLE 4   Find the inequality whose solution is graphed in Figure 7.38.

Solution   We start by finding the equation of the boundary line. We can compute the slope of the line by substituting (4, 0) and (0, −4) into the slope formula:

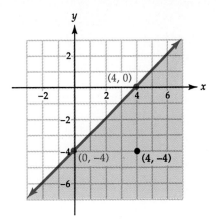

FIGURE 7.38

$$m = \frac{4 - 0}{0 - (-4)}$$

$$= \frac{4}{4}$$

$$= 1$$

The line passes through the point $(0, -4)$, so it has y intercept -4. Now we can substitute $m = 1$ and $b = -4$ into the slope-intercept form of the equation of a line. This gives us

$$y = mx + b$$

$$y = x - 4$$

The equation of the boundary line is $y = x - 4$. Since the boundary is drawn as a solid line, the inequality must be either $\leq$ or $\geq$. To determine which one, we choose a point (solution) within the shaded half-plane, say $(4, -4)$, and find which inequality is satisfied. That is, we substitute $x = 4$ and $y = -4$ into the inequalities $y \leq x - 4$ and $y \geq x - 4$ to find which one is a true statement.

$y \leq x - 4$		$y \geq x - 4$	
$-4 \leq 4 - 4$		$-4 \geq 4 - 4$	
$-4 \leq 0$	True	$-4 \geq 0$	False

Therefore the required inequality is $y \leq x - 4$.   ■

DISCUSSION QUESTIONS 7.5

1. What is a linear inequality? What is its solution?

2. What is the graph of a linear inequality?

3. What is a test point? How is it used to graph the solution of a linear inequality?

EXERCISES 7.5

In Exercises 1–24, graph the given linear inequality.

1. $y < 2x + 7$ 2. $y \le 3x - 5$ 3. $y \ge 5x - 1$ 4. $y > 4x + 2$

5. $y > -8x$ 6. $y \ge 0.25x$ 7. $x \ge -2$ 8. $y < 4$

9. $x \ge 4y$ 10. $x \le -y$ 11. $x - y \le 0$ 12. $x - 3y > 0$

13. $y < 7$ 14. $x \ge -1$ 15. $0.2x + 0.5y > -1$ 16. $x + y \ge 1$

17. $x + y \le 3$ 18. $\dfrac{x}{4} + \dfrac{y}{6} < 2$ 19. $4x + y > 8$ 20. $3x - y \ge 6$

21. $6x + 8y \ge 24$ 22. $5x - 6y > -30$ 23. $2x - 3y > 5$ 24. $3x + 5y \ge 2$

In Exercises 25–32, find the inequality whose solution is represented by the given graph.

25.

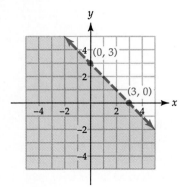

FIGURE 7.39

26.

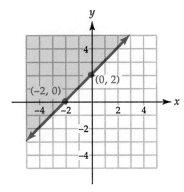

FIGURE 7.40

27.

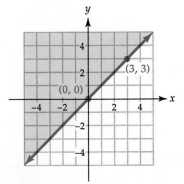

FIGURE 7.41

28.

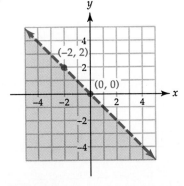

FIGURE 7.42

29.

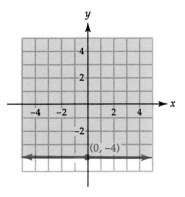

FIGURE 7.43

30.

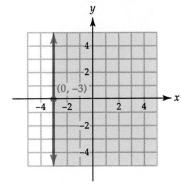

FIGURE 7.44

31.

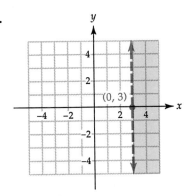

FIGURE 7.45

32.

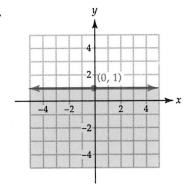

FIGURE 7.46

33. An electronics company makes regular calculators and scientific calculators. It takes 4 hours (h) of labor to make a regular calculator and 6 h of labor to make a scientific calculator. The company has at most 108 h of labor per day to manufacture the calculators. Let x = number of regular calculators made per day and y = number of scientific calculators made per day.
  a. Write an inequality that describes the company's situation.
  b. Graph the inequality of part a.
  c. Is (10, 18) a solution of the inequality? Interpret your answer in terms of the given situation.

34. A dietician in a hospital must create a special diet with at least 120 units of vitamin A, using two foods. Food C has 10 units of vitamin A per ounce, and food D has 30 units of vitamin A per ounce. Let x = ounces of food C in the diet and y = ounces of food D in the diet.
  a. Write a linear inequality that describes the dietician's situation.
  b. Graph the inequality of part a.
  c. Is (6, 2) a solution of the inequality? Interpret your answer in terms of the given situation.

Preparing for Section 8.1

In Exercises 35–38, show that the ordered pair (5, 2) satisfies both the given equations.

35. $x + y = 7$ and $2x - y = 8$

36. $x + 2y = 9$ and $x - y = 3$

37. $y = 3x - 13$ and $y = 0.4x$

38. $y = -4x + 22$ and $x = 2.5y$

In Exercises 39–42, graph the given pair of equations on the same coordinate system.

39. $y = 3x$ and $y = x + 4$

40. $y = -2x$ and $y = x - 3$

41. $x - y = 5$ and $4x + 3y = 6$

42. $x + y = 8$ and $2x - 3y = 26$

Chapter Summary

Important Ideas

Slope of a line $m = \dfrac{y_2 - y_1}{x_2 - x_1}$

Slope-intercept form of the equation of a line $y = mx + b$

Chapter Review

Section 7.1

In Exercises 1–4, determine whether the ordered pair is a solution of the given equation.

1. $(3, 4)$; $2x + y = 10$ **2.** $(2, -3)$; $x - 2y = 8$ **3.** $(8, 2)$; $y = 4x$ **4.** $(15, 3)$; $x = 5y$

In Exercises 5–8, complete the ordered pair to make it a solution of the given equation.

5. $(-1, \ )$; $3x - 2y = 1$ **6.** $(\ , -2)$; $4x + 3y = -2$ **7.** $(\ , -5)$; $x = -3$ **8.** $(-4, \ )$; $y = 6$

In Exercises 9–10, complete the table of values for the given equation.

9.

x	y
0	
2	
	0
	-3

$3x + 4y = 12$

10.

x	y
	0
	-2
0	
6	

$5x - 10y = 20$

11. **a.** Give three ordered pairs (x, y) that satisfy the following condition: The y coordinate is 3 more than 6 times the x coordinate.
    b. Write a linear equation that describes the linear relationship between x and y given by the condition in part a.

12. The cost of a regular checking account at a local bank is $5.00 per month plus $0.20 per check written. Complete the table below:

Checks Written	Monthly Cost
5	
10	
15	
20	
30	
50	

If x = number of checks written and y = monthly cost of the checking account, write an equation that describes the relationship between checks written and monthly cost. Is it a linear relationship? Explain.

Section 7.2

In Exercises 13–16, plot the point and determine the quadrant in which it lies.

13. $(2, -5)$ **14.** $(-6, 3)$ **15.** $(-1, -4)$ **16.** $(0, 5)$

In Exercises 17–20, write the ordered pair that corresponds to the given point as plotted in Figure 7.47.

17. A **18.** B **19.** C **20.** D

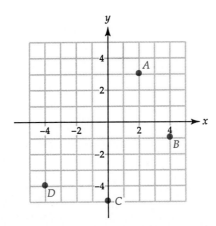

FIGURE 7.47

In Exercises 21–22, complete the table of values for the given equation. Graph the equation.

21.

x	y
0	
3	
−1	

$y = 4x + 8$

22.

x	y
0	
4	
−2	

$3x − 6y = 12$

In Exercises 23–29, graph the given equation and find its intercepts.

23. $5x − y = 10$ **24.** $2x − 7y = 14$ **25.** $8x + 4y = 16$ **26.** $x = −3y$ **27.** $y = 7x$

28. $y = 4$ **29.** $x = −3$

Section 7.3

In Exercises 30–33, find the slope of the line that passes through the given points.

30. $(3, −1), (6, 8)$ **31.** $(−2, 9), (4, 3)$ **32.** $(−7, 3), (−1, 3)$ **33.** $(6, −3), (6, 2)$

In Exercises 34–35, find the slope of the given line.

34.

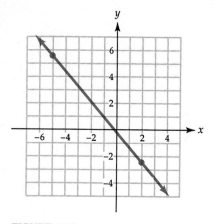

FIGURE 7.48

35.

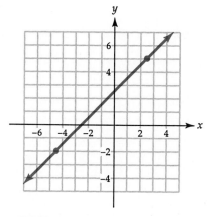

FIGURE 7.49

In Exercises 36–40, graph the line that passes through the given point and has the given slope.

36. $(0, −3), m = 4$ **37.** $(−2, −5), m = \dfrac{3}{8}$ **38.** $(5, 1), m = −\dfrac{7}{5}$

39. (2, 6), m is undefined   **40.** (−1, −6), $m = 0$

41. The height of a candle y (measured in centimeters) after burning for x hours is given by the equation $y = 15 − 2x$.
   a. Graph the equation $y = 15 − 2x$ and find the slope of the line.
   b. If the height of the candle is currently 10 centimeters (cm), what will be its height 4 h from now?

Section 7.4

In Exercises 42–47, find the slope and y intercept of the line which has the given equation.

42. $y = 4x + 10$ **43.** $2x + 4y = 16$ **44.** $3x + 6y = 0$

45. $x = 2y$ **46.** $x = 2$ **47.** $y − 1 = 0$

In Exercises 48–54, write an equation of the line that passes through the given point and has the given slope.

48. (0, 0), $m = 3$ **49.** (0, 4), $m = −2$ **50.** (−8, 0), $m = 1.75$

51. (−2, 3), $m = −\dfrac{1}{2}$ **52.** (5, −8), $m = \dfrac{7}{4}$ **53.** (1, −3), $m = 0$

54. (−7, 3), m is undefined

In Exercises 55–61, write an equation of the line that passes through the given points.

55. (3, 2), (5, 10)   **56.** (6, 8), (−3, 5)   **57.** (−3, 1), (4, −6)   **58.** (3, −9), (−1, −4)

59. (6, −1), (6, 2)   **60.** (−2, 4), (3, 4)   **61.** $\left(-1, \dfrac{3}{5}\right), \left(\dfrac{3}{4}, 2\right)$

62. The income tax for taxpayers in a certain bracket is $7000 plus 28% of the amount of income over $40,000. Let x = amount of income over $40,000 and y = income tax, and assume that a linear relationship exists between amount of income over $40,000 and income tax.
   a. What is the income tax for a taxpayer who earns $60,000?
   b. Write an equation that describes the linear relationship between amount of income over $40,000 and income tax.
   c. Graph the equation of part b.

Section 7.5

In Exercises 63–70, graph the given linear inequality.

63. $y \geq 3x - 5$ **64.** $y < -4x + 3$ **65.** $2x + 6y < 12$ **66.** $4x - 2y \geq 6$

67. $8x - 4y \leq 0$ **68.** $5x + 15y > 0$ **69.** $y > -2$ **70.** $x \leq 9$

In Exercises 71–74, write the inequality represented by the given graph.

71.

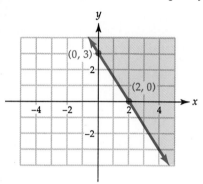

FIGURE 7.50

72.

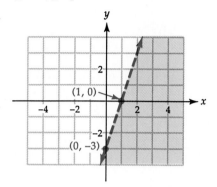

FIGURE 7.51

73.

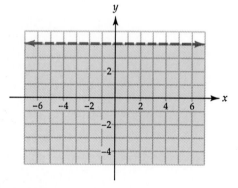

FIGURE 7.52

74.

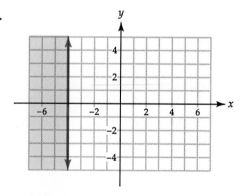

FIGURE 7.53

75. An elevator has a capacity of 3600 lb. Suppose that the average weight of a woman is 120 lb and the average weight of a man is 180 lb. Let $x =$ the number of women on the elevator and $y =$ the number of men on the elevator.
   a. Write a linear inequality that indicates the elevator is overloaded.
   b. Graph the inequality of part a.
   c. Is (14, 8) a solution of the inequality? Interpret your answer in terms of the given situation.

Chapter 7 Test

In Problems 1–2, determine whether the ordered pair is a solution of the equation $8x + 3y = 11$.

1. $(4, -7)$ **2.** $(-2, 9)$

In Problems 3–4, complete the ordered pair to make it a solution of the equation $5x - 6y = 18$.

3. $(0,)$ **4.** $(, -2)$

In Problems 5–6, plot the point and determine the quadrant in which it lies.

5. $(2, -5)$ **6.** $(-1, 0)$

7. Complete the table of values for the equation $4x + 2y = 10$. Graph the equation.

x	y
	0
3	
	-5
-1	

In Problems 8–10, graph the given equation and find its intercepts.

8. $x - 4y = 8$ **9.** $3x + 6y = 0$ **10.** $y = 5$

In Problems 11–12, find the slope of the line that passes through the given points.

11. $(5, -2), (-3, 6)$ **12.** $(0, 4), (-7, 1)$

In Problems 13–14, graph the line that passes through the given point and has the given slope.

13. $(8, 0), m = -3$ **14.** $(2, -6), m = \dfrac{3}{7}$

In Problems 15–16, find the slope and y intercept of the line which has the given equation.

15. $5x - y = 4$ **16.** $3x + 5y = 18$

In Problems 17–18, write an equation of the line that passes through the given point and has the given slope.

17. $(0, 5), m = -\dfrac{7}{4}$ **18.** $(-3, 2), m = 1$

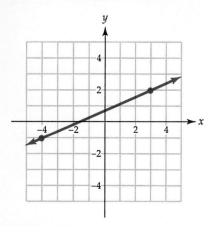

FIGURE 7.54   **FIGURE 7.55**

19. Find the slope and write an equation of the line whose graph is shown in Figure 7.54.

In Problems 20–22, graph the given linear inequality.

20. $x + 6y \leq 3$ 21. $8x > 4y$ 22. $x + 8 \geq 0$

23. Find the inequality whose solution is graphed in Figure 7.55.

24. Biologists have determined that there is a linear relationship between the number of chirps per minute made by a certain species of cricket and the temperature. Suppose that crickets chirp 124 times per minute at 68°F, and 172 times per minute at 80°F.
    a. Let x = chirps per minute and y = temperature. Write an equation that describes the linear relationship between chirps per minute and temperature.
    b. Graph the equation of part a.
    c. What is the temperature when there is no chirping?
    d. If the number of chirps per minute increases by 16, what has happened to the temperature?

25. A farmer has at most $2400 to spend on corn and soybean seed. Corn seed costs $30 per acre, and soybean seed costs $20 per acre. Let x = the number of acres of corn planted and y = the number of acres of soybeans planted.
    a. Write a linear inequality that describes the farmer's situation.
    b. Graph the inequality of part a.
    c. Is (45, 30) a solution of the inequality? Interpret your answer in terms of the given situation.

8 Systems of Linear Equations and Inequalities

8.1 Solution by Graphing

FOCUS

We solve systems of two linear equations in two variables by graphing.

A bank offers a regular checking account and a special checking account to its customers. The regular checking account has a monthly fee of $3.00 plus a charge of $0.10 for each check written. The special checking account has no monthly fee, but there is a charge of $0.20 for each check written. Which account is the better deal for the customer?

We can determine which is the better deal by comparing the monthly costs of the two checking accounts. We first note that the monthly cost of each account depends on the number of checks written in a month, so we let x = number of checks written in a month and y = monthly cost in dollars. Then the monthly cost of the regular checking account is given by the linear equation

$$y = 3.00 + 0.10x$$

Similarly, the monthly cost of the special checking account is given by the linear equation

$$y = 0.20x$$

To compare the monthly costs of the two checking accounts, we can graph these equations on the same coordinate system. This is done in Figure 8.1, which shows that the two lines intersect at the point (30, 6). Since

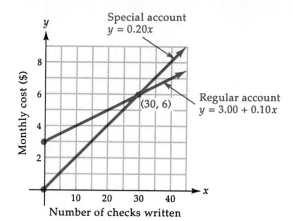

FIGURE 8.1

the point (30, 6) lies on both lines, it is a solution of both equations. There-fore, when 30 checks are written in a month, the monthly cost is $6 for each checking account. We can check this result by substituting $x = 30$ and $y = 6$ into each equation:

For the regular account:	For the special account:
$y = 3.00 + 0.10x$	$y = 0.20x$
$6 = 3.00 + 0.10(30)$	$6 = 0.20(30)$
$6 = 3.00 + 3.00$	$6 = 6$ True
$6 = 6$ True	

Now, which account is a better deal? That depends on how many checks are written each month. To see which is the better deal when, for example, 20 checks are written, we can compute the monthly cost of each account when 20 checks are written in a month. We do that by substituting $x = 20$ into each equation:

For the regular account:	For the special account:
$y = 3.00 + 0.10x$	$y = 0.20x$
$y = 3.00 + 0.10(20)$	$y = 0.20(20)$
$y = 3.00 + 2.00$	$y = 4.00$
$y = 5.00$	

The special checking account is the better deal when 20 checks are written in a month. In fact, since the graph of $y = 0.20x$ is below the graph of $y = 3.00 + 0.10x$ when $x < 30$, the special account costs less when fewer than 30 checks are written in a month. Similarly, you can check that the regular checking account costs less when more than 30 checks are written in a month.

The two equations used in the checking account problem are called a **system of linear equations** in two variables. The **solution of a system of linear equations** consists of all ordered pairs that *satisfy both equations;* that is, all ordered pairs that make both equations true statements. Graphing the equations on the same coordinate system is one way to determine the solution of the system.

EXAMPLE 1

Solve the following system of equations by graphing:

$$x + y = 6$$
$$x - y = -2$$

Solution

The graphs of the equations are shown in Figure 8.2. The intersection point (2, 4) is the only point that satisfies both equations, since it is the only point that lies on both lines. Therefore, the solution of the system is the ordered pair (2, 4). We can check this solution by substituting $x = 2$ and $y = 4$ into each equation:

$$x + y = 6 \qquad\qquad x - y = -2$$
$$2 + 4 = 6 \qquad\qquad 2 - 4 = -2$$
$$6 = 6 \quad \text{True} \qquad\qquad -2 = -2 \quad \text{True}$$

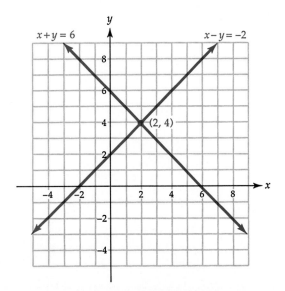

FIGURE 8.2

■

A system of linear equations that has exactly one solution is called a **consistent system.**

EXAMPLE 2

Solve the following system of equations by graphing:

$$2x - y = 4$$

$$2x - y = 2$$

Solution

The graphs of the equations are shown in Figure 8.3. The two lines are parallel, and because parallel lines do not intersect, there is no solution to the system.

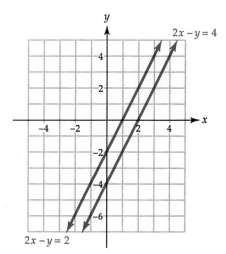

FIGURE 8.3

A system of linear equations with no solution is called an **inconsistent system.**

By writing the two equations in Example 2 in slope-intercept form, we can determine the slope and y intercept of each line:

$2x - y = 4$	$2x - y = 2$
$-y = -2x + 4$	$-y = -2x + 2$ Adding $2x$ to both sides
$y = 2x - 4$	$y = 2x - 2$ Multiplying both sides by -1

The two lines have the same slope ($m = 2$), but different y intercepts. Therefore the lines do not intersect, that is, they are parallel. *Two lines with the same slope and different y intercepts are parallel.*

EXAMPLE 3 Solve the following system of equations by graphing:

$$x + y = 3$$
$$2x + 2y = 6$$

Solution Writing these equations in slope-intercept form, we see that they have the same slope ($m = -1$) and the same y intercept ($b = 3$):

$$x + y = 3 \qquad\qquad 2x + 2y = 6$$
$$y = -x + 3 \qquad\qquad 2y = -2x + 6$$
$$y = -\frac{2}{2}x + \frac{6}{2}$$
$$y = -x + 3$$

Hence they have the same graph—the line shown in Figure 8.4. Every point on that line must satisfy both equations, so the solution of the system is an infinite number of ordered pairs.

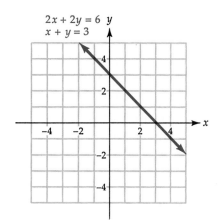

FIGURE 8.4 ■

A system of linear equations with an infinite number of solutions is called a **dependent system**.

Examples 1 to 3 illustrate the three possibilities for the solution of a system of two linear equations in two variables. These possibilities are shown in Figure 8.5.

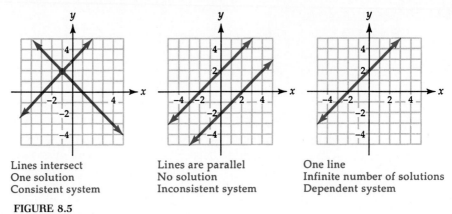

Lines intersect
One solution
Consistent system

Lines are parallel
No solution
Inconsistent system

One line
Infinite number of solutions
Dependent system

FIGURE 8.5

We can summarize the graphing method of solution as follows:

To solve a system of two linear equations by graphing

1. Carefully graph the two equations on the same coordinate system.
2. If the graphs intersect at one point, the system is consistent. The coordinates of the intersection point give the solution. Check the solution in each equation.
3. If the graphs are parallel lines, the system is inconsistent. It has no solution.
4. If the graphs are the same line, the system is dependent. Every point on the line represents a solution. The system has an infinite number of solutions.

DISCUSSION QUESTIONS 8.1

1. What do we mean by the solution of a system of linear equations?

2. Describe the graph of a consistent system. How many solutions are there?

3. Describe the graph of an inconsistent system. How many solutions are there?

4. Describe the graph of a dependent system. How many solutions are there?

5. In your own words, state how to find the solution of a system of linear equations by graphing.

EXERCISES 8.1

In Exercises 1–32, solve the given system of equations by graphing. Check the result.

1. $x + y = 10$
    $x - y = 6$

2. $x - y = 5$
    $x + y = 7$

3. $y = x - 4$
    $2x + y = 8$

4. $x + 2y = 6$
    $y = x + 3$

5. $y = -3x$
    $y = 2x + 10$

6. $y = -4x + 6$
    $y = 2x$

7. $2x + 3y = 7$
    $2x + y = -3$

8. $2x + 6y = 6$
    $-x + y = 5$

9. $x - 2y = 14$
    $5x + 2y = 10$

10. $4x + 3y = -9$
    $x - 3y = -6$

11. $x + 3y = 2$
    $3x - y = 6$

12. $x - 5y = 1$
    $5x + 2y = 32$

13. $\frac{1}{2}x - y = -3$
    $3x + 2y + 10 = 0$

14. $y = 5 - 2x$
    $x - \frac{4}{3}y = 8$

15. $2x - y = 8$
    $y = 2x - 1$

16. $y = -3x + 7$
    $3x + y = 9$

17. $3x + 4y = 6$
    $x - 2y = 2$

18. $4x - 2y = 8$
    $3x + y = 6$

19. $y = 4x + 3$
    $8x - 2y = -6$

20. $6x + 3y = 15$
    $y = -2x + 5$

21. $2x + y = 0$
    $y = 2x$

22. $y = 4x$
    $x + 2y = 0$

23. $2x - 4y = 8$
    $2x - y = -1$

24. $3x - 5y = 15$
    $2x - y = -4$

25. $\frac{2}{3}x - \frac{1}{3}y = 1$
    $x - \frac{1}{2}y = \frac{5}{4}$

26. $\frac{3}{4}x - y = 4$
    $\frac{1}{6}x - \frac{1}{3}y = 1$

27. $3x + y = -6$
    $2y = -6x - 12$

28. $4x - 2y = 12$
    $y = 2x - 6$

29. $3x - y = 12$
    $x = -2$

30. $x + 3y = 6$
    $y = 1$

31. $y = 3$
    $6x + 4y = 24$

32. $3x + 4y = 12$
    $x = -3$

In Exercises 33–38, answer the following questions for the given system:

a. What are the slope and the y intercept of each equation in the system?

b. Without graphing the equations, state whether the system is consistent, inconsistent, or dependent.

33. $3x + y = 3$
    $y = -3x + 6$

34. $4x - y = 4$
    $y = 4x - 2$

35. $4x - 3y = 12$
    $2x + 3y = 15$

36. $2x + 5y = 10$
    $3x - 7y = 14$

37. $x - 6y = 8$
    $-3x + 18y = -24$

38. $5x + 3y = -1$
    $10x + 6y = -2$

39. A record club offers two membership options. The first option requires a \$10 entry fee and \$5 for each record purchased. The second option has no entry fee but requires a payment of \$7 for each record purchased.

    a. For each membership option, write an equation that gives the total cost y of purchasing x records.

    b. Graph the system of part a.

c. How many records can be purchased before the second option becomes more expensive than the first?

40. A lot currently worth $40,000 is increasing in value at the rate of $4,000 per year. A second lot currently worth $36,000 is increasing in value at a rate of $4,800 per year.

   a. For each lot, write an equation that gives the value y of the lot after x years.
   b. Graph the system of part a.
   c. When will the two lots have the same value?

41. A shoe company sells a special running shoe for $50 per pair. The company has fixed costs of $120,000 (rent, utilities, insurance, and so on) plus a variable cost of $20 per pair of shoes made.

   a. Write an equation that gives the revenue y from selling x pairs of shoes.
   b. Write an equation that gives the total cost y to make x pairs of shoes.
   c. By graphing the equations of part a and part b, find the number of pairs of shoes that the company must make and sell in order to break even (that is, to have total cost equal revenue).

42. An electronics company sells a basic calculator for $5. The company has fixed costs of $60,000 plus a variable cost of $2 per calculator made.

   a. Write an equation that gives the revenue y from selling x calculators.
   b. Write an equation that gives the total cost y to make x calculators.
   c. By graphing the equations of part a and part b, find the number of calculators that the company must make and sell in order to break even.

Preparing for Section 8.2

In Exercises 43–48, add the given expressions.

43. $\begin{array}{r} x + 2y \\ 2x - y \\ \hline \end{array}$ **44.** $\begin{array}{r} 3x - 2y \\ x + 5y \\ \hline \end{array}$ **45.** $\begin{array}{r} 4x + 3y \\ 6x + 4y \\ \hline \end{array}$ **46.** $\begin{array}{r} 5x - 8y \\ 7x - 6y \\ \hline \end{array}$ **47.** $\begin{array}{r} x + y \\ x - y \\ \hline \end{array}$ **48.** $\begin{array}{r} x - 2y \\ 3x + 2y \\ \hline \end{array}$

8.2   Solution by Elimination

FOCUS

We solve systems of equations by eliminating one of the variables from the system. This method does not require graphing the system.

Graphical solutions of systems of linear equations are not always accurate, because they require the estimation of coordinates on a graph. For example, consider the following system:

$$x + 2y = 4$$
$$x - 2y = 1$$

As shown in Figure 8.6, the system has one solution, but the exact coordinates of the point of intersection are difficult to estimate.

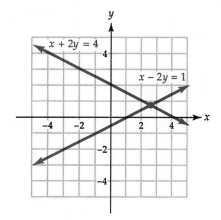

FIGURE 8.6

In this section and the next, we will discuss two algebraic methods of solving systems of linear equations. These methods do not require graphing. The method of this section, called the **elimination method,** is based on the following property of equality:

$$\text{If } A = B \text{ and } C = D, \text{ then } A + C = B + D.$$

This property says that if equal quantities are added to equal quantities, then the sums are equal. It is often more useful to write this property as follows:

$$\begin{aligned}
&\text{If} & A &= B \\
&\text{and} & C &= D \\
\hline
&\text{then} & A + C &= B + D \qquad \text{Adding left sides and right sides}
\end{aligned}$$

The goal of the elimination method is to eliminate one variable, obtaining a single equation by adding the equations of the system. We illustrate the procedure in the next several examples.

EXAMPLE 1   Solve the following system of equations by the elimination method:

$$x + 2y = 4$$
$$x - 2y = 1$$

Solution   The term $2y$ in the first equation and the term $-2y$ in the second equation are opposites. Therefore the variable y can be eliminated from the system by adding the corresponding sides of the two equations:

$$
\begin{array}{l}
x + 2y = 4 \\
\underline{x - 2y = 1} \\
2x \quad\;\; = 5
\end{array}
\qquad \text{Adding left sides and right sides}
$$

We can now solve the resulting equation, $2x = 5$, for x:

$$2x = 5$$
$$x = \frac{5}{2} \text{ or } 2.5 \qquad \text{Dividing both sides by 2}$$

Figure 8.6 is the graph of the two original equations. The x coordinate of the point of intersection of the two lines is $\dfrac{5}{2}$ or 2.5. To find the y coordinate we can substitute $x = 2.5$ into either of the original equations and solve for y. If we choose the first equation, we get

$$x + 2y = 4$$
$$2.5 + 2y = 4 \qquad \text{Substituting } x = 2.5$$
$$2y = 1.5 \qquad \text{Subtracting 2.5 from both sides}$$
$$y = 0.75 \text{ or } \frac{3}{4} \qquad \text{Dividing both sides by 2}$$

Therefore the solution of the system is the ordered pair (2.5, 0.75) or $\left(\dfrac{5}{2}, \dfrac{3}{4}\right)$. We check this result by substituting $x = 2.5$ and $y = 0.75$ into each equation:

$$
\begin{array}{ll}
x + 2y = 4 & \qquad x - 2y = 1 \\
2.5 + 2(0.75) = 4 & \qquad 2.5 - 2(0.75) = 1 \\
2.5 + 1.5 = 4 & \qquad 2.5 - 1.5 = 1 \\
4 = 4 \quad \text{True} & \qquad 1 = 1 \quad \text{True} \qquad \blacksquare
\end{array}
$$

In Example 1, we were able to eliminate the variable y merely by adding the left sides and right sides of the equations, because the y terms in the equations were opposites. But most systems of equations do not contain opposite variable terms. Then we have to work on one or both of the equations to produce opposite terms *before* we add the two equations. This procedure is illustrated in the next two examples.

EXAMPLE 2

Solve the following system of equations by the elimination method:

$$x + y = 7$$
$$4x - 5y = 1$$

Solution

Adding the left and right sides of the two equations does not eliminate one of the variables from the system:

$$x + y = 7$$
$$\underline{4x - 5y = 1}$$
$$5x - 4y = 8 \qquad \text{Adding left sides and right sides}$$

However, we can produce opposite terms in two ways. If we want to eliminate x, we can multiply both sides of the first equation by -4; this produces the term $-4x$ in the first equation, which is the opposite of the term $4x$ in the second equation. If we want to eliminate y, we can multiply both sides of the first equation by 5; this produces the term $5y$ in the first equation, which is the opposite of the term $-5y$ in the second equation.

We choose to eliminate y, because it involves multiplying by a positive number. Multiplying the first equation by 5 gives us

$$5(x + y) = 5(7) \quad \rightarrow \quad 5x + 5y = 35$$
$$\underline{4x - 5y = 1}$$
$$9x \qquad = 36 \qquad \text{Adding left sides and right sides}$$

We can now solve the equation $9x = 36$ for x:

$$9x = 36$$
$$x = 4 \qquad \text{Dividing both sides by 9}$$

To find y we can substitute $x = 4$ into either of the original equations and solve for y. If we use the first equation, we get

$$x + y = 7$$
$$4 + y = 7$$
$$y = 3$$

Therefore the solution of the system is the ordered pair (4, 3). You should check this solution in each of the original equations.   ■

EXAMPLE 3   Solve the following system of equations by the elimination method:

$$2x - 3y = 1$$
$$3x - 4y = 7$$

Solution   Here, to produce opposite terms we must work on both equations. We shall eliminate x by multiplying the first equation by -3 and the second equation by 2; this will produce the term $-6x$ in the first equation and the term $6x$ in the second equation. This gives us

$$-3(2x - 3y) = -3(1) \rightarrow \quad -6x + 9y = -3$$
$$2(3x - 4y) = 2(7) \rightarrow \quad \underline{6x - 8y = 14}$$
$$y = 11 \qquad \text{Adding the left and right sides}$$

Now, to find x we substitute $y = 11$ into the second equation:

$$3x - 4y = 7$$
$$3x - 4(11) = 7$$
$$3x - 44 = 7$$
$$3x = 51$$
$$x = 17$$

Therefore the solution of the system is the ordered pair (17, 11). You should check this solution in each of the original equations.   ■

The next two examples show how the elimination method can be used to determine whether a system of equations is inconsistent or dependent.

EXAMPLE 4   Solve the following system of equations by the elimination method:

$$\frac{1}{3}x + \frac{2}{3}y = 1$$
$$\frac{1}{2}x + y = \frac{1}{4}$$

Solution

We first simplify the system by clearing the equations of fractions. To do so, we multiply the first equation by 3 and the second equation by 4. This produces the system

$$3\left(\frac{1}{3}x + \frac{2}{3}y\right) = 3(1) \rightarrow \quad x + 2y = 3$$

$$4\left(\frac{1}{2}x + y\right) = 4\left(\frac{1}{4}\right) \rightarrow \quad 2x + 4y = 1$$

We choose to eliminate x by multiplying the first equation by -2. After multiplying, we add the resulting equation and the second equation:

$$-2(x + 2y) = -2(3) \rightarrow \quad -2x - 4y = -6$$
$$\underline{\qquad\qquad\qquad\qquad\quad 2x + 4y = \quad 1}$$
$$\qquad\qquad\qquad\qquad\qquad\quad 0 = -5 \qquad \text{False}$$

We obtain the false statement $0 = -5$, which indicates that the system has no solution. Therefore the system is inconsistent. The graphs of the equations are parallel lines, as shown in Figure 8.7.

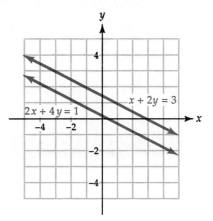

FIGURE 8.7

EXAMPLE 5

Solve the following system of equations by the elimination method:

$$4x - 3y = 6$$
$$6y = 8x - 12$$

Solution

To use the elimination method, we must write the second equation as $-8x + 6y = -12$. We choose to eliminate y by multiplying the first equation by 2. After multiplying, we add the resulting equation and the second equation:

$$2(4x - 3y) = 2(6) \rightarrow \quad 8x - 6y = \quad 12$$
$$-8x + 6y = -12$$
$$\overline{\ 0 = 0} \qquad \text{True}$$

The identity $0 = 0$, which is a true statement for all values of x and y, indicates that the system has an infinite number of solutions. Therefore the system is dependent. The graphs of the equations are the same line, as shown in Figure 8.8.

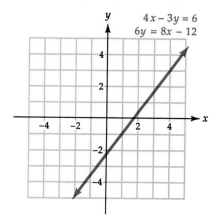

FIGURE 8.8   ∎

We can summarize the elimination method as follows:

To solve a system of two linear equations by elimination

1. Write both equations in the form $ax + by = c$.
2. Decide which variable to eliminate. In some cases one variable will be easier to eliminate than the other.
3. Multiply one or both equations by the appropriate numbers to produce opposite terms in the variable selected in step 2.
4. Add the two equations produced in step 3, and solve the resulting equation in one variable. (If the result is a false statement, the system has no solution; it is an inconsistent system. If the result is an identity, the system has an infinite number of solutions; it is a dependent system.)
5. Substitute the value found in step 4 into either of the original equations and solve for the other variable.
6. Check the solution in each of the original equations.

DISCUSSION QUESTIONS 8.2

1. What is the advantage of the elimination method over the graphing method when solving a system of equations?

2. When using the elimination method, what indicates that the system is inconsistent?

3. When using the elimination method, what indicates that the system is dependent?

EXERCISES 8.2

In Exercises 1–42, solve the given system of equations by the elimination method.

1. $x + y = 5$
$x - y = 9$

2. $x - y = 4$
$x + y = 6$

3. $x + y = 10$
$-x + y = 2$

4. $-x + y = 7$
$x + y = 1$

5. $2x + y = 9$
$3x - y = 1$

6. $3x + y = 2$
$4x - y = 5$

7. $4x + 2y = 8$
$2x + y = 2$

8. $x + 3y = 6$
$2x + 6y = 10$

9. $x + 4y = 17$
$-x + 7y = -6$

10. $-x + 5y = 4$
$x + 3y = 12$

11. $3x - 2y = 5$
$6x + 2y = 4$

12. $6x + 3y = 12$
$-6x + 2y = -2$

13. $3x + 2y = 5$
$-3x + 5y = 9$

14. $5x - 4y = 12$
$3x + 4y = 4$

15. $x + 3y = 14$
$4x - y = 4$

16. $x - 5y = 2$
$3x + y = 14$

17. $3x - y = 7$
$10x - 5y = 25$

18. $5x + 3y = 11$
$2x + y = 5$

19. $0.2x + 0.3y = 1.3$
$x - 0.1y = -1.5$

20. $3.5x + y = 30$
$x - 0.5y = 1.5$

21. $\frac{2}{3}x - y = 1$
$x + \frac{3}{5}y = -9$

22. $x - \frac{1}{4}y = \frac{3}{2}$
$x + \frac{1}{3}y = \frac{1}{3}$

23. $3x + 4y = 24$
$2x = 11 - y$

24. $x = 2y - 5$
$4x - 3y = -20$

25. $\frac{1}{6}x + \frac{1}{2}y = 1$
$\frac{1}{4}x + \frac{1}{2}y = 1$

26. $\frac{1}{4}x - \frac{1}{8}y = 1$
$\frac{1}{3}x - \frac{1}{9}y = 1$

27. $y = -5x + 12$
$10x + 2y = 6$

28. $8x - 16y = -5$
$2x = 4y + 3$

29. $1.5x - y = -8.5$
$0.2x + 0.4y = 1$

30. $x - 0.75y = 0.25$
$0.75x + y = -4.5$

31. $6x - 7y = 12$
$5x - 4y = 10$

32. $8x - 3y = 15$
$7x - 4y = 20$

33. $\frac{3}{4}x + \frac{5}{8}y = \frac{13}{8}$
$\frac{1}{2}x + \frac{1}{3}y = \frac{2}{3}$

34. $\frac{1}{3}x + \frac{11}{21}y = \frac{5}{3}$
$\frac{3}{4}x - \frac{3}{2}y = \frac{15}{4}$

35. $4x - 5y = -19$
$3x + 7y = 18$

36. $6x - 5y = 28$
$4x + 9y = -6$

37. $2x + 3y = -11$
$5x + 3y = 22$

38. $3x + 5y = -7$
$5x + 4y = 10$

39. $3x - 7y = 0$
$4x = -5y$

40. $3y = -4x$ **41.** $20x + 25 = 10y$ **42.** $6x - 18y = 24$
 $5x - 2y = 0$ $4x - 2y = -5$ $36y - 48 = -12x$

In Exercises 43–46, solve the given problem by using a system of equations and the elimination method.

43.   The sum of two numbers is 9. Their difference is -1. Find the two numbers.

44.   The sum of two numbers is -5. Their difference is 7. Find the two numbers.

45.   One belt and three ties cost $55. Two belts and one tie cost $35. Find the cost of one belt and one tie.

46.   Three pounds of peanuts and two pounds of cashews cost $10.20. Two pounds of peanuts and five pounds of cashews cost $15.60. Find the cost of one pound of peanuts and one pound of cashews.

Preparing for Section 8.3

In Exercises 47–52, solve the given equation.

47. $3x + (2x - 1) = 5$ **48.** $(3y - 5) + 4y = 16$ **49.** $-(6w + 7) + 4w = -13$

50. $6z - (9z + 2) = -5$ **51.** $5p - 2(p - 3) = -6$ **52.** $-8(q - 7) + 2q = 14$

8.3   Solution by Substitution

FOCUS

We discuss and use another algebraic method for solving systems of linear equations.

Recall the checking account problem at the beginning of Section 8.1: The regular checking account involves a monthly fee of $3.00 plus a charge of $0.10 for each check written; the special checking account has no monthly fee, but there is a charge of $0.20 for each check written. We let x = number of checks written in a month and y = monthly cost, and obtained the system of equations

$$y = 3.00 + 0.10x$$

$$y = 0.20x$$

The second equation tells us that $0.20x$ is equal to y. This means we can substitute $0.20x$ for y in the first equation, obtaining

$$0.20x = 3.00 + 0.10x$$

We can now solve this equation for x:

$0.20x = 3.00 + 0.10x$

$0.10x = 3.00$ Subtracting $0.10x$ from both sides

$x = 30$ Dividing both sides by 0.10

To find y we can substitute $x = 30$ into either of the original equations and solve for y. This produces $y = 6$. As before, the solution of the system is the ordered pair $(30, 6)$, which indicates that each checking account has a monthly cost of \$6 when 30 checks are written in a month.

    This method of solving a system of two linear equations is called the **substitution method.** To use it, we must be able to solve one of the equations for one of the variables.

EXAMPLE 1

Solve the following system of equations by the substitution method:

$$x = 12 - y$$
$$x + 3y = 8$$

Solution

The first equation is already solved for x. Therefore we can replace x in the second equation with $12 - y$ and solve the resulting equation for y:

$x + 3y = 8$

$12 - y + 3y = 8$ Substituting $x = 12 - y$

$12 + 2y = 8$ Simplifying

$2y = -4$ Subtracting 12 from both sides

$y = -2$ Dividing both sides by 2

To find x we can substitute $y = -2$ into either of the original equations and solve for x. We choose the first equation because it is already solved for x:

$$x = 12 - y$$
$$= 12 - (-2)$$
$$= 14$$

Therefore the solution of the system is the ordered pair $(14, -2)$. You should check this solution in each of the original equations. ∎

BE CAREFUL!

When one equation in a system is solved for one of the variables, the result must be substituted into the *other* equation. Consider the following *incorrect* solution of Example 1:

$$x = 12 - y \qquad \text{First equation}$$
$$12 - y = 12 - y \qquad \text{Substituting for } x \text{ in the first equation}$$
$$0 = 0$$

The expression for x (or y) in the one equation must replace x (or y) in the *other* equation to produce the solution.

EXAMPLE 2

Solve the following system of equations by the substitution method:

$$3x + y = 9$$
$$x - 2y = 10$$

Solution

In this system, we can easily solve the first equation for y or the second equation for x. We choose to solve the first equation for y:

$$3x + y = 9$$
$$y = 9 - 3x$$

We can now replace y in the second equation with $9 - 3x$ and solve for x:

$$x - 2y = 10$$
$$x - 2(9 - 3x) = 10 \qquad \text{Substituting } y = 9 - 3x$$
$$x - 18 + 6x = 10 \qquad \text{By the distributive property}$$
$$7x = 28 \qquad \text{Simplifying}$$
$$x = 4 \qquad \text{Dividing both sides by 7}$$

To find y, we substitute $x = 4$ into the equation $y = 9 - 3x$:

$$y = 9 - 3x$$
$$= 9 - 3(4)$$
$$= 9 - 12$$
$$= -3$$

Therefore the solution of the system is the ordered pair $(4, -3)$. You should check this solution in each of the original equations. ∎

In the solution of Example 2 it was relatively easy to solve the first equation for y. We could have solved the first equation for x, but the result would have been a fraction:

$$3x + y = 9$$
$$3x = 9 - y$$
$$x = \frac{9 - y}{3}$$

Replacing x in the second equation with $\frac{9 - y}{3}$ produces a messy equation to solve. It is always a good idea to avoid fractions when using the substitution method. If fractions cannot be avoided, it might be better to try the elimination method.

EXAMPLE 3   Solve the following system of equations by the substitution method:

$$0.7x + 0.2y = 3.4$$
$$0.5x - 0.1y = 0$$

Solution   Before solving one of the equations for one of the variables, we can simplify the system by multiplying each equation by 10. This produces integer coefficients and constants in the resulting system:

$$10(0.7x + 0.2y) = 10(3.4) \rightarrow 7x + 2y = 34$$
$$10(0.5x - 0.1y) = 10(0) \quad \rightarrow \quad 5x - y = 0$$

To avoid fractions, we solve the second equation for y:

$$5x - y = 0$$
$$-y = -5x$$
$$y = 5x$$

We can now replace y in the first equation with $5x$ and solve for x:

$$7x + 2y = 34$$
$$7x + 2(5x) = 34$$
$$7x + 10x = 34$$
$$17x = 34$$
$$x = 2$$

To find y, we can substitute $x = 2$ into the equation $y = 5x$:

$$y = 5x$$

$$y = 5(2)$$

$$y = 10$$

Therefore the solution of the system is the ordered pair (2, 10). **The check is left to you.** ∎

EXAMPLE 4   Solve the following system of equations by the substitution method:

$$x - 2y = -5$$

$$2x - 4y = 7$$

Solution   To avoid fractions, we solve the first equation for x:

$$x - 2y = -5$$

$$x = 2y - 5$$

We can now replace x in the second equation with $2y - 5$ and solve for y:

$$2x - 4y = 7$$

$$2(2y - 5) - 4y = 7$$

$$4y - 10 - 4y = 7$$

$$-10 = 7 \qquad \text{False}$$

The false statement $-10 = 7$ indicates that the system has no solution. The graphs of the two equations are parallel lines, and the system is inconsistent. ∎

We can summarize the substitution method as follows:

To solve a system of two linear equations by substitution

1. If necessary, solve one of the equations for one of the variables. Try to avoid fractions.
2. Substitute the result of step 1 into the other equation, obtaining an equation in one variable.
3. Solve the equation produced in step 2.
4. To find the value of the other variable, substitute the value obtained in step 3 into the equation produced in step 1.
5. Check the solution in each of the original equations.

DISCUSSION QUESTIONS 8.3

1. When should the substitution method generally not be used to solve a system of equations?

2. Rework Example 2 by solving the second equation for x.

3. To use substitution to solve the system below, which variable would you solve for in which equation? Explain.

$$3x - 2y = 12$$
$$5x + y = 10$$

EXERCISES 8.3

In Exercises 1–32, solve the given system of equations by the substitution method.

1. $x + y = 12$
   $y = x + 10$

2. $y = x - 5$
   $x + y = 3$

3. $4x - y = 6$
   $y = 4 - x$

4. $y = 6 - x$
   $3x - 2y = -2$

5. $x = -2y$
   $2x - 3y = 14$

6. $0.3x - 0.7y = 0.8$
   $0.1x = 0.5y$

7. $0.2x + 0.5y = 2.2$
   $0.1y = 0.4x$

8. $y = -3x$
   $6x + 4y = 18$

9. $\frac{1}{3}x - y = \frac{10}{3}$
   $\frac{1}{2}x + \frac{y}{4} = \frac{3}{2}$

10. $\frac{1}{2}x + y = \frac{1}{2}$
    $\frac{1}{4}x - \frac{1}{4}y = -2$

11. $2x + y = 1$
    $x - 2y = -1$

12. $x - 3y = 10$
    $2x + y = 6$

13. $x - 4y = 9$
    $3x + y = 14$

14. $2x + y = 1$
    $x - 5y = 17$

15. $4x + 2y = 10$
    $2x + y = 5$

16. $x - 6y = 3$
    $3x - 18y = 9$

17. $2x + y = 0$
    $2y = -x$

18. $x - 4y = 0$
    $y = 3x$

19. $0.5x - 0.1y = 0.3$
    $0.3x - 0.1y = 1.1$

20. $x + 5y = 3$
    $x - 2y = 10$

21. $2x - y = 0$
    $4x + 3y = 30$

22. $0.5x - 0.2y = 1.1$
    $0.3x + 0.1y = 0$

23. $7x + y = 4$
    $14x + 2y = 8$

24. $3x - 9y = 6$
    $x - 3y = 3$

25. $x + \frac{1}{5}y = \frac{14}{5}$
    $\frac{1}{5}x + \frac{1}{10}y = \frac{1}{2}$

26. $\frac{1}{2}x - \frac{3}{4}y = 1$
    $x + \frac{1}{3}y = \frac{1}{3}$

27. $4x + 3y = 31$
    $2x - y = 7$

28. $3x - y = -2$
    $4x + 5y = 48$

29. $x + 2y = 3$
    $4x + 8y = 7$

30. $2x - 5y = 9$
    $x - 4y = 3$

31. $24x - 3y = -15$
    $8x - y = -5$

32. $x + 9y = 2$
    $2x + 18y = 4$

33. The quantity q of a product that consumers are willing to buy at a price of p dollars is given by the demand equation, $q = 20 - 5p$. The quantity q of the product that producers are willing to sell at price p is given by

the supply equation $q = 12p - 14$. The two equations form a system. Solve the system by substitution to obtain what economists call the equilibrium price; that is, the price of the product when the quantity supplied equals the quantity demanded.

34. The quantity q of a product that consumers are willing to buy at a price of p dollars is given by the demand equation, $q = 640 - 10p$. The quantity q of the product that producers are willing to sell at price p is given by the supply equation, $q = 5p - 50$. Find the price of the product when the quantity supplied equals the quantity demanded.

35. A bicycle company sells a mountain bike for $180. The company has fixed costs of $200,000 plus a variable cost of $100 for each bike made.
    a. Write an equation that gives the revenue y from selling x mountain bikes. Write an equation that gives the total cost y to make x mountain bikes.
    b. The break-even point occurs when revenue equals total cost. Find the break-even point by using substitution to solve the system of equations in part a.
    c. How many mountain bikes must be sold for the company to make a profit?

36. A shoe company sells shoelaces for $0.40 per pair. The company has fixed costs of $200 plus a variable cost of $0.20 for each pair of shoelaces made.
    a. Write an equation that gives the revenue y from selling x pairs of shoelaces. Write an equation that gives the total cost y to make x pairs of shoelaces.
    b. Find the break-even point by using substitution to solve the system of equations in part a.
    c. How many pairs of shoelaces must be sold for the company to make a profit?

Preparing for Section 8.4

In Exercises 37–40, write a linear equation in two variables to describe the given condition(s).

37. The difference of two numbers is 20.

38. The perimeter of a rectangle is 36 inches (in.).

39. A shirt costs x dollars and a tie costs y dollars. The total cost of 3 shirts and 2 ties is $85.

40. Sabrina invested x dollars at 7% simple interest and y dollars at 9% simple interest. She earned a total of $500 interest the first year.

In Exercises 41–44, solve the given equation.

41. $7x + 2(5 - 4x) = 11$

42. $3y - 6(8y + 1) = 39$

43. $2.50w - 1.25(400 + w) = 1000$

44. $0.80z + 0.60(40 - z) = 30$

8.4   Applications Involving Systems of Linear Equations

FOCUS

We use systems of linear equations to solve a variety of word problems.

As we saw in the checking account problem at the beginning of this chapter, systems of linear equations can be used to solve real-world problems. In this section we solve a variety of word problems by using systems of linear equations. First, though, we must modify the word-problem solution procedure of Section 3.6 to account for the two unknown quantities and corresponding system of equations.

> **To solve word problems using a system of equations**
>
> 1. Read the problem carefully. Draw a diagram or make a table if it will help you understand the problem.
> 2. Represent each unknown quantity with a different variable. If necessary, represent other quantities in terms of these variables.
> 3. Use the given conditions and numerical values, and perhaps a formula, to write a system of equations that describes the problem situation.
> 4. Solve the system of equations.
> 5. Check the solution in each equation, and against the conditions of the problem.

We illustrate these steps in the following examples.

EXAMPLE 1

Shao and Jennifer won $57,000 in the state lottery. They agreed that Jennifer's share would be twice Shao's share. How much money does each receive?

Solution

We begin by choosing two variables to represent Jennifer's and Shao's share of the winnings:

$$x = \text{Jennifer's share}$$
$$y = \text{Shao's share}$$

Since they won a total of $57,000, we have the equation

Jennifer's share	+	Shao's share	=	Total winnings
x	+	y	=	57,000

Since Jennifer's share is twice as much as Shao's share, we have a second equation:

Jennifer's share	=	Twice Shao's share
x	=	$2y$

This gives us the system

$$x + y = 57,000$$
$$x = 2y$$

We can solve this system by substituting $2y$ for x in the first equation and then solving for y:

$$x + y = 57,000$$
$$2y + y = 57,000 \qquad \text{Substituting } x = 2y$$
$$3y = 57,000 \qquad \text{Combining like terms}$$
$$y = 19,000 \qquad \text{Dividing both sides by 3}$$

Shao's share of the winnings is $19,000. To find Jennifer's share, we can substitute $y = 19,000$ into either of the original equations and solve for x:

$$x = 2y$$
$$x = 2(19,000)$$
$$x = 38,000$$

Jennifer's share of the winnings is $38,000. We check the solution (38,000, 19,000) in the original system:

$$x + y = 57{,}000 \qquad\qquad\qquad x = 2y$$

$$38{,}000 + 19{,}000 = 57{,}000 \qquad\qquad 38{,}000 = 2(19{,}000)$$

$$57{,}000 = 57{,}000 \quad \text{True} \qquad 38{,}000 = 38{,}000 \quad \text{True} \qquad ■$$

EXAMPLE 2

At a school play, 280 tickets were sold, producing total revenue of $1155. The cost of an adult ticket is $5.00, and the cost of a child's ticket is $1.50. How many adult tickets and how many child tickets were sold?

Solution

We are asked to find the number of adult tickets sold and the number of child tickets sold. We use a different variable to represent each of these quantities:

$$x = \text{number of adult tickets sold}$$

$$y = \text{number of child tickets sold}$$

Since the total number of tickets sold is 280, we immediately have the equation

$$
\begin{array}{ccccc}
\text{Adult tickets} & + & \text{Child tickets} & = & \text{Total tickets} \\
x & + & y & = & 280
\end{array}
$$

The total revenue produced from selling the tickets is $1155. Since each adult ticket costs $5.00, the revenue from selling x adult tickets is $5.00x$. Similarly, the revenue from selling y child tickets is $1.50y$. This gives a second equation:

$$
\begin{array}{ccccc}
\text{Revenue from} & + & \text{Revenue from} & = & \text{Total revenue} \\
\text{adult tickets} & & \text{child tickets} & & \\
5.00x & + & 1.50y & = & 1155
\end{array}
$$

Therefore we have the system

$$x + y = 280$$

$$5.00x + 1.50y = 1155$$

To solve this system, we shall solve the first equation for y (we could just as easily solve for x) and substitute the result into the second equation:

$$x + y = 280$$

$$y = 280 - x \qquad \text{Subtracting } x \text{ from both sides}$$

So the second equation becomes

$$5.00x + 1.50(280 - x) = 1155 \qquad \text{Substituting } y = 280 - x$$

We can now solve the last equation for x:

$$5.00x + 1.50(280 - x) = 1155$$
$$5.00x + 420 - 1.50x = 1155 \qquad \text{By the distributive property}$$
$$3.50x = 735 \qquad \text{Simplifying}$$
$$x = 210 \qquad \text{Dividing both sides by 3.50}$$

Hence 210 adult tickets were sold. To find the number of child tickets sold, we can substitute $x = 210$ into either of the original equations and solve for y:

$$x + y = 280$$
$$210 + y = 280 \qquad \text{Substituting } x = 210$$
$$y = 70 \qquad \text{Subtracting 210 from both sides}$$

Thus 70 child tickets were sold. We check the solution (210, 70) in the original system:

$$x + y = 280 \qquad\qquad\qquad 5.00x + 1.50y = 1155$$
$$210 + 70 = 280 \qquad\qquad\qquad 5.00(210) + 1.50(70) = 1155$$
$$280 = 280 \quad \text{True} \qquad\qquad 1050 + 105 = 1155$$
$$1155 = 1155 \quad \text{True} \quad \blacksquare$$

The next example illustrates the use of the distance formula $d = rt$, where d is the distance traveled, r is the rate of travel, and t is the time of travel.

EXAMPLE 3

Rita decided to row her kayak on a gently flowing river. In 3 hours (h) she rowed 18 miles (mi) downstream; then she turned around but was able to row only 12 mi upstream in the same time. Find Rita's speed in still water, and the speed of the river current.

Solution

We use two variables to represent the unknown quantities:

$$x = \text{Rita's speed in still water}$$
$$y = \text{speed of the current}$$

Rita makes two trips, one downstream and one upstream. When she rows downstream, the current is pushing the kayak, so her net speed downstream is $x + y$ (faster than in still water). When she rows upstream, the current is acting against the kayak, so her net speed upstream is $x - y$ (slower than in still water). We can use a table to organize the relevant information for each direction:

	d	r	t
Downstream	18	$x + y$	3
Upstream	12	$x - y$	3

We can now use the formula $d = rt$ to write two equations:

Downstream	$18 = (x + y)3$
Upstream	$12 = (x - y)3$

After simplifying the right side of each equation, we have

$$18 = 3x + 3y$$
$$12 = 3x - 3y$$

The term $3y$ in the first equation and the term $-3y$ in the second equation are opposites, so we can easily solve the system with the elimination method:

$$18 = 3x + 3y$$
$$\underline{12 = 3x - 3y}$$

$30 = 6x$	Adding left sides and right sides
$5 = x$	Dividing both sides by 6

Rita's speed in still water is 5 mph. To find the speed of the current, we substitute $x = 5$ into either of the original equations and solve for y:

$$18 = 3x + 3y$$
$$18 = 3(5) + 3y$$
$$18 = 15 + 3y$$
$$3 = 3y$$
$$1 = y$$

The speed of the current is 1 mph. You should check that $x = 5$ and $y = 1$ satisfy the original system. ∎

EXAMPLE 4

According to information on the labels, Pizzaz cereal contains 30% sugar and Shazam cereal contains 50% sugar. How many ounces of each cereal are needed to make 20 ounces (oz) of a mixture containing 45% sugar?

Solution

We begin by defining two variables:

$$x = \text{ounces of Pizzaz in the mixture}$$

$$y = \text{ounces of Shazam in the mixture}$$

We have data regarding the ounces of the mixture and the percentages of sugar. Since Pizzaz contains 30% sugar, in x ounces of Pizzaz there are $0.30x$ (30% of x) ounces of sugar. Similarly, in y ounces of Shazam there are $0.50y$ ounces of sugar, and in 20 oz of the mixture there are $0.45(20) = 9$ oz of sugar. Before we try to write a system of equations, we should organize the relevant data in a table:

	Ounces of Cereal	Ounces of Sugar	
Pizzaz	x	$0.30x$	30% of Pizzaz is sugar
Shazam	y	$0.50y$	50% of Shazam is sugar
Mixture	20	$0.45(20) = 9$	45% of 20 oz of mixture is sugar

Since Pizzaz and Shazam are combined to make the mixture, we can write two equations:

Ounces of Pizzaz	+	Ounces of Shazam	=	Ounces of mixture
x	+	y	=	20

and

Ounces of sugar in the Pizzaz	+	Ounces of sugar in the Shazam	=	Ounces of sugar in the mixture
$0.30x$	+	$0.50y$	=	9

Therefore we have the system

$$x + y = 20$$

$$0.30x + 0.50y = 9$$

Before solving the system, we simplify the second equation by multiplying both sides by 10. This produces the system

$$x + y = 20$$

$$10(0.30x + 0.50y) = 10(9) \rightarrow 3x + 5y = 90$$

To solve this system, we multiply both sides of the first equation by -3 and add the result to the second equation:

$$-3x - 3y = -60 \qquad \text{Multiplying both sides by } -3$$

$$\underline{3x + 5y = 90}$$

$$2y = 30 \qquad \text{Adding left sides and right sides}$$

$$y = 15 \qquad \text{Dividing both sides by 2}$$

There are 15 oz of Shazam in the mixture. To find the amount of Pizzaz in the mixture, we substitute $y = 15$ into either of the original equations and solve for x:

$$x + y = 20$$

$$x + 15 = 20$$

$$x = 5$$

There are 5 oz of Pizzaz in the mixture. You should check that the solution (5, 15) satisfies the conditions of the problem.   ∎

DISCUSSION QUESTIONS 8.4

1. In your own words, state how to solve word problems using a system of equations.

2. A baseball team charges $9 for a box seat and $6 for a reserved seat. How much revenue is obtained from selling x box seats and y reserved seats? Explain.

3. An airplane's speed in still air is x miles per hour. The wind is blowing y miles per hour. What is the plane's net speed with the wind? Against the wind? Explain your answers.

EXERCISES 8.4

In Exercises 1–4, match the given set of conditions with the appropriate system of equations (a–e). You may use a system more than once. Do not solve the system.

1. The difference between twice a certain number x and a smaller number y is 18. The sum of twice the larger number and twice the smaller number is 24.

2. The perimeter of a rectangle is 24 in. The difference between twice its length x and its width y is 18 in.

3. The sum of Jenny's age y and twice Mary's age x is 24. The sum of Mary's age and twice Jenny's age is 18.

4. Twice a certain number x is 18 more than a second number y. The sum of twice the first number and the second number is 24.

   a. $2x + 2y = 24$
   $2x - y = 18$

   b. $2x = y + 18$
   $2x + y = 24$

   c. $x - 2y = 18$
   $2x + 2y = 24$

   d. $2x + y = 24$
   $x + 2y = 18$

   e. $2x + 2y = 24$
   $2x - 2y = 18$

In Exercises 5–34, use a system of equations to solve the problem.

5. The difference between two numbers is 8. The larger number is 2 less than 3 times the smaller number. Find the two numbers.

6. The sum of two numbers is 27. The larger number is 3 more than 5 times the smaller number. Find the two numbers.

7. The perimeter of a rectangle is 40 centimeters (cm). The length of the rectangle is 3 times the width. Find the length and the width of the rectangle.

8. The perimeter of a rectangle is 60 in. The width of the rectangle is one-fifth the length. Find the length and the width of the rectangle.

9. A man named his wife and sister as beneficiaries of a $300,000 insurance policy. His wife is to be paid 4 times the amount paid to his sister. How much will each be paid?

10. Jan and Mike contribute a total of $1000 to charities each year. Jan contributes $200 less than twice Mike's contribution. How much does each contribute?

11. For the grand opening of Davies Symphony Hall, a total of 3500 tickets were sold. An orchestra ticket cost $75, and a balcony ticket cost $50. The revenue obtained from selling the tickets was $225,000. How many orchestra tickets and how many balcony tickets were sold?

12. The total revenue from selling 750 tickets to a high school football game was $2650. An adult ticket cost $5, and a student ticket cost $3. How many adult tickets and how many student tickets were sold?

13. The cost of driving Larry's Buick is $0.55 per mile. The cost of driving his Honda is $0.28 per mile. Last month the cars were driven a total of 900 mi at a cost of $360. How many miles was each car driven?

14. Anne and Gerhard work at a local supermarket. Anne earns $12 per hour as a checker, and Gerhard earns $7 per hour as a bagger. They worked a total of 75 h last week and were paid a total of $700. How much did each earn?

15. A child's painting kit costs $7.00 and contains 2 brushes and 5 tubes of paint. The standard kit contains 4 brushes and 9 tubes of paint and costs $13.10. Find the cost of a brush and the cost of a tube of paint.

16. A gas station sold 500 gal of unleaded gas and 425 gal of regular gas for a total of $842.50. Unleaded gas costs $0.02 per gallon more than regular gas. Find the cost of a gallon of unleaded gas and the cost of a gallon of regular gas.

17. A plane can fly 1000 mi in 2.5 h against the wind and 1250 mi in the same time with the wind. Find the speed of the plane in still air and the speed of the wind.

18. A boat travels 90 mi downstream in 4 h and 70 mi upstream in the same time. Find the speed of the boat in still water and the speed of the current.

19. A walker and a jogger are 18 mi apart. They start toward each other, the jogger traveling twice as fast as the walker. If they meet in 2 h, find the speed of each.

20. A scooter and a car are 320 mi apart. The car's speed is 3 times as fast as the scooter's. If they meet in 4 h, find the speed of each.

21. An enemy missile traveling 6000 miles per hour (mph) is detected on radar to be 5000 mi from a friendly missile base. Immediately, an anti-missile missile with a speed of 9000 mph is launched from the base. How far does each missile travel before they meet?

22. Two couriers driving toward each other are 375 mi apart. One is traveling at 55 mph and the other at 70 mph. How far does each courier travel before they meet?

23. A landscape gardener has two types of grass seed. One type is 40% bluegrass, and the other type is 70% bluegrass. How many pounds of each type are needed to make 500 pounds (lb) of a mixture that is 52% bluegrass?

24. Old silver coins are 90% pure silver. New silver coins are 60% pure silver. How many ounces of each kind of coin must be melted together to produce 250 oz of an alloy that is 72% pure silver?

25. Carbon monoxide gas is 43% carbon. Carbon dioxide gas is only 27% carbon. The Environmental Protection Agency analyzed 1600 milligrams (mg) of auto exhaust and found it was 32% carbon. How many milligrams of carbon monoxide and how many milligrams of carbon dioxide were in the exhaust?

26. The octane number of gasoline is the percentage of the compound octane in the gas. A wholesaler has a supply of 82-octane gas and

92-octane gas. How many gallons of each kind of gas must be mixed to produce 5000 gal of 86-octane gas?

27. Animals in a scientific experiment must receive a daily diet that contains 20 grams (g) of protein and 6 g of fat. One food mix is 10% protein and 6% fat. A second food mix is 20% protein and 2% fat. How many grams of each mix are needed for the animals to get the proper diet?

28. Fruit Grain consists of dried fruit and cereal. Each box of Fruit Grain contains 24 g of protein and 330 g of carbohydrate. The dried fruit is 5% protein and 75% carbohydrate, whereas the cereal is 10% protein and 85% carbohydrate. How many grams of dried fruit and how many grams of cereal are in a box of Fruit Grain?

29. Kona coffee sells for $4.50 per pound, and Dark Roast coffee sells for $6.00 per pound. John spent $52.50 for a 10-lb mixture of the two coffees. How many pounds of each coffee were in the mixture?

30. A candy store sells dark chocolates for $4.30 per pound, and caramels for $2.80 per pound. Tina spent $6.35 on a 2-lb mixture of dark chocolates and caramels. How many pounds of each candy were in the mixture?

31. Kevin invested a total of $10,000 in two accounts. One account pays 7% simple interest, and the other account pays 12% simple interest. How much is invested in each account if Kevin earned a total of $900 interest the first year?

32. Barbara invested a total of $20,000 in two accounts. One account pays 5% simple interest, and the other account pays 7% simple interest. How much is invested in each account if Barbara earned a total of $1120 interest the first year?

33. In a 4-point grading system, an A is worth 4 points per semester hour, a B is worth 3 points, a C is worth 2 points, a D is worth 1 point, and an F is worth 0 points. Last semester, Maureen earned 46 points for 14 semester hours. If she made only A's and B's, how many semester hours of each grade did she earn?

34. One year Ernie Banks hit 18 more home runs than Hank Aaron. However, when the two digits in Aaron's total are interchanged, the result is Banks' total. If the sum of the digits in each total is 6, how many home runs did each player hit?

Preparing for Section 8.5

In Exercises 35–42, graph the solution of the given inequality.

35. $x + y \geq 3$
36. $x - y < -7$
37. $2x - y > -4$
38. $x + 3y \geq 9$
39. $4x + 10y \leq 20$
40. $6x - 2y \leq -12$
41. $y < -5$
42. $x > 8$

8.5   Systems of Linear Inequalities

FOCUS

We solve systems of linear inequalities by graphing.

In Section 7.5 we saw how to graph linear inequalities. For example, to graph the inequality $x + y < 5$, we first graph the boundary line $x + y = 5$ (see Figure 8.9). We draw a broken line, since the equality is not included in the inequality symbol $<$. We then use the origin $(0, 0)$ as a test point to determine which half-plane contains the solution of the inequality. We substitute $x = 0$ and $y = 0$ into the inequality:

$$x + y < 5$$
$$0 + 0 < 5$$
$$0 < 5 \qquad \text{True}$$

Since the ordered pair $(0, 0)$ makes the inequality $x + y < 5$ a true statement, it is a part of the solution. We shade the half-plane containing $(0, 0)$, as shown in the figure. The shaded half-plane is the graph of the solution of $x + y < 5$ because all the points in this half-plane satisfy the inequality.

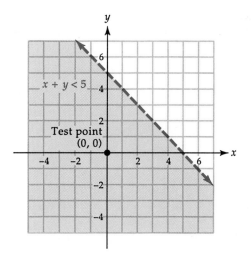

FIGURE 8.9

A **system of linear inequalities** consists of two or more linear inequalities. The **solution of a system of linear inequalities** contains all ordered pairs that satisfy all the inequalities simultaneously; that is, the solution includes all ordered pairs that make all the inequalities true. We graph the solution of a system of linear inequalities by graphing all the inequalities. The region common to all the inequalities is the solution of the system.

EXAMPLE 1

Graph the solution of the system:

$$x + y < 5$$
$$x - y \geq 2$$

Solution

We must graph the inequality $x + y < 5$ and the inequality $x - y \geq 2$ on the same coordinate system. We have already graphed $x + y < 5$ (in the example above). It is shown along with the graph of $x - y \geq 2$ in Figure 8.10. The most heavily shaded region in the figure contains all points that satisfy both inequalities. Therefore, it is the solution of the system. Note that the graph of the solution contains part of the boundary line $x - y = 2$.

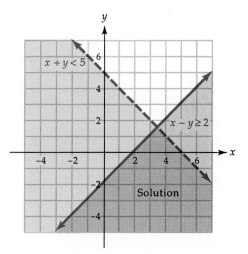

FIGURE 8.10   ■

EXAMPLE 2

Graph the solution of the system:

$$y \geq 4x$$
$$2x + y \geq 6$$

Solution

The inequality $y \geq 4x$ is graphed in Figure 8.11a. The graph of the solution of the system is the most heavily shaded region in Figure 8.11b. It includes portions of the boundary lines $y = 4x$ and $2x + y = 6$.

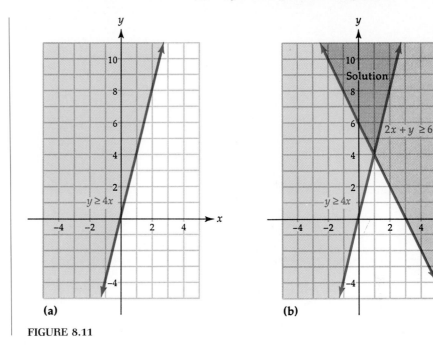

(a)   **(b)**

FIGURE 8.11   ∎

EXAMPLE 3

Tom's dad is more than twice as old as Tom. The difference in their ages is less than 25 years. What are their possible ages?

Solution

We will write a system of inequalities that describes the situation. We begin by defining the variables:

$$x = \text{Dad's age}$$
$$y = \text{Tom's age}$$

Since Tom's dad is more than twice as old as Tom, we have the inequality

$$x > 2y$$

Since the difference in their ages is less than 25 years, we have a second inequality:

$$x - y < 25$$

There are two additional inequalities because their ages are positive numbers:

$$x > 0 \quad \text{and} \quad y > 0$$

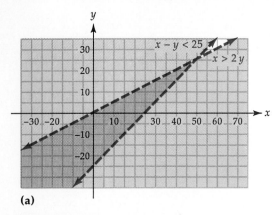

(a)

(b)

FIGURE 8.12

Therefore we have the following system:

$$x > 2y$$
$$x - y < 25$$
$$x > 0$$
$$y > 0$$

The graphs of $x > 2y$ and $x - y < 25$ are shown in Figure 8.12. The graph of the solution of the system is the most heavily shaded region in Figure 8.12b. Note that the last two inequalities restrict the graph to points in quadrant I. The graph contains all ordered pairs (possible ages) that satisfy all four inequalities. Among others, the ordered pair (35, 15), meaning that Dad is 35 years old and Tom is 15 years old, is part of the solution. ∎

DISCUSSION QUESTIONS 8.5

1. What constitutes the solution of a system of linear inequalities?

2. In your own words, state how to solve a system of inequalities graphically.

3. Find a second ordered pair that is contained in the solution of Example 3.

EXERCISES 8.5

In Exercises 1–28, graph the solution of the given system of inequalities.

1. $x + y > 3$
   $y \le x - 1$

2. $x - y < 5$
   $y \ge 3 - x$

3. $3x + y \ge 5$
   $y < x$

4. $x \le -3y$
   $3x + y > -6$

5. $x + y > 6$
   $x - y < 3$

6. $x - y < 5$
   $x + y > 4$

7. $x + y < 2$
   $2x - y \ge -6$

8. $2x + y \le 8$
   $x - y > 1$

9. $x + 2y > 6$
    $2x + 4y < 8$

10. $3x - y < -6$
    $6x - 2y > -18$

11. $2x + y \geq 4$
    $x > 3$

12. $x - 3y < 6$
    $y \leq -1$

13. $x - 2y \geq 6$
    $y \leq -2$

14. $x \geq 5$
    $6x + y \geq 18$

15. $2x + 5y > 10$
    $5x - 3y < 18$

16. $3x - 2y < 12$
    $2x + 4y > 8$

17. $4x - 8y > 12$
    $x - 2y > 3$

18. $5x - 10y < 25$
    $-x + 2y > -5$

19. $2x + 3y \leq 6$
    $x > 0$
    $y > 0$

20. $3x + 4y \leq 24$
    $x < 0$
    $y < 0$

21. $4x - 8y > 8$
    $x \geq 3$
    $y > 2$

22. $3x - 6y > 6$
    $x > -1$
    $y \leq 3$

23. $2x + 3y \leq 12$
    $x + 3y < 15$
    $y \geq 4$

24. $2x + y \leq 14$
    $4x + 3y > 12$
    $x < 5$

25. $2x + y \geq 10$
    $x + 2y \leq 8$
    $x \geq 0$
    $y \leq 0$

26. $3x + y \leq 9$
    $x + 3y \geq 3$
    $x \geq 0$
    $y \leq 0$

27. $2x + 6y < 24$
    $2x - 4y \geq 16$
    $x > 0$
    $y < -2$

28. $3x - 5y \leq 15$
    $2x + 3y > 24$
    $x < 10$
    $y > 0$

29. A sow has a litter of at most 10 pigs. At least 4 of them are males and at least 3 are females. Let x = the number of male pigs and y = the number of female pigs.
    a. Write a system of inequalities that describes the situation above.
    b. Graph the solution of the system in part a.
    c. List two ordered pairs that are contained in the solution of the system.

30. At a certain college, students may enroll in no more than five lecture courses and no more than three lab courses in a given semester. In addition, the number of lab courses can be no more than the number of lecture courses. Let x = the number of lecture courses and y = the number of lab courses in which a student may enroll.
    a. Write a system of inequalities that describes the situation above.
    b. Graph the solution of the system in part a.
    c. List two ordered pairs that are contained in the solution of the system.

31. An investment firm has at most $270,000 to invest in two stocks. Stock A sells for $45 per share, and stock B sells for $60 per share. The company decides that the total number of shares of both stocks to be purchased cannot exceed 5000. Let x = the number of shares of stock A purchased and y = the number of shares of stock B purchased.
    a. Write a system of inequalities that describes the situation above.
    b. Graph the solution of the system in part a.
    c. List two ordered pairs that are contained in the solution of the system.

Preparing for Section 9.1

In Exercises 32–34, evaluate the given expression.

32. -6^3 33. $(-3)^5$ 34. $-(-2)^4$

In Exercises 35–38, solve the given equation.

35. $x^2 - 100 = 0$ **36.** $y^2 - 49 = 0$ **37.** $w^2 = 256$ **38.** $z^2 = 64$

Chapter Summary

Important Ideas

To solve word problems using a system of equations

a. Read the problem carefully. Draw a diagram or make a table if it will help you understand the problem.

b. Represent each unknown quantity with a different variable. If necessary, represent other quantities in terms of these variables.

c. Use the given conditions and numerical values, and perhaps a formula, to write a system of equations that describes the problem situation.

d. Solve the system of equations.

e. Check the solution in each equation, and against the conditions of the problem.

Chapter Review

Section 8.1

In Exercises 1–4, determine whether the given ordered pair is a solution of a given system of equations.

1. (2, 7) $2x - y = -3$
              $3x + 5y = 41$

2. (3, −4) $x + 2y = -5$
               $4x + 3y = 0$

3. (−1, 5) $4x + 3y = 11$
               $7x - 2y = 3$

4. (−6, −2) $2x - 7y = 2$
                $6x - 14y = -8$

In Exercises 5–16, solve the given system of equations by graphing.

5. $x + y = 8$
     $2x - y = 10$

6. $x - 6y = 12$
     $x - y = 2$

7. $6x + 2y = 4$
     $y = -3x + 12$

8. $y = 4x - 3$
     $6x - 3y = -15$

9. $x = -4y$
     $4x - 2y = 36$

10. $2x + 4y = 20$
      $x - 3y = 0$

11. $6x - 3y = 6$
      $y = 2x - 2$

12. $3x + 3y = -6$
      $12x + 6y = 12$

13. $0.3x + 0.2y = -0.4$
      $0.1x - 0.2y = 0.4$

14. $0.3x = 0.6$
      $0.1y - 0.5 = 0$

15. $x - \dfrac{1}{2}y = -\dfrac{1}{2}$
      $\dfrac{1}{4}x - \dfrac{1}{2}y = 1$

16. $x + \dfrac{3}{2}y = 1$
      $\dfrac{1}{8}x + \dfrac{1}{4}y = 1$

Section 8.2

In Exercises 17–28, solve the given system of equations by the elimination method.

17. $x - y = 9$
$x + y = 7$

18. $5x + 3y = 4$
$x - 3y = -10$

19. $6x + 4y = 8$
$6x + 2y = -2$

20. $3x - 6y = 15$
$2x - y = -2$

21. $3x - 5y = 17$
$2x + 4y = 4$

22. $6x + 5y = 3$
$4x - 8y = -32$

23. $7x + 3y = 33$
$5x + 2y = 23$

24. $3x - 7y = 26$
$5x - 6y = 15$

25. $0.5x - 0.2y = 1.3$
$0.3x - 0.2y = 0.7$

26. $0.4x + 0.3y = 1.1$
$0.5x - 0.6y = 0.4$

27. $\dfrac{1}{3}x + y = 1$
$\dfrac{1}{3}x - \dfrac{1}{6}y = 1$

28. $\dfrac{3}{2}x - \dfrac{1}{4}y = -2$
$\dfrac{1}{4}x + \dfrac{1}{8}y = -2$

Section 8.3

In Exercises 29–38, solve the given system of equations by the substitution method.

29. $y = 2x - 1$
$x + y = 8$

30. $x - y = -6$
$x = 4y - 18$

31. $2x + 3y = 8$
$5x + y = -6$

32. $x - 3y = 10$
$3x - 5y = 18$

33. $2x - 3y = 14$
$x + 4y = -26$

34. $4x - 8y = 4$
$7x + y = 22$

35. $0.7x + 0.1y = 1.4$
$0.2x - 0.5y = -3.3$

36. $0.3x - 0.7y = 2.8$
$0.1x + 0.4y = 0.3$

37. $\dfrac{1}{4}x - y = -\dfrac{1}{2}$
$\dfrac{1}{6}x - y = \dfrac{4}{3}$

38. $\dfrac{1}{5}x + \dfrac{1}{2}y = \dfrac{1}{10}$
$-\dfrac{1}{8}x + \dfrac{3}{4}y = 1$

Section 8.4

In Exercises 39–44, use a system of equations to solve the given problem.

39. The sum of two numbers is 7. The smaller number is 23 less than the larger number. Find the two numbers.

40. The perimeter of a rectangle is 44 cm. The length of the rectangle is 1 cm more than twice the width. Find the length and the width of the rectangle.

41. Century City issued $1000 and $500 school bonds to build a new high school. There were 2750 bonds sold with a total value of $1,875,000. How many $1000 bonds and how many $500 bonds were sold?

42. A coin collector has 45 dimes and quarters with a face value of $9.45. (They are actually worth a lot more.) How many dimes and how many quarters does she have?

43. A chemist has one solution that is 10% acid and another solution that is 15% acid. How many liters of each solution should he use to make 20 liters of a solution that is 12% acid?

44. An auto mechanic has a supply of 90% antifreeze and 75% antifreeze. How many gallons of each should she mix together to make 25 gal of 84% antifreeze?

Section 8.5

In Exercises 45–54, graph the solution of the given system of inequalities.

45. $y \geq 3x - 10$
$x - y > 2$

46. $x + 5y > 10$
$y \leq 2x + 6$

47. $x < 8y$
$2x + y > 6$

48. $y \geq -3x$
$x + y < 4$

49. $x - 4y \leq 8$
$y \geq -1$

50. $x \leq 1$
$6x - 3y \leq 12$

51. $3x + 2y < 6$
$y \geq -1.5x + 1$

52. $x > \dfrac{3}{4}y - 2$
$-8x + 6y < 16$

53. $3x + 4y < 12$
$x \geq 0$
$y \geq 0$

54. $4x + 2y > 8$
$y \leq 4x$
$x > 1$

55. A university's basketball arena seats 12,000. Student tickets cost $3, whereas tickets for the general public cost $9. The university must have at least $4500 revenue from ticket sales to earn a profit. Let $x =$ number of student tickets sold and $y =$ number of general public tickets sold.
   a. Write a system of inequalities that describes the situation above. (Assume that the university makes a profit.)
   b. Graph the solution of the system of inequalities in part a.
   c. List two ordered pairs that are contained in the solution of the system.

Chapter 8 Test

In Problems 1–2, solve the given system by graphing.

1. $3x - y = 3$
$y = x + 1$

2. $y = -0.5x$
$2x + 8y = 8$

In Problems 3–6, solve the given system by the elimination method.

3. $x - 5y = 8$
$x + 5y = -2$

4. $3x + y = -11$
$x + 2y = -12$

5. $2x - y = 1$
$4x - 3y = -5$

6. $2x + 3y = 4$
$5x - 4y = 33$

In Problems 7–9, solve the given system by the substitution method.

7. $x = y + 3$ **8.** $4x - y = 0$ **9.** $3x - 4y = -15$
        $5x - 2y = 3$ $x + 2y = 18$ $4x - y = 6$

In Problems 10–12, solve the given system by any method.

10. $2x + 3y = 7$ **11.** $4x + 5y = 2$ **12.** $x + \dfrac{2y}{3} = -3$
          $6x + y = -19$ $3x - 6y = 21$

$\qquad\qquad\qquad\qquad\qquad\qquad\qquad\qquad\qquad\qquad \dfrac{x}{2} - \dfrac{y}{4} = 2$

In Problems 13–16, solve each problem using a system of equations.

13.   A bus has 48 passengers. The number of adults on the bus is 4 less than 3 times the number of children. How many adults and how many children are on the bus?

14.   An appliance repairer charges a fixed amount plus an hourly rate to service a house call. The repairer charged $113 to repair a dishwasher that required 2 h of labor and $95 to repair a stove that required 1.5 h of labor. Find the appliance repairer's fixed charge and hourly rate.

15.   A submarine travels submerged for 10 h against a current. It returns with the current in 6 h. If the submarine travels 150 mi on each trip, find its speed in still water and find the speed of the current.

16.   A biologist has a supply of 10% formaldehyde and 30% formaldehyde. How many liters of each must she mix together to make 200 liters of 15% formaldehyde?

In Problems 17–19, graph the solution of the given system of inequalities.

17. $y \ge x$ **18.** $x + y > 5$ **19.** $x > -1$
          $x - 2y > 12$ $2x + y \le 10$ $y > x - 2$
$\qquad\qquad\qquad\qquad\qquad\qquad\qquad\qquad\qquad\qquad\qquad\qquad x + 3y \le 6$

20.   A bookstore has a maximum of $6000 to spend on business consultants. A consulting firm charges $200 per day for a computer consultant and $150 per day for a technical consultant. The firm requires at least 10 days to do the work requested by the bookstore. Let x = number of days the computer consultant works and y = number of days the technical consultant works to complete the job for the bookstore.
   a. Write a system of inequalities that describes the situation above.
   b. Graph the solution of the system in part a.
   c. List two ordered pairs that are contained in the solution of the system.

9 *Radicals*

9.1 Square Roots and Radicals

FOCUS

We review the concept of the square root of a number (which we first introduced in Chapter 1), and use square roots in several ways.

Recall from Chapter 1 that a is a *square root* of b if $a^2 = b$. If a is not negative, we abbreviate the phrase "a is a square root of b" by writing $a = \sqrt{b}$. The "$\sqrt{}$" symbol is called a **radical symbol.**

In Chapter 1, we also noted that if a is positive, then a can be thought of as the length of the side of a square with area b.

EXAMPLE 1

A man wants to put a fence around a part of his backyard to keep his dog from terrorizing the neighborhood. The part he wants to enclose is a square with an area of 140 square feet (ft²). What total length of fence does the man need? (Ignore waste.)

Solution

We let s be the length of a side of the square to be enclosed, as in Figure 9.1. Since the area of the square is 140 ft², we know that $s^2 = 140$. The first thing we need to find is s, which is the square root of 140. We use the $\boxed{\sqrt{}}$ key on a calculator.

Expression	Key Sequence	Display	Rounded Result
$\sqrt{140}$	140 $\boxed{\sqrt{}}$	11.83216	11.8

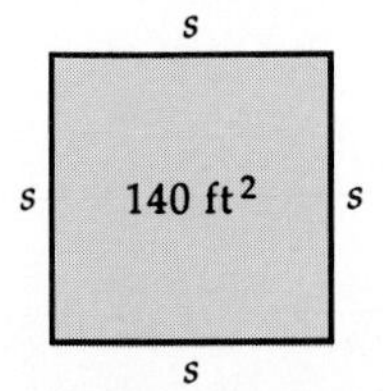

FIGURE 9.1

Hence, each of the four sides of the square is slightly longer than 11.8 ft. To enclose the backyard, the man needs a length of fence equal to its perimeter.

$$P = 4s$$

$$\approx 4 \cdot 11.8$$

$$= 47.2$$

The man needs about 47.2 ft of fence.   ■

In the previous example, our calculator display for $\sqrt{140}$ was 11.83216, but this is only an approximation to five decimal places. The exact decimal equivalent of $\sqrt{140}$ is an infinite decimal. In general, for most numbers n, the decimal equivalent of $\sqrt{n}$ has an infinite number of places. Furthermore, $\sqrt{n}$ usually cannot be written as a fraction $\dfrac{a}{b}$, in which a and b are both whole numbers. Thus, for most values of n, $\sqrt{n}$ is an *irrational number*. There are exceptions, however, like 4 and 9.

A rational number whose square root is rational is called a **perfect square.** For example, the square root of 16 is 4 because $4^2 = 16$. Thus, 16 is a perfect square, as are 1, 4, 9, 25, 36, 49, 64, 81, and 100. In addition, fractions like $\dfrac{4}{9}, \dfrac{16}{25}$, and $\dfrac{81}{49}$ are perfect squares. You should know how to find square roots of these numbers without using your calculator.

EXAMPLE 2   Find the square root of each number without using a calculator or table of square roots.

a.   121   **b.** $\dfrac{4}{9}$ **c.**   2.25

Solution   **a.**   The square root of 121 must be greater than 10 because $10^2 = 100$. A reasonable guess would be 11, which we check by squaring: $11^2 = 11 \cdot 11 = 121$. Hence, $\sqrt{121} = 11$

b.   We must find a positive number whose square is $\dfrac{4}{9}$. Remembering that $2^2 = 4$ and $3^2 = 9$ leads us to guess that $\dfrac{2}{3}$ might be the square root of $\dfrac{4}{9}$. In fact,

$$\left(\frac{2}{3}\right)^2 = \frac{2}{3} \cdot \frac{2}{3} = \frac{4}{9}$$

Hence, $\sqrt{\dfrac{4}{9}} = \dfrac{2}{3}$

c.   The square root of 2.25 must be a number between 1 and 2. (Why?) A reasonable first guess would be 1.5. Squaring, we see that

$$(1.5)^2 = (1.5)(1.5) = 2.25$$

Hence, $\sqrt{2.25} = 1.5$ ■

Principal Square Root

In our examples, the results have all been positive numbers. But just as $4^2 = 16$, it is also true that $(-4)^2 = 16$. Hence, according to our definition, both 4 and -4 are square roots of 16. We say that 4 is the *positive* square root of 16, and we write it as $\sqrt{16}$, so $\sqrt{16} = 4$. The *negative* square root of 16 is -4, and it is written $-\sqrt{16}$; hence, $-\sqrt{16} = -4$. The positive square root of a number is sometimes called its **principal square root.**

EXAMPLE 3

Find the negative square roots of these numbers.

a.   81   b. $\dfrac{9}{25}$ c.   1.96

Solution

a. $-\sqrt{81} = -9$ because $(-9)^2 = 81$.

b. $-\sqrt{\dfrac{9}{25}} = -\dfrac{3}{5}$ because $\left(-\dfrac{3}{5}\right)^2 = \dfrac{9}{25}$.

c.   This one is not so obvious. However, $\sqrt{1.96}$ must be a little less than $\sqrt{2.25} = 1.5$, which we found in Example 1. We try -1.4 and note that $(-1.4)^2 = 1.96$. Hence, $-\sqrt{1.96} = -1.4$ ■

Our examples have only involved finding square roots of positive numbers. Do negative numbers have real-number square roots? One way to find out is to try to use your calculator to find $\sqrt{-9}$. But if you enter 9 $\boxed{+/-}$ $\boxed{\sqrt{}}$, you will receive an error message. To see why, observe what happens when we square a number. If the number is positive, its square is positive; if the number is negative, its square is again positive; and if it is zero, its square is zero. Thus,

$$(+3)^2 = +9 \qquad (-3)^2 = +9 \qquad (0)^2 = 0$$

Hence, the square of every real number is positive or zero. So *only* positive numbers and zero have real-number square roots; negative numbers do not have real square roots. (In Chapter 10, we define another set of numbers, called the *complex numbers*, and show that negative numbers do have square roots in that set.)

EXAMPLE 4   Indicate whether each of these numbers is a real number. If so, write the number as a decimal to the nearest tenth.

a. $\sqrt{-5}$ **b.** $-\sqrt{87}$ **c.** $-\sqrt{-14}$

Solution   **a.**   The square root of a negative number, -5, is not a real number.

b.   This is a real number. To find it, we find $\sqrt{87}$ and write its negative. On a calculator,

Expression	Key Sequence	Display	Rounded Result
$-\sqrt{87}$	87 $\boxed{\sqrt{}}$ $\boxed{+/-}$	-9.3273791	-9.3

So, $-\sqrt{87} \approx -9.3$

c.   The square root of a negative number, -14, is not a real number. Hence, $\sqrt{-14}$ is not real, and the negative of $\sqrt{-14}$ is not real either.   ■

Graph of $y = \sqrt{x}$

The graph of $y = \sqrt{x}$, where $x \geq 0$, provides another way to represent the square roots of nonnegative numbers. To plot it, we develop the following table of square roots, which you can verify with your calculator.

x	0	1	2	3	4	5	6	7	8	9	10
$y = \sqrt{x}$	0	1	1.4	1.7	2	2.2	2.4	2.6	2.8	3	3.2

Next, we plot these ordered pairs and join them with a smooth curve, as in Figure 9.2. Now we can use the graph to find the square roots of numbers from 0 to 10.

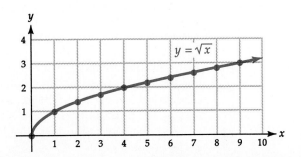

FIGURE 9.2

EXAMPLE 5   Use the graph of $y = \sqrt{x}$ to estimate each of these square roots.

a. $\sqrt{7.3}$ **b.** $\sqrt{1.5}$ **c.** $\sqrt{4.4}$

Solution   In each case, we locate the given number on the x axis and estimate the corresponding y value (see Figure 9.3). The results are

a. $\sqrt{7.3} \approx 2.7$ **b.** $\sqrt{1.5} \approx 1.2$ **c.** $\sqrt{4.4} \approx 2.1$ ■

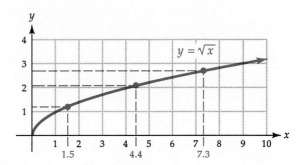

FIGURE 9.3

The idea of a square root can be generalized to the nth roots of numbers.

A number a is an **nth root** of b if $a^n = b$, where a and b are any real numbers and n is a positive integer.

If $a \geq 0$, we write $a = \sqrt[n]{b}$; if $a < 0$, we write $a = -\sqrt[n]{b}$. In $\sqrt[n]{b}$, n is called the **index** and b is called the **radicand.** The entire expression $\sqrt[n]{b}$ is called a **radical** or a **radical expression.**

For example,

$$\sqrt[4]{16} = 2 \text{ because } 2^4 = (2)(2)(2)(2) = 16$$
$$\sqrt[3]{-27} = -3 \text{ because } (-3)^3 = (-3)(-3)(-3) = -27$$

Also, in $\sqrt[4]{16}$, the index is 4 and the radicand is 16. In square roots, such as $\sqrt{9}$, the index is understood to be 2; however, it is perfectly correct to write $\sqrt[2]{9}$.

Recall that a cube is a rectangular solid with equal length, width, and height. The formula for the volume of a cube of side s is $V = s^3$. Thus, if a cube has volume V, then the length of a side of the cube is $s = \sqrt[3]{V}$. For this reason, the third root of b is also called the **cube root** of b. A rational number whose cube root is rational is called a **perfect cube.** For example, 27 is a perfect cube, because $\sqrt[3]{27} = 3$.

EXAMPLE 6   Find to the nearest tenth the length of a side of a cube whose volume is 137.5 cubic inches (in.^{3}).

Solution   As we noted above, the length of the side will be $\sqrt[3]{137.5}$. To compute this, we use a calculator. Most calculators have a $\boxed{\sqrt{}}$ key.

Expression	Key Sequence	Display	Rounded Result
$\sqrt[3]{137.5}$	137.5 $\boxed{\sqrt{}}$	5.1614006	5.2

To the nearest tenth, the length of the side is 5.2 in.   ∎

DISCUSSION QUESTIONS 9.1

1. Which numbers have real-number square roots? Which do not? Explain.

2. Just above Example 2, we listed the perfect squares up to 100. Name the next five perfect squares.

3. Does $\sqrt{81}$ equal 9 or -9? Justify your answer.

4. Does $-\sqrt{4} = \sqrt{-4}$? Justify your answer.

EXERCISES 9.1

1. Identify the real numbers below.

   a. $-\sqrt{2}$ **b.** $\sqrt{-25}$ **c.** $-\sqrt{52}$ **d.** $\sqrt{-36}$

2. Name the index and the radicand in each of the following:

   a. $\sqrt{42}$ **b.** $\sqrt{13}$ **c.** $\sqrt[3]{-67}$ **d.** $\sqrt[5]{0}$

In Exercises 3–14, use the graph in Figure 9.4 to estimate the square root of each number to the nearest tenth.

3. 8   4. 56

5. 17   6. 89

7. 42   8. 32

9. 66   10. 73

11. 24   12. 91

13. 9.5   14. 37.8

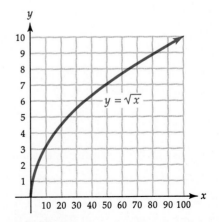

FIGURE 9.4

In Exercises 15–28, use a calculator to compute the square root of each number to the nearest tenth.

15. 7.8   **16.** 15   **17.** 74   **18.** $5\frac{1}{3}$ **19.** $17\frac{3}{4}$ **20.** 112

21. 175   **22.** 45.6   **23.** 135.9   **24.** 348   **25.** 659   **26.** 382.6

27. 12,685   **28.** 87,520

In Exercises 29–38, find, without the aid of a calculator, the exact square root of each number.

29. 144   **30.** 225   **31.** $\dfrac{25}{64}$ **32.** 3.61   **33.** $\dfrac{100}{9}$

34. 256   **35.** $\dfrac{81}{169}$ **36.** 4.41   **37.** $\dfrac{900}{25}$ **38.** 729

In Exercises 39–56, determine whether the statement is true, and justify your answer.

39. $\sqrt[3]{8} = 2$ **40.** $\sqrt[3]{64} = 7.6$ **41.** $\sqrt[3]{12} = 3.4$ **42.** $\sqrt[3]{29} = 3$

43. $\sqrt[3]{125} = 5$ **44.** $\sqrt[3]{200} < 6$ **45.** $\sqrt[3]{-1} = -1$ **46.** $\sqrt[3]{-27} = -3$

47. $\sqrt[3]{-120} > -5$ **48.** $\sqrt[3]{-258} < -6$ **49.** $\sqrt{3} = 1.732$ **50.** $-\sqrt{81} = -9$

51. $\sqrt{11} < 3.3$ **52.** $\sqrt{20} > 4.5$ **53.** $\sqrt{-25} = -5$ **54.** $\sqrt{1444} = 38$

55. $11.6 < \sqrt{135} < 11.7$ **56.** $\sqrt{3.24} < 1.8$

In Exercises 57–64, determine whether each statement is true when $N = 16$ and when $N = -16$. Do you think the statement is true for all positive values of N?

57. $\sqrt{N^2} = N$ **58.** $\sqrt{-N} = \sqrt{N}$ **59.** $(\sqrt{N})^2 = N$ **60.** $N = \sqrt{(-N)^2}$

61. $\sqrt{-N^2} = -N$ **62.** $(-\sqrt{N})^2 = N$ **63.** $\sqrt{N} = -\sqrt{N}$ **64.** $\sqrt{-N} \cdot \sqrt{N} = -N$

65. The area of a square is 346 in.^{2}. Find its perimeter to the nearest hundredth of an inch.

66. Find the perimeter of the figure in Figure 9.5 if its area is 150 square centimeters (cm^2).

Preparing for Section 9.2

In Exercises 67–72, compute both values and compare them.

67. $\sqrt{5} \times \sqrt{10}$ and $\sqrt{50}$ **68.** $\sqrt{8} \times \sqrt{20}$ and $\sqrt{160}$

69. $\sqrt{14} \div \sqrt{2}$ and $\sqrt{7}$ **70.** $\sqrt{68} \div \sqrt{17}$ and $\sqrt{4}$

71. $\sqrt{140} \div \sqrt{70}$ and $\sqrt{2}$ **72.** $\sqrt{24} \div \sqrt{5}$ and $\sqrt{4.8}$

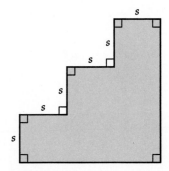

FIGURE 9.5

9.2 Multiplying Radicals

FOCUS

We multiply and simplify radical expressions by removing perfect-square factors from radicands.

By definition, we have $(\sqrt{a})^2 = a$. For example, since 2 is the square root of 4, we have $2^2 = 4$. Now consider the product of two square roots, $\sqrt{a}\sqrt{b}$. Is this product equal to $\sqrt{ab}$? To answer this question, we square $\sqrt{a}\sqrt{b}$.

$$(\sqrt{a}\sqrt{b})^2 = (\sqrt{a})^2(\sqrt{b})^2 = ab$$

Since the square of $\sqrt{a}\sqrt{b}$ is ab, it follows that $\sqrt{a}\sqrt{b} = \sqrt{ab}$. This property also holds for nth roots.

> **Multiplication property for radicals**
>
> If $\sqrt[n]{a}$ and $\sqrt[n]{b}$ are defined, then
>
> $$\sqrt[n]{a}\sqrt[n]{b} = \sqrt[n]{ab}$$

EXAMPLE 1

Show that $\sqrt{a}\sqrt{b} = \sqrt{ab}$ holds for $a = 9$ and $b = 25$.

Solution

$$\sqrt{a}\sqrt{b} = \sqrt{9}\sqrt{25} = 3 \cdot 5 = 15$$

and

$$\sqrt{ab} = \sqrt{9 \cdot 25} = \sqrt{225} = 15$$

Thus, $\sqrt{a}\sqrt{b} = \sqrt{ab} = 15$ for these values of a and b. ∎

The multiplication property for radicals can be used to remove perfect-square factors from the radicand of a square root. This is the first of several steps in *simplifying* a radical expression.

EXAMPLE 2

Simplify by removing perfect-square factors from the radicand.

a. $\sqrt{18}$ **b.** $\sqrt{20}$ **c.** $\sqrt{75}$

Solution

a. We find a factor of 18 that is a perfect square, namely 9. Then we use the multiplication property, $\sqrt{ab} = \sqrt{a}\sqrt{b}$.

$$\sqrt{18} = \sqrt{9 \cdot 2} \qquad \text{Factoring 18}$$

$$= \sqrt{9}\sqrt{2} \qquad \text{Using the multiplication property}$$

$$= 3\sqrt{2} \qquad \text{Taking the square root of 9}$$

You can check with your calculator to see that $\sqrt{18} = 3\sqrt{2}$.

b. We note that 4 is a perfect square and $20 = 4 \cdot 5$. Hence,

$$\sqrt{20} = \sqrt{4 \cdot 5} \qquad \text{Factoring 20}$$

$$= \sqrt{4}\sqrt{5} \qquad \text{Using the multiplication property}$$

$$= 2\sqrt{5} \qquad \text{Taking the square root of 4}$$

c. A perfect-square factor of 75 is 25.

$$\sqrt{75} = \sqrt{25 \cdot 3} \qquad \text{Factoring 75}$$

$$= \sqrt{25}\sqrt{3} \qquad \text{Using the multiplication property}$$

$$= 5\sqrt{3} \qquad \text{Taking the square root of 25} \qquad \blacksquare$$

The multiplication property also allows us to compute and simplify products of radicals.

EXAMPLE 3

Simplify by removing perfect-square factors from the resulting radicands.

a. $\sqrt{2}\sqrt{18}$ **b.** $\sqrt{20}\sqrt{75}$ **c.** $\sqrt{36}\sqrt{18}$

Solution

a. We use the multiplication property.

$$\sqrt{2}\sqrt{18} = \sqrt{2 \cdot 18} \qquad \text{Using the multiplication property}$$

$$= \sqrt{36} \qquad \text{Multiplying 18 by 2}$$

$$= 6 \qquad \text{Taking the square root of 36}$$

b. One way to approach this problem is to write

$$\sqrt{20}\sqrt{75} = \sqrt{20 \cdot 75} \qquad \text{Using the multiplication property}$$

$$= \sqrt{1500} \qquad \text{Multiplying 20 by 75}$$

We then must look for the greatest perfect-square factor of 1500, which is 100.

$$\sqrt{1500} = \sqrt{100 \cdot 15} \qquad \text{Factoring 1500}$$

$$= \sqrt{100}\sqrt{15} \qquad \text{Using the multiplication property}$$

$$= 10\sqrt{15} \qquad \text{Taking the square root of 100}$$

c. We could begin by multiplying the two radicands, as we did in part b. Instead, we will simplify the radicals before multiplying.

$$\sqrt{36}\sqrt{18} = 6\sqrt{18} \qquad \text{Taking the square root of 36}$$

$$= 6\sqrt{9 \cdot 2} \qquad \text{Factoring 18}$$

$$= 6\sqrt{9}\sqrt{2} \qquad \text{Using the multiplication property}$$

$$= (6 \cdot 3)\sqrt{2} \qquad \text{Taking the square root of 9}$$

$$= 18\sqrt{2} \qquad \text{Multiplying 6 by 3} \qquad \blacksquare$$

The multiplication property for radicals can be used to multiply radicals and remove factors involving higher-order roots, too.

EXAMPLE 4

a. Multiply $\sqrt[3]{4}\sqrt[3]{2}$.

b. Write $\sqrt[3]{54}$ with no perfect-cube factors in the radicand.

Solution

a. Apply the multiplication property.

$$\sqrt[3]{4}\sqrt[3]{2} = \sqrt[3]{4 \cdot 2} \qquad \text{Using the multiplication property}$$

$$= \sqrt[3]{8} \qquad \text{Multiplying 4 by 2}$$

$$= 2 \qquad \text{Taking the cube root of 8}$$

Use your calculator's $\boxed{\sqrt{}}$ key to verify this equation.

b. We note that $27 = 3^3$ is a factor of 54.

$$\sqrt[3]{54} = \sqrt[3]{27 \cdot 2} \qquad \text{Factoring 54}$$

$$= \sqrt[3]{27}\sqrt[3]{2} \qquad \text{Using the multiplication property}$$

$$= 3\sqrt[3]{2} \qquad \text{Taking the cube root of 27} \qquad \blacksquare$$

Some radical expressions involving variables in the radicand can be simplified using the following property.

If $\sqrt[n]{a}$ is defined, then

$$\sqrt[n]{a^n} = a$$

We apply this property for square roots in the following example.

EXAMPLE 5

Simplify each of the radicals. Variables represent nonnegative real numbers.

a. $\sqrt{x^2 y^4}$ **b.** $\sqrt{16x^3}$ **c.** $\sqrt{45x^5}$

Solution

a. We note that $y^4 = (y^2)^2$ and use the multiplication property for radicals.

$$
\begin{aligned}
\sqrt{x^2 y^4} &= \sqrt{x^2}\sqrt{y^4} && \text{Using the multiplication property}\\
&= \sqrt{x^2}\sqrt{(y^2)^2} && \text{Substituting } (y^2)^2 \text{ for } y^4\\
&= xy^2 && \text{Taking the square roots of } x^2 \text{ and } (y^2)^2
\end{aligned}
$$

b. We see that $16 = 4^2$ and $x^3 = x^2 \cdot x$, and both 16 and x^2 are perfect squares.

$$
\begin{aligned}
\sqrt{16x^3} &= \sqrt{(16x^2)(x)} && \text{Factoring } 16x^3\\
&= \sqrt{16}\sqrt{x^2}\sqrt{x} && \text{Using the multiplication property}\\
&= 4x\sqrt{x} && \text{Taking the square roots of 16 and } x^2
\end{aligned}
$$

c. The perfect-square factors we need are 9 and x^4, since $45 = (9)(5)$ and $x^5 = (x^4)(x)$.

$$
\begin{aligned}
\sqrt{45x^5} &= \sqrt{(9)(5)(x^4)(x)} && \text{Factoring } 45x^5\\
&= \sqrt{9}\sqrt{x^4}\sqrt{5x} && \text{Using the multiplication property}\\
&= 3x^2\sqrt{5x} && \text{Taking the square roots of 9 and } x^4 \quad \blacksquare
\end{aligned}
$$

BE CAREFUL!

1. The radical sign is *not* distributive over sums or differences. For example:

$$\sqrt{16+9} \neq \sqrt{16}+\sqrt{9} \quad \text{and} \quad \sqrt{25-9} \neq \sqrt{25}-\sqrt{9}$$

   (Compute each side to verify that the expressions are not equal.)
2. In expressions like $2\sqrt{5}$ you *cannot* multiply 2 times $\sqrt{5}$ by just multiplying 2 times 5, since 2 is not under a radical sign. On the other hand, $\sqrt{2}\sqrt{5} = \sqrt{10}$.

DISCUSSION QUESTIONS 9.2

1. Find the perfect-cube whole numbers between 1 and 1000.

2. Is $a\sqrt{b} + c$ equal to $\sqrt{ab + ac}$? Is $\sqrt{a}\sqrt{b} + c$ equal to $\sqrt{ab + ac}$? Give reasons for your answers.

3. If you compute $\sqrt{20} \times \sqrt{5}$ on a calculator, the display reads 10. Explain why.

EXERCISES 9.2

In Exercises 1–5, all variables represent nonnegative numbers. Find each root.

1. $\sqrt{x^4}$
2. $\sqrt{x^2y^4}$
3. $\sqrt{x^4y^6}$
4. $\sqrt[3]{-27x^3y^6}$
5. $\sqrt[3]{8(x + 1)^3}$

In Exercises 6–13, indicate whether you think the two given expressions are equal or unequal. Check your answer by computing each on a calculator.

6. $\sqrt{7 - 2}$; $\sqrt{7} - \sqrt{2}$
7. $\sqrt{5}\sqrt{3}$; $\sqrt{15}$
8. $\sqrt{7}\sqrt{11}$; $\sqrt{77}$
9. $6\sqrt{5}$; $\sqrt{30}$
10. $7\sqrt{2}$; $\sqrt{14}$
11. $2\sqrt{3}$; $\sqrt{12}$
12. $5\sqrt{2}$; $\sqrt{50}$
13. $\sqrt[3]{3}\sqrt[3]{6}$; $\sqrt[3]{18}$

In Exercises 14–28, multiply and write each radical expression with no perfect-square factors in the radicand.

14. $\sqrt{2}\sqrt{5}$
15. $\sqrt{6}\sqrt{3}$
16. $\sqrt{8}\sqrt{3}$
17. $\sqrt{19}\sqrt{3}$
18. $\sqrt{5}\sqrt{15}$
19. $\sqrt{2}\sqrt{32}$
20. $\sqrt{6}\sqrt{12}$
21. $\sqrt{10}\sqrt{15}$
22. $\sqrt[3]{2}\sqrt[3]{3}$
23. $\sqrt[3]{6}\sqrt[3]{4}$
24. $2\sqrt{3}\sqrt{3}$
25. $3\sqrt{5}\sqrt{5}$
26. $(3\sqrt{3})(5\sqrt{2})$
27. $(2\sqrt{8})(3\sqrt{2})$
28. $(3\sqrt{7})(7\sqrt{3})$

In Exercises 29–38, simplify each radical by factoring.

29. $\sqrt{117}$
30. $\sqrt{180}$
31. $\sqrt{112}$
32. $\sqrt{245}$
33. $\sqrt{192}$
34. $\sqrt{147}$
35. $\sqrt{162}$
36. $\sqrt{128}$
37. $\sqrt[3]{32}$
38. $\sqrt[3]{81}$

In Exercises 39–46, simplify the expression. Assume that all expressions denote real numbers.

39. $\sqrt{x}\sqrt{y}$
40. $\sqrt{y^2}\sqrt{y^4}$
41. $(2\sqrt{x})(-5\sqrt{x})$
42. $(3\sqrt{y^3})(-2\sqrt{y})$
43. $(2\sqrt{xy^3})(3\sqrt{x^2y^2})$
44. $(7\sqrt{r^3s})(-\sqrt{r^4s^2})$
45. $\sqrt[3]{x^2}\sqrt[3]{x}$
46. $\sqrt[3]{a^5}\sqrt[3]{a^2}$

47. The length of a rectangular solid is $3\sqrt{5}$ cm, its width is $\sqrt{5}$ cm, and its height is $2\sqrt{5}$ cm. Find (a) its volume, (b) its lateral surface area, and (c) its total surface area. Leave your answers in simplest radical form.

48. The area of one square is 24, and the area of a second square is twice that of the first. Find the length of a side of each square. Is the side of the larger square twice that of the smaller one?

49. The length of the diagonal d of a square of side s is given by $d = s\sqrt{2}$. Find to the nearest meter the length of the diagonal of a square field that is 5000 meters (m) on a side.

Preparing for Section 9.3

In Exercises 50–52, use a calculator to compute each expression and compare the results.

50. $\sqrt{10} \div \sqrt{5}$; $\sqrt{2}$ **51.** $\sqrt{18} \div \sqrt{3}$; $\sqrt{6}$ **52.** $\sqrt{28} \div \sqrt{7}$; $\sqrt{4}$

In Exercises 53–56, write each fraction in simplest form.

53. $\dfrac{12}{16}$ **54.** $\dfrac{4}{18}$ **55.** $\dfrac{8}{26}$ **56.** $\dfrac{16}{82}$

9.3   Dividing Radicals

FOCUS

We divide and simplify radical expressions by rationalizing denominators.

In Section 9.2, we showed that the product of two square roots is the square root of the product, that is, $\sqrt{a}\sqrt{b} = \sqrt{ab}$. We also showed that $\sqrt{a + b} \neq \sqrt{a} + \sqrt{b}$ and $\sqrt{a - b} \neq \sqrt{a} - \sqrt{b}$. What about quotients of radicals? Does $\dfrac{\sqrt{a}}{\sqrt{b}} = \sqrt{\dfrac{a}{b}}$? Yes, because the square of the left side is $\dfrac{a}{b}$.

That is,

$$\left(\frac{\sqrt{a}}{\sqrt{b}}\right)^2 = \frac{(\sqrt{a})^2}{(\sqrt{b})^2} = \frac{a}{b}$$

By the definition of square root, then, $\dfrac{\sqrt{a}}{\sqrt{b}} = \sqrt{\dfrac{a}{b}}$. Formally,

---

Division property for radicals

If $\sqrt[n]{a}$ and $\sqrt[n]{b}$ are defined and $b \neq 0$, then

$$\frac{\sqrt[n]{a}}{\sqrt[n]{b}} = \sqrt[n]{\frac{a}{b}}$$

EXAMPLE 1   Write these quotients in simplified form.

a. $\dfrac{\sqrt{48}}{\sqrt{6}}$ b. $\sqrt{128} \div \sqrt{8}$

Solution   We apply the division property in both parts.

a. $\dfrac{\sqrt{48}}{\sqrt{6}} = \sqrt{\dfrac{48}{6}}$ Using the division property

$= \sqrt{8}$ Dividing 48 by 6

$= \sqrt{4 \cdot 2}$ Factoring 8

$= \sqrt{4}\sqrt{2}$ Using the multiplication property

$= 2\sqrt{2}$ Taking the square root of 4

b. $\sqrt{128} \div \sqrt{8} = \sqrt{\dfrac{128}{8}}$ Using the division property

$= \sqrt{16}$ Dividing 128 by 8

$= 4$ Taking the square root of 16   ■

An expression with a radical in the denominator can be rewritten as an equivalent expression with no radical in the denominator. This process is called **rationalizing the denominator.**

EXAMPLE 2   Rationalize the denominator in each.

a. $\sqrt{\dfrac{14}{9}}$ b. $\dfrac{\sqrt{5}}{\sqrt{3}}$ c. $\dfrac{3\sqrt{3}}{2\sqrt{2}}$

Solution   a. $\sqrt{\dfrac{14}{9}} = \dfrac{\sqrt{14}}{\sqrt{9}}$ Using the division property

$= \dfrac{\sqrt{14}}{3}$ Taking the square root of 9

b.   Since the denominator is $\sqrt{3}$, we multiply by $\dfrac{\sqrt{3}}{\sqrt{3}}$, which is another form of 1. This will remove the radical from the denominator.

$\dfrac{\sqrt{5}}{\sqrt{3}} = \dfrac{\sqrt{5}}{\sqrt{3}} \cdot \dfrac{\sqrt{3}}{\sqrt{3}}$ Multiplying by $\dfrac{\sqrt{3}}{\sqrt{3}}$

$= \dfrac{\sqrt{5}\sqrt{3}}{\sqrt{3}\sqrt{3}}$ Multiplying fractions

$= \dfrac{\sqrt{15}}{3}$ Multiplying $\sqrt{3}$ by $\sqrt{3}$

c.   We multiply both the numerator and the denominator of the given fraction by $\sqrt{2}$, which is the same as multiplying by $\dfrac{\sqrt{2}}{\sqrt{2}}$ or 1. This removes the radical from the denominator.

$$\frac{3\sqrt{3}}{2\sqrt{2}} = \frac{3\sqrt{3}\sqrt{2}}{2\sqrt{2}\sqrt{2}} \qquad \text{Multiplying numerator and denominator by } \sqrt{2}$$

$$= \frac{3\sqrt{6}}{2\sqrt{4}} \qquad \text{Using the multiplication property}$$

$$= \frac{3\sqrt{6}}{4} \qquad \text{Taking the square root of 4}$$ ∎

One advantage of a fraction with a rational denominator is that it is easier to convert to a decimal. As an example, by rationalizing we can show that

$$\frac{1}{\sqrt{2}} = \frac{\sqrt{2}}{2}$$

Now $\sqrt{2}$ is approximately 1.414. Hence, to evaluate $\dfrac{1}{\sqrt{2}}$, we could calculate either

$$\frac{1}{\sqrt{2}} \approx \frac{1}{1.414} \quad \text{or} \quad \frac{\sqrt{2}}{2} \approx \frac{1.414}{2}$$

It is easy to see that the second quotient is 0.707, but the computation in the first case is much more difficult.

We can (and usually should) also rationalize denominators in radical expressions involving variables.

EXAMPLE 3

Rationalize each of the following. Assume that all expressions are real numbers.

a. $\dfrac{5}{\sqrt{x}}$ b. $\sqrt{\dfrac{3x}{4y}}$ c. $\sqrt[3]{\dfrac{8x^2}{27y^3}}$

Solution

a. $\dfrac{5}{\sqrt{x}} = \dfrac{5}{\sqrt{x}} \cdot \dfrac{\sqrt{x}}{\sqrt{x}} \qquad \text{Multiplying by } \dfrac{\sqrt{x}}{\sqrt{x}}$

$= \dfrac{5\sqrt{x}}{x} \qquad \text{Multiplying}$

b. $\sqrt{\dfrac{3x}{4y}} = \dfrac{\sqrt{3x}}{\sqrt{4y}}$ Using the division property

$\qquad\quad = \dfrac{\sqrt{3x}}{\sqrt{4y}} \cdot \dfrac{\sqrt{4y}}{\sqrt{4y}}$ Multiplying by $\dfrac{\sqrt{4y}}{\sqrt{4y}}$

$\qquad\quad = \dfrac{\sqrt{12xy}}{4y}$ Multiplying fractions

$\qquad\quad = \dfrac{\sqrt{4 \cdot 3xy}}{4y}$ Factoring $12xy$

$\qquad\quad = \dfrac{2\sqrt{3xy}}{4y}$ Taking the square root of 4

$\qquad\quad = \dfrac{\sqrt{3xy}}{2y}$ Dividing numerator and denominator by 2

c. This example requires the use of both the multiplication and division properties for radicals.

$\sqrt[3]{\dfrac{8x^2}{27y^3}} = \dfrac{\sqrt[3]{8x^2}}{\sqrt[3]{27y^3}}$ Using the division property

$\qquad\quad = \dfrac{\sqrt[3]{8}\sqrt[3]{x^2}}{\sqrt[3]{27}\sqrt[3]{y^3}}$ Using the multiplication property

$\qquad\quad = \dfrac{2\sqrt[3]{x^2}}{3y}$ Taking the cube root of 8, 27, and y^3 ■

An expression involving radicals is said to be in **simplest radical form** if there are no perfect-square factors in the radicand, there is no radical in the denominator, and the numerator and denominator have no common factors. The steps for simplifying a radical are as follows:

To write a radical in simplest radical form

1. Remove all perfect-square factors from the radicand.
2. Rationalize the denominator.
3. Reduce the fraction.

DISCUSSION QUESTIONS 9.3

1. If you compute $\sqrt{20} \div \sqrt{5}$ on a calculator, the display reads 2. Explain why.

2. What does it mean to rationalize the denominator in a radical expression?

3. Give one reason why an expression with a rational denominator is simpler than one with a radical in the denominator.

EXERCISES 9.3

In Exercises 1–4, all variables represent nonnegative numbers. Find each root.

1. $\dfrac{\sqrt{x^2}}{\sqrt[4]{y^4}}$

2. $\sqrt{\dfrac{4a^2}{9}}$

3. $\sqrt{\dfrac{x^6}{16y^2}}$

4. $\sqrt[3]{\dfrac{x^6}{y^3}}$

In Exercises 5–14, indicate whether you think the two given expressions are equal or unequal. Check your answer by computing each on a calculator.

5. $\dfrac{\sqrt{10}}{2}; \sqrt{5}$

6. $\dfrac{\sqrt{21}}{3}; \sqrt{7}$

7. $\dfrac{1}{\sqrt{3}}; \dfrac{\sqrt{3}}{3}$

8. $\dfrac{1}{\sqrt{5}}; \dfrac{\sqrt{5}}{5}$

9. $\dfrac{\sqrt{10}}{\sqrt{2}}; \sqrt{5}$

10. $\dfrac{\sqrt{21}}{\sqrt{3}}; \sqrt{7}$

11. $\dfrac{8}{\sqrt{2}}; 2$

12. $\dfrac{15}{\sqrt{5}}; \sqrt{3}$

13. $\dfrac{3\sqrt{5}}{6}; \dfrac{\sqrt{5}}{2}$

14. $\dfrac{8\sqrt{7}}{4}; 2\sqrt{7}$

In Exercises 15–37, write each radical expression in simplest radical form.

15. $\sqrt{\dfrac{14}{4}}$

16. $\sqrt{\dfrac{7}{36}}$

17. $\sqrt{\dfrac{18}{25}}$

18. $\sqrt{\dfrac{12}{36}}$

19. $\sqrt{\dfrac{24}{64}}$

20. $\sqrt{\dfrac{5}{8}}$

21. $\dfrac{\sqrt{40}}{\sqrt{81}}$

22. $\dfrac{\sqrt{5}}{\sqrt{12}}$

23. $\dfrac{1}{\sqrt{5}}$

24. $\dfrac{2}{\sqrt{2}}$

25. $\dfrac{28}{2\sqrt{7}}$

26. $\dfrac{15}{\sqrt{6}}$

27. $\sqrt{2\dfrac{1}{8}}$

28. $\sqrt{5\dfrac{3}{4}}$

29. $\dfrac{\sqrt{11}}{\sqrt{44}}$

30. $\dfrac{\sqrt{98}}{\sqrt{2}}$

31. $\dfrac{25\sqrt{3}}{3\sqrt{5}}$

32. $\dfrac{10\sqrt{2}}{3\sqrt{20}}$

33. $\dfrac{8\sqrt{2}}{6\sqrt{24}}$

34. $\dfrac{6\sqrt{7}}{7\sqrt{6}}$

35. $\dfrac{1}{2}\sqrt{162}$

36. $\dfrac{2}{3}\sqrt{180}$

37. $\dfrac{5\sqrt[3]{32}}{8\sqrt[3]{2}}$

In Exercises 38–47, simplify the expression. Assume that all expressions denote real numbers.

38. $\sqrt{\dfrac{1}{r}}$

39. $\sqrt{\dfrac{2}{x}}$

40. $\sqrt{\dfrac{2}{3x}}$

41. $\sqrt{\dfrac{5}{x^2}}$

42. $\sqrt{\dfrac{5x}{3y}}$

43. $\sqrt{\dfrac{2y}{5x}}$

44. $\sqrt{\dfrac{x^3}{y}}$

45. $\dfrac{\sqrt{3a^5}}{\sqrt{8b}}$

46. $\dfrac{\sqrt{3x}}{\sqrt{x^3}}$

47. $\dfrac{\sqrt{5y}}{\sqrt{y^5}}$

48. The length of the diagonal d of a square is $\sqrt{2}$ times the length of its side s. Thus, $d = \sqrt{2}s$. Solving for s, we have $s = \dfrac{d}{\sqrt{2}}$. Find to the nearest tenth the length of the side of a square with the given diagonal.

   a. 10 cm   **b.** 15 cm   **c.** 26 m   **d.** 135 m

49. The area of one square is 24, and the area of a second square is 32. Find the ratio of the perimeter of the smaller square to that of the larger square (in simplified radical form).

50. The pendulum of a clock makes one cycle in the time $T = 2\pi\dfrac{\sqrt{L}}{\sqrt{32}}$, where L is the length of the pendulum in feet, and T is the time in seconds.
a. Find T as a product of π and an expression with no radical in the denominator.
b. Find T to the nearest tenth of a second for a pendulum that is 28 ft long.

Preparing for Section 9.4

In Exercises 51–54, combine like terms. Assume all variables represent positive real numbers.

51. $5x + 8x$ **52.** $3y^2 - 9y^2$ **53.** $2\sqrt{3} + 5\sqrt{3}$ **54.** $5\sqrt{x} - 12\sqrt{x}$

9.4 **Combining Like Radicals**

FOCUS

We use the distributive property to add and subtract radicals that have the same index and the same radicand.

Like radicals are radicals that have the same index and the same radicand. Thus the expressions $7\sqrt{13}$ and $4\sqrt{13}$ are like radicals because both contain $\sqrt{13}$. But $7\sqrt{13}$ and $4\sqrt[3]{13}$ are not like radicals because $\sqrt{13}$ has index 2 and $\sqrt[3]{13}$ has index 3.

The distributive property can be used to combine *like* radicals. For example,

$$7\sqrt{13} + 4\sqrt{13} = (7 + 4)\sqrt{13} \qquad \text{By the distributive property}$$

$$= 11\sqrt{13}$$

and $7\sqrt{13} - 4\sqrt{13} = (7 - 4)\sqrt{13} \qquad \text{By the distributive property}$

$$= 3\sqrt{13}$$

Unlike radicals cannot be simplified in this way.

EXAMPLE 1   Perform the indicated operation. Assume that all variables represent non-negative numbers.

    a. $4\sqrt{3} + \sqrt{3}$ **b.** $2\sqrt{5} - 9\sqrt{5}$

    c. $3\sqrt{x} + 6\sqrt{x}$ **d.** $5\sqrt[3]{10y^2} - 4\sqrt[3]{10y^2}$

Solution   We use the distributive property to add or subtract the like radicals in each case.

    a. $4\sqrt{3} + \sqrt{3} = 4\sqrt{3} + 1\sqrt{3}$

$$= (4 + 1)\sqrt{3} \qquad \text{By the distributive property}$$

$$= 5\sqrt{3}$$

    b. $2\sqrt{5} - 9\sqrt{5} = (2 - 9)\sqrt{5} \qquad \text{By the distributive property}$

$$= -7\sqrt{5}$$

    c. $3\sqrt{x} + 6\sqrt{x} = (3 + 6)\sqrt{x}$

$$= 9\sqrt{x}$$

    d. $5\sqrt[3]{10y^2} - 4\sqrt[3]{10y^2} = (5 - 4)\sqrt[3]{10y^2}$

$$= \sqrt[3]{10y^2} \qquad\qquad\qquad ■$$

    Sometimes, by simplifying unlike radicals, we produce like radicals, which then can be combined. To simplify square roots, we remove perfect-square factors from the radicand. Similarly, we remove perfect-cube factors from the radicand to simplify cube roots.

EXAMPLE 2   Perform the indicated operation. Assume that all variables represent non-negative numbers.

    a. $\sqrt{2} + \sqrt{8}$ **b.** $\sqrt{75} - \sqrt{27}$

    c. $3\sqrt{20z^2} + \sqrt{45z^2}$ **d.** $2w\sqrt[3]{54w} - 5\sqrt[3]{2w^4}$

Solution   **a.** We first simplify $\sqrt{8}$ and then combine like radicals:

$$\sqrt{2} + \sqrt{8} = \sqrt{2} + \sqrt{4 \cdot 2} \qquad \text{Factoring 8}$$

$$= \sqrt{2} + \sqrt{4}\sqrt{2} \qquad \text{Using the multiplication property}$$

$$= \sqrt{2} + 2\sqrt{2} \qquad \text{Taking the square root of 4}$$

$$= 3\sqrt{2} \qquad\qquad \text{Combining like radicals}$$

b. We first simplify $\sqrt{75}$ and $\sqrt{27}$ and then combine like radicals:

$$\sqrt{75} - \sqrt{27} = \sqrt{25 \cdot 3} - \sqrt{9 \cdot 3} \qquad \text{Factoring 75 and 27}$$
$$= \sqrt{25}\sqrt{3} - \sqrt{9}\sqrt{3} \qquad \text{Using the multiplication property}$$
$$= 5\sqrt{3} - 3\sqrt{3} \qquad \text{Taking the square roots of 25 and 9}$$
$$= 2\sqrt{3} \qquad \text{Combining like radicals}$$

c. We first simplify $\sqrt{20z^2}$ and $\sqrt{45z^2}$ and then combine like radicals:

$$3\sqrt{20z^2} + \sqrt{45z^2}$$
$$= 3\sqrt{20}\sqrt{z^2} + \sqrt{45}\sqrt{z^2} \qquad \text{Using the multiplication property}$$
$$= 3\sqrt{4}\sqrt{5}\sqrt{z^2} + \sqrt{9}\sqrt{5}\sqrt{z^2} \qquad \text{Simplifying } \sqrt{20} \text{ and } \sqrt{45}$$
$$= 3 \cdot 2\sqrt{5} \cdot z + 3\sqrt{5} \cdot z \qquad \text{Taking the square roots of 4, 9, and } z^2$$
$$= 6z\sqrt{5} + 3z\sqrt{5} \qquad \text{Simplifying}$$
$$= 9z\sqrt{5} \qquad \text{Combining like radicals}$$

d. Since the radicals are cube roots, we factor the radicands on the right side of the equation to produce perfect-cube factors:

$$2w\sqrt[3]{54w} - 5\sqrt[3]{2w^4}$$
$$= 2w\sqrt[3]{27 \cdot 2w} - 5\sqrt[3]{w^3 \cdot 2w} \qquad \text{Factoring } 54w \text{ and } 2w^4$$

Applying the multiplication property of radicals to $\sqrt[3]{27 \cdot 2w}$ and $\sqrt[3]{w^3 \cdot 2w}$, we get

$$2w\sqrt[3]{54w} - 5\sqrt[3]{2w^4}$$
$$= 2w\sqrt[3]{27}\sqrt[3]{2w} - 5\sqrt[3]{w^3}\sqrt[3]{2w}$$
$$= 2w \cdot 3\sqrt[3]{2w} - 5 \cdot w\sqrt[3]{2w} \qquad \text{Taking the cube roots of 27 and } w^3$$

Using the commutative and associative properties to simplify the right side of the equation, we obtain

$$2w\sqrt[3]{54w} - 5\sqrt[3]{2w^4}$$
$$= 6w\sqrt[3]{2w} - 5w\sqrt[3]{2w} \qquad \text{Using the multiplication property}$$
$$= w\sqrt[3]{2w} \qquad \text{Combining like radicals} \qquad \blacksquare$$

As illustrated in Example 2, simplifying radicals can sometimes be a formidable task. However, with sufficient practice, you will probably discover some shortcuts, thereby eliminating some of the steps required to obtain a solution. For example, in part c of Example 2, one alternative solution is to write the radicals as

$$\sqrt{20z^2} = \sqrt{4z^2 \cdot 5} = \sqrt{(2z)^2 \cdot 5}$$

and

$$\sqrt{45z^2} = \sqrt{9z^2 \cdot 5} = \sqrt{(3z)^2 \cdot 5}$$

and then apply the multiplication property of radicals.

EXAMPLE 3

Simplify each expression. Assume that all variables represent nonnegative numbers.

a. $2\sqrt{28} - \sqrt{18} + 4\sqrt{7}$ b. $4\sqrt{2k} + 7\sqrt[3]{2k} - 5\sqrt[3]{2k}$

Solution

a. We start by simplifying $2\sqrt{28}$ and $\sqrt{18}$:

$2\sqrt{28} - \sqrt{18} + 4\sqrt{7}$

$\quad = 2\sqrt{4 \cdot 7} - \sqrt{9 \cdot 2} + 4\sqrt{7}$ Factoring 28 and 18

$\quad = 2\sqrt{4}\sqrt{7} - \sqrt{9}\sqrt{2} + 4\sqrt{7}$ Using the multiplication property

$\quad = 2 \cdot 2\sqrt{7} - 3\sqrt{2} + 4\sqrt{7}$ Taking the square roots of 4 and 9

$\quad = 4\sqrt{7} - 3\sqrt{2} + 4\sqrt{7}$

Now we can combine like radicals:

$2\sqrt{28} - \sqrt{18} + 4\sqrt{7} = 8\sqrt{7} - 3\sqrt{2}$ Combining like radicals

This result cannot be simplified further because the radicands are different.

b. $4\sqrt{2k} + 7\sqrt[3]{2k} - 5\sqrt[3]{2k} = 4\sqrt{2k} + 2\sqrt[3]{2k}$ Combining like radicals

This result cannot be simplified further because the indices are different. ∎

DISCUSSION QUESTIONS 9.4

1. What are like radicals?

2. How are like radicals combined?

3. Does $\sqrt{a^2} + \sqrt{b^2} = \sqrt{a^2 + b^2}$? Explain.

EXERCISES 9.4

In Exercises 1–43, perform the indicated operation(s). Assume that all variables represent nonnegative numbers.

1. $3\sqrt{5} + 8\sqrt{5}$ **2.** $4\sqrt{3} + 7\sqrt{3}$ **3.** $4\sqrt{13} + \sqrt{13}$ **4.** $\sqrt{17} + 2\sqrt{17}$

5. $\sqrt{23} - 5\sqrt{23}$ **6.** $8\sqrt{19} - \sqrt{19}$ **7.** $\sqrt{8} + 3\sqrt{2}$ **8.** $4\sqrt{5} + \sqrt{20}$

9. $4\sqrt{12} + 6\sqrt{75}$ **10.** $3\sqrt{24} - 5\sqrt{54}$ **11.** $2\sqrt{18} - 3\sqrt{8}$ **12.** $4\sqrt{32} + 6\sqrt{72}$

13. $\sqrt[3]{100} + \sqrt[3]{100}$ **14.** $\sqrt[3]{27} + \sqrt{27}$ **15.** $6\sqrt[3]{40} - 5\sqrt[3]{135}$ **16.** $2\sqrt[3]{24} - 4\sqrt[3]{81}$

17. $7\sqrt[4]{32} + 9\sqrt[4]{162}$ **18.** $5\sqrt[4]{48} + 8\sqrt[4]{243}$ **19.** $\sqrt{3m} - 2\sqrt{3m}$ **20.** $\sqrt{7n} + \sqrt{7n}$

21. $3\sqrt{5q} + 7\sqrt{5q}$ **22.** $3\sqrt{13p} - 9\sqrt{13p}$ **23.** $w\sqrt{300} - \sqrt{27w^2}$ **24.** $\sqrt{28z^2} - z\sqrt{63}$

25. $\sqrt{20p^2} + 8\sqrt{45p^2}$ **26.** $6\sqrt{40q^2} + \sqrt{90q^2}$ **27.** $6\sqrt[3]{5m^3} + \sqrt[3]{40m^3}$ **28.** $\sqrt[3]{54n^3} + 7\sqrt[3]{3n^3}$

29. $\sqrt[3]{54k} - \sqrt[3]{16k^4}$ **30.** $\sqrt[3]{256h^5} - \sqrt[3]{32h^2}$ **31.** $5\sqrt[3]{81x^5} + 4\sqrt[3]{24x^5}$

32. $8\sqrt[3]{135y^4} - 3\sqrt[3]{40y^4}$ **33.** $4\sqrt{12} + 5\sqrt{8} - 2\sqrt{50}$ **34.** $4\sqrt{28} + 6\sqrt{48} - \sqrt{63}$

35. $8\sqrt{75} - 7\sqrt{12} + 4\sqrt{8}$ **36.** $3\sqrt{48} - 9\sqrt{18} + 7\sqrt{50}$ **37.** $2\sqrt[3]{192} + 9\sqrt[3]{24} - 9\sqrt[3]{3}$

38. $\sqrt[3]{81} - 5\sqrt[3]{192} + 2\sqrt[3]{24}$ **39.** $3\sqrt{6w} - 6\sqrt{w} + \sqrt{6w}$ **40.** $\sqrt[3]{4z} - 3\sqrt{4z} - 2\sqrt{4z}$

41. $\sqrt{8} + \sqrt[3]{16} - 3\sqrt{2}$ **42.** $\sqrt[4]{6y} + \sqrt[3]{6y} - 2\sqrt[4]{6y}$ **43.** $4\sqrt{7p^2q} - 7\sqrt{28pq^2} + 12\sqrt{63p}$

Preparing for Section 9.5

In Exercises 44–47, find the product.

44. $(2v - 3)(v - 9)$ **45.** $(x - 15)^2$ **46.** $(5y + 8)^2$ **47.** $(4m - 13)(4m + 13)$

9.5   Simplifying Radical Expressions

FOCUS

We simplify products of radical expressions that contain more than one term. We also discuss a method for rationalizing binomial denominators.

Now that we know how to combine like radicals, we can simplify products and quotients of radical expressions that contain more than one term. In the next two examples, we simplify products of radical expressions by first using the distributive property, the FOIL method, and special products, and then combining the resulting like radicals.

EXAMPLE 1   Simplify each expression. Assume that x is a nonnegative number.

a. $\sqrt{3}(\sqrt{18} + \sqrt{32})$ **b.** $(\sqrt{5} + 4)(\sqrt{5} + 2)$ **c.** $(3\sqrt{x} - 2)(\sqrt{x} + 5)$

Solution   **a.** We begin by using the distributive property to find the product $\sqrt{3}(\sqrt{18} + \sqrt{32})$:

$$\sqrt{3}(\sqrt{18} + \sqrt{32}) = \sqrt{3}\sqrt{18} + \sqrt{3}\sqrt{32} \qquad \text{By the distributive property}$$

$$= \sqrt{54} + \sqrt{96} \qquad \text{Using the multiplication property}$$

Next we simplify $\sqrt{54}$ and $\sqrt{96}$ and then combine the resulting like radicals:

$$\sqrt{3}(\sqrt{18} + \sqrt{32}) = \sqrt{9 \cdot 6} + \sqrt{16 \cdot 6} \qquad \text{Factoring 54 and 96}$$

$$= \sqrt{9}\sqrt{6} + \sqrt{16}\sqrt{6} \qquad \text{Using the multiplication property}$$

$$= 3\sqrt{6} + 4\sqrt{6} \qquad \text{Taking the square roots of 9 and 16}$$

$$= 7\sqrt{6} \qquad \text{Combining like radicals}$$

An alternative solution is to first simplify the radicals inside the parentheses and then combine the resulting like radicals.

b. We can use the FOIL method to find the product $(\sqrt{5} + 4)(\sqrt{5} + 2)$:

$$(\sqrt{5} + 4)(\sqrt{5} + 2)$$

$$\overset{\text{F}\qquad\quad\text{O}\qquad\quad\text{I}\qquad\quad\text{L}}{= \sqrt{5} \cdot \sqrt{5} + \sqrt{5} \cdot 2 + 4 \cdot \sqrt{5} + 4 \cdot 2}$$

$$= \sqrt{25} + 2\sqrt{5} + 4\sqrt{5} + 8 \qquad \text{Using the multiplication property}$$

$$= 5 + 6\sqrt{5} + 8 \qquad \text{Taking the square root of 25 and combining like radicals}$$

$$= 13 + 6\sqrt{5}$$

c. $(3\sqrt{x} - 2)(\sqrt{x} + 5)$

$$= 3\sqrt{x} \cdot \sqrt{x} + 3\sqrt{x} \cdot 5 - 2 \cdot \sqrt{x} - 2 \cdot 5 \qquad \text{Multiplying by the FOIL method}$$

$$= 3\sqrt{x^2} + 15\sqrt{x} - 2\sqrt{x} - 10 \qquad \text{Using the multiplication property}$$

$$= 3x - 13\sqrt{x} - 10 \qquad \text{Taking the square root of } x^2 \text{ and combining like radicals} \quad \blacksquare$$

Sometimes we can use the formulas for special products given in Chapter 4 to simplify products of radical expressions.

EXAMPLE 2   Simplify each expression.

    a. $(5 + \sqrt{2})(5 - \sqrt{2})$ **b.** $(\sqrt{p} + 6)^2$ **c.** $(\sqrt{w} - \sqrt{z})^2$

Solution   **a.** We can use the FOIL method, or the formula for the special product $(a + b)(a - b) = a^2 - b^2$, to find $(5 + \sqrt{2})(5 - \sqrt{2})$:

$$(5 + \sqrt{2})(5 - \sqrt{2}) = 5^2 - (\sqrt{2})^2 \qquad \text{Using } (a + b)(a - b) = a^2 - b^2$$
$$= 25 - 2$$
$$= 23$$

    b. We can use the formula for squaring a binomial $(a + b)^2 = a^2 + 2ab + b^2$ to simplify $(\sqrt{p} + 6)^2$:

$$(\sqrt{p} + 6)^2 = (\sqrt{p})^2 + 2 \cdot \sqrt{p} \cdot 6 + 6^2$$
$$= p + 12\sqrt{p} + 36$$

    c. We can use the formula for squaring a binomial $(a - b)^2 = a^2 - 2ab + b^2$ to simplify $(\sqrt{w} - \sqrt{z})^2$:

$$(\sqrt{w} - \sqrt{z})^2 = (\sqrt{w})^2 - 2\sqrt{w}\sqrt{z} + (\sqrt{z})^2$$

$$w - 2\sqrt{wz} + z \qquad\qquad \text{Using the multiplication}$$
$$\text{property} \qquad\qquad\qquad\blacksquare$$

Rationalizing Binomial Denominators

Binomial expressions of the form $a + b$ and $a - b$, where either a or b (or both) are radicals, are called **conjugates.** For example, the binomials $5 + \sqrt{2}$ and $5 - \sqrt{2}$ are conjugates. The result of part a of Example 2 shows that the product of a pair of conjugates does not contain a radical. This observation provides a way to rationalize a denominator that contains a sum or difference of radicals. We simply multiply the numerator and the denominator by the conjugate of the denominator.

EXAMPLE 3   Simplify the expression $\dfrac{3}{5 + \sqrt{2}}$.

Solution   We multiply the numerator and the denominator of $\dfrac{3}{5 + \sqrt{2}}$ by $5 - \sqrt{2}$ (the conjugate of the denominator). This "multiplication by 1" eliminates the radical from the denominator:

$$\frac{3}{5 + \sqrt{2}} = \frac{3}{5 + \sqrt{2}} \cdot \frac{5 - \sqrt{2}}{5 - \sqrt{2}}$$

$$= \frac{3(5 - \sqrt{2})}{5^2 - (\sqrt{2})^2}$$

Multiplying fractions and using the formula $(a + b)(a - b) = a^2 - b^2$

$$= \frac{15 - 3\sqrt{2}}{25 - 2}$$

By the distributive property and by squaring

$$= \frac{15 - 3\sqrt{2}}{23}$$

∎

EXAMPLE 4

Simplify the expression $\dfrac{\sqrt{10}}{\sqrt{10} - 4}$.

Solution

The conjugate of $\sqrt{10} - 4$ is $\sqrt{10} + 4$. Therefore we multiply the numerator and the denominator of $\dfrac{\sqrt{10}}{\sqrt{10} - 4}$ by $\sqrt{10} + 4$ to eliminate the radical from the denominator:

$$\frac{\sqrt{10}}{\sqrt{10} - 4} = \frac{\sqrt{10}}{\sqrt{10} - 4} \cdot \frac{\sqrt{10} + 4}{\sqrt{10} + 4}$$

$$= \frac{\sqrt{10}(\sqrt{10} + 4)}{(\sqrt{10})^2 - 4^2}$$

Multiplying fractions and using the formula $(a + b)(a - b) = a^2 - b^2$

$$= \frac{10 + 4\sqrt{10}}{10 - 16}$$

By the distributive property and by squaring

$$= \frac{10 + 4\sqrt{10}}{-6}$$

We can reduce $\dfrac{10 + 4\sqrt{10}}{-6}$ to lowest terms by factoring the numerator and then dividing out the common factor 2 from the numerator and the denominator. Continuing, we have

$$\frac{\sqrt{10}}{\sqrt{10} - 4} = \frac{2(5 + 2\sqrt{10})}{-6}$$

Factoring

$$= \frac{5 + 2\sqrt{10}}{-3}$$

Dividing numerator and denominator by 2

$$= -\frac{5 + 2\sqrt{10}}{3}$$

∎

EXAMPLE 5   Simplify the expression $\dfrac{\sqrt{5}-\sqrt{3}}{\sqrt{5}+\sqrt{3}}$.

Solution   We multiply the numerator and the denominator of $\dfrac{\sqrt{5}-\sqrt{3}}{\sqrt{5}+\sqrt{3}}$ by $\sqrt{5}-\sqrt{3}$ to eliminate the radical from the denominator:

$$\frac{\sqrt{5}-\sqrt{3}}{\sqrt{5}+\sqrt{3}} = \frac{\sqrt{5}-\sqrt{3}}{\sqrt{5}+\sqrt{3}} \cdot \frac{\sqrt{5}-\sqrt{3}}{\sqrt{5}-\sqrt{3}}$$

$$= \frac{(\sqrt{5}-\sqrt{3})(\sqrt{5}-\sqrt{3})}{(\sqrt{5})^2 - (\sqrt{3})^2}$$

Multiplying fractions and using the formula $(a+b)(a-b)=a^2-b^2$

We can use the FOIL method, or the formula for the special product $(a-b)^2 = a^2 - 2ab + b^2$, to find $(\sqrt{5}-\sqrt{3})(\sqrt{5}-\sqrt{3})$:

$$\frac{\sqrt{5}-\sqrt{3}}{\sqrt{5}+\sqrt{3}} = \frac{(\sqrt{5})^2 - 2\sqrt{5}\sqrt{3} + (\sqrt{3})^2}{(\sqrt{5})^2 - (\sqrt{3})^2}$$

Using $(a-b)^2 = a^2 - 2ab + b^2$

$$= \frac{5 - 2\sqrt{15} + 3}{5 - 3}$$

Using the multiplication property and squaring

$$= \frac{8 - 2\sqrt{15}}{2}$$

$$= \frac{2(4 - \sqrt{15})}{2}$$

Factoring

$$= 4 - \sqrt{15}$$

Dividing numerator and denominator by 2   ∎

DISCUSSION QUESTIONS 9.5

1.   What are conjugates?

2.   How are conjugates used to rationalize denominators that contain a sum or difference of radicals?

3.   Does the product of conjugates contain a radical? Why?

EXERCISES 9.5

In Exercises 1–46, simplify the given expression.

1. $3(2\sqrt{5}+4)$

2. $7(8+4\sqrt{3})$

3. $\sqrt{8}(\sqrt{2}+\sqrt{8})$

4. $\sqrt{12}(\sqrt{3}+\sqrt{12})$

5. $\sqrt{3}(2\sqrt{50}-4\sqrt{72})$

6. $\sqrt{6}(3\sqrt{10}-2\sqrt{15})$

7. $\sqrt{x}(3\sqrt{x} + 7)$ **8.** $2\sqrt{y}(4 - 8\sqrt{y})$ **9.** $(\sqrt{3} + 5)(\sqrt{3} + 2)$

10. $(7 + \sqrt{5})(4 + \sqrt{5})$ **11.** $(1 - \sqrt{2})(7 + \sqrt{2})$ **12.** $(2 + \sqrt{3})(9 - \sqrt{3})$

13. $(\sqrt{q} + 6)(\sqrt{q} - 9)$ **14.** $(2\sqrt{p} + 1)(\sqrt{p} + 10)$ **15.** $(3\sqrt{r} - 4)(2\sqrt{r} - 7)$

16. $(4\sqrt{s} + 5)(6\sqrt{s} - 1)$ **17.** $(\sqrt{2h} + k)(\sqrt{2h} - 5k)$ **18.** $(\sqrt{13} + 6)^2$

19. $(2 + \sqrt{11})^2$ **20.** $(7 - \sqrt{10})^2$ **21.** $(\sqrt{y} - 9)^2$ **22.** $(\sqrt{x} + 4)^2$

23. $(\sqrt{6} + \sqrt{8})^2$ **24.** $(\sqrt{18} - \sqrt{6})^2$ **25.** $(\sqrt{3p} + 5\sqrt{2q})^2$

26. $(3\sqrt{k} - 7\sqrt{2h})^2$ **27.** $(2 + \sqrt{3})(2 - \sqrt{3})$ **28.** $(\sqrt{7} - 5)(\sqrt{7} + 5)$

29. $(\sqrt{z} + \sqrt{8})(\sqrt{z} - \sqrt{8})$ **30.** $(\sqrt{11} - \sqrt{w})(\sqrt{11} + \sqrt{w})$ **31.** $(2\sqrt{6} - \sqrt{q})(2\sqrt{6} + \sqrt{q})$

32. $(\sqrt{p} + 3\sqrt{8})(\sqrt{p} - 3\sqrt{8})$ **33.** $(\sqrt{2r} + 3\sqrt{t})(\sqrt{2r} - 3\sqrt{t})$ **34.** $(x\sqrt{y} + \sqrt{5xy})(x\sqrt{y} - \sqrt{5xy})$

35. $\dfrac{18}{\sqrt{15} - 3}$ **36.** $\dfrac{50}{\sqrt{30} - 5}$ **37.** $\dfrac{\sqrt{8}}{\sqrt{7} - \sqrt{5}}$ **38.** $\dfrac{\sqrt{18}}{\sqrt{6} - \sqrt{3}}$

39. $\dfrac{\sqrt{11} - \sqrt{3}}{\sqrt{11} + \sqrt{3}}$ **40.** $\dfrac{\sqrt{13} - \sqrt{10}}{\sqrt{13} + \sqrt{10}}$ **41.** $\dfrac{\sqrt{k} + 2}{\sqrt{k} - 2}$ **42.** $\dfrac{5 + \sqrt{h}}{5 - \sqrt{h}}$

43. $\dfrac{\sqrt{r} - \sqrt{s}}{\sqrt{r} + \sqrt{s}}$ **44.** $\dfrac{\sqrt{z} - \sqrt{w}}{\sqrt{z} + \sqrt{w}}$ **45.** $\dfrac{7\sqrt{z + w}}{7 + \sqrt{z + w}}$ **46.** $\dfrac{\sqrt{s} + \sqrt{s + 6}}{\sqrt{s} - \sqrt{s + 6}}$

Preparing for Section 9.6

In Exercises 47–51, simplify each expression. Write each answer with positive exponents only.

47. $x^2 \cdot x^{-3}$ **48.** $(p^{-4})^{-1}$ **49.** $\dfrac{m^{-4}}{m^{-5} \cdot m^2}$ **50.** $\left(\dfrac{w}{5}\right)^{-3}$ **51.** $\left(\dfrac{2z^{-4}}{z^3}\right)^{-2}$

9.6   Fractional Exponents

FOCUS

We define fractional exponents and apply the properties of exponents to them.

In Chapter 4 we defined integer exponents and developed the properties of exponents. But we said nothing about exponents that are fractions. As an example, $3^{1/2}$ has a fractional exponent; we should define $3^{1/2}$ so that all the properties of exponents hold. Suppose, for example, that we apply the power-to-a-power property of exponents by squaring $3^{1/2}$. We get

$$(3^{1/2})^2 = 3^{1/2 \cdot 2}$$
$$= 3^1$$
$$= 3$$

More generally, suppose a is a nonnegative number and n is a positive integer. If the power-to-a-power property is to hold, we should have

$$(a^{1/n})^n = a^{n/n}$$
$$= a^1$$
$$= a$$

Since the nth power of $a^{1/n}$ is a, it follows by our definition in Section 9.1 that $a^{1/n}$ must be an nth root of a. Thus, we have the following definition of the **fractional exponent $\dfrac{1}{n}$.**

If n is a positive integer and a is a real number such that $\sqrt[n]{a}$ is defined, then

$$a^{1/n} = \sqrt[n]{a}$$

EXAMPLE 1

Evaluate each expression.

a. $49^{1/2}$ **b.** $8^{1/3}$ **c.** $(x^{1/5})^5$ **d.** $(x^3y^3)^{1/3}$

Solution

We use the definition of the fractional exponent $\dfrac{1}{n}$ to evaluate each expression.

a. $49^{1/2} = \sqrt{49} = 7$ Because $7^2 = 49$

b. $8^{1/3} = \sqrt[3]{8} = 2$ Because $2^3 = 8$

c. $(x^{1/5})^5 = (\sqrt[5]{x})^5$ Definition of $a^{1/n}$

$\qquad\quad = x$ Definition of $\sqrt[5]{x}$

d. $(x^3y^3)^{1/3} = \sqrt[3]{x^3y^3}$ Definition of $a^{1/n}$

$\qquad\qquad = \sqrt[3]{x^3}\sqrt[3]{y^3}$ Using the multiplication property

$\qquad\qquad = xy$ Because $\sqrt[3]{a^3} = a$ ■

For $a^{1/n}$ as we have defined it, all the properties of exponents actually do hold true. Hence we can use them to evaluate expressions like those in parts c and d of Example 1, without rewriting the powers as roots. For part c, for example,

$$(x^3y^3)^{1/3} = (x^3)^{1/3}(y^3)^{1/3} \qquad \text{By the product-to-a-power property}$$
$$= x^{3/3}y^{3/3} \qquad \text{By the power-to-a-power property}$$
$$= x^1y^1 \qquad \text{Simplifying}$$
$$= xy$$

Powers with fractional exponents m/n can also be defined in a way that is consistent with the properties of exponents. If the properties of exponents are to hold, then we must have

$$a^{m/n} = (a^{1/n})^m$$

Since we have already defined $a^{1/n}$, we have the following definition.

If m/n is a rational number, n is a positive integer, and a is a real number such that $\sqrt[n]{a}$ is defined, then

$$a^{m/n} = (\sqrt[n]{a})^m = \sqrt[n]{a^m}$$

Either of the two forms in the definition may be used to compute a fractional power. However, the first form is often easier to use, since it involves finding roots of smaller numbers.

EXAMPLE 2   Compute each of the following as $(\sqrt[n]{a})^m$ and as $\sqrt[n]{a^m}$.

a. $27^{2/3}$ **b.** $(-8)^{4/3}$ **c.** $16^{-3/4}$

Solution   We compute $(\sqrt[n]{a})^m$ on the left, and $\sqrt[n]{a^m}$ on the right.

a. $27^{2/3} = (\sqrt[3]{27})^2$ and $27^{2/3} = \sqrt[3]{27^2}$
$$= 3^2 \qquad\qquad\qquad\qquad = \sqrt[3]{729}$$
$$= 9 \qquad\qquad\qquad\qquad\quad\; = 9$$

b. $(-8)^{4/3} = (\sqrt[3]{-8})^4$ and $(-8)^{4/3} = \sqrt[3]{(-8)^4}$
$$= (-2)^4 \qquad\qquad\qquad\quad = \sqrt[3]{4096}$$
$$= 16 \qquad\qquad\qquad\qquad\;\; = 16$$

c. $16^{-3/4} = (\sqrt[4]{16})^{-3}$ and $16^{-3/4} = \sqrt[4]{16^{-3}}$

$$= 2^{-3} \qquad\qquad\qquad = \sqrt[4]{\dfrac{1}{16^3}}$$

$$= \left(\dfrac{1}{2}\right)^3 \qquad\qquad = \sqrt[4]{\dfrac{1}{4096}}$$

$$= \dfrac{1}{8} \qquad\qquad\qquad = \dfrac{1}{8}$$

∎

You can also use your calculator to evaluate numerical expressions that contain a fractional exponent. For example, to evaluate these expressions, use the following keying sequences.

Expression	*Key Sequence*	*Display*
$125^{1/3}$	125 $\boxed{y^x}$ $\boxed{(}$ 3 $\boxed{1/x}$ $\boxed{)}$ $\boxed{=}$	5
$38.5^{3/4}$	38.5 $\boxed{y^x}$ $\boxed{(}$ 3 $\boxed{\div}$ 4 $\boxed{)}$ $\boxed{=}$	15.455944

BE CAREFUL!

If the base of a power is a negative number, the power is not defined for some fractional exponents. For example,

$(-8)^{1/4}$ **or** $\sqrt[4]{-8}$ **is not defined, because there is no number** n **for which** $n^4 = -8$**.**

However, $(-8)^{2/3} = (\sqrt[3]{-8})^2 = (-2)^2 = 4$

All the properties of exponents in Chapter 4 hold for fractional exponents.

EXAMPLE 3

Simplify and express the result using only positive exponents. Assume that all expressions represent real numbers.

a. $x^{2/3}x^{1/4}$ **b.** $(x^3y^6)^{2/3}$ **c.** $(3x^{1/4}y^{-5/6})^3$ **d.** $\dfrac{-16r^{2/3}}{6r^{-1/4}}$

Solution

a. $x^{2/3}x^{1/4} = x^{2/3\,+\,1/4}$ By the product property of exponents

$\qquad\qquad\quad = x^{11/12}$ Adding fractions

b. $(x^3y^6)^{2/3} = (x^3)^{2/3}(y^6)^{2/3}$ By the product-to-a-power property

$\qquad\qquad\quad = x^{3\,\cdot\,2/3}y^{6\,\cdot\,2/3}$ By the power-to-a-power property

$\qquad\qquad\quad = x^2y^4$ Multiplying fractions

c. $(3x^{1/4}y^{-5/6})^3 = 3^3(x^{1/4})^3(y^{-5/6})^3$ By the product-to-a-power property

$$= 27x^{3 \cdot 1/4}y^{3(-5/6)}$$ By the power-to-a-power property

$$= 27x^{3/4}y^{-5/2}$$ Multiplying fractions

$$= \frac{27x^{3/4}}{y^{5/2}}$$ By the definition of negative exponents

d. $\dfrac{-16r^{2/3}}{6r^{-1/4}} = \dfrac{-16}{6}(r^{2/3 - (-1/4)})$ By the quotient property of exponents

$$= -\frac{8}{3}r^{11/12}$$ Subtracting and simplifying fractions   ∎

Sometimes we can simplify a radical by writing it in exponential form, as in the next example.

EXAMPLE 4   Write each radical in exponential form, simplify, and then write the simplified result in radical form.

a. $\sqrt[6]{16^3}$ b. $(\sqrt[4]{x})^2$

Solution   a. $\sqrt[6]{16^3} = (16^3)^{1/6}$ By the definition of fractional exponents

$$= 16^{3/6}$$ By the power-to-a-power property

$$= 16^{1/2}$$ Simplifying $\dfrac{3}{6}$

$$= \sqrt{16}$$ By the definition of fractional exponents

$$= 4$$ Taking the square root of 16

b. $(\sqrt[4]{x})^2 = (x^{1/4})^2$ By the definition of fractional exponents

$$= x^{2 \cdot 1/4}$$ By the power-to-a-power property

$$= x^{2/4}$$ Multiplying fractions

$$= x^{1/2}$$ Simplifying $\dfrac{2}{4}$

$$= \sqrt{x}$$ By the definition of fractional exponents

(This assumes that $x \geq 0$.)   ∎

DISCUSSION QUESTIONS 9.6

1. Does $-4^{3/2} = (-4)^{3/2}$? Explain.

2. Does $\left(\dfrac{2}{3}\right)^{-3/4} = \left(\dfrac{3}{2}\right)^{3/4}$? Explain.

3. Does $-8^{-2/3} = (-8)^{-2/3}$? Explain.

EXERCISES 9.6

In Exercises 1–10, use your calculator to evaluate the numerical expression to the nearest thousandth. If the expression is not a real number, explain why.

1. $\sqrt{68}$ **2.** $\sqrt{8^3}$ **3.** $5^{1/3}$ **4.** $(-5)^{1/3}$ **5.** $5^{-1/3}$

6. $23.5^{-2/5}$ **7.** $-18.56^{4/3}$ **8.** $(-3)^{3/4}$ **9.** $(-53.89)^{-3/5}$ **10.** $(-3.6)^{1/6}$

In Exercises 11–34 evaluate the given numerical expression.

11. $-\sqrt{100}$ **12.** $\sqrt[3]{\dfrac{-27}{64}}$ **13.** $\sqrt[4]{\dfrac{81}{16}}$ **14.** $\sqrt[5]{-243}$ **15.** $36^{1/2}$

16. $(-8)^{1/3}$ **17.** $625^{1/4}$ **18.** $(-32)^{1/5}$ **19.** $64^{2/3}$ **20.** $81^{3/2}$

21. $9^{5/2}$ **22.** $49^{-1/2}$ **23.** $(-216)^{-1/3}$ **24.** $(-64)^{-4/3}$ **25.** $\left(\dfrac{3}{4}\right)^{-2}$

26. $\left(\dfrac{9}{49}\right)^{-1/2}$ **27.** $\left(\dfrac{-1}{27}\right)^{-2/3}$ **28.** $\left(\dfrac{1}{32}\right)^{-4/5}$ **29.** $(3^{1/4})(3^{3/4})$ **30.** $(5^{2/3})(5^{4/3})$

31. $(3^{4/3})(3^{-1/3})$ **32.** $(4^{7/2}) \div (4^{1/2})$ **33.** $\left(\dfrac{1}{3}\right)^{1/4} \div \left(\dfrac{1}{3}\right)^{5/4}$ **34.** $\left(-\dfrac{3}{4}\right)^{1/3} \div \left(\dfrac{3}{4}\right)^{-2/3}$

In Exercises 35–48, simplify and express the result using only positive exponents. Assume that all expressions represent real numbers.

35. $x^{4/5} \cdot x^{3/5}$ **36.** $y^{-5/6} \cdot y^{1/6}$ **37.** $(k^{-2/3})^{-3/8}$ **38.** $(h^{-1/6})^{3/2}$ **39.** $(27p)^{1/3}$

40. $(16q)^{-1/2}$ **41.** $(16m^2)^{-3/4}$ **42.** $(64n^{-2/3})^{-4}$ **43.** $\dfrac{w^{-4/3}}{w^{-2/3}}$ **44.** $\left(\dfrac{t^{2/3}}{t^{-2/3}}\right)^{-1/2}$

45. $\dfrac{z^{3/4} \cdot z^{-5/4}}{z^{-1/4}}$ **46.** $\dfrac{w^{-5/8}}{w^{3/8} \cdot w^{-1/8}}$ **47.** $\left(\dfrac{4p^{-1/4}}{p^{3/8} \cdot p^{-1/8}}\right)^{-1/2}$ **48.** $\left(\dfrac{q^{-5/3} \cdot q^{-2/3}}{64q^{2/3}}\right)^{1/6}$

In Exercises 49–54, write each radical in exponential form, simplify, and then write the simplified result in radical form.

49. $\sqrt[6]{h^2}$ **50.** $\sqrt[8]{k^6}$ **51.** $(\sqrt[8]{m})^6$ **52.** $(\sqrt[4]{n})^8$ **53.** $\sqrt[3]{\dfrac{p^9}{q^6}}$ **54.** $\sqrt[4]{\left(\dfrac{1}{n}\right)^{-12}}$

55. The amount of light A that passes through a certain type of tinted glass m centimeters thick is given by the formula $A = (0.81)^m$. Find the amount of light that passes through tinted glass $\dfrac{1}{2}$ cm thick.

56. The wind speed v in mph needed to produce p watts of power from a windmill is given by the formula $v = \left(\dfrac{p}{0.015}\right)^{1/3}$. How much wind speed is needed to produce 120 watts (W) of power?

57. Medical researchers have determined that a person h inches tall has an average pulse rate of approximately p beats per minute, where $p = 600h^{-1/2}$. What is the approximate pulse rate of a person 6 ft tall?

Preparing for Section 9.7

In Exercises 58–63, solve the given equation.

58. $7y + 3 = 17$

59. $3z + 10 = 7z - 6$

60. $5w - 13 = 6w + 1$

61. $p^2 = 5p + 6$

62. $q^2 - 12 = 4$

63. $(s - 1)^2 = s - 1$

9.7   Solving Equations Containing Radicals

FOCUS

We discuss a method for solving equations containing radicals. The solution strategy is to isolate the radical on one side of the equation and then square both sides to eliminate the radical.

Scientists use an equation containing a radical to study the effect of gravity on the motion of a "free falling" body. If air resistance is negligible, an object dropped from d feet above the earth takes t seconds to hit the ground, where

$$t = \frac{\sqrt{d}}{4}$$

If a ball dropped from a rooftop takes 1 second (s) to hit the ground, what is the height of the roof?

To answer this question we can substitute $t = 1$ into the equation above and solve for d:

$$1 = \frac{\sqrt{d}}{4}$$

$$4 = \sqrt{d} \qquad \text{Multiplying both sides by 4}$$

To solve this equation, we can apply the following property of equality:

> If $A = B$, then $A^2 = B^2$.

This property says that if two expressions are equal, then the squares of the expressions are also equal. We can solve the equation $4 = \sqrt{d}$ for d by squaring both sides:

$$4^2 = (\sqrt{d})^2 \qquad \text{Squaring both sides}$$

$$16 = d$$

The roof is 16 ft high. We can check this solution by substituting $d = 16$ into the equation $1 = \dfrac{\sqrt{d}}{4}$:

$$1 = \frac{\sqrt{16}}{4}$$

$$1 = \frac{4}{4}$$

$$1 = 1 \qquad \text{True}$$

In this example, we squared both sides of the equation to eliminate the radical and produce the solution. However, sometimes squaring both sides of an equation produces solutions that do not check in the original equation. These solutions are called **extraneous solutions** and must be discarded. For example, to solve the equation $\sqrt{x + 1} = -1$, we square both sides and solve for x:

$$\sqrt{x + 1} = -1$$

$$(\sqrt{x + 1})^2 = (-1)^2 \qquad \text{Squaring both sides}$$

$$x + 1 = 1$$

$$x = 0$$

We check this solution by substituting $x = 0$ into the original equation:

$$\sqrt{x + 1} = -1$$

$$\sqrt{0 + 1} = -1$$

$$1 = -1 \qquad \text{False}$$

Since the solution does not check, $x = 0$ is an extraneous solution. In fact, the equation $\sqrt{x + 1} = -1$ has no solutions. (Note that $\sqrt{x + 1}$ represents a nonnegative square root, so it cannot equal -1, a *negative* number.)

EXAMPLE 1

Solve the equation $2\sqrt{y} = \sqrt{3y + 5}$.

Solution

We can eliminate the radicals by squaring both sides of the equation $2\sqrt{y} = \sqrt{3y + 5}$:

$$2\sqrt{y} = \sqrt{3y + 5}$$

$$(2\sqrt{y})^2 = (\sqrt{3y + 5})^2 \qquad \text{Squaring both sides}$$

$$2^2(\sqrt{y})^2 = (\sqrt{3y + 5})^2 \qquad \text{By the product-to-a-power property of exponents}$$

$$4y = 3y + 5$$

We can now solve the last equation for y:

$$4y = 3y + 5$$

$$y = 5 \qquad \text{Subtracting } 3y \text{ from both sides}$$

We check the solution by substituting $y = 5$ into the original equation:

$$2\sqrt{y} = \sqrt{3y + 5}$$

$$2\sqrt{5} = \sqrt{20}$$

$$= \sqrt{4 \cdot 5}$$

$$= 2\sqrt{5} \qquad \text{True} \qquad \blacksquare$$

Sometimes squaring both sides of an equation does not eliminate the radical. For example, squaring both sides of the equation $\sqrt{w} + 3 = 4$ does not eliminate the radical:

$$\sqrt{w} + 3 = 4$$

$$(\sqrt{w} + 3)^2 = 4^2 \qquad \text{Squaring both sides}$$

$$(\sqrt{w})^2 + 2 \cdot \sqrt{w} \cdot 3 + 3^2 = 16 \qquad \text{Using the formula } (a + b)^2 = a^2 + 2ab + b^2$$

$$w + 6\sqrt{w} + 9 = 16$$

The last equation is more complicated than the original equation, and it still contains a radical. A better strategy—and the one you should use—is to *isolate the radical* on one side of the equation before squaring. That is, get the radical on one side of the equation and all the other terms on the other side.

EXAMPLE 2   Solve the equation $\sqrt{w} + 3 = 4$.

Solution   We can isolate the radical on the left side of the equation by subtracting 3 from both sides:

$$\sqrt{w} + 3 = 4$$

$$\sqrt{w} = 1 \qquad \text{Subtracting 3 from both sides}$$

Now squaring both sides of the equation eliminates the radical:

$$(\sqrt{w})^2 = 1^2 \qquad \text{Squaring both sides}$$

$$w = 1$$

We check the solution by substituting $w = 1$ into the original equation:

$$\sqrt{w} + 3 = 4$$
$$1 + 3 = 4$$
$$4 = 4 \qquad \text{True} \qquad\blacksquare$$

EXAMPLE 3

Solve the equation $p = \sqrt{p - 2} + 4$.

Solution

We can isolate the radical on the right side of the equation by subtracting 4 from both sides. We get

$$p - 4 = \sqrt{p - 2}$$

We now square both sides of the last equation and solve for p. To square the left side of the equation, we can use the formula $(a - b)^2 = a^2 - 2ab + b^2$ for squaring a binomial:

$$(p - 4)^2 = (\sqrt{p - 2})^2$$
$$p^2 - 8p + 16 = p - 2$$

The last equation is a quadratic equation. We write it in standard form and solve by factoring:

$$p^2 - 9p + 18 = 0 \qquad \text{Subtracting } p \text{ from and adding 2 to both sides}$$
$$(p - 3)(p - 6) = 0 \qquad \text{Factoring}$$
$$p - 3 = 0 \quad \text{or} \quad p - 6 = 0 \qquad \text{By the zero product property}$$
$$p = 3 \quad \text{or} \qquad p = 6$$

We check these solutions by substituting $p = 3$ and $p = 6$ into the original equation:

$$p = \sqrt{p - 2} + 4 \qquad\qquad p = \sqrt{p - 2} + 4$$
$$3 = \sqrt{3 - 2} + 4 \qquad\qquad 6 = \sqrt{6 - 2} + 4$$
$$3 = \sqrt{1} + 4 \qquad\qquad 6 = \sqrt{4} + 4$$
$$3 = 1 + 4 \qquad\qquad 6 = 2 + 4$$
$$3 = 5 \quad \text{False} \qquad\qquad 6 = 6 \quad \text{True}$$

The solution $p = 3$ is extraneous because it produces a false statement. Therefore, $p = 6$ is the only solution of the equation. $\qquad\blacksquare$

The next example shows that sometimes we must square both sides more than once to solve an equation containing radicals.

EXAMPLE 4   Solve the equation $\sqrt{z + 7} - \sqrt{z} = 1$.

Solution   We can isolate $\sqrt{z + 7}$ on the left side of the equation by adding $\sqrt{z}$ to both sides. We get

$$\sqrt{z + 7} = 1 + \sqrt{z}$$

We now square both sides, treating the right side as a binomial, and then simplify the result:

$(\sqrt{z + 7})^2 = (1 + \sqrt{z})^2$	Squaring both sides
$z + 7 = 1 + 2\sqrt{z} + z$	Using the formula $(a + b)^2 = a^2 + 2ab + b^2$
$6 = 2\sqrt{z}$	Subtracting 1 and z from both sides
$3 = \sqrt{z}$	Dividing both sides by 2

The last equation still contains a radical, so we square again:

$3^2 = (\sqrt{z})^2$	Squaring both sides
$9 = z$	

We check the solution by substituting $z = 9$ into the original equation:

$$\sqrt{z + 7} - \sqrt{z} = 1$$
$$\sqrt{16} - \sqrt{9} = 1$$
$$4 - 3 = 1$$
$$1 = 1 \quad \text{True} \qquad ■$$

BE CAREFUL!

Squaring a binomial like $1 + \sqrt{z}$ is often done *incorrectly* by squaring each term:

$$(1 + \sqrt{z})^2 \neq 1^2 + (\sqrt{z})^2$$

Do not square a binomial by squaring each term. Instead, use the formulas for squaring a binomial:

$$(a + b)^2 = a^2 + 2ab + b^2 \quad \text{and} \quad (a - b)^2 = a^2 - 2ab + b^2$$

We can summarize the procedure for solving equations containing radicals as follows:

To solve an equation containing radicals

1. Isolate one of the radicals on one side of the equation.
2. Square both sides of the equation.
3. If the equation still contains a radical, simplify and repeat steps 1 and 2.
4. Solve the resulting equation.
5. Check the solutions in the original equation.

DISCUSSION QUESTIONS 9.7

1. What are extraneous solutions?

2. If an equation contains a radical, does squaring both sides always eliminate the radical? Explain.

3. Does $(\sqrt{a} + \sqrt{b})^2 = (\sqrt{a})^2 + (\sqrt{b})^2 = a + b$? Explain.

EXERCISES 9.7

In Exercises 1–38, solve the given equation.

1. $\sqrt{x + 2} = 6$
2. $\sqrt{y + 6} = 4$
3. $\sqrt{3p + 1} = 7$
4. $\sqrt{6q + 4} = 8$

5. $\sqrt{m + 9} = 14$
6. $\sqrt{n + 7} = 16$
7. $\sqrt{5s + 1} - 3 = 4$
8. $\sqrt{4t + 9} - 2 = 5$

9. $\sqrt{3u - 6} + 8 = 5$
10. $\sqrt{2v - 9} + 3 = 1$
11. $\sqrt{2y + 3} = \sqrt{y + 4}$

12. $\sqrt{x + 4} = \sqrt{4x - 2}$
13. $\sqrt{6z + 7} = \sqrt{5z + 1}$
14. $\sqrt{4w - 7} = \sqrt{3w + 5}$

15. $\sqrt{8n} = 2\sqrt{3n - 4}$
16. $3\sqrt{2m - 4} = \sqrt{6m}$
17. $\sqrt{3t - 2} = t$

18. $s = \sqrt{4s + 5}$
19. $\sqrt{h^2 - 7} = 3$
20. $\sqrt{k^2 - 9} = 4$

21. $v = \sqrt{v^2 - 5v + 20}$
22. $\sqrt{u^2 - 11u - 12} = u$
23. $\sqrt{x + 6} = x$

24. $y = \sqrt{2y} + 4$
25. $z + 5 = \sqrt{z^2 + 5}$
26. $\sqrt{w^2 - 8} = w + 4$

27. $\sqrt{6 - p} = p - 4$
28. $q + 3 = \sqrt{q + 5}$
29. $n = \sqrt{2n - 3} + 3$

30. $\sqrt{4m - 3} + 2 = m$
31. $\sqrt{x} = \sqrt{x + 16} - 2$
32. $\sqrt{y + 12} = 2 + \sqrt{y}$

33. $3\sqrt{h + 13} = h + 9$
34. $k - 1 = 2\sqrt{k + 7}$
35. $\sqrt{t + 8} - \sqrt{t} = 2$

36. $\sqrt{r} - \sqrt{r + 5} = 1$
37. $\sqrt{3u + 1} - \sqrt{u} = 1$
38. $\sqrt{2v} - \sqrt{v + 7} = 1$

39. A ball dropped from the top of the Transamerica Pyramid building in San Francisco takes 7.5 s to hit the ground. How tall is the building? (Hint: Use the formula $t = \dfrac{\sqrt{d}}{4}$.)

40. The time t in seconds needed for a pendulum of length L feet to complete one cycle is given by the formula $t = 2\pi\sqrt{\dfrac{L}{32}}$. What length of pendulum will complete one cycle in 4 s?

41. The formula $v = \sqrt{24L}$ is used by police to estimate the speed v of a car (in miles per hour) from the length L (in feet) of its skid marks on dry pavement. How long are the skid marks of a car going 60 miles per hour (mph)?

42. The distance d in miles to the horizon from an altitude of h feet above the surface of the earth is given by the formula $d = \sqrt{1.5h}$. How far is it to the horizon from a glider flying at an altitude of 2000 ft?

Preparing for Section 10.1

In Exercises 43–46, solve the given equation.

43. $4h - 9 = 11$ 44. $8 - 3k = -7$ 45. $x^2 = 121$ 46. $y^2 = 169$

In Exercises 47–50, factor the given trinomial.

47. $z^2 - 11z + 30$ 48. $w^2 - 3w - 54$ 49. $3p^2 - 16p - 35$ 50. $10q^2 + q - 24$

Chapter Summary

Important Ideas

Rules for working with radicals

$\sqrt[n]{a}\,\sqrt[n]{b} = \sqrt[n]{ab}$ *Multiplication property for radicals*

$\dfrac{\sqrt[n]{a}}{\sqrt[n]{b}} = \sqrt[n]{\dfrac{a}{b}}$ *Division property for radicals*

$a^{1/n} = \sqrt[n]{a}$

$a^{m/n} = (\sqrt[n]{a})^m = \sqrt[n]{a^m}$

Conjugates
a. Binomial expressions of the form $a + b$ and $a - b$, where either a or b (or both) are radicals, are called conjugates.
b. The product of conjugates does not contain a radical.

Chapter Review

Section 9.1

In Exercises 1–6, use a calculator to find the square root to the nearest hundredth.

1. $\sqrt{11.4}$
2. $\sqrt{6.95}$
3. $\sqrt{318}\sqrt{51}$
4. $\sqrt{425}\sqrt{78}$
5. $-\sqrt{79}$
6. $-\sqrt{56}$

7. The area of a square is 236 cm². Find to the nearest hundredth of a centimeter the length of a side of the square.

8. The area of a square lawn is 950 ft². Find to the nearest foot the perimeter of the lawn.

Section 9.2

In Exercises 9–18, multiply and simplify where possible. Assume all variables represent nonnegative numbers.

9. $\sqrt{8}\sqrt{5}$
10. $\sqrt{10}\sqrt{5}$
11. $\sqrt{6}\sqrt{18}$
12. $\sqrt{x}\sqrt{x^5}$

13. $(3\sqrt{x})(5\sqrt{x^2})$
14. $(2\sqrt{x})(3\sqrt{x^3})$
15. $\sqrt[3]{-27(x-1)^6}$
16. $\sqrt[4]{16(x^2+3)^4}$

17. $(3\sqrt{xy^3})(-5\sqrt{x^3y})$
18. $(-6\sqrt[3]{a^2b^4})(-2\sqrt[3]{ab^2})$

19. If the side of an equilateral triangle is s, then its height h is given by the equation, $h = 0.5s\sqrt{3}$. Find to the nearest tenth the height of an equilateral triangle with side 18.

Section 9.3

In Exercises 20–25, write each radical expression in simplest radical form.

20. $\sqrt{\dfrac{20}{6}}$
21. $\dfrac{1}{\sqrt{7}}$
22. $\dfrac{4\sqrt{3}}{2\sqrt{6}}$
23. $\sqrt{\dfrac{3}{2x}}$
24. $\sqrt{\dfrac{x^5}{y}}$
25. $\sqrt{\dfrac{x}{y^3}}$

26. The area of a rectangle is $15\sqrt{2}$ in.², and its width is 5 in. Find its length to the nearest tenth of an inch.

27. Find to the nearest tenth of a cm the length of a side of an equilateral triangle whose height is 12.3 cm. See Exercise 19.

Section 9.4

In Exercises 28–37, simplify the given expression. Assume that all variables represent nonnegative numbers.

28. $\sqrt{14} + 7\sqrt{14}$
29. $x\sqrt{15} + 2x\sqrt{15}$
30. $5\sqrt{8} - 3\sqrt{18}$
31. $2\sqrt{27} - 4\sqrt{12}$

32. $4\sqrt{w} + \sqrt{4w}$ **33.** $\sqrt{300y} - 5\sqrt{27y}$ **34.** $5\sqrt{75x^2} - 2x\sqrt{27}$ **35.** $4\sqrt[3]{16} - 2\sqrt[3]{128}$

36. $\sqrt{50} + 6\sqrt{12} - 3\sqrt{32}$ **37.** $\sqrt{24p^2q} + 2p\sqrt{8q} - 5\sqrt{32pq^2}$

Section 9.5

In Exercises 38–48, simplify the given expression. Assume that all variables represent positive real numbers.

38. $\sqrt{3}(\sqrt{2} + \sqrt{12})$ **39.** $(2 - \sqrt{6})(3 + \sqrt{2})$

40. $(\sqrt{7} - 4)^2$ **41.** $(5 + 2\sqrt{y})^2$

42. $(\sqrt{10w} + 5)(\sqrt{10w} - 5)$ **43.** $(\sqrt{2k} - 5\sqrt{h})(\sqrt{2k} + 5\sqrt{h})$

44. $\dfrac{1}{\sqrt{17} - 4}$ **45.** $\dfrac{\sqrt{8}}{\sqrt{5} + 1}$ **46.** $\dfrac{\sqrt{3} + 1}{3 - \sqrt{6}}$ **47.** $\dfrac{\sqrt{12}}{\sqrt{10} - \sqrt{7}}$ **48.** $\dfrac{\sqrt{2} - \sqrt{3}}{\sqrt{3} - \sqrt{2}}$

Section 9.6

In Exercises 49–56, simplify the given expression and write the result with positive exponents only. Assume that all variables represent positive real numbers.

49. $8^{-1/3}$ **50.** $256^{-3/4}$ **51.** $(-1000)^{2/3}$ **52.** $8^{2/3} \cdot 8^{1/3}$

53. $(81w^{-2})^{3/4}$ **54.** $\dfrac{16^{-5/8}}{16^{1/8}}$ **55.** $\dfrac{3^{-3/5} \cdot 3^{-4/5}}{3^{-6/5}}$ **56.** $\left(\dfrac{4p^{-1/2}}{p^{1/4}}\right)^{-3/2}$

Section 9.7

In Exercises 57–63, solve the given equation.

57. $\sqrt{3p - 8} + 5 = 6$ **58.** $\sqrt{n - 2} = 3\sqrt{n + 6}$ **59.** $\sqrt{m + 6} = m$ **60.** $h - 6 = \sqrt{3h}$

61. $\sqrt{2 - k} - 4 = k$ **62.** $\sqrt{5v + 1} = 1 + \sqrt{3v}$ **63.** $\sqrt{3u + 1} - 1 = \sqrt{3u - 8}$

64. The voltage v needed to operate a hair dryer that uses 1000 amperes (A) of power and has a resistance of r ohms is given by the formula $v = \sqrt{1000r}$. If the hair dryer needs 100 volts (V) to operate properly, what is the resistance?

65. After falling d feet, an object's speed in feet per second is given by the formula $v = 8\sqrt{d}$. If an object is dropped from a plane, how far has it fallen when its speed is 896 ft/s?

Chapter 9 Test

In Problems 1–4, use the graph in Figure 9.4, Section 9.1, to estimate the square root of each number to the nearest tenth.

1. 11   2. 39   3. 54   4. 78

5. Use a calculator to compute $\sqrt[3]{7}$ to the nearest hundredth.

In Problems 6–26, simplify the expression. Assume that all variables represent positive numbers.

6. $\sqrt{48}$ 7. $\sqrt{6}\sqrt{12}$ 8. $\sqrt{x^2 y^3}$ 9. $(3\sqrt{x})(3\sqrt{x^4})$

10. $\sqrt{24x^5}$ 11. $\dfrac{1}{2\sqrt{8}}$ 12. $\sqrt{\dfrac{7}{12}}$ 13. $\sqrt{\dfrac{x^3}{4y}}$

14. $\sqrt{48} - \sqrt{12}$ 15. $2\sqrt[3]{24} + 5\sqrt[3]{3}$ 16. $3x\sqrt{18} - 8\sqrt{2x^2}$ 17. $\sqrt{5}(\sqrt{10} + \sqrt{15})$

18. $(2 - \sqrt{3})(4\sqrt{6} + 1)$ 19. $(5 + 3\sqrt{y})^2$ 20. $\dfrac{8}{\sqrt{6} + 2}$ 21. $\dfrac{\sqrt{w}}{3 - \sqrt{w}}$

22. $\dfrac{\sqrt{10} - \sqrt{8}}{\sqrt{10} + \sqrt{8}}$ 23. $(-27)^{-4/3}$ 24. $\dfrac{z^{2/3} \cdot z^{5/6}}{z^{-1/3}}$ 25. $(100p^{-4})^{1/2}$

26. $\left(\dfrac{q^{3/4}}{81q^{-1/4}}\right)^{-1/2}$

In Problems 27–29, solve the given equation.

27. $\sqrt{8m - 4} + 6 = 2$ 28. $h = \sqrt{2h - 3} + 3$ 29. $\sqrt{2k + 1} + \sqrt{k} = 1$

30. Find to the nearest tenth the length of the side of a square whose diagonal is 14.

10 *Quadratic Equations*

10.1 Solving Quadratic Equations

FOCUS

We define quadratic equations and discuss some ways to solve them.

We have already encountered quadratic equations in Chapters 5 and 9. A quadratic equation in one variable is a second-degree equation; that is, it contains one or more terms with the variable squared—for example, x^2 or $7y^2$—and no variable terms of higher degree. More precisely, a **quadratic equation** in x is any equation that can be written in the **standard form** $ax^2 + bx + c = 0$, where $a \neq 0$.

EXAMPLE 1

Which of the following are quadratic equations?

a. $2z^2 = 1 - 6z^2$ **b.** $3x(x - 2) = 1$ **c.** $y^2 = y(y^2 - 1)$

Solution

We check for the highest exponent of a variable, which must be 2 in a quadratic equation.

a. $2z^2 = 1 - 6z^2$ is a quadratic equation, since two terms contain z^2 but none contain a higher power of z.
b. Multiplying through by $3x$ to eliminate the parentheses gives $3x^2 - 6x = 1$. This is a quadratic equation because it has the second-degree term $3x^2$ and no variable terms of higher degree.
c. Simplifying gives $y^2 = y^3 - y$, which is *not* a quadratic equation because it contains the third-degree term y^3. ■

In Chapter 5, we worked extensively on factoring and, in particular, on solving equations by factoring. Many of the equations we solved were quadratic equations, for which we used the following procedure.

> **To solve a quadratic equation by factoring**
>
> 1. Write the equation in standard form.
> 2. Factor it completely.
> 3. Set each factor equal to zero, and solve the resulting equations.
> 4. Check each solution in the original equation.

EXAMPLE 2

Solve each equation by factoring.

a. $y^2 - 6y - 27 = 0$ **b.** $5w = w(w - 1)$ **c.** $3x^2 + 4 = 13x$

Solution

a. This equation is in standard form; that is, the right side is zero and the exponents of y are in descending order. Hence, we can begin by factoring the left side of the equation:

$$y^2 - 6y - 27 = 0$$
$$(y - 9)(y + 3) = 0$$

Next, we set each factor equal to zero, and solve the resulting equations:

$$y - 9 = 0 \text{ or } y + 3 = 0$$
$$y = 9 \qquad y = -3$$

The solutions are $y = 9$ and $y = -3$. We'll check $y = 9$:

$$y^2 - 6y - 27 = 0$$
$$9^2 - 6 \cdot 9 - 27 = 0 \qquad \text{Substituting 9 for } y$$
$$81 - 54 - 27 = 0$$
$$27 - 27 = 0 \qquad \text{True}$$

The check of $y = -3$ is left as an exercise.

b. This equation needs to be simplified by multiplying through to remove the parentheses, and then transformed to standard form:

$$5w = w(w - 1)$$
$$5w = w^2 - w \qquad \text{Multiplying and removing parentheses}$$
$$5w - w^2 + w = 0 \qquad \text{Adding } -w^2 + w \text{ to both sides}$$

$$6w - w^2 = 0 \qquad \text{Combining like terms}$$

$$w(6 - w) = 0 \qquad \text{Factoring the left side}$$

$$w = 0 \ \text{ or } \ 6 - w = 0 \qquad \text{Setting each factor equal to 0}$$

$$w = 0 \qquad w = 6 \qquad \text{Solving each equation}$$

Check these solutions before continuing. They both are correct.

c.   We write this equation in standard form, factor, and solve:

$$3x^2 - 13x + 4 = 0 \qquad \text{Writing in standard form}$$

$$(3x - 1)(x - 4) = 0 \qquad \text{Factoring the left side}$$

$$3x - 1 = 0 \ \text{ or } \ x - 4 = 0 \qquad \text{Setting each factor equal to 0}$$

$$x = \frac{1}{3} \qquad x = 4 \qquad \text{Solving each equation}$$

You should check these solutions as well. ∎

BE CAREFUL!

Be sure to write a quadratic equation in standard form before trying to solve it by factoring. For example, you *cannot* solve

$$x^2 + 5x + 6 = 12$$

by factoring first and then setting each factor equal to 12.

Next we consider quadratic equations that have no first-degree term, such as $s^2 = 140$. Can this quadratic equation be solved by factoring? The answer is *no* because $s^2 - 140 = 0$ cannot be factored as $(s + a)(s + b)$ with a and b integers. However, the equation $s^2 = 140$ may be viewed as a problem involving square roots. The solution is then simple, because the positive and negative square roots of 140 satisfy the equation $s^2 = 140$. The solutions are $s = \sqrt{140} = 2\sqrt{35}$ and $s = -\sqrt{140} = -2\sqrt{35}$.

EXAMPLE 3   Solve each equation.

a. $x^2 - 8 = 0$ **b.** $4y^2 + 3 = 18$ **c.** $(t + 1)^2 = 5$

Solution   **a.**   You can check that $x^2 - 8 = 0$ cannot be solved by factoring. Hence, we consider the equation $x^2 = 8$, whose roots must be $x = \sqrt{8} = 2\sqrt{2}$ $x = -\sqrt{8} = -2\sqrt{2}$. We'll check $x = 2\sqrt{2}$:

$$x^2 = 8$$

$$(2\sqrt{2})^2 = 8 \qquad \text{Substituting } 2\sqrt{2} \text{ for } x$$

$$2\sqrt{2} \cdot 2\sqrt{2} = 8$$

$$4\sqrt{4} = 8$$

$$8 = 8 \qquad \text{True}$$

You should check the solution $x = -2\sqrt{2}$.

b.   We will solve $4y^2 + 3 = 18$ by using the same reasoning as in part a:

$$4y^2 + 3 = 18$$

$$4y^2 = 15 \qquad \text{Subtracting 3 from both sides}$$

$$y^2 = \frac{15}{4} \qquad \text{Dividing both sides by 4}$$

$$y = \sqrt{\frac{15}{4}} \text{ or } y = -\sqrt{\frac{15}{4}}$$

Those solutions, when simplified, are

$$y = \frac{\sqrt{15}}{2} \text{ or } y = -\frac{\sqrt{15}}{2}$$

Sometimes we write this as

$$y = \pm\frac{\sqrt{15}}{2}$$

c.   In one sense the equation $(t + 1)^2 = 5$ is really no different from the equation $x^2 = 5$. We are looking for a number, which we are calling $t + 1$, whose square is 5. That number must be a square root of 5:

$$t + 1 = \sqrt{5} \text{ or } t + 1 = -\sqrt{5}$$

By subtracting 1 from both sides of the above equations, we find that

$$t = -1 + \sqrt{5} \text{ or } t = -1 - \sqrt{5}$$

For short, we can write $t = -1 \pm \sqrt{5}$. Checking is done just as for other solutions. We check $t = -1 + \sqrt{5}$:

$$(t + 1)^2 = 5$$

$$(-1 + \sqrt{5} + 1)^2 = 5 \qquad \text{Substituting } -1 + \sqrt{5} \text{ for } t$$

$$(\sqrt{5})^2 = 5$$

$$5 = 5 \qquad \text{True}$$

The check of $t = -1 - \sqrt{5}$ is left as an exercise.   ∎

Some equations are quadratic though at first they may not appear to be.

EXAMPLE 4   Show that each equation is quadratic and solve it by factoring.

a. $y + 2 = \dfrac{18}{y - 5}$ b. $\dfrac{6}{x} + 3 = x - 2$

Solution   a.   Note that if $y = 5$, then $y - 5 = 0$. But $y - 5$ is the denominator of a fraction, so y cannot be 5. We multiply both sides of the equation by $y - 5$:

$$(y + 2)(y - 5) = \frac{18}{y - 5}(y - 5)$$

$$y^2 - 3y - 10 = 18$$

This is clearly a quadratic equation. We write it in standard form, factor, and solve:

$$y^2 - 3y - 28 = 0 \qquad \text{Writing in standard form}$$

$$(y - 7)(y + 4) = 0 \qquad \text{Factoring the left side}$$

$$y - 7 = 0 \text{ or } y + 4 = 0 \qquad \text{Setting each factor equal to 0}$$

$$y = 7 \qquad y = -4 \qquad \text{Solving each equation}$$

Neither of these solutions is 5, so they are acceptable. Nevertheless, you should check them in the usual way.

b.   Note that x cannot be 0. Why? For $x \neq 0$, we multiply both sides of the equation by x:

$$\left(\frac{6}{x} + 3\right)(x) = (x - 2)(x)$$

$$6 + 3x = x^2 - 2x$$

We now see that this is a quadratic equation. We write it in standard form, factor, and solve:

$$x^2 - 5x - 6 = 0 \qquad \text{Writing in standard form}$$

$$(x - 6)(x + 1) = 0 \qquad \text{Factoring the left side}$$

$$x - 6 = 0 \text{ or } x + 1 = 0 \qquad \text{Setting each factor equal to 0}$$

$$x = 6 \qquad x = -1 \qquad \text{Solving each equation}$$

Check these solutions as usual.   ∎

Quadratic equations arise in many applications. The next example shows one type.

EXAMPLE 5   A rectangular field is 25 feet (ft) longer than it is wide. Its area is 3750 ft². Find the dimensions of the field.

Solution   We let w represent the width in feet of the field; then its length is $w + 25$. The area of the field is length times width, so we have the following equation:

$$(w + 25)(w) = 3750$$

We solve this equation for w:

$$w^2 + 25w = 3750 \qquad \text{Removing parentheses}$$

$$w^2 + 25w - 3750 = 0 \qquad \text{Subtracting 3750 from both sides}$$

$$(w + 75)(w - 50) = 0 \qquad \text{Factoring}$$

$$w + 75 = 0 \text{ or } w - 50 = 0 \qquad \text{Setting each factor equal to zero}$$

$$w = -75 \qquad w = 50 \qquad \text{Solving each equation}$$

Since w is the width of a rectangle, it cannot be negative. Hence, w cannot be −75, so w is 50 ft. Since the length is $w + 25$, we have

Width: 50 ft   Length: 75 ft

To check, find the area of a rectangle that is 50 ft by 75 ft. Is it 3750 ft²? ∎

DISCUSSION QUESTIONS 10.1

1. Is $xy = 6x^2$ a quadratic equation in one variable? Justify your answer.

2. Which of the following equations are in standard form?
    a. $x^2 = 16$ **b.** $3x^2 - x = 0$ **c.** $5x^2 + 2x = 10$

3. What is wrong with the following solution to $x^2 - 4x = 3$? Since $x^2 - 4x$ can be factored, we have

$$x(x - 4) = 3$$
$$x = 3 \text{ or } x - 4 = 3$$
$$x = 3 \text{ or } x = 7$$

EXERCISES 10.1

1. Check the solution in Example 2, part a.

2. Check the solution in Example 3, part c.

In Exercises 3–26, solve each quadratic equation by factoring.

3. $x^2 - 36 = 0$ 4. $y^2 - 45 = 19$ 5. $y^2 - 2y - 3 = 0$ 6. $x^2 + 8x + 12 = 0$

7. $t^2 = 4(3 - t)$ 8. $z^2 - 7z = -10$ 9. $30 + r = r^2$ 10. $x^2 + 28 = 11x$

11. $9a^2 - 49 = 0$ 12. $2x^2 - 7x = 0$ 13. $2w^2 - 5w + 2 = 0$ 14. $7u^2 - 3 = 4u$

15. $3k^2 = 8 - 2k$ 16. $2y = 5y^2 - 4y$ 17. $29m^2 = 169 - 7m^2$ 18. $n(n - 3) = 70$

19. $(x + 7)(x + 3) = 5$ 20. $\dfrac{x^2}{3} - \dfrac{7x}{3} = -3\dfrac{1}{3}$ 21. $\dfrac{5x^2}{3} = 3x + \dfrac{2}{3}$ 22. $0.5y = 0.3 - 0.2y^2$

23. $6.3x^2 - 23.1x = 8.4$ 24. $\dfrac{3}{5 - 2z} = z + 1$ 25. $\dfrac{-12}{x - 10} = x - 2$ 26. $\dfrac{2}{x} - 4 = x - 5$

In Exercises 27–42, solve each quadratic equation by using square roots. Write your answers in simplified radical form.

27. $x^2 - 49 = 0$ 28. $y^2 - 15 = 0$ 29. $w^2 = \dfrac{9}{16}$ 30. $6w^2 = \dfrac{1}{5}$ 31. $\dfrac{1}{2}t^2 = 72$

32. $9z^2 - 64 = 0$ 33. $121m^2 = 1$ 34. $v^2 = 20 - v^2$ 35. $\dfrac{x^2}{3} = 15$ 36. $\dfrac{7}{y^2} = 36$

37. $1.8y^2 - 20.5 = 33.5$ 38. $\dfrac{2t}{3}\left(t - \dfrac{3}{8}\right) = \dfrac{14}{3} - \dfrac{t}{4}$ 39. $\dfrac{k + 60}{5} = \dfrac{5}{k} + 12$

40. $(p + 1)^2 = 4$ 41. $(2z + 3)^2 = 6$ 42. $(7x - 1)^2 - 16 = 0$

43. The vertical distance d that a dropped object falls in t seconds is calculated from the formula $d = 16t^2$, where d is the number of feet fallen in t seconds. How long would it take a pebble to reach the ground if it is dropped from an airplane at an altitude of 6400 ft?

44. The base of a triangle is 9 centimeters (cm) longer than its height; its area is 56 cm². Find the base and height of the triangle.

45. The sum of the squares of two consecutive integers is 113. What are the integers?

46. A projectile is hurled vertically upward with a velocity of 128 feet per second (ft/s). The height h of the object after t seconds is given by the formula $h = -16t^2 + 128t$. In how many seconds will the object be 192 ft above the ground? When will it hit the ground?

47. A number minus its reciprocal equals $\dfrac{7}{12}$. Find the number. (*Hint:* The reciprocal of n is $\dfrac{1}{n}$.)

48. A farmer cuts a strip of grain around the border of a rectangular field that is 60 hectometers (hm) by 120 hm, as in Figure 10.1. Find the width w of the strip if he cuts 3200 hm^2 of grain.

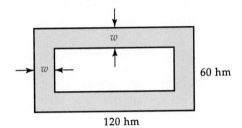

FIGURE 10.1

49. If the sides of a square are increased by 3 cm, its area is increased by 45 cm^2. Find the length of a side of the original square.

Preparing for Section 10.2

In Exercises 50–52, factor each expression that can be factored. If it cannot be factored, indicate that.

50. $2x^2 - 10x - 12$ **51.** $y^2 - y + 3$ **52.** $2x^2 - 6x - 4$

In Exercises 53–56, use the given numbers to evaluate the expression $b^2 - 4ac$.

53. $a = 2, b = 4, c = 0$ **54.** $a = 1, b = 3, c = -1$

55. $a = 2, b = -4, c = 2$ **56.** $a = 3, b = 1, c = -2$

10.2   The Quadratic Formula

FOCUS

We discuss the quadratic formula and use it to find the real-number roots of quadratic equations.

You have learned, thus far, two methods of solving quadratic equations: factoring and finding square roots. But many quadratic equations *cannot* be solved by either of these methods; for example, the equation $x^2 - 6x + 2 = 0$ cannot be solved by factoring and cannot be written in the form $x^2 = N$. The *quadratic formula* gives us a way of solving this equation and many others that cannot be solved by other methods.

The quadratic formula is developed by applying a solution method known as *completing the square*. A thorough understanding of this method is not necessary for our purposes, so we will just look at one example of its use and then proceed to the quadratic formula.

EXAMPLE 1

Solve $x^2 - 6x + 2 = 0$ by completing the square.

Solution

First, we write the equation as

$$x^2 - 6x = -2 \qquad \text{Subtracting 2 from both sides}$$

Then we take half the coefficient of x, square it, and add the result to both sides of the equation. Since one-half of 6 is 3 and 3^2 is 9, we add 9 to both sides of the equation:

$$x^2 - 6x + 9 = -2 + 9 \qquad \text{Adding 9 to both sides}$$
$$x^2 - 6x + 9 = 7 \qquad \text{Adding like terms}$$

By adding 9 we have "completed the square" on the left side; that is, we have produced the perfect-square trinomial $x^2 - 6x + 9$, which can now be factored:

$$x^2 - 6x + 9 = (x - 3)(x - 3)$$
$$= (x - 3)^2$$

The equation to be solved is now

$$(x - 3)^2 = 7$$

But we know how to solve this type of equation by the method of square roots (see Example 3, part c in Section 10.1).

$$x - 3 = +\sqrt{7} \text{ or } x - 3 = -\sqrt{7} \qquad \text{Taking the square root of both sides}$$
$$x = 3 + \sqrt{7} \text{ or } x = 3 - \sqrt{7} \qquad \text{Adding 3 to both sides}$$

The check is left to you.   ∎

As you can see, completing the square is not an easy method to use. On the other hand, the quadratic formula provides a simple means for solving *any* quadratic equation. The formula is found by completing the square in the quadratic equation, $ax^2 + bx + c = 0$, where $a \neq 0$. We will not derive it here (the derivation can be found in the Appendix), but you will be expected to know how to *apply* the formula to solve quadratic equations in this text and elsewhere.

The quadratic formula

The solutions of the quadratic equation $ax^2 + bx + c = 0$, where $a \neq 0$, are

$$x = \frac{-b + \sqrt{b^2 - 4ac}}{2a} \quad \text{and} \quad x = \frac{-b - \sqrt{b^2 - 4ac}}{2a}$$

Often the solutions determined by the **quadratic formula** are written in the shortened form

$$x = \frac{-b \pm \sqrt{b^2 - 4ac}}{2a}$$

To use the quadratic formula, we first find the numbers a, b, and c, and then substitute them into the formula.

EXAMPLE 2

Solve each quadratic equation by using the quadratic formula.

a. $x^2 + 3x - 10 = 0$ **b.** $2k^2 = 6k - 3$ **c.** $z^2 - 2z + 2 = 0$

Solution

a.   The equation $x^2 + 3x - 10 = 0$ is already in standard form. If we write it as

$$ax^2 + bx + c = 0$$
$$1x^2 + 3x + (-10) = 0$$

we see that $a = 1$, $b = 3$, and $c = -10$. Substituting these values into the quadratic formula gives

$$x = \frac{-3 \pm \sqrt{3^2 - 4 \cdot 1 \cdot (-10)}}{2 \cdot 1}$$

$$= \frac{-3 \pm \sqrt{9 + 40}}{2}$$

$$= \frac{-3 \pm \sqrt{49}}{2}$$

$$= \frac{-3 \pm 7}{2}$$

This last equation actually stands for two equations, which we solve:

$$x = \frac{-3 + 7}{2} \quad \text{and} \quad x = \frac{-3 - 7}{2}$$

$$= \frac{4}{2} \qquad\qquad\qquad = -\frac{10}{2}$$

$$= 2 \qquad\qquad\qquad\quad = -5$$

The two solutions are 2 and −5. Both should be checked in the original equation.

b. First we write $2k^2 = 6k - 3$ in standard form by adding $-6k + 3$ to both sides:

$$2k^2 - 6k + 3 = 0$$

or

$$2k^2 + (-6)k + 3 = 0$$

From this equation we see that $a = 2$, $b = -6$, and $c = 3$, and we can continue as follows:

$$k = \frac{-(-6) \pm \sqrt{(-6)^2 - 4 \cdot 2 \cdot 3}}{2 \cdot 2}$$

$$= \frac{6 \pm \sqrt{36 - 24}}{4}$$

$$= \frac{6 \pm \sqrt{12}}{4}$$

$$= \frac{6 \pm 2\sqrt{3}}{4}$$

The two solutions are $k = \dfrac{6 + 2\sqrt{3}}{4}$ and $k = \dfrac{6 - 2\sqrt{3}}{4}$.

Using a calculator, you can write these as approximate decimals: $k \approx 2.37$ and $k \approx 0.63$. The solutions can also be written in simplified radical form as follows:

$$k = \frac{6 \pm 2\sqrt{3}}{4}$$

$$= \frac{3 \pm \sqrt{3}}{2} \qquad \text{Dividing numerator and denominator by 2}$$

To check these solutions, substitute each into both sides of the original equation for k. We complete the check for $k = \dfrac{3 - \sqrt{3}}{2}$:

$$2k^2 = 6k - 3$$

$$2\left(\frac{3 - \sqrt{3}}{2}\right)^2 = 6\left(\frac{3 - \sqrt{3}}{2}\right) - 3$$

$$2\left(\frac{9 - 6\sqrt{3} + 3}{4}\right) = \frac{18 - 6\sqrt{3}}{2} - 3$$

$$\left(\frac{12 - 6\sqrt{3}}{2}\right) = 9 - 3\sqrt{3} - 3$$

$$6 - 3\sqrt{3} = 6 - 3\sqrt{3} \qquad \text{True}$$

c.   The equation $z^2 - 2z + 2 = 0$ is in standard form. We see that $a = 1$, $b = -2$, and $c = 2$. Then

$$z = \frac{-(-2) \pm \sqrt{(-2)^2 - 4 \cdot 1 \cdot 2}}{2 \cdot 1}$$

$$= \frac{2 \pm \sqrt{4 - 8}}{2}$$

$$= \frac{2 \pm \sqrt{-4}}{2}$$

These solutions are not real numbers because $\sqrt{-4}$, the square root of a negative number, is not a real number. Hence the quadratic equation $z^2 - 2z + 2 = 0$ has no real-number solutions ∎

The situation in part c of the preceding example is not unusual. In general, a quadratic equation does not have real-number solutions if $b^2 - 4ac$ is negative, for then $\sqrt{b^2 - 4ac}$ is not a real number. The quantity $b^2 - 4ac$ is called the **discriminant** of the quadratic equation $ax^2 + bx + c = 0$, and it provides important information about the roots of the equation:

Roots of quadratic equations

A quadratic equation, $ax^2 + bx + c = 0$, has real-number solutions (roots) as follows:
a.   Two real roots if $b^2 - 4ac > 0$.
b.   One real root if $b^2 - 4ac = 0$.
c.   No real roots if $b^2 - 4ac < 0$.

EXAMPLE 3

Without solving, determine the number of real roots of each quadratic equation.

a. $3x^2 - 4 = 6x$ **b.** $5 = 6y - 2y^2$ **c.** $x^2 - 6x = -9$

Solution

We write each equation in standard form and then calculate the discriminant.

a. The standard form of $3x^2 - 4 = 6x$ is

$$3x^2 - 6x - 4 = 0$$

or

$$3x^2 + (-6)x + (-4) = 0$$

Hence, $a = 3$, $b = -6$, and $c = -4$, and the discriminant is

$$b^2 - 4ac = (-6)^2 - 4(3)(-4)$$
$$= 36 + 48$$
$$= 84$$

Since the discriminant is positive, $3x^2 - 4 = 6x$ has two distinct real solutions.

b. The standard form of $5 = 6y - 2y^2$ is

$$2y^2 - 6y + 5 = 0$$

We evaluate the discriminant with $a = 2$, $b = -6$, and $c = 5$:

$$b^2 - 4ac = (-6)^2 - 4(2)(5)$$
$$= 36 - 40$$
$$= -4$$

Since the discriminant is negative, $5 = 6y - 2y^2$ does not have a real solution.

c. The standard form of $x^2 - 6x = -9$ is

$$x^2 - 6x + 9 = 0$$

We evaluate the discriminant with $a = 1$, $b = -6$, and $c = 9$:

$$b^2 - 4ac = (-6)^2 - 4(1)(9)$$
$$= 36 - 36$$
$$= 0$$

Since the discriminant is zero, $x^2 - 6x = -9$ has one real solution. ∎

DISCUSSION QUESTIONS 10.2

1. To use the quadratic formula to solve $7 - 3x + 2x^2 = 0$, what values would you use for a, b, and c?

2. What is the discriminant of $ax^2 + bx + c = 0$? Why is the discriminant useful?

3. Can the equation in Example 3, part a, be solved by factoring (with integer coefficients) as well as by the quadratic formula?

EXERCISES 10.2

In Exercises 1–35, solve each equation by using the quadratic formula. Give solutions in simplified radical form. If there are no real roots, indicate that.

1. $x^2 - 6x + 5 = 0$
2. $x^2 - 7x + 10 = 0$
3. $w^2 + 5w + 2 = 0$
4. $x^2 - 3x + 1 = 0$

5. $5y^2 - 9 = 0$
6. $3t^2 - 4 = 0$
7. $s^2 - 10 = 3s$
8. $6z = z^2 + 1$

9. $6x = 5x^2$
10. $7v^2 = 2v$
11. $3 - 4x + x^2 = 0$
12. $8 + 3x - x^2 = 0$

13. $2y^2 - 3y + 2 = 0$
14. $3x^2 + 2x - 2 = 0$
15. $3x - 4 = 3x^2$
16. $2z + 1 = 4z^2$

17. $3w^2 = 12 - 5w$
18. $6s^2 + 5s = 0$
19. $3t^2 - 5t = 0$
20. $1 = x(1 - x)$

21. $y(2 + y) = 3$
22. $5x^2 - 9x + 8 = 0$
23. $6m^2 - 13m + 6 = 0$
24. $p + 4 = \dfrac{21}{p}$

25. $\dfrac{3}{x} = x - 5$
26. $(y - 3)(3y + 1) = 2y$
27. $(2k + 1)(k - 3) = 5k$

28. $\dfrac{4n^2}{7} = 1$
29. $3 = \dfrac{6x^2}{5}$
30. $\dfrac{r - 1}{r + 2} = 2r + 1$

31. $\dfrac{3x^2 - x}{2x - 1} = x - 1$
32. $3y^2 = 4y - 2$
33. $7x = 5 - x^2$

34. $y(y + 2) - (y - 2)(2y - 1) = 5$
35. $(2z + 1)(z + 3) - z(z - 4) = 3$

In Exercises 36–42, solve the equations by any method. Give solutions correct to the nearest hundredth.

36. $x^2 + 5x + 9 = 0$
37. $\dfrac{1}{3}w^2 - w - \dfrac{1}{2} = 0$

38. $\dfrac{1}{2}y^2 - \dfrac{3}{5}y - \dfrac{1}{4} = 0$
39. $\dfrac{2}{3}z^2 - \dfrac{1}{4}z - \dfrac{3}{4} = 0$

40. $12.9w^2 + 22.2w + 1.4 = 0$
41. $1.2x^2 + 0.88x - 5.3 = 0$

42. $2.4t^2 + 7.2t + 3.8 = 0$

43. **a.** Find the discriminant of each equation in Exercises 36–42.
   b. Which equations have a positive discriminant? How many real solutions do these equations have?
   c. Which equations have zero for their discriminant? How many real solutions do these equations have?
   d. Which equations have a negative discriminant? How many real solutions do these equations have?

44. The formula $H = 2.5d^2n$, where d is the diameter of a cylinder in centimeters and n is the number of cylinders, gives the rated horsepower H of a typical gasoline engine. A certain four-cylinder gasoline engine develops 130 horsepower. What is the diameter of the cylinders (to the nearest thousandth)?

45. The amount of water that flows through a pipe varies directly as the square of the diameter of the pipe, provided the pressure remains constant. (*Hint:* Use a proportion.)
   a. If a pipe 2 inches (in.) in diameter delivers 3 gallons per minute (gal/min), how much will a pipe 5 in. in diameter deliver in 1 min?
   b. A plumber wants a pipe to deliver 5 gal/min of water. What must be the diameter of the pipe? (Use the constant in part a and give the answer to the nearest tenth of an inch.)

46. The velocity v in centimeters per second of blood flowing in an arterial capillary is given approximately by the formula

$$v = 1.2 - 2000r^2 \text{ cm/s}$$

where r is the radius of the capillary.
   a. What is the speed of blood flowing in an arterial capillary of radius 0.004 cm?
   b. A researcher measured the speed of blood in an arterial capillary to be 0.7 cm/s. What is the radius of the capillary (to the nearest hundredth of a centimeter)?

Preparing for Section 10.3

47. Define a right angle.

In Exercises 48–51, give the formula for the area of each figure.

48. Rectangle with length x and width y.

49. Triangle with base p and height q.

50. Parallelogram with base r and height s.

51. What is the perimeter of a rectangle with base 6 cm and height 10 cm?

<table>
<tr><td>**10.3**</td><td></td></tr>
</table>

Applications of Quadratic Equations

FOCUS

We apply quadratic equations in various settings in the examples and exercises. We discuss the Pythagorean Theorem, since it is useful in some of these settings, too.

Rate Problems

In the two previous sections, we have used quadratic equations to solve various applications. Some rate problems, similar to those we solved with linear equations, give rise to quadratic equations.

EXAMPLE 1

Two buses drove from Iowa City to Indianapolis, a distance of 300 miles (mi), to a basketball tournament. The first bus arrived 1 hour (h) before the second because its average speed was 10 miles per hour (mph) faster than that of the second. Assuming that the two buses started at the same time, find the average speed of each.

Solution

As in most rate problems, we are concerned with two trips. It is always helpful to organize the information we are given and the information we are seeking in a table. We'll let r represent the average rate of the first bus; then the rate of the second is $r - 10$. We can find an expression for the time of each bus trip by using the formula

$$\text{Time} = \frac{\text{Distance}}{\text{Rate}} \quad \text{or} \quad t = \frac{d}{r}$$

	Distance (mi)	Rate (mph)	Time (h)
First bus	300	r	$\dfrac{300}{r}$
Second bus	300	$r - 10$	$\dfrac{300}{r - 10}$

We are told that the first bus arrived 1 h before the second, even though they started the trip at the same time. It follows that

$$\text{Time of first bus} + 1 = \text{Time of second bus}$$

We substitute the expressions in the table for the times and get

$$\frac{300}{r} + 1 = \frac{300}{r - 10}$$

Noting that the LCD of the two rational fractions is $r(r - 10)$, we write

$$r(r - 10)\left(\frac{300}{r} + 1\right) = r(r - 10)\left(\frac{300}{r - 10}\right)$$ Multiplying both by the LCD

$$\frac{300r(r - 10)}{r} + r(r - 10) = \frac{300r(r - 10)}{(r - 10)}$$ Removing parentheses

$$300(r - 10) + r(r - 10) = 300r$$ Simplifying fractions

$$300r - 3000 + r^2 - 10r = 300r$$ Removing parentheses

$$r^2 - 10r - 3000 = 0$$ Adding $-300r$

The solution of this quadratic equation is left to you.   ■

Pythagorean Theorem

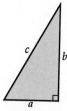

FIGURE 10.2

An important figure in geometry is a **right triangle,** that is, a triangle with one 90° angle. (A 90° angle is called a *right* angle.) The sides that form the 90° angle are called the **legs** of the right triangle. The longest side (which is always opposite the right angle) is called the **hypotenuse** of the right triangle. In Figure 10.2, the lengths of the legs are a units and b units and the length of the hypotenuse is c units.

In previous chapters we used diagrams to show that some algebraic relations are true. For example, to show that $(a + b)^2 = a^2 + 2ab + b^2$, we represented a and b with line segments as in Figure 10.3 and then drew Figure 10.4. In the next example, a theorem for right triangles is suggested by a pictorial argument.

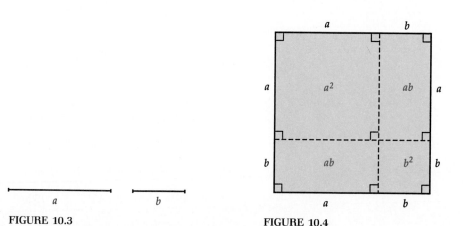

FIGURE 10.3

FIGURE 10.4

EXAMPLE 2

The area of the large square in Figure 10.4 is $(a + b)(a + b) = (a + b)^2$, but its area is also the sum of the areas of its four parts. Hence $(a + b)^2 = a^2 + 2ab + b^2$. The area of the large square in Figure 10.5 is also $(a + b)^2$, but it is divided into parts in a different way.

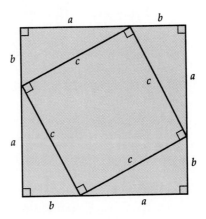

FIGURE 10.5

a. Write the sum of the areas of the parts in Figure 10.5, assuming that the inner figure is a square.

b. Set your result in part a equal to $a^2 + 2ab + b^2$. [This is valid, since both expressions are equal to $(a + b)^2$.] Then show that the resulting equation reduces to $a^2 + b^2 = c^2$.

Solution

a. Figure 10.5 consists of four right triangles and a square with side c. The area of each triangle is $\frac{1}{2}ab$. Hence the area of all four right triangles is

$$4\left(\frac{1}{2}ab\right) = 2ab$$

The inner figure is a square, so its area is c^2. Hence, the area of the large square, which is the sum of the areas of its parts, is

$$c^2 + 2ab$$

b. We are to set the expression we found in part a equal to $a^2 + 2ab + b^2$. This gives the equation

$$c^2 + 2ab = a^2 + 2ab + b^2$$

Subtracting $2ab$ from both sides then gives

$$a^2 + b^2 = c^2$$

■

Example 2 suggests an important relationship between the legs and hypotenuse of any right triangle. This relationship is called the **Pythagorean Theorem** in honor of the Greek mathematician Pythagoras.

Pythagorean Theorem

In a right triangle, the sum of the squares of the lengths of the legs is equal to the square of the length of the hypotenuse. In symbols: If a and b are the lengths of the legs and c is the length of the hypotenuse of a right triangle, we have

$$a^2 + b^2 = c^2$$

EXAMPLE 3

Find x in each triangle in Figure 10.6. Give square roots to the nearest tenth.

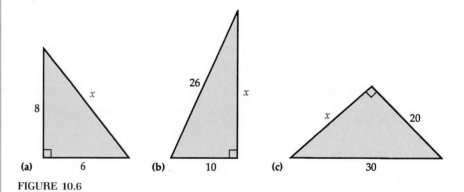

(a)   6   (b)   10   (c)   30

FIGURE 10.6

Solution

a. Apply the Pythagorean Theorem directly:

$$x^2 = 6^2 + 8^2$$
$$= 36 + 64$$
$$= 100$$
$$x = \sqrt{100}$$
$$= 10$$

b. Here x is the length of a leg, not the length of the hypotenuse. The hypotenuse is 26. Hence,

$$26^2 = x^2 + 10^2$$
$$676 = x^2 + 100$$
$$576 = x^2$$
$$x = 24$$

c.   Again x is the length of a leg:

$$x^2 + 20^2 = 30^2$$
$$x^2 + 400 = 900$$
$$x^2 = 500$$
$$x \approx 22.4$$ ∎

BE CAREFUL!

The Pythagorean Theorem applies to *right* triangles only. Do not try to use it on triangles that have no right angle.

The Pythagorean Theorem is useful for solving many practical problems, as the next example illustrates.

EXAMPLE 4

A pole for a circus act is 20 meters (m) high. Each of three wire braces run from the top of the pole to a point 5 m from the base of the pole, as in Figure 10.7. Find the total length of the three braces.

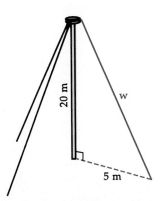

FIGURE 10.7

Solution

We assume that the pole stands straight. In other words, it makes a 90° angle with the ground. We can then think of the pole and each wire brace as forming a right triangle with the ground as the third side. The length w of the wire brace is the hypotenuse of the right triangle, as in Figure 10.7. We can find w by applying the Pythagorean Theorem:

$$w^2 = 5^2 + 20^2$$
$$= 25 + 400$$
$$= 425$$
$$w \approx 20.6 \text{ m}$$

Since there are three wires all the same length, their total length is about $3 \cdot 20.6 = 61.8$ m. ∎

DISCUSSION QUESTIONS 10.3

1. What do we call the sides of a right triangle that form the right angle? What do we call the side that is opposite the right angle? What is the longest side of a right triangle?

2. Name the legs and the hypotenuse of the right triangle in Figure 10.8.

3. State the Pythagorean Theorem in terms of Figure 10.8.

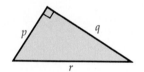

FIGURE 10.8

EXERCISES 10.3

In Exercises 1–6, sides a and b are the lengths of the legs and c is the length of the hypotenuse of a right triangle. Find the unknown value correct to the nearest tenth.

1. $a = 5, b = 12, c = ?$

2. $a = 10, b = 15, c = ?$

3. $a = ?, b = 16, c = 20$

4. $a = ?, b = 9, c = 11$

5. $a = 4, b = ?, c = 7$

6. $a = 6, b = ?, c = 12$

In Exercises 7–12, the length, width, and diagonal of a rectangle are l, w, and d, respectively. Find the unknown value correct to the nearest tenth.

7. $l = 14, w = 10, d = ?$

8. $l = 3.5, w = 5.2, d = ?$

9. $l = 9.1, w = ?, d = 13.4$

10. $l = 3, w = ?, d = 6$

11. $l = ?, w = 5.5, d = 8.3$

12. $l = ?, w = 23, d = 35$

13. Find the area and perimeter of the triangle in Figure 10.9. Round to the nearest tenth.

FIGURE 10.9

14. Find the area and perimeter of the triangle in Figure 10.10. Round to the nearest tenth.

15. Find the area and perimeter of the parallelogram in Figure 10.11.

16. Find the area and perimeter of the figure in Figure 10.12.

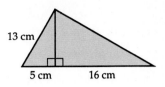

FIGURE 10.10

17. The number of meters in the perimeter of a square is equal to the number of square meters in its area. Find the length of a side of the square.

18. A crew rows 4 mi downstream and then back in $1\frac{1}{2}$ h. The speed of the current is 4 mph. How fast could the crew row in still water?

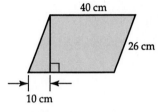

FIGURE 10.11

19. If the top of a 16-ft ladder just reaches the edge of a roof 12 ft above the ground, how far from the house is the base of the ladder?

20. A ladder must reach over a 3-ft-wide flower bed beside a house to the edge of a roof 12 ft above the ground. What is the shortest ladder that can be used?

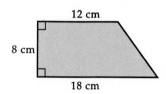

FIGURE 10.12

21. After traveling 60 mi at a certain speed, a motorist increases his speed by 5 mph and travels 48 mi farther. If he took 4 h to drive the entire 108 mi, find his speed during the first 60 mi.

22. A car traveling at r miles per hour can be stopped in a minimum of d feet, where d is given by the formula $d = 0.045r^2 + 1.1r$. What is the maximum speed at which a car can be driven if it must be stopped within 50 ft?

23. A boat has a speed in still water of 9 mph. It goes 25 mi downstream with the current and returns against the current in a total of $6\frac{1}{4}$ h. Find the speed of the current.

24. The pressure p (in pounds per square foot) of a wind blowing at v miles per hour is $p = 0.003v^2$. What velocity will produce a pressure of 30 pounds per square foot (lb/ft^2)?

25. A surveyor wants to measure the distance from A to B across a lake. She makes the measurements marked in Figure 10.13. What is the distance AB across the lake?

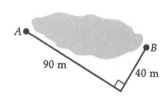

FIGURE 10.13

26. The relation between the measured area A and the effective area E of a chimney is given by $E = A - 0.6\sqrt{A}$. How large should the measured area be to produce an effective area of 30 ft^2? (*Hint:* Let $x = \sqrt{A}$. What is x^2?)

27. In a rectangular auditorium there are 616 seats. The number of rows of seats is 6 less than the number of seats in each row. How many rows of seats are there, and how many seats in each row?

In Exercises 28–30, use the fact that a baseball infield is in the shape of a square 90 ft on a side, as in Figure 10.14.

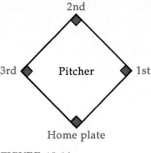

28. What is the distance from home plate to second base?

29. The pitching rubber lies on a straight line between home plate and second base. It is 60.5 ft from home plate. Does it lie on the line joining first base and third base? If not, is it in front of the line (closer to home plate) or behind the line?

30. How long is the throw from third base to first base? From a point exactly midway between second and third base to first base?

FIGURE 10.14

In Exercises 31–32, use the fact that the rectangle in Figure 10.15, which is called a *golden rectangle,* has

$$\frac{1}{x} = \frac{x}{x + 1}$$

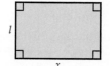

31. Solve for x in simplest radical form.

FIGURE 10.15

32. The value of the ratio $\dfrac{1}{x}$ is called the *golden ratio.* Find its value to the nearest hundredth.

Exercises 33–35 refer to this situation:

A manufacturer of precision ink pens determined that his total cost C is given by the equation $C = x^2 + 12x - 160$ where x is the number of pens produced.

33. How many pens should he produce to keep his cost at $1873?

34. What is the average cost per pen of producing x pens?

35. How many pens must be produced to attain an average cost of $18 per pen?

Preparing for Section 10.4

36. Complete this table, where $y = -x^2$.

x	0	1	−1	2	−2	3	−3	4	−4
y	0	−1	−1	?	?	?	?	?	?

37. The table in Exercise 36 gives ordered pairs (x, y) that satisfy the equation $y = -x^2$. Plot the pairs on a coordinate system and sketch the curve that appears to be the graph of this equation.

38. Using the graph in Exercise 37, graph $y = x^2$. (*Hint:* Change the sign of the y coordinate in each of the ordered pairs in Exercise 36.)

10.4   Graphing $y = ax^2 + bx + c$

FOCUS

We graph equations of the form $y = ax^2 + bx + c$ and examine the properties of these graphs, especially their relationship to the solution of quadratic equations.

Graphing Parabolas

An equation of the form $y = ax^2 + bx + c$, where $a \neq 0$, is called a *quadratic function*. The simplest quadratic function we will consider is $y = x^2$, in which we have $a = 1$, $b = 0$, and $c = 0$. The graph of this equation is related to the graphs of other, more complicated equations of the same form.

EXAMPLE 1

Draw the graph of each of these equations.

a. $y = x^2$ **b.** $y = -x^2$ **c.** $y = 2x^2$ **d.** $y = -0.5x^2$

Solution

a. We first develop some ordered pairs that satisfy $y = x^2$. These are

x	0	1	−1	2	2.5	−2.5	−2	3	−3	4	−4
$y = x^2$	0	1	1	4	6.25	6.25	4	9	9	16	16

Next, we plot these ordered pairs on a set of axes, and join them with a smooth curve. This is done in Figure 10.16.

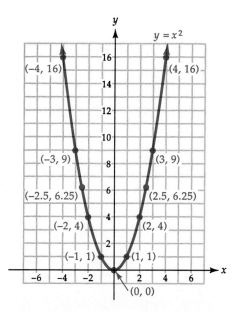

FIGURE 10.16

b. If we change the sign of each y coordinate in the table of part a, the resulting pairs satisfy the equation $y = -x^2$. These pairs are plotted and joined with a curve in Figure 10.17.

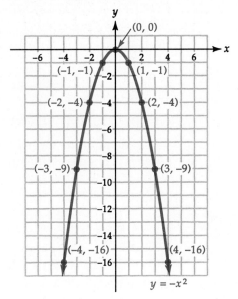

FIGURE 10.17

c. If we multiply the y coordinate in each ordered pair in part a by 2, the resulting pairs satisfy $y = 2x^2$. These pairs are plotted and joined with a smooth curve in Figure 10.18.

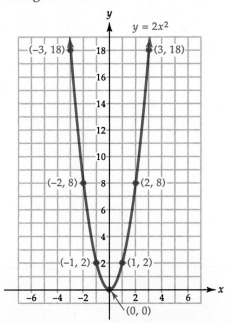

FIGURE 10.18

d. Multiplying the y coordinate in each ordered pair in part a by -0.5 gives us pairs that satisfy $y = -0.5x^2$. These pairs are plotted and joined with a curve in Figure 10.19.

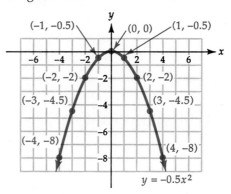

FIGURE 10.19   ∎

The curves in Figures 10.16 to 10.19 are called parabolas. In fact, the graph of every equation of the form $y = ax^2 + bx + c$ is a **parabola.** The graph of $y = x^2$ can be thought of as the "basic" parabola, and, as in Example 1, graphs of other quadratic equations are variations of it. The following example further illustrates this idea.

EXAMPLE 2   Graph the following equations.

a. $y = x^2 - 2$ **b.** $y = -x^2 + 2x + 3$ **c.** $y = 2x^2 + x + 1$

Solution   **a.** By subtracting 2 from the y coordinate of each ordered pair in Example 1, part a, we obtain ordered pairs that satisfy $y = x^2 - 2$. These ordered pairs and the parabola that contains them are drawn in Figure 10.20. It

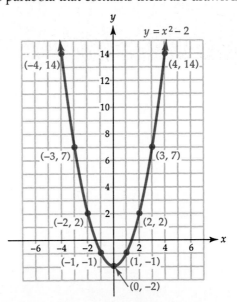

FIGURE 10.20

has the same size and shape as the graph of $y = x^2$, but is moved down 2 units.

b. The connection between the equations of parts b and c and the equations in Example 1 is not so clear. We will compute some ordered pairs that satisfy each equation, and then graph the ordered pairs.

    To find y when x is, say, -2, we substitute -2 for x in the given equation and solve for y. Hence, we get

$$y = -x^2 + 2x + 3$$
$$= -(-2)^2 + 2(-2) + 3$$
$$= -4 + (-4) + 3$$
$$= -5$$

Proceeding in the same way for other values of x, we obtain the following table of pairs where $y = -x^2 + 2x + 3$:

x	0	1	-1	2	-2	3	-3	4	-4	5
y	3	4	0	3	-5	0	-12	-5	-21	-12

In Figure 10.21, we graph these ordered pairs and draw the parabola. Notice that this graph opens downward like the graph of $y = -x^2$ in Figure 10.17. In fact, except for its location, it is exactly the same as the graph of $y = -x^2$.

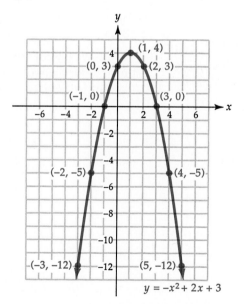

FIGURE 10.21

$y = -x^2 + 2x + 3$

c. You should verify that the following pairs satisfy $y = 2x^2 + x + 1$:

x	0	1	-1	2	-2
y	1	4	2	11	7

In Figure 10.22, we plot these pairs and draw the parabola. This graph has exactly the same shape as the graph of $y = 2x^2$ in Figure 10.18.

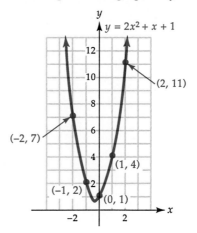

FIGURE 10.22   ■

Each graph in Examples 1 and 2 is a parabola that opens either upward (as in Figure 10.16) or downward (as in Figure 10.17). You can verify that each parabola opens upward when a is positive, and downward when a is negative.

x Intercepts and Solutions

Another important property of the graph of the equation $y = ax^2 + bx + c$ has to do with the x coordinates of the points where the graph intersects the x axis (the x *intercepts*). As the three parabolas in Example 2 illustrate, the graph of $y = ax^2 + bx + c$ may intersect the x axis at one point, two points, or no points. The number of points of intersection is equal to the number of real solutions of the quadratic equation $0 = ax^2 + bx + c$. Moreover, each intercept gives a real solution to that equation, as we see in the next example.

EXAMPLE 3

Find the solutions of the equations in parts a, b, and c of Example 2, with y set to zero. Compare the solutions with the x intercepts in Figures 10.20, 10.21, and 10.22, respectively.

Solution

a.   With y set to zero, the equation becomes $0 = x^2 - 2$. We first solve for x^2 and then take the square root of both sides:

$$0 = x^2 - 2$$

$$x^2 = 2 \qquad \text{Adding 2 to both sides}$$

$$x = \pm\sqrt{2} \qquad \text{Taking square roots of both sides}$$

The graph of $y = x^2 - 2$ in Figure 10.20 crosses the x axis at about $x = 1.4$ and $x = -1.4$, which are approximations for $x = \pm\sqrt{2}$.

b. We solve by factoring, beginning with

$$0 = -x^2 + 2x + 3$$

$$x^2 - 2x - 3 = 0 \qquad\qquad \text{Adding } x^2 - 2x - 3 \text{ to both sides}$$

$$(x - 3)(x + 1) = 0 \qquad\qquad \text{Factoring}$$

$$x - 3 = 0 \quad \text{or} \quad x + 1 = 0 \qquad \text{Setting each factor to zero}$$

$$x = 3 \qquad\qquad x = -1 \qquad \text{Solving for } x$$

The graph of $y = -x^2 + 2x + 3$ in Figure 10.21 crosses the x axis at $x = 3$ and $x = -1$.

c. We use the quadratic formula on

$$0 = 2x^2 + x + 1$$

or
$$2x^2 + x + 1 = 0$$

In the quadratic formula, we let $a = 2$, $b = 1$, and $c = 3$, obtaining

$$x = \frac{-1 \pm \sqrt{1^2 - 4(2)(3)}}{2(2)}$$

$$x = \frac{-1 \pm \sqrt{-23}}{4}$$

Because the discriminant, $b^2 - 4ac$ is less than 0, there are no real solutions. This is confirmed by the graph of $y = 2x^2 + x + 1$ in Figure 10.22, which does not cross the x axis.   ∎

We summarize the important properties illustrated in this section:

Properties of $y = ax^2 + bx + c$

1. The graph of $y = ax^2 + bx + c$ is a parabola.
2. The parabola opens upward if a is positive, and downward if a is negative.
3. The number of points at which the parabola intersects the x axis is the same as the number of solutions of $0 = ax^2 + bx + c$.
4. The x intercepts of the parabola are precisely the solutions of $0 = ax^2 + bx + c$.

Applications

EXAMPLE 4

When an object is shot upward with an initial velocity of 100 ft/s, its height h in feet after t seconds is given by the equation, $h = -16t^2 + 100t$.
  a. Graph this equation.
  b. Use the graph to determine the height of the object at $t = 1.5$ s.
  c. How long does it take the object to come back to earth?

Solution

a. From its equation, we know that the graph is a parabola that opens downward. We assign nonnegative values to t and then determine the corresponding values of h. (Why would we not use negative values of t?) For example, for $t = 0$, we have

$$h = -16 \cdot 0^2 + 100 \cdot 0$$
$$= 0$$

For $t = 1$, we obtain

$$h = -16 \cdot 1^2 + 100 \cdot 1$$
$$= -16 + 100$$
$$= 84$$

Continuing in this manner yields the following table:

t	0	1	2	3	4	5	6
h	0	84	136	156	144	100	24

These pairs of numbers are plotted in Figure 10.23, and connected with a smooth curve. This is the graph of $h = -16t^2 + 100t$.

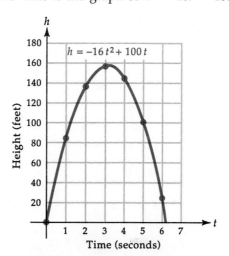

FIGURE 10.23

b. We want the value of h when $t = 1.5$. On the graph, the height h corresponding to 1.5 s appears to be just less than 120, say 118.

c. On the graph, the time t at which the height is 0 appears to be about 6.4 s. How could you find the exact value?   ■

DISCUSSION QUESTIONS 10.4

1. What does the number a tell us about the graph of $y = ax^2 + bx + c$?

2. What is the value of y at the point (x, y) at which a parabola intersects the x axis? How is this point related to the equation $0 = ax^2 + bx + c$?

3. Match each of the following statements with the appropriate graph of $y = ax^2 + bx + c$ in Figure 10.24. Some statements apply to more than one graph.

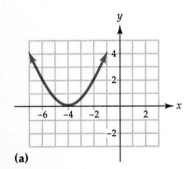

(a)

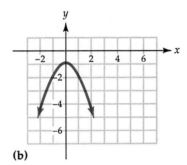

(b)

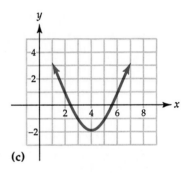

(c)

FIGURE 10.24.

a. The graph contains the point $(4, -2)$.
b. $a > 0$.
c. $0 = ax^2 + bx + c$ has no real solutions.
d. The graph contains the point $(-4, 0)$.
e. The graph contains the point $(0, -1)$.

f. $0 = ax^2 + bx + c$ has two real solutions.
g. $a < 0$.
h. $0 = ax^2 + bx + c$ has one real solution.
i. The graph contains the point $(6, 1)$.

EXERCISES 10.4

In Exercises 1–8, without graphing, determine (a) whether the graph of the equation opens upward or downward and (b) at how many points it crosses the x axis. (*Hint:* For part b, just compute the discriminant.)

1. $y = x^2 - 4x - 5$
2. $y = -3x^2 + 12x + 8$
3. $y = -2x^2 + 3x - 1$

4. $y = -x^2 - 41$
5. $y = 3x^2 + 10$
6. $y = 2x^2 - 3x$

7. $y = -x^2 + 2x$
8. $y = 2x^2 - x - 1$

In Exercises 9–16, graph the equation. Estimate the x intercepts, if any, of the graph.

9. $y = x^2 - 7x$ **10.** $y = 14 + 5x - x^2$ **11.** $y = 2x^2 + x - 3$ **12.** $2x^2 - y = 3x + 12$

13. $0 = 1 - \dfrac{x^2}{49} - y$ **14.** $8x - 5 = 3x^2 + y$ **15.** $2x(x - 1) = 1 + y$ **16.** $y = 2x^2 - 3x + 5$

In Exercises 17–24, set $y = 0$ in the equation of the given exercise and solve the result for x. Give solutions correct to the nearest hundredth. Your solutions will be the exact x intercepts of the graph.

17. Exercise 9   **18.** Exercise 10   **19.** Exercise 11   **20.** Exercise 12

21. Exercise 13   **22.** Exercise 14   **23.** Exercise 15   **24.** Exercise 16

Exercises 25–31 refer to this situation:

The Flying Kite Company has discovered that its daily profit y (in dollars) can be determined from the number of kites produced per day, x, by the equation, $y = -x^2 + 500x - 52{,}500$.

25. If the company produces 200 kites per day, how much profit is made?

26. If the company produces 100 kites per day, how much profit is made? Explain the significance of your answer.

27. Make a graph of $y = -x^2 + 500x - 52{,}500$.

28. What is the minimum number of kites the company must produce per day in order to break even (have no profit or loss)?

29. How many kites should the company produce per day to maximize its profit?

30. What is the maximum profit attainable?

31. Would it be profitable for the company to produce more than 500 kites per day? Justify your answer.

Exercises 32–34 refer to this situation:

The approximate cost C (in dollars per hour) of operating a car at speed x (in miles per hour) is given by the equation, $C = 0.0040x^2 - 0.40x + 20.30$ where x is between 35 and 780, inclusive.

32. Graph this equation.

33. At what speed in this range is cost a minimum?

34. What is the minimum operating cost per hour?

Preparing for Section 10.5

In Exercises 35–37, perform the indicated operations.

35. $(1 + 3i) + (2 - i)$ **36.** $(7 + 5i) - (-2 - 4i)$ **37.** $(-1 - i)(3 + 7i)$

In Exercises 38–41, use the properties of square roots to simplify each expression.

38. $\sqrt{5} \cdot \sqrt{6}$ **39.** $5\sqrt{3} \cdot \sqrt{3}$ **40.** $\sqrt{320}$ **41.** $\sqrt{108}$

10.5 Complex Numbers

FOCUS

We introduce complex numbers and compute their sums, differences, products, and quotients. We then use the quadratic formula to find the complex-number roots of quadratic equations.

Imaginary Numbers and Complex Numbers

Some quadratic equations have no real-number solutions. For example, $x^2 + 1 = 0$ has no real-number solution, since there is no real number x such that $x^2 = -1$. Mathematicians, however, have defined a set of numbers called the *complex numbers* in which *every* quadratic equation with real-number coefficients has solutions. The complex numbers are defined in terms of the **imaginary unit** $i = \sqrt{-1}$. Thus, $i^2 = -1$.

Obviously, i is not a real number, but to a certain extent it can be treated like other square roots. It is also used to define the square roots of negative numbers.

Square root of a negative number

If $a \geq 0$, then $\sqrt{-a} = \sqrt{a}\sqrt{-1} = \sqrt{a}i$.

EXAMPLE 1 | Simplify each of the following.

a. i^2 **b.** i^5 **c.** $\sqrt{-25}$ **d.** $\sqrt{-12}$

Solution   **a.** Since $i = \sqrt{-1}$, we have

$$i^2 = (\sqrt{-1})^2 = -1$$

b. Now with $i^2 = -1$, we have

$$i^3 = i^2 \cdot i = -1 \cdot i = -i$$
$$i^4 = i^2 \cdot i^2 = (-1)(-1) = 1$$

and finally
$$i^5 = i^4 \cdot i = (1)(i) = i$$

c. $\sqrt{-25} = \sqrt{25}\sqrt{-1} = 5i$

d. $\sqrt{-12} = \sqrt{12}\sqrt{-1}$

$$= \sqrt{4(3)}i$$
$$= 2\sqrt{3}i$$ ■

Numbers of the form bi, where b is a real number, are called **imaginary numbers.** A complex number is the sum of a real number and an imaginary number.

> A **complex number** is a number that can be written in the form $a + bi$, where $i = \sqrt{-1}$ and a and b are real numbers.

The form $a + bi$ is called the **standard form** of a complex number. Note that any real number a is a complex number, since

$$a = a + 0i$$

Any imaginary number bi is also a complex number, since

$$bi = 0 + bi$$

It is also important to note that $a + bi = c + di$ *if and only if* $a = c$ and $b = d$.

Addition and Subtraction of Complex Numbers

Addition and subtraction of complex numbers are defined so that the properties of real numbers hold.

> **Sum and difference of complex numbers**
>
> The sum and difference of complex numbers $a + bi$ and $c + di$ are defined, respectively, as follows:
>
> $$(a + bi) + (c + di) = (a + c) + (b + d)i$$
> $$(a + bi) - (c + di) = (a - c) + (b - d)i$$

In other words, to add (or subtract) complex numbers, we add (or subtract) the real-number parts and the imaginary-number parts.

EXAMPLE 2   Compute.

a. $(-3 + 6i) + (2 - i)$ **b.** $(2 - 3i) - (-4 - 2i)$

Solution   **a.** $(-3 + 6i) + (2 - i) = (-3 + 2) + (6 - 1)i$

$$= -1 + 5i$$

b. $(2 - 3i) - (-4 - 2i) = (2 - (-4)) + (-3 - (-2))i$

$$= 6 + (-1)i$$

$$= 6 - i$$ ∎

BE CAREFUL!

The square root of a negative number is an imaginary number, but the negative of the square root of a positive number is a negative real number. For example,

$$\sqrt{-1} = i \quad \text{but} \quad -\sqrt{1} = -1$$
$$\sqrt{-49} = 7i \quad \text{but} \quad -\sqrt{49} = -7$$

Multiplication of Complex Numbers

Multiplication of complex numbers is also defined so that the properties of real numbers hold.

EXAMPLE 3   Find the product $(2 + i)(3 + 2i)$.

Solution   We apply the distributive property as we would to multiply, say, $(2 + x)$ · $(3 + 2x)$.

$$(2 + i)(3 + 2i) = 2(3 + 2i) + i(3 + 2i)$$
$$= 6 + 4i + 3i + 2i^2$$
$$= 6 + (4 + 3)i + 2(-1) \qquad \text{Because } i^2 = -1$$
$$= 6 + 7i - 2$$
$$= 4 + 7i$$

Another way we could have found this product is with the FOIL method. ∎

This example suggests the following definition for the product of two complex numbers.

Product of complex numbers

The product of two complex numbers $a + bi$ and $c + di$ is

$$(a + bi)(c + di) = (ac - bd) + (ad + bc)i$$

You could memorize the definition, but it is easier to use the FOIL method as we do in the next example.

EXAMPLE 4

Find each product.

a. $(5 - 2i)(6 + 3i)$ **b.** $(4 + 2i)(4 - 2i)$

Solution

$$\qquad\qquad\qquad \text{F} \quad \text{O} \quad \text{I} \quad \text{L}$$

a. $(5 - 2i)(6 + 3i) = 30 + 15i - 12i - 6i^2$
$$= 30 + 3i - 6(-1)$$
$$= 36 + 3i$$

b. $(4 + 2i)(4 - 2i) = 16 - 8i + 8i - 4i^2$
$$= 16 - 4(-1)$$
$$= 20 \qquad\qquad\qquad\qquad ∎$$

In part b of Example 4, the product of two complex numbers turns out to be a real number. This will always be the case if the two complex numbers are conjugates. Complex numbers of the form $a + bi$ and $a - bi$ are called **complex conjugates** or just **conjugates**.

EXAMPLE 5 Find the conjugate of the given number. Then multiply the given number by its conjugate.

a. $-3 + i$ **b.** $3 - i$ **c.** 5 **d.** $4i$

Solution **a.** The conjugate is $-3 - i$. And,

$$(-3 + i)(-3 - i) = 9 - 3i + 3i - i^2$$
$$= 9 - (-1)$$
$$= 10$$

b. The conjugate is $3 + i$. And,

$$(3 - i)(3 + i) = 9 - 3i + 3i - i^2$$
$$= 9 - (-1)$$
$$= 10$$

c. The conjugate of 5 (or $5 + 0i$) is $5 - 0i$, which is 5. And,

$$(5 + 0i)(5 - 0i) = 25$$

d. The conjugate of $4i$ (or $0 + 4i$) is $0 - 4i$, or simply $-4i$. Also,

$$(4i)(-4i) = -16i^2$$
$$= -16(-1)$$
$$= 16 \qquad \blacksquare$$

Division of Complex Numbers

To divide two complex numbers, we write the division as a fraction and then multiply both the numerator and the denominator by the conjugate of the denominator.

Quotient of complex numbers

The quotient of two complex numbers $a + bi$ and $c + di$ is

$$\frac{a + bi}{c + di} = \frac{a + bi}{c + di} \cdot \frac{c - di}{c - di} = \frac{(ac + bd) + (bc - ad)i}{c^2 + d^2}$$

Again, it is easier to do the work than to memorize the definition.

EXAMPLE 6   Find each quotient.

a. $(4 - 5i) \div 3i$ **b.** $20 \div (5 - i)$ **c.** $(2 + 3i) \div (1 + i)$

Solution   In each case, we write the quotient as a fraction and then multiply numerator and denominator by the conjugate of the denominator.

a. $(4 - 5i) \div 3i = \dfrac{4 - 5i}{3i}$ Writing as a fraction

$\qquad = \dfrac{4 - 5i}{3i} \cdot \dfrac{-3i}{-3i}$ Multiplying by conjugate

$\qquad = \dfrac{-12i + 15i^2}{-9i^2}$ Multiplying

$\qquad = \dfrac{-12i - 15}{9}$ Substituting -1 for i^2

$\qquad = -\dfrac{5}{3} - \dfrac{4}{3}i$ Simplifying

b. $20 \div (5 - i) = \dfrac{20}{5 - i}$ Writing as a fraction

$\qquad = \dfrac{20}{5 - i} \cdot \dfrac{5 + i}{5 + i}$ Multiplying by conjugate

$\qquad = \dfrac{100 + 20i}{26}$ Multiplying

$\qquad = \dfrac{50}{13} + \dfrac{10}{13}i$ Simplifying

c. $(2 + 3i) \div (1 + i) = \dfrac{2 + 3i}{1 + i}$ Writing as a fraction

$\qquad = \dfrac{2 + 3i}{1 + i} \cdot \dfrac{1 - i}{1 - i}$ Multiplying by conjugate

$\qquad = \dfrac{5 + i}{2}$ Multiplying

$\qquad = \dfrac{5}{2} + \dfrac{1}{2}i$ Simplifying

Complex Numbers and Equations

Linear equations that involve complex numbers can be solved in much the same way as we solved such equations involving real numbers.

EXAMPLE 7   Solve for z.

$$z - (4 - 5i) = 3 + i$$

Solution   Since $4 - 5i$ is subtracted from z, we add it to both sides of the equation.

$$z - (4 - 5i) = 3 + i$$
$$[z - (4 - 5i)] + (4 - 5i) = (3 + i) + (4 - 5i) \qquad \text{Adding } 4 - 5i$$
$$z = (3 + 4) + (1 - 5)i \qquad \text{Combining}$$
$$= 7 - 4i \qquad\qquad\blacksquare$$

If the discriminant $b^2 - 4ac$ of a quadratic equation $ax^2 + bx + c = 0$ is negative, then the equation has no real solutions. However, such an equation has two distinct complex roots. They are found in the same way as real roots—by factoring, taking square roots, or using the quadratic formula.

EXAMPLE 8   Find all complex-number roots of these equations.

a. $x^2 + 3 = 0$ **b.** $x^2 - x + 1 = 0$

Solution   **a.**   We solve this equation by taking square roots.

$$x^2 + 3 = 0$$
$$x^2 = -3 \qquad \text{Adding } -3 \text{ to both sides}$$
$$x = \pm\sqrt{-3} \qquad \text{Taking the square root of } -3$$
$$= \pm\sqrt{3}i \qquad \text{Writing in terms of } i$$

We check one of the roots, $-\sqrt{3}i$, by substitution.

$$(-\sqrt{3}i)^2 + 3 = 0$$
$$(-\sqrt{3})^2 i^2 + 3 = 0$$
$$(3)(-1) + 3 = 0$$
$$-3 + 3 = 0 \qquad \text{True}$$

You should check the other root.

b. We use the quadratic formula with $a = 1$, $b = -1$, and $c = 1$ to solve this equation.

$$x = \frac{-(-1) \pm \sqrt{(-1)^2 - 4(1)(1)}}{2(1)}$$

$$= \frac{1 \pm \sqrt{-3}}{2}$$

$$= \frac{1}{2} \pm \frac{\sqrt{3}}{2}i$$

The solutions are $\frac{1}{2} + \frac{\sqrt{3}}{2}i$ and $\frac{1}{2} - \frac{\sqrt{3}}{2}i$.

We check to be sure $x = \frac{1}{2} + \frac{\sqrt{3}}{2}i$ satisfies the equation.

$$x^2 - x + 1 = 0$$

$$\left(\frac{1}{2} + \frac{\sqrt{3}}{2}i\right)^2 - \left(\frac{1}{2} + \frac{\sqrt{3}}{2}i\right) + 1 = 0$$

$$\frac{1}{4} + \frac{\sqrt{3}}{4}i + \frac{\sqrt{3}}{4}i + \frac{3}{4}i^2 - \frac{1}{2} - \frac{\sqrt{3}}{2}i + 1 = 0$$

$$\left(\frac{1}{4} + \frac{3}{4}(-1) - \frac{1}{2} + 1\right) + \left(\frac{\sqrt{3}}{4} + \frac{\sqrt{3}}{4} - \frac{\sqrt{3}}{2}\right)i = 0$$

$$\left(\frac{5}{4} - \frac{5}{4}\right) + \left(\frac{\sqrt{3}}{2} - \frac{\sqrt{3}}{2}\right)i = 0$$

$$0 + 0i = 0 \qquad \text{True} \qquad \blacksquare$$

DISCUSSION QUESTIONS 10.5

1. Write -6 as a complex number in standard form. Do the same for $4i$.

2. Write the conjugate of each number in standard form.
   a. 6   **b.** $-5i$ **c.** $2 - 3i$

3. If $x + 2i = -3 + yi$, find x and y.

4. Suppose the graph of $y = ax^2 + bx + c$ lies entirely below the x axis. Is its discriminant positive, zero, or negative?

EXERCISES 10.5

In Exercises 1–10, write each expression as a complex number in standard form. Classify each as imaginary, real, or neither. Then write its conjugate in standard form.

1. i^8 **2.** $-i^7$ **3.** $i^6 - 1$ **4.** $i^{25} - 3$ **5.** $i^3 + i$

6. $\sqrt{-81}$ **7.** $-\sqrt{16}$ **8.** $\sqrt{-100}$ **9.** $-\sqrt{100}$ **10.** $\sqrt{-63}$

In Exercises 11–37, perform the indicated operation.

11. $6i + 4i$ **12.** $-5i + 2i$ **13.** $(-6 - 5i) + (-5 - 3i)$

14. $(-4 - 2i) + (-2 - i)$ **15.** $(7 - 8i) + 3i$ **16.** $(8 + 2i) + 5i$

17. $9i - (-3i)$ **18.** $-5i - (-i)$ **19.** $(10 + 2i) - (4 + i)$

20. $(6 - 5i) - (3 + 3i)$ **21.** $(7 - 3i) - 5i$ **22.** $8 - (-2 + 10i)$

23. $(2i)(-3i)$ **24.** $(5i)(7i)$ **25.** $(-2i)(1 + 3i)$

26. $(4i)(-2 - 5i)$ **27.** $(2 - 4i)(-3 + 3i)$ **28.** $(3 + 4i)(3 - 4i)$

29. $(-2 - 3i)(-2 + 3i)$ **30.** $(6 - 3i) \div 6$ **31.** $(-12 + 5i) \div 3i$

32. $(1 + 2i) \div (1 - 2i)$ **33.** $(-3 - i) \div (3 + 4i)$ **34.** $2i \div (5 - i)$

35. $-7i \div (-6 + 8i)$ **36.** $\left(\dfrac{1}{3} + 2i\right) \div \left(\dfrac{4}{3} - 9i\right)$ **37.** $\left(\dfrac{3}{4} - \dfrac{1}{5}i\right) \div \left(\dfrac{3}{5} + \dfrac{2}{5}i\right)$

In Exercises 38–43, solve each equation for the complex number z. Express it in standard form.

38. $z + (2 + i) = 12 + 6i$ **39.** $z + (5 - 3i) = -2 + 6i$ **40.** $z - (9 + 3i) = 6 + 5i$

41. $z - (-7 + 2i) = 15 - 8i$ **42.** $9 + z = 5 - 2i$ **43.** $7i + z = 6 + 8i$

In Exercises 44–55, find all the complex-number roots of each equation. Express the roots in standard form and check.

44. $x^2 + 9 = 0$ **45.** $x^2 + 18 = 0$ **46.** $x^2 - 7x = 0$ **47.** $-x^2 - 3x = 0$

48. $x^2 + 2x + 6 = 0$ **49.** $x^2 + 2x - 6 = 0$ **50.** $x^2 + 3x + 5 = 0$ **51.** $x^2 - 3x + 5 = 0$

52. $x^2 - x + 4 = 0$ **53.** $3x^2 - 15 = 0$ **54.** $2x^2 - 3x + 6 = 0$ **55.** $6x^2 - x + 1 = 0$

56. One root of $x^2 + bx + c = 0$ is $1 + i$. Based on your answers in Exercises 44–55, what is the other root?

57. One root of $x^2 - 2x + c = 0$ is $-1 - 2i$. Find the second root, and find c.

In Exercises 58 and 59, use this information:

In terms of the resistance R and the reactance X, the impedance Z of an electric circuit is given by $Z = R + Xi$, where quantities are measured in ohms.

58. If the resistance of an electric circuit is 10 ohms (Ω) and the reactance is 3 Ω, find its impedance.

59. If the impedance of a circuit is $15 + 8i$ Ω, find the resistance and the reactance.

In Exercises 60 and 61, use this information:

> The voltage E, current I, and resistance R in an electric circuit are related by Ohm's law, $E = IR$, where these quantities are measured in volts, amperes, and ohms, respectively.

60. Find E if $I = 2 - 3i$ amperes (A) and $R = 3 + 4i$ Ω.

61. Find E if $I = 4 - 5i$ A and $R = 3 - 2i$ Ω.

Chapter Summary

Important Ideas

The quadratic formula

$$x = \frac{-b + \sqrt{b^2 - 4ac}}{2a} \quad \text{and} \quad x = \frac{-b - \sqrt{b^2 - 4ac}}{2a}$$

Pythagorean Theorem

$$a^2 + b^2 = c^2$$

Sum and difference of complex numbers

$$(a + bi) + (c + di) = (a + c) + (b + d)i$$
$$(a + bi) - (c + di) = (a - c) + (b - d)i$$

Product of complex numbers

$$(a + bi)(c + di) = (ac - bd) + (ad + bc)i$$

Quotient of complex numbers

$$\frac{a + bi}{c + di} = \frac{a + bi}{c + di} \cdot \frac{c - di}{c - di} = \frac{(ac + bd) + (bc - ad)i}{c^2 + d^2}$$

A *complex number* is a number that can be written in the form $a + bi$, where $i = \sqrt{-1}$ and a and b are real numbers.

Chapter Review

Section 10.1

In Exercises 1–12, solve each equation by factoring or finding square roots.

1. $x^2 + 5x + 4 = 0$
2. $x^2 + 3x + 2 = 0$
3. $x^2 - 7x + 12 = 0$
4. $y^2 - y = -6$

5. $2x^2 = 4x - 2$
6. $s - s^2 = -16 + s$
7. $x^2 = 6x - 8$
8. $(w + 3)^2 = 25$

9. $3x + 2 = 2x^2$
10. $y(y - 5) = -6$
11. $x(x + 10) = 24$
12. $(x - 3)^2 = 49$

13. The sum of a number and its reciprocal is $\dfrac{25}{12}$. Find the number.

14. If the sides of a square are decreased by 3 cm, its area is decreased by 63 cm^2. Find the length of a side of the original square.

Section 10.2

In Exercises 15–26, solve the equation using the quadratic formula. Give solutions in simplified radical form. If there are no real roots, indicate that.

15. $3x^2 - x - 1 = 0$
16. $y^2 + 4y + 3 = 0$
17. $x^2 + 3x + 1 = 0$
18. $y^2 - y - 3 = 0$

19. $2x^2 = 5 - x$
20. $2x^2 + 3x = 10$
21. $x - 3x^2 = 8$
22. $10 - x = x^2 + 10$

23. $4x - x^2 = 3x - 6$
24. $3x^2 + x = 4x + 1$
25. $10 + 2x = 2x + x^2$
26. $6s^2 = 5s$

27. A rectangular wall is 8 ft longer than it is high. If its area is 172 ft^2, find its height and length correct to the nearest tenth of a foot.

28. A quadratic equation has no real-number roots if its discriminant, $b^2 - 4ac$, is less than zero. Find the values of c for which $x^2 + 3x + c = 0$ has no real roots.

Section 10.3

In Exercises 29–34, sides a and b are the legs and c is the hypotenuse of a right triangle. Find the unknown value correct to the nearest tenth.

29. $a = 6, b = 9, c = \underline{\ ?\ }$
30. $a = 4.5, b = 7.4, c = \underline{\ ?\ }$
31. $a = 5, b = 8, c = \underline{\ ?\ }$

32. $a = 3.5, b = 7.8, c = \underline{\ ?\ }$
33. $a = 8, b = \underline{\ ?\ }, c = 12$
34. $a = \underline{\ ?\ }, b = 6.9, c = 8.7$

35. Find the width of a rectangle whose length and diagonal are 15 ft and 25 ft, respectively.

36. A rectangular picture is 5 in. wide and 7 in. long. It is surrounded by a frame of uniform width. If the area of the picture and frame together is 80 in.^{2}, find the width of the frame.

Section 10.4

In Exercises 37–40, graph each equation. Estimate the x intercepts, if any, of the graph.

37. $y = -x^2 + 2$ **38.** $y = x^2 - x$ **39.** $y = 2x^2 + x + 1$ **40.** $y = -x^2 - 2x - 2$

In Exercises 41–44, set $y = 0$ in the equation of the given exercise and solve the result for x. Give solutions correct to the nearest hundredth. Your solutions will be the x intercepts of the graph.

41. Exercise 37 **42.** Exercise 38 **43.** Exercise 39 **44.** Exercise 40

45. A fire department's hose shoots water in a path described by the equation $h = -0.025x^2 + 2x + 0.5$, where x is the horizontal distance in meters from the nozzle and h is the height in meters. Graph this equation and use the graph to estimate the distance from the nozzle at which the water reaches its maximum height.

46. For what value of c will the graph of $y = 2x^2 - x + c$ have just one x intercept? *Hint:* The discriminant must be zero.

Section 10.5

In Exercises 47–50, write each expression as a complex number in standard form. Classify each as imaginary, real, or neither.

47. i^9 **48.** $i^{12} + 4$ **49.** $-\sqrt{-8}$ **50.** $-\sqrt{25} + i$

In Exercises 51–62, perform the indicated operations.

51. $-3i + i$ **52.** $(2 - i) + (-5i)$ **53.** $(-5 - 9i) + (-4 + 7i)$

54. $\left(\frac{3}{4} + \frac{1}{4}i\right) + \left(-\frac{1}{5} - 2i\right)$ **55.** $(2 + i) - (-1 - 3i)$ **56.** $(5.3 + 6.4i) - (-2.4 + 3.1i)$

57. $(-5)(4 - 7i)$ **58.** $(-7i)(8 - 3i)$ **59.** $(-2 - 3i)(-4 - i)$

60. $4 \div (1 + i)$ **61.** $(2 - i) \div (-1 + 2i)$ **62.** $\left(\frac{1}{2} + \frac{1}{2}i\right) \div \left(\frac{1}{3}i\right)$

In Exercises 63 and 64, verify that the given complex number is a root of the equation. Give the second root of the equation.

63. $x^2 + 12 = 3$; $-3i$ **64.** $x^2 - 4x + 5 = 0$; $2 - i$

In Exercises 65–71, find all the complex roots of each equation. Express answers in standard form.

65. $x^2 - 3x = 0$ **66.** $-3x^2 + x - 5 = 0$ **67.** $x^2 - 2x + 2 = 0$

68. $3x^2 - 2x + 1 = 0$ **69.** $5x^2 - 6x + 2 = 0$ **70.** $3x^2 - x + 2 = 0$

71. $4.2x^2 - 5.4x + 3.7 = 0$

Chapter 10 Test

In Problems 1–4, find all complex roots. Express answers in standard form, simplifying all radicals.

1. $x^2 - 2x - 4 = 0$ **2.** $x^2 = 20 + 4x$ **3.** $2x^2 + x + 1 = 0$ **4.** $3x^2 = 2x + 2$

In Problems 5–10, simplify and write the resulting expression as a complex number in standard form.

5. i^{28}

6. $\sqrt{-64}$

7. $(3 - 2i) - (-4 + i)$

8. $-5i \div (2 - 6i)$

9. $2i(3 + 5i)$

10. $(3 - i)(-2 + 4i)$

In Problems 11 and 12, graph the equation. Estimate the x intercepts, if any.

11. $y = -x^2 + 1$

12. $y = x^2 + x + 1$

13. The length of a rectangle is 7 cm and its diagonal is 9 cm. Find to the nearest tenth the width of the rectangle.

14. A 30-ft ladder leans against the side of a building so that the top of the ladder touches the building 20 ft above the ground. How far is the foot of the ladder from the bottom of the building?

15. A rectangular living room is 3 ft longer than it is wide, and it has 180 ft^2 of space. Find the length and width of the room.

16. A rectangular billboard is 6 m long and 4 m high. The area of the billboard is increased by 36 m^2 by increasing the length and width each by an amount x. Find the length and width of the enlarged billboard.

17. A rectangular garden is 4 m longer than it is wide, and its area is 117 m^2. Find the length and width of the garden.

18. A block of wood is in the shape of a rectangular solid. Its width is 1 less than its height, and its length is 2 more than its height. If the total surface area of the block is 12 square units, find its length, width, and height.

19. The stopping distance d, in feet, of a car traveling r mph is

$$d = 0.06r^2 + 0.1r$$

How fast is a car traveling if it takes 5 ft to stop?

20. In Problem 19, how fast is a car traveling if the rate r is numerically equal to the distance d?

Glossary

Abscissa (3.1)   See *x coordinate*.

Absolute value (2.1)   The absolute value of a real number is its distance from zero on the number line.

Addition property (3.2)   Any number or expression can be added to both sides of an equation without changing the relationship between the sides.

Additive identity (2.2)   For real numbers, this is zero.

Additive inverse (2.2)   For a real number a, this is $-a$.

Algebraic expression (1.1)   A string of numbers, operations, and variables.

Area (1.3)   The measure of the size of an enclosed region.

Arithmetic mean (1.2)   The sum of a set of numbers divided by the number of numbers in the set (also called the *average*).

Associative property for addition (1.4)   If a, b, and c are any real numbers, then $(a + b) + c = a + (b + c)$.

Associative property for multiplication (1.4)   If a, b, and c are any real numbers, then $(ab)c = a(bc)$.

Asymmetric property (3.4)   If $a < b$, then $b > a$. If $a > b$, then $b < a$. If $a \leq b$, then $b \geq a$. If $a \geq b$, then $b \leq a$.

Average (1.2)   See *arithmetic mean*.

Axes (1.5)   The pair of perpendicular number lines in a rectangular coordinate system.

Base (1.1)   In a^n, a is called the base.

Binomial (4.4)   A polynomial that contains two terms.

Circumference (1.3)   The length or distance around a circle.

Clearing an equation of fractions (6.5)   Multiplying both sides of an equation by the least common denominator of all the fractions and rational expressions in the equation.

Commutative property for addition (1.4)   If a and b are any real numbers, then $a + b = b + a$.

Commutative property for multiplication (1.4)   If a and b are any real numbers, then $ab = ba$.

Complex conjugates (10.5)   Complex numbers of the form $a + bi$ and $a - bi$.

Complex fraction (6.4)   A fraction that contains fractions or rational expressions in its numerator or denominator or both.

Complex number (10.5)   A number that can be written in the form $a + bi$, where $i = \sqrt{-1}$ and a and b are real numbers.

Conditional equation (3.1)   An equation that is not true for all values of the variables, but may be true for some values.

Conjugates (9.5)   Binomial expressions of the form $a + b$ and $a - b$, where either a or b (or both) are radicals.

Consistent system (8.1)   A system of equations that has exactly one solution.

Constant of proportionality (6.6)   See *constant of variation*.

Constant of variation (6.6)   The constant k in a direct or inverse variation equation (also called a *constant of proportionality*).

Cross products (3.6)   The product of the numerator of one ratio in a proportion and the denominator of the other.

Cube root (9.1)   The third root of a number.

Degree of a polynomial (4.4)   The highest degree of the polynomial's terms.

Denominator (1.1) b in a fraction $\frac{a}{b}$.

Dependent system (8.1)   A system of equations with an infinite number of solutions.

Difference of two squares (5.4)   A binomial of the form $a^2 - b^2$.

Directly proportional to (6.6)   See *varies directly as*.

G1

Discriminant (10.2) For a quadratic equation in standard form, $b^2 - 4ac$.

Distributive property (1.4) If a, b, and c are any real numbers, then $a(b + c) = ab + ac$ and $(b + c)a = ba + ca$.

Division property (3.3) Both sides of an equation can be divided by the same nonzero number without changing the relationship of the sides.

Divisors (5.1) See *factors*.

Equation (1.1) A statement of the form $a = b$, which means that the expression on the left of the equal sign represents the same number as the expression on the right.

Equivalent equations (3.2) Two equations that have the same solutions.

Exponent (1.1) In a^n, n is called the exponent.

Exponentiation (1.1) The process of raising a number to a power.

Extraneous solution (9.7) A solution resulting from squaring both sides of an equation that does not check in the original equation.

Factoring (5.1) Writing an expression as a multiplication of expressions.

Factors (5.1) If $a = bc$, then b and c are factors of a (also called *divisors*).

First-degree equation (3.1) An equation in which the variable (or each variable) has exponent 1 (also called a *linear equation*).

FOIL method (4.5) To find the product of two binomials, find the product of the first terms, the outer terms, the inner terms, and finally the last terms, and then write the sum of these products.

Fractional exponent (9.6) If n is a positive integer and a is a real number such that $\sqrt[n]{a}$ is defined, then $a^{1/n} = \sqrt[n]{a}$.

Graph (1.1) A geometric representation of a set of numbers as points on a number line or in a coordinate plane.

Graph of a linear equation (7.2) All ordered pairs that satisfy the equation, all of which lie on a line.

Graph of a linear inequality (7.5) The region in the coordinate plane that contains all ordered-pair solutions of the inequality.

Greater than (1.1) If a is less than b (i.e., a appears to the left of b on the number line), we also say b is greater than a.

Greatest common factor (5.1) The largest number that is a factor of two or more integers.

Guess-and-test method (3.7) Guessing an answer for a word problem, computing the result of that answer, and then adjusting the guess to better fit the other data in the problem.

Half-plane (7.5) The region on one side of a given line in the plane.

Hypotenuse (10.3) The side of a right triangle that lies opposite the right angle.

Identity (3.1) An equation that is true for all real-number values of the variable or variables.

Imaginary number (10.5) A number of the form bi, where b is a real number and $i = \sqrt{-1}$.

Imaginary unit (10.5) $i = \sqrt{-1}$ is called the imaginary unit.

Inconsistent system (8.1) A system of equations that has no solution.

Index (9.1) n in $\sqrt[n]{b}$.

Inequality (3.4) A statement saying that two numbers or expressions are, or may be, unequal.

Integers (2.1) The whole numbers and their negatives.

Inversely proportional to (6.6) See *varies inversely as*.

Lateral surface area (right circular cylinder) (1.3) The lateral surface area is $2\pi rh$, where the height is h and the radius of the base is r.

Least common denominator (6.3) The least common multiple of a set of denominators.

Legs of a right triangle (10.3) The sides that form the right angle.

Less than (1.1) Whole number a is less than whole number b if a appears to the left of b on a number line.

Like radicals (9.4) Radicals that have the same index and the same radicand.

Like terms (1.4) Terms in which the same variables are raised to the same powers.

Linear equation (3.1) See *first-degree equation*.

Linear equation in two variables (7.1) An equation that can be expressed in the form $ax + by + c = 0$, where a, b, and c are real numbers and a and b are not both zero.

Linear inequality in two variables (7.5) An inequality that can be expressed in the form $ax + by < c$, where a, b, and c are real numbers and a and b are not both zero. ("<" may be replaced by $>$, $\leq$, or $\geq$.)

Monomial (4.4) A polynomial that consists of just one term.

Multiplication property (3.3) Both sides of an equation can be multiplied by the same number without changing the relationship of the sides.

Multiplicative identity (2.4) For real numbers, this is 1.

Multiplicative inverse (2.4) For a nonzero real number a, this is $\dfrac{1}{a}$ (also called *reciprocal*).

Negative real numbers (2.1) All real numbers that are less than zero.

nth power of a (1.1) If n is a whole number, $a^n = \underbrace{a \cdot a \cdot a \cdot \ldots \cdot a}_{n \text{ factors}}.$

nth root (9.1) A number a is an nth root of b if $a^n = b$, where a and b are real numbers and n is a positive integer.

Number line (1.1) A line with a correspondence between its points and a set of real numbers.

Numerator (1.1) a in a fraction $\dfrac{a}{b}$.

Numerical coefficient (1.4) The number that multiplies the variable(s) in a term.

Opposite (2.1) The opposite of a number is the number that is the same distance from zero on the number line but in the opposite direction.

Opposite of a polynomial (4.4) The polynomial gotten by changing the sign of each term in a given polynomial.

Ordered (1.1) A set of numbers is ordered if for any two numbers a and b, exactly one of the following holds: $a = b$, $a < b$, or $a > b$.

Ordered pair (7.1) A pair of real numbers in a specific order, usually written in the form (x, y).

Ordinate (3.1) See *y coordinate*.

Origin (1.5) The point of intersection of the axes in a rectangular coordinate system.

Parabola (10.4) The graph of an equation of the form $y = ax^2 + bx + c$ where $a \neq 0$.

Perfect cube (9.1) A rational number whose cube root is rational.

Perfect square (9.1) A rational number whose square root is rational.

Perfect square trinomial (5.4) A trinomial that can be written in the form $a^2 + 2ab + b^2$ or $a^2 - 2ab + b^2$.

Perimeter (1.3) The total distance around an enclosed two-dimensional figure.

Polynomial (4.4) Either a single term or the sum of a finite number of terms.

Positive real numbers (2.1) All real numbers that are greater than zero.

Prime factored form (5.1) A number written as a product of prime numbers.

Prime number (5.1) A positive integer greater than 1 whose only whole-number factors are itself and 1.

Prime polynomial (5.2) A polynomial that cannot be factored with integer factors other than 1.

Principal square root (9.1) The positive square root of a number.

Proportion (3.6) An equation stating that two ratios are equal.

Pythagorean theorem (10.3) If a and b are the lengths of the legs and c is the length of the hypotenuse of a right triangle, then $a^2 + b^2 = c^2$.

Quadrants (7.2) The four regions into which the plane is divided by a rectangular coordinate system.

Quadratic equation (5.5) An equation that can be written in the form $ax^2 + bx + c = 0$, where a, b, and c are real numbers and $a \neq 0$.

Quadratic formula (10.2) The solution of the quadratic equation in standard form is
$$x = \frac{-b \pm \sqrt{b^2 - 4ac}}{2a}.$$

Radical (9.1) An expression of the form $\sqrt[n]{b}$ (also called a *radical expression*).

Radical expression (9.1) See *radical*.

Radical symbol (9.1) The symbol $\sqrt{}$.

Radicand (9.1) The number or expression under a radical sign.

Rate of change (7.3) For a linear relationship between x and y, the change in y for a unit change in x.

Rate of work (6.6) The fractional part of a job that can be done in one hour.

Ratio (3.6) The ratio of a to b is the fraction $\frac{a}{b}$.

Rational expression (6.1) An expression of the form $\frac{A}{B}$, where A and B are polynomials and $B \neq 0$.

Rational numbers (1.1) Numbers that can be written in the form $\frac{a}{b}$, where a and b are integers with $b \neq 0$.

Rationalizing the denominator (9.3) The process of rewriting a radical expression so that the denominator is a rational number.

Real numbers (1.1) All numbers that correspond to a point on a number line.

Reciprocal (2.4) See *multiplicative inverse*.

Rectangular coordinate system (1.5) A plane containing a pair of perpendicular number lines that allow us to correspond ordered pairs of real numbers to points in the plane.

Reduced to lowest terms (6.1) A rational expression in which the numerator and denominator contain no common factors.

Right triangle (10.3) A triangle that has a right angle.

Rise (7.3) A vertical change from one point to another.

Root (3.1) A replacement of the variable or variables that make a conditional equation true (also called *solution*).

Run (7.3) A horizontal change from one point to another.

Scientific notation (4.3) A positive real number written in the form $a \cdot 10^n$, where $1 \leq a < 10$ and n is an integer.

Signed number (2.1) A number that carries a plus or minus sign.

Simplest radical form (9.3) A radical expression with no perfect squares (or factors with rational nth roots, if nth roots are involved) in a radicand, no radical in a denominator, and no common factors in the numerator and denominator of a fraction.

Simplified or reduced form (1.1) A fraction $\frac{a}{b}$ in which there are no integral numbers that are divisors of both a and b.

Slope of a line (7.3) The ratio $\frac{y_2 - y_1}{x_2 - x_1}$ where (x_1, y_1) and (x_2, y_2) are any two points on the line.

Slope-intercept form (7.4) The equation of a line in the form $y = mx + b$, where m is the slope and b is the y intercept.

Solution (3.1) See *root*.

Solution of a linear equation (7.1) An ordered pair of numbers that makes the equation a true statement.

Solution of a linear inequality (7.5) An ordered pair of numbers that makes the inequality a true statement.

Solution of a system of equations (8.1) All ordered pairs that satisfy all equations in a system simultaneously.

Solution of a system of inequalities (8.5) All ordered pairs that satisfy all inequalities in a system simultaneously.

Square root (1.3) A square root of a nonnegative number N is x if $x^2 = N$.

Standard form (of a complex number) (10.5) The form is $a + bi$.

Standard form (of a quadratic equation) (10.1) The form is $ax^2 + bx + c = 0$, where $a \neq 0$. See also *Standard Form* in Section 5.5.

Subscripts (7.3) A number written below and to the right of a variable, such as 1 in a_1.

Subtraction property (3.2) Any number or expression can be subtracted from both sides of an equation without changing the relationship between the sides.

Symmetric property (3.1) For real numbers a and b, if $a = b$, then $b = a$.

System of linear equations (8.1) Two or more linear equations considered simultaneously.

System of linear inequalities (8.5) Two or more linear inequalities considered simultaneously.

Table of values (7.1) A list of ordered pairs that satisfy an equation in two variables.

Test point (7.5) A point used to determine which half-plane is included in the graph of a linear inequality.

Total surface area (right circular cylinder) (1.3) The total surface area is $2\pi rh + 2\pi r^2$, where the height is h and the radius of the base is r.

Trinomial (4.4) A polynomial that contains three terms.

Variable (1.1) A symbol, usually a letter, that represents any number or, sometimes, a particular unknown number.

Varies directly as (6.6) A relationship between x and y, $y = kx$ where k is a constant (also called *directly proportional to*).

Varies inversely as (6.6) A relationship between x and y, $y = \dfrac{k}{x}$ where k is a constant (also called *inversely proportional to*).

Volume (1.3) A measure of the amount of space in the interior of a closed three-dimensional figure.

Whole numbers (1.1) These are the numbers: 0, 1, 2, 3, ...

x axis (7.2) The horizontal axis in a rectangular coordinate system.

x coordinate (3.1) The first coordinate in an ordered pair (also called *abscissa*).

x intercept (7.2) The x coordinate of the point where a graph intersects the x axis.

y axis (7.2) The vertical axis in a rectangular coordinate system.

y coordinate (3.1) The second coordinate in an ordered pair (also called *ordinate*).

y intercept (7.2) The y coordinate of the point where a graph intersects the y axis.

Zero product property (5.5) If a and b are real numbers such that $ab = 0$, then either $a = 0$ or $b = 0$ or both.

APPENDIX: The Quadratic Formula

The quadratic formula is derived by applying the method called *completing the square* to solve the general quadratic equation

$$ax^2 + bx + c = 0 \quad (a \neq 0)$$

This method is illustrated in Example 1 of Section 10.2. The first step is to add $-c$ to both sides of the equation:

$$ax^2 + bx = -c$$

Next we multiply both sides of the equation by $\frac{1}{a}$:

$$x^2 + \frac{bx}{a} = -\frac{c}{a}$$

Our next goal is to add to both sides of the equation the constant that will make the left side a perfect-square trinomial. This is done by squaring one-half of $\frac{b}{a}$, the coefficient of x:

$$\frac{1}{2} \cdot \frac{b}{a} = \frac{b}{2a} \quad \text{and} \quad \left(\frac{b}{2a}\right)^2 = \frac{b^2}{4a^2}$$

We add this to both sides of the equation:

$$x^2 + \frac{bx}{a} + \frac{b^2}{4a^2} = -\frac{c}{a} + \frac{b^2}{4a^2}$$

The left side is now a perfect-square trinomial. We write it in factored form and simplify the right side:

$$\left(x + \frac{b}{2a}\right)^2 = \frac{b^2 - 4ac}{4a^2}$$

We now solve by the method of square roots. Remember that both the positive and the negative square roots must be considered.

$$x + \frac{b}{2a} = \pm\sqrt{\frac{b^2 - 4ac}{4a^2}}$$

But the right side can be simplified:

$$x + \frac{b}{2a} = \pm \frac{\sqrt{b^2 - 4ac}}{2a}$$

Finally, we add $-\dfrac{b}{2a}$ to both sides:

$$x = \frac{-b \pm \sqrt{b^2 - 4ac}}{2a}$$

This last equation, of course, is the quadratic formula.

Answers to Odd-Numbered Exercises

Chapter 1

EXERCISES 1.1 *(page 11)*

1. C **3.** D **5.** E **7.** 500 **9.** 1000 **11.** 68,400 **13.** 0.275 **15.** 1.333
17. $\frac{4}{7}$ **19.** $3\frac{8}{9}$ **21.** $\frac{3}{4}$ **23.** $\frac{3}{4}$ **25.** $\frac{4}{5}$ **27.** $14\frac{7}{12}$ **29.** $\frac{9}{40}$ **31.** $\frac{7}{10}$ **33.** $1\frac{5}{6}$

35. *Common Fraction:* $\frac{1}{6}, \frac{3}{8}, \frac{5}{8}, \frac{5}{6}, \frac{1}{1}$; *Decimal Fraction:* 0.167, 0.125, 0.625, 0.875, 0.833; *Percent:* 12.5%, 37.5%,
87.5%, 100% **37.** 210% **39.** 30% **41.** 19.5% **43.** 140% **45.** 70% **47.** 0.36 **49.** 0.005
51. 0.009 **53.** 0.0523 **55.** 1.82 **57.** 209.67 **59.** 18.9 **61.** 38.4 **63.** 64 **65.** 0.49

67. 100,000 **69.** 11,881,376 **71.** 440 mi, 3.6 hr, 50 mph, 65t mi, 12r mi, $\frac{d}{68}$ hr, $\frac{d}{9}$ mph, rt mi

Preparing for Section 1.2 *(page 13)*
73. 5.205 **75.** 0.9384 **77.** sum **79.** 5, 3

EXERCISES 1.2 *(page 17)*

1. 1024 **3.** 2.89 **5.** 22 **7.** $\frac{3}{40}$ **9.** 0 **11.** $\frac{3}{8}$ **13.** 1.21 **15.** 8.75 **17.** 83,569

19. 11.18 **21.** 8 **23.** 2 **25.** $8\frac{1}{3}$ **27.** 99 **29.** 8.2 **31.** 0.357 **33.** 18 **35.** 0.3

37. 100 **39.** c **41.** a **43.** b **45.** a **47.** 288 **49.** b **51.** a **53.** b

Preparing for Section 1.3 *(page 19)*

55.   **57.** 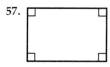 **59.**

EXERCISES 1.3 *(page 28)*

1. 35 in.^{2}, 24 in. **3.** 20 m^2 or 200,000 cm^2, 18 m or 1800 cm **5.** $\frac{9}{16}$ in.^{2}, 3 in. **7.** 1.0 m^2, 4 m

9. 207 in.^{2}, 68 in. **11.** 50.7 ft^2, 31.6 ft **13.** $17\frac{1}{2}$ in.^{2}, $17\frac{1}{2}$ in. **15.** 59.2 ft^2 **17.** 2.6 cm^2

19. $\frac{11}{16}$ yd^2 or $6\frac{3}{16}$ ft^2 **21.** 20 in. **23.** $6\frac{1}{2}$ ft **25.** 13 cm **27.** $4\frac{1}{10}$ ft **29.** 25.1 cm, 50.3 cm^2
31. 20.4 in, 33.2 in.^{2} **33.** 8.8 yd, 6.2 yd^2 **35.** 41.9 ft, 139.6 ft^2 **37.** 84.8 in., 572.6 in.^{2}
39. 120 ft^3, 148 ft^2 **41.** 1747.9 in.^{2}, 4804.4 in.^{3} **43.** 30.1 ft^2 or 4338 in.^{2}, 10.7 ft^3 or 18,468 in.^{3}
45. 54.9 m^2 or 549,200 cm^2, 26.4 m^3 or 26,392,500 cm^3 **47.** 4188.8 cm^3, 1256.6 cm^2 **49.** 7 in.
51. 7.6 ft **53.** 1.4 ft **55.** 39 ft^2 **57.** 21 cm^2 **59.** 21 cm^2 **61.** 5400 yd^2

63. 250 cm, 2,090 cm², 1744 cm²   **65.** 300 ft²   **67.** 448.9 in.³, 283.5 in.²   **69.** 14,834 cm³ or 3355.1 m²
71. 3.2 cm   **73. a.** 120 ft²   **b.** 0.5 gallons   **c.** 2 gallons   **d.** Not enough information is given.
We need to know the woman's rate of work.

Preparing for Section 1.4   *(page 32)*
75. A variable is a symbol (usually a letter) that represents any number or sometimes particular unknown
numbers.   **77.** 72   **79.** 39   **81.** 60

EXERCISES 1.4   *(page 40)*

1. 3, 7, 5, 1, 1, 8, 2, 7   **3.** 7   **5.** $1\frac{7}{12}$ **7.** $21\frac{1}{3}$ **9.** 65   **11.** 40.936

13. 8, commutative property for multiplication   **15.** 6, distributive property   **17.** $0.9n$, commutative
property for addition   **19.** 7, distributive property   **21.** 3, commutative property for multiplication
23. 4, distributive property   **25.** 7, associative property for addition   **27.** 3, distributive property

29. 8.4, distributive property   **31.** $4x + 24$ **33.** $10 + 6x$ **35.** $8x + 4$ **37.** $\frac{3}{10}x^2 + \frac{1}{5}x$ **39.** $8xy + 8x^2$

41. $11x + 2$ **43.** $8 + 12y$ **45.** $3x + 11.5$ **47.** $3x^2 + 3x$ **49.** $\frac{5}{8}x + \frac{11}{60}$ **51.** $24x^5 + 28$

53. $24x^2 + 40x + 60$ **55.** $16x + 40$ **57.** $35 + 10x + 5x^2$ **59.** $16x + 10$ **61.** $12x^2 + 2x + 2$
63. $28x + 174$ **65.** $36x^2 + 36x + 160$ **67.** $5x$ **69.** $x + 1$ **71.** $2x + 6$ **73.** $(x + 2)(x + 7)$
75. $2h + f + d$, $500 - (2h + f + d)$ **77.** $28t$

Preparing for Section 1.5   *(page 42)*

79. 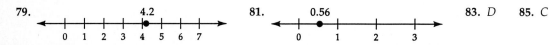   **81.**   **83.** *D*   **85.** *C*

EXERCISES 1.5   *(page 48)*

1. *D*   **3.** *J*   **5.** *C*   **7.** *E*

9.

x	0	1	2	3	4
y	0	1	2	3	4

11.

x	0	1	2	3	4
y	0	2	4	6	8

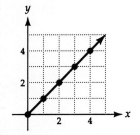

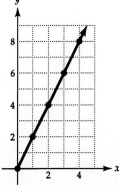

13.

x	0	1	2	3	4
y	10	9	8	7	6

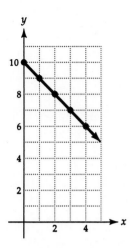

15.

x	0	1	2	3	4
y	0	1	4	9	16

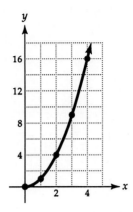

17. Figure 1.32   **19.** Figure 1.34   **21.** Figure 1.35   **23.** $y = x$ **25.** $y < x$ **27.** About 1.6

29. A

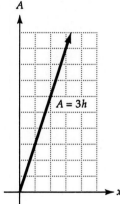

$A = 3h$

31. a. d **b.** 57 mph

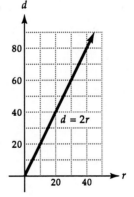

$d = 2r$

33. 100 ft   **35.** 2.5 seconds   **37.** $A = 2b^2$ **39.** About 1.7, about 2.2, about 2.6, about 3.2

41.

x	0	2	4	6	8	10	12
y	12	10	8	6	4	2	0

Preparing for Section 2.1   *(page 51)*
43. a. +1   **b.** −1   **c.** +12   **d.** −6

45. a. 8°   **b.** 8°   **c.** 12°   **d.** 12°

CHAPTER REVIEW   *(page 53)*

1.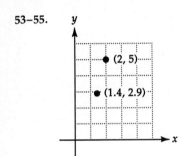

$\frac{2}{9}$ 1.26 $\frac{8}{5}$ π

3. 0.00032   **5.** $\frac{27}{40}$ **7.** 12   **9.** 98.58   **11.** 59.1

13. $\frac{173}{500}$ **15.** $1.32, $.48, $1.56, 8, $1.08, $2.28, 12, $2.16, m, $1.08 + (m - 3)(.24)$, $(x - 3)(.24)$, $1.08 + (x - 3)(.24)$

17. 34   **19.** 38   **21.** 13   **23.** c   **25.** 42 in.^{2} **27.** 63 ft^2 or 7 yd^2 **29.** 8 m

31. 29.8 cm^3, 57.7 cm^2 **33.** 10 in.   **35.** 134 in.^{2}, 134 in.^{2} **37.** 20 ft   **39.** 6,048 cm^2, 72,576 cm^3

41. 1   **43.** 6, commutative property for addition   **45.** $3x$, commutative property for addition

47. $\frac{5}{2}x + \frac{7}{4}$ **49.** $2r \div (r + 1)$ **51.** $n(n + 2)$

53–55.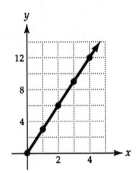

57.

x	0	1	2	3	4
y	0	3	6	9	12

59.

x	0	1	2	3	4
y	3	4	7	12	19

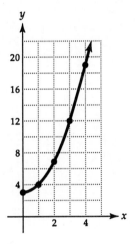

61. yes   **63.** yes   **65.** About 0.7 units

Chapter 2

EXERCISES 2.1   *(page 64)*

1. $\left|-\frac{1}{4}\right| = \left|\frac{1}{4}\right| = \frac{1}{4}$, $|1.5| = |-1.5| = 1.5$, $|3| = |-3| = 3$

3. $|-3.5| = 3.5$,   **5.** $|0| = 0$,

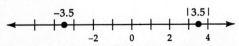

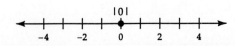

7. $|-\pi| = \pi$    **9.** $-12°F$ **11.** $+4$ yards   **13.** -43.6 m

15. -10 points   **17.** $8\frac{1}{3}$ **19.** $-61\frac{2}{9}$ **21.** 0   **23.** 16   **25.** 11   **27.** $\frac{8}{3}$ or $2\frac{2}{3}$ **29.** 3   **31.** 2

33. 3   **35.** 2   **37.** 2   **39.** 15   **41.** 5   **43.** 6   **45.** 8   **47.** 8   **49.** 2   **51.** $\frac{1}{2}$ **53.** 10

55. 52

Preparing for Section 2.2   *(page 65)*

57. $\frac{13}{10}$ or $1\frac{3}{10}$ **59.** 2.05

61.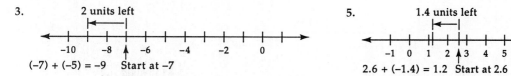

63.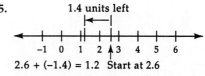

EXERCISES 2.2   *(page 73)*

1.

3. 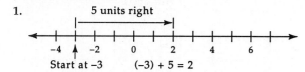   **5.**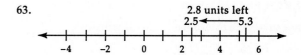

7. $-\frac{5}{7}$ **9.** -12 **11.** $-\frac{5}{5}$ or -1 **13.** -18 **15.** -2 **17.** -30 **19.** $\frac{1}{8}$ **21.** -712 **23.** 2.03

25. -1 **27.** 4   **29.** 6   **31.** 4   **33.** 132   **35.** 54   **37.** 58.6   **39.** -421.34

41. 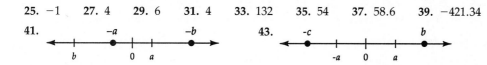   **43.**

45. $500 + (-120) = 380$; going up 380 feet   **47.** $(-50) + 40 = -10$; withdrawing \$10
49. $(-13) + 5 + (-8) = -16$; temperature falling 16°   **51.** $(-12) + t + 2t = 3t - 12$; temperature
changing $(3t - 12)°$ **53.** 17 yards gained   **55.** yes, \$689   **57.** No. The equation
should be 280 ft + 1350 ft = 1630 ft because the balloon rose 280 ft to get out of Death Valley.

Preparing for Section 2.3   *(page 75)*

59. 5, 5; equal   **61.** $\frac{5}{12}$, $\frac{5}{12}$; equal

63.

65.

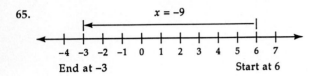

67.

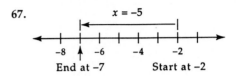

EXERCISES 2.3   *(page 80)*

1. negative   **3.** negative   **5.** negative

7.

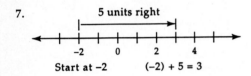

9.

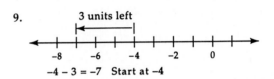

11.

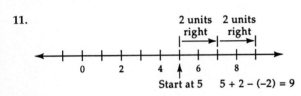

13. $|-3 - (-5.5)| = |2.5| = 2.5$

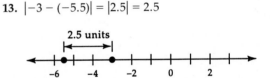

15. $|4.2 - (-4.3)| = |8.5 = 8.5$

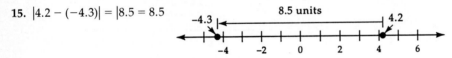

17. -1 **19.** -55 **21.** $-\dfrac{2}{3}$ **23.** $-\dfrac{9}{17}$ **25.** -1352 **27.** 28   **29.** 2,583   **31.** 16   **33.** 24.9

35. -5 **37.** 1   **39.** -9 **41.** 3   **43.** -9 **45.** -13 **47.** 8   **49.** 21.8   **51.** -1156
53. $-1,668$ **55.** 10   **57.** 93, 93   **59.** $-259.14°C$ **61.** 3521 m above sea level   **63.** $-\$57, \$73$

Preparing for Section 2.4   *(page 82)*
65. a. 1, 1   **b.** 4, 4   **67. a.** 8, 8   **b.** 5, 5   **69.** $(5)(2) + (5)(-2)$

EXERCISES 2.4   *(page 88)*

1. -120 **3.** -63 **5.** $\dfrac{2}{3}$ **7.** -13.86 **9.** -1.4 **11.** $\dfrac{39}{8}$ or $4\dfrac{7}{8}$ **13.** -144 **15.** -20.976

17. -4 **19.** -40 **21.** -7.2 **23.** 12   **25.** 6   **27.** -14 **29.** $-\dfrac{64}{7}$ or $-9\dfrac{1}{7}$ **31.** 0   **33.** 0.47

35. 12,129.214 **37.** $-\dfrac{10}{19}$ **39.** $\dfrac{5}{9}$ **41.** 1 **43.** 4 **45.** 4 **47.** 16 **49.** $-\dfrac{1}{2}$ **51.** 3 **53.** -1

55. $\dfrac{1}{3}$ **57.** -48 **59.** -28 **61. a.** $-4°F$ **b.** $24°F$ **c.** $38°F$

Preparing for Section 2.5 *(page 89)*
63. $5x + 5$ **65.** $20s + 14$ **67.** $3(x + y)$

EXERCISES 2.5 *(page 94)*

1. $-x$ **3.** $-(x + y)$ **5.** $3x + 2$ **7.** $8x - 3y$ **9.** $-4; -4$ **11.** $-2; -2$ **13.** $-14; -14$ **15.** $16x$

17. $2d + 3$ **19.** $z + 5$ **21.** $0.4a + 2.7b$ **23.** $\dfrac{7}{9}u + \dfrac{1}{2}v$ **25.** $3x^2 - 2x - 3$ **27.** $12n - 6$

29. $5k^3 - 3k - 3$ **31.** $12q^2 - 5q - 4p$ **33.** $-12w + 8$ **35.** $2g + 12$ **37.** $-mn + n + 1$ **39.** $-n + 8$
41. $-4.9664x^2 - 21.8652$ **43.** $-0.12z + 1.1$ **45.** $6xy - 22$ **47.** $-13z + 2$ **49.** $15w + 15$

51. $32x - 132$ **53. a.** n **b.** $2n$ **c.** $2n - 100$ **d.** $\dfrac{1}{2}(2n - 100)$ **e.** $n - \dfrac{1}{2}(2n - 100)$

55. $n - n + 50 = 50$ **57. a.** $\dfrac{1}{8}x$ **b.** $\dfrac{1}{12}y$ **59. a.** $7.25\left(\dfrac{7}{16}x\right)$ **b.** $16\left(\dfrac{11}{12}y\right)$

Preparing for Section 3.1 *(page 96)*

61.

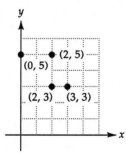

63–69.

CHAPTER REVIEW *(page 97)*

1.

3. -2 **5.** $-\dfrac{19}{8}$ or $-2\dfrac{3}{8}$ **7.** 8.6 **9.** 16.4

11. a loss of 4 pounds; no change in weight **13.** 1.9 **15.** -3.09 **17.** $-9\dfrac{3}{10}$ **19.** 75 **21.** 37.464

23. 11 **25.** 23 **27.** $-\dfrac{41}{210}$ **29.** 15,743 **31.** -6 **33.** 7 **35.** 305.58 **37** $12°$ **39.** $\dfrac{9}{16}$

41. 0 **43.** $-\dfrac{6}{41}$ or about -0.1463414 **45.** 189 **47.** -33 **49.** 12.6 **51.** 10 yards gained

53. $-16x + 50$ **55.** $4c - 6$ **57.** $-13z + 2$ **59.** $\dfrac{5}{2}x - \dfrac{29}{36}$ **61.** $322x + 33702$ **63.** $4w - 8$

65. $((-6) + x + 5 + 3x) \div 4;\ x - \dfrac{1}{4}$

Chapter 3

EXERCISES 3.1   *(page 107)*

1–15.

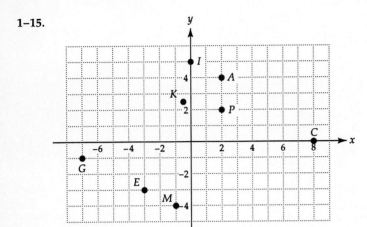

17. *D*   **19.** *A*   **21.** *G*   **23.** *J*   **25.** *C*

27. $x = -2$ **29.** $x = 10$

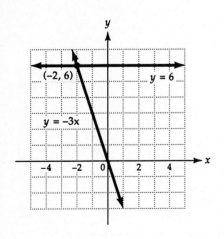

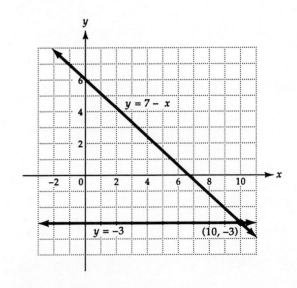

31. $x = 1$

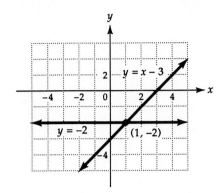

33. $x = 1$

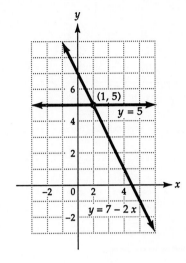

35. $x = 2$

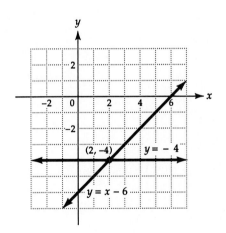

37. $x = -\dfrac{3}{2}$

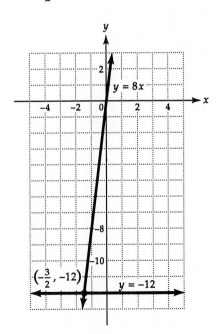

39. a.

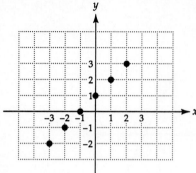

b. $y = x + 1$

41. a.

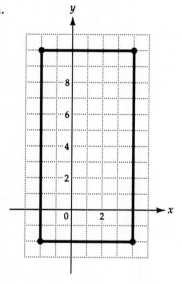

b. rectangle   **c.** $A = 78$, $p = 38$

43.

45. Points with negative coordinates make no sense, since time and distance must be nonnegative.
47. 20 km   **49.** $d = 16t$

Preparing for Section 3.2   *(page 108)*

51. $2(5) + 6 \neq 10 - 3(5)$ since $16 \neq -5$; no

53. $-3.2, -3, -2.9, 0, 2\frac{3}{4}, 2.8, 3.1, \pi$

EXERCISES 3.2   *(page 115)*

1. a. $2 = 10 - x$ **b.** $x - 8 = 0$ **c.** $2 - x = 10 - 2x$ **d.** $x = 8$; d is simplest.

3. $x = \dfrac{41}{6}$

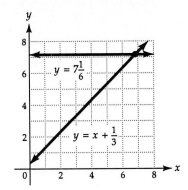

5. $x = \dfrac{8}{3}$

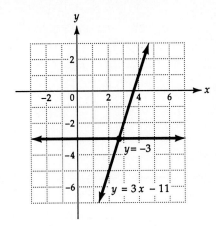

7. 22   **9.** -5 **11.** -6.8 **13.** 3   **15.** $\dfrac{1}{2}$ **17.** 3   **19.** 8   **21.** -9 **23.** -9 **25.** 0.17

27. 13   **29.** 5   **31.** 10   **33.** 9   **35.** $y + 11 = 25; 14$ **37.** $2x = x - 8; -8$ **39.** $9x + 1 = 8x - 1; -2$

41. $L - 12 = 142; 154$ lb   **43.** $x + 1.13 = 4.21; 3.08$

Preparing for Section 3.3   *(page 116)*

45. -48 **47.** 24   **49.** 3   **51.** -15

EXERCISES 3.3   *(page 120)*

1. a. $8x + 48 = 0$ **b.** $x = -6$ **c.** $8x - 8 = 56$ **d.** $-\dfrac{1}{6}x = 1$ **b** results in the simplest equation

3.
$$5 - 3b = -2.5 \qquad \text{Subtracting } b \text{ from both sides}$$
$$-3b = -7.5 \qquad \text{Subtracting 5 from both sides}$$
$$b = 2.5 \qquad \text{Dividing both sides by } -3$$

5. -7 **7.** 4   **9.** $\dfrac{13}{8}$ or $1\dfrac{5}{8}$ **11.** 6   **13.** No solution   **15.** $-\dfrac{7}{3}$ or $-2\dfrac{1}{3}$ **17.** 7   **19.** 2   **21.** 8

23. No solution   **25.** -2 **27.** -5 **29.** $\dfrac{11}{4}$ or $2\dfrac{3}{4}$ **31.** -1 **33.** $\dfrac{8}{3}$ or $2\dfrac{2}{3}$

35. $x = 1$

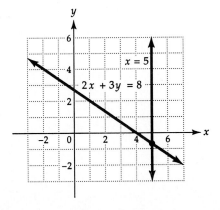

37. $y = 3.5$

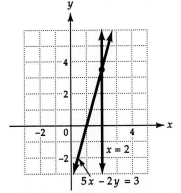

39. $\frac{1}{4}x + 6 = 0$; $-24°$ **41.** $2w + 2(2w + 55) = 350$; length $= 135$ ft, width $= 40$ ft

Preparing for Section 3.4   *(page 122)*

43. $-6, -4, 0, 3, 7$ **45.** $\frac{2}{3}, \frac{3}{4}, \frac{4}{5}, \frac{4}{3}, \frac{3}{2}$ **47.** 2, 3, 4, for example

EXERCISES 3.4   *(page 128)*

1. $x < -2$;
3. $x \le 7$;
5. $n < 10$;

7. $w \ge 7$;
9. $u > 7$;
11. $y < 5$;

13. $v \le -7.5$;
15. $s > -4.9$;

17. $h < -12.5$;

19. $3x - 4 + 2x > 4 - 8x + 5$ Removing parentheses
　　　$5x - 4 > -8x + 9$ Combining like terms
　　　$13x - 4 > 9$ Adding $8x$ to both sides
　　　$13x > 13$ Adding 4 to both sides
　　　$x > 1$ Dividing both sides by 13

21. $k > 5$ **23.** $y < \frac{13}{11}$ **25.** $t < \frac{9}{10}$ **27.** $y < -5\frac{1}{4}$ **29.** $q \le 2$ **31.** $z \ge -18$

Preparing for Section 3.5   *(page 129)*

33. 6   **35.** $3\frac{1}{7}$ **37.** $l = \frac{A}{W}$

EXERCISES 3.5   *(page 133)*

1. $x = 10 - y$ **3.** $x = \frac{1}{2}(y + 41)$ **5.** $a = \frac{1}{5} + \frac{1}{4}b$ **7.** $y = \frac{2}{15}x - \frac{7}{5}$ **9.** $x = \frac{1}{a}(c - by)$ **11.** $x = \frac{1}{m}(y - b)$

13. $y = \frac{12x + 2}{3x^2}$ **15.** $a = \frac{bx}{b - y}$ **17.** $m = \frac{y - a}{x - b}$ **19.** $b = \frac{mx - y + a}{m}$ **21.** $x = \frac{ry}{5s}$ **23.** $r = \frac{5xs}{y}$

25. a. $s = \frac{P}{4}$ **b.** 36   **27. a.** $W = \frac{A}{l}$ **b.** 5   **29. a.** $r = \frac{C}{2\pi}$ **b.** 12.73   **31. a.** $a = \frac{2A - hb}{h}$ **b.** 32

33. a. $R = \frac{E}{I}$ **b.** 38   **35. a.** $m = \frac{E}{d^2}$ **b.** 2   **37. a.** $h = \frac{A - 2\pi r^2}{2\pi r}$ **b.** 3.96   **39.** 17 miles per gallon

41. \$279.96   **43.** Acme Rentals; \$70.71

Preparing for Section 3.6   *(page 134)*
45. 0.483   **47.** 31.2   **49.** $x = 13.5$ **51.** 16-ounce can

EXERCISES 3.6 *(page 142)*

1. $\dfrac{3}{8}$ **3.** $\dfrac{9}{5}$ **5.** $\dfrac{5}{14}$ **7.** $2\dfrac{68}{105}$ **9.** $\dfrac{1}{3}$ **11.** 2.5 **13.** 17.33 **15.** 0.36 **17.** −2.36 **19.** −7.11

21. yes **23.** yes **25.** yes **27.** no **29.** yes **31.** 27 **33.** 27 **35.** $\dfrac{8}{3}$ **37.** 4.33 **39.** 2.47

41. $\dfrac{1}{2}(m+n)$ **43.** $\dfrac{a}{b}=\dfrac{y}{x}$; $\dfrac{x}{b}=\dfrac{y}{a}$; $\dfrac{a}{y}=\dfrac{b}{x}$; $\dfrac{x}{y}=\dfrac{b}{a}$ **45. a.** 2:3; 1:4 **b.** 3:7 **c.** no; $\dfrac{2}{3}+\dfrac{1}{4}=\dfrac{11}{12}$ **d.** 0.429

47. 3.125; 3.75; 4; 3.6 **49.** 7.5 **51.** about 337 **53.** $5150

Preparing for Section 3.7 *(page 144)*
55. a. $2x$ or $45-x$ **b.** $2x=45-x$ **c.** 15

EXERCISES 3.7 *(page 151)*

1. Suppose there were 14 apples in each of the three original piles. That would be $3\cdot14$ or 42 apples in all. If the man threw 8 away, that would leave $42-8$ or 34 apples, and that makes 2 piles of 17 apples each.
3. Suppose the flagpole is 13 meters long, and the height of the building is $4\cdot13$ or 52 meters. Then the top of the pole would be $13+52$ or 65 meters above the level of the ground. **5.** a, b, c, e, f **7. a.** $n+6$ **b.** $5n$
c. $10(n+6)$ **d.** $5n+10(n+6)$ or $15n+60$ **e.** $15n+60=225$; 11 **f.** 11 nickels and 17 dimes
g. Eleven nickels are worth 55 cents and 17 dimes are worth $1.70. So the value of the collection would be

$1.70 + $.55 or $2.25. **9. a.** 55 mph **b.** 60 mph **c.** 1040 miles **11.** $4\dfrac{1}{6}$ hours **13.** 140 miles **15.** 72

17. 8 hours, 57 minutes of darkness; 15 hours, 3 minutes of daylight **19.** 1: 12 million Kw; 2: 6 million Kw; 3 and 4 each: 5 million Kw **21.** $9011.76 **23.** $396 **25.** 35 games **27.** 60 cents **29.** 35°, 70°, 75°
31. $10,000, $1000 **33.** 4 hours after takeoff

Preparing for Section 4.1 *(page 154)*
35. 6; 5 **37.** 243 **39.** 5 **41.** 98

CHAPTER REVIEW *(page 156)*

1–7.

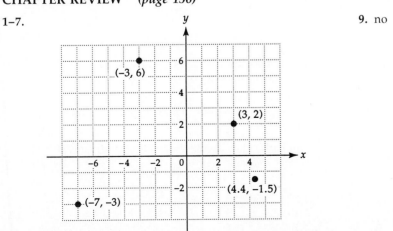

9. no

11. $x = 3\dfrac{1}{2}$

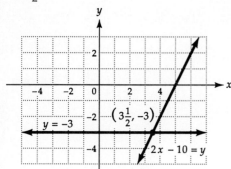

13. $x = 5.5$

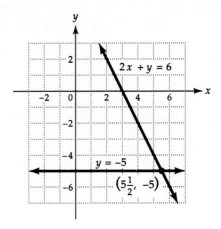

15. $x = 3\dfrac{1}{16}$

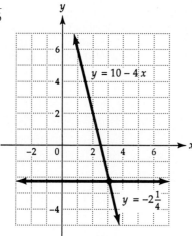

17. 10   **19.** $\dfrac{7}{6}$ or $1\dfrac{1}{6}$ **21.** 18   **23.** $6\dfrac{1}{8}$

25. $2\dfrac{8}{13}$ **27.** $2x + 7 = 19$; 6   **29.** $8x = 40$ Adding 3 to both sides   **31.** -2 **33.** $1\dfrac{6}{29}$ **35.** $-3\dfrac{5}{6}$
$x = 5$ Dividing both sides by 8

37. $1\dfrac{22}{37}$ **39.** -13 **41.** 42.02 in.   **43.** $x \geq 4.7$ ←———⊕———→   **45.** remove parenthesis,
**4.7**

combine like terms, add $10x$ to both sides, subtract 8 from both sides, divide both sides by 14, simplify

47. $x > -1$ **49.** $x \leq -\dfrac{4}{7}$ **51.** $x \geq 7.04$ **53. a.** $a = \dfrac{Mg - T}{M}$ **b.** 30   **55. a.** $x = \dfrac{21 - 2y}{5}$ **b.** $3\dfrac{4}{15}$

57. a. $y = \dfrac{-c - ax}{b}$ **b.** $y = \dfrac{2a - c}{b}$ **59.** 140 ft   **61.** $\dfrac{5}{3}$ **63.** $11\dfrac{1}{4}$ **65.** $-6\dfrac{1}{4}$ **67.** 6 cans

69. $r = 24.5$ cm, $A = 1887.26$ cm^2 **71.** \$3600   **73.** 6 miles

Chapter 4

EXERCISES 4.1   *(page 168)*

1. 5^3; base 5, exponent 3   **3.** -8^2; base 8, exponent 2   **5.** $-(-6)^5$; base -6, exponent 5
7. x^6; base x, exponent 6   **9.** $-(-m)^2$; base $-m$, exponent 2   **11.** 8   **13.** -243 **15.** -16 **17.** -9
19. 224   **21.** 135   **23.** -2592 **25.** -3888 **27.** $5^5 = 3125$ **29.** $\left(\dfrac{3}{4}\right)^4 = \dfrac{81}{256}$ **31.** $(-8)^9$ **33.** x^8

35. $(-w)^{14}$ **37.** h^9 **39.** n^{21} **41.** 4^6 **43.** $\left(\dfrac{1}{4}\right)^{15}$ **45.** t^{54} **47.** p^{16} **49.** $81x^4$ **51.** $-64z^3$

53. $32p^5q^5$ **55.** $\dfrac{9}{25}h^2k^2$ **57.** $15q^9$ **59.** $-32u^{13}$ **61.** m^{28} **63.** $784x^6$ **65.** $-45p^8$ **67.** $-864m^{22}$
69. $96w^8z^9$ **71.** $-8000r^{21}t^{12}$ **73. a.** $6x^2$ **b.** 216   **75. a.** $52w^2$ **b.** $24w^3$ **c.** $Sa = 1053$, $V = 2187$

Preparing for Section 4.2   *(page 169)*

77. -3 **79.** $-\dfrac{1}{12}$ **81.** $\dfrac{1}{9}$ **83.** -27

EXERCISES 4.2   *(page 179)*

1. 1   **3.** 1   **5.** -1 **7.** 1   **9.** $\dfrac{1}{9}$ **11.** $-\dfrac{1}{64}$ **13.** $-\dfrac{1}{7}$ **15.** 625   **17.** $-\dfrac{243}{32}$

19. approximately 0.055   **21.** $\dfrac{15}{k^5}$ **23.** $-\dfrac{2}{m^8}$ **25.** $\dfrac{s^6}{3}$ **27.** $\dfrac{2}{u} - \dfrac{1}{2u} = \dfrac{3}{2u}$ **29.** $\dfrac{3^2}{8^2} = \dfrac{9}{64}$ **31.** $\dfrac{2^3}{5^3} = \dfrac{8}{125}$

33. $\dfrac{y^4}{2^4} = \dfrac{y^4}{16}$ **35.** $\dfrac{z^5}{w^5}$ **37.** $3^2 = 9$ **39.** $\dfrac{1}{5^5} = \dfrac{1}{3125}$ **41.** $4^3 = 64$ **43.** $\dfrac{1}{h^5}$ **45.** m^9 **47.** $\dfrac{1}{t^2}$ **49.** $\dfrac{v^5}{4}$

51. $-\dfrac{9y^{16}}{2}$ **53.** $\dfrac{1}{729}$ **55.** $\dfrac{1}{m^5}$ **57.** $\dfrac{1}{4}$ **59.** $-\dfrac{25}{3t^{14}}$ **61.** $16u^{16}$ **63.** -27 **65.** $\dfrac{3n^{12}}{m^7}$

Preparing for Section 4.3   *(page 180)*
67. 25   **69.** 0.61   **71.** 0.00729

EXERCISES 4.3   *(page 185)*

1. $7.0 \cdot 10^3$ **3.** $9.246 \cdot 10^{-3}$ **5.** $2.4 \cdot 10^8$ **7.** $5.0 \cdot 10^{-4}$ **9.** $3.8 \cdot 10^{-8}$ **11.** 98,000   **13.** 0.00006
15. 2,478,000,000   **17.** 7.8034   **19.** 0.000001004   **21.** $8 \cdot 10^{15}$ **23.** $2 \cdot 10^4$ **25.** $3.15 \cdot 10^7$
27. $1.4 \cdot 10^{11}$ **29.** $4.371 \cdot 10^{-7}$ **31.** $5.19 \cdot 10^{-15}$ **33.** $6.592 \cdot 10^{-13}$ **35.** $1.16 \cdot 10^8$ **37.** $3.36 \cdot 10^{-7}$
39. $9.64 \cdot 10^{-7}$ **41.** $1.61 \cdot 10^{10}$ miles   **43.** $3.456 \cdot 10^{12}$ computations   **45.** $5.3125 \cdot 10^9$ dollars
47. $3.75 \cdot 10^{-2}$ square kilometers

Preparing for Section 4.4 (page 187)
49. $12x$ **51.** $3w$ **53.** $8k^2$ **55.** $-m^4$

EXERCISES 4.4   *(page 194)*

1. $u^5 + 4u$ **3.** $-4x^5 + 4x^3$ **5.** $9w^6 - 4w^3 + 7w^2$ **7.** $-5p^7 + 10p^4 - 5p^3$ **9.** $4m^6 + 2m^5 - 5m^4 - 3m^2$
11. $-k^9 - 5k^7 + 6k^5$ **13.** $8r^3 - 8r$; degree 3, binomial   **15.** $-h^5$; degree 5, monomial   **17.** $-3x^5 + x^2$;
degree 5, binomial   **19.** $-6p^7 + 11p^4$; degree 7, binomial   **21.** $-2m^{10} - 4m^8 + 7m^3$; degree 10, trinomial
23. $-4r^2s + 8r^2 + 7s$; degree 3, trinomial   **25.** $-2x^2y^5 - 6x^4y^2 + 10xy^3 - 5x^4$; degree 7
27. $7h^3 + 3h$ **29.** $-3x^2 + 8$ **31.** $7m^3 + 5m^2 - m + 6$ **33.** $-5p^5 + 2p^4 - 6p^3 - 3p$
35. $-2t^6 + 4t^5 - 3t^4 - 3t^2 - 1$ **37.** $5m - 3$ **39.** $2x^4 + 11x^3$ **41.** $-8p^3 - 9p^2 + 4p - 18$
43. $15r^8 + 4r^7 - 8r^6 + 6r^5 - r^3 + 6r^2$ **45.** $3m^7 + 2m^6 - 8m^5 + 2m^3 - 4m + 9$ **47.** $10p^2q^2 + 11pq + 5q - 6$
49. $6h^3k + 17h^2k - 4hk - 2hk^2$ **51.** \$18,000   **53.** 5,100 units   **55.** $9x - 2$

57. a. **(i)** $4x$; monomial, degree 1   **59. a.** **(i)** $6z^2$; monomial, degree 2
  (ii) $2x + 2y$; binomial, degree 1   **(ii)** $4xz + 2z^2$; binomial, degree 2
  (iii) $2x + 2y$; binomial, degree 1   **(iii)** $2xy + 2xz + 2yz$; trinomial, degree 2
  (iv) $4y$; monomial, degree 1

   b. **(i)** x^2; monomial, degree 2   **b.** **(i)** z^3; monomial, degree 3
  (ii) xy; monomial, degree 2   **(ii)** xz^2; monomial, degree 3
  (iii) xy; monomial, degree 2   **(iii)** xyz; monomial, degree 3
  (iv) y^2; monomial, degree 2

61. a. 1   **b.** 2   **c.** 3

Preparing for Section 4.5   *(page 197)*
63. $-16y + 12$ **65.** $-28z^3 + 4z - 8$ **67.** $-8q^{12}$ **69.** $14k^{13}$

EXERCISES 4.5   *(page 205)*

1. $-15w^3 + 20w$ **3.** $8p^3 - 16p^2 + 24p$ **5.** $6t^8 - 24t^5 - 12t^3$ **7.** $8x^4y + 14x^3y^2 - 2x^2y^3$
9. $-54w^5z^3 + 27w^5z^4 - 45w^4z^5$ **11.** $h^3 + 7h^2 + 18h + 18$ **13.** $2x^3 - 17x^2 + 36x - 15$
15. $12w^3 - 25w^2 + w + 2$ **17.** $24p^5 - 3p^4 - 17p^3 - 17p^2 - 20p + 28$ **19.** $n^2 + 3n + 2$ **21.** $-h^2 + h + 20$
23. $x^2 - 14x + 48$ **25.** $2w^2 + 9w + 4$ **27.** $4p^2 + 21p - 18$ **29.** $6m^2 - 13m - 5$ **31.** $63r^2 - 50r + 8$
33. $6h^4 + 29h^2 + 35$ **35.** $8x^3 + 42x^2 - 11x$ **37.** $16w^2 + 66wz - 27z^2$ **39.** $x^2 + 16x + 64$ **41.** $4 - p^2$
43. $16m^2 - 40m + 25$ **45.** $4h^2 - 81$ **47.** $36x^2 + 84xy + 49y^2$ **49.** $64p^2 - 48pq + 9q^2$
51. $18x^3 + 129x^2 + 21x$ **53.** $14w^3 + 7w^2 - 252w$ **55.** $5p^3 + 100p^2 + 500p$ **57.** $-36m^3 + 192m^2 - 256m$
59. $150r^3 - 96r$ **61.** $8h^3 + 44h^2 + 68h + 24$ **63.** $42x^3 - 261x^2 + 48x + 36$ **65.** $9w^3 + 3w^2 - 8w - 4$
67. $r^3 + 3r^2 + 3r + 1$ **69.** $h^3 - 12h^2 + 48h - 64$ **71.** $21x^3 + 43x^2 - 11x + 7$ **73.** $3p^2 + \dfrac{1}{2}pq - \dfrac{1}{6}q^2$
75. $\dfrac{4}{25}r^2 - \dfrac{3}{10}rt + \dfrac{9}{64}t^2$ **77.** $12x^2 - 8x$ **79. a.** $34z^2 - 4z - 6$ **b.** $10z^3 + 4z^2 - 6z$ **81.** $5m^2 - m - 3$

Preparing for Section 4.6   *(page 207)*
83. $3x$ **85.** $\dfrac{3}{w^4}$ **87.** $-\dfrac{5p^3}{3}$ **89.** $-\dfrac{3}{5m^3}$

EXERCISES 4.6   *(page 214)*

1. $2x + 6$ **3.** $-5x^4 + 4x^2 - 7x$ **5.** $-2y^2 + 5$ **7.** $-6y^5 + 10y^2 + \dfrac{14}{y}$ **9.** $5z^5 - 7z + 3$

11. $16z^4 - 20z^2 + 11 - \dfrac{9}{z}$ **13.** $-2k^5w + 5k^3$ **15.** $7k^4w^5 - kw^4 + 3w^2$ **17.** $h + 5$ **19.** $q + 2 + \dfrac{2}{q + 3}$

21. $h + 1 + \dfrac{21}{h - 7}$ **23.** $s - 4 - \dfrac{10}{s - 1}$ **25.** $3x + 5$ **27.** $3p - 3 + \dfrac{13}{3p + 2}$ **29.** $u + 3$ **31.** $5k + 3$

33. $3m + 4 + \dfrac{11}{4m - 3}$ **35.** $w^2 + 2w - 4$ **37.** $z^2 + 3z + 1$ **39.** $4p + 3 - \dfrac{2}{p^2 - 1}$ **41.** $4m^2 + 2m + 1$

43. $3h^2 + 2 + \dfrac{2}{3h^2 + 2}$ **45.** $5x^2 + x + 3 + \dfrac{x + 7}{3x^2 - 1}$ **47.** $r^2 + r - 2 + \dfrac{10}{r^2 - r - 2}$

49. $u^2 - 2u + 1 + \dfrac{8}{u^2 + 2u - 4}$ **51.** $x + 5$ **53.** $8w - 2$

Preparing for Section 5.1   *(page 216)*
55. 1, 2, 3, 6, 9, 18   **57.** 1, 2, 3, 6, 7, 14, 21, 42   **59.** $12p^2 + 54p$ **61.** $-18r^5 + 12r^3 + 30r^2$

CHAPTER REVIEW   *(page 217)*

1. x^4; 16   **3.** $(-x)^3$; -8 **5.** y^7 **7.** $81z^2$ **9.** $64t^{21}$ **11.** $81r^{10}$ **13.** $-1125m^{19}n^5$ **15.** -1 **17.** $\dfrac{1}{8}$

19. 50,000   **21.** h^5 **23.** $-\dfrac{3}{4t^4}$ **25.** $\dfrac{3}{2z^2}$ **27.** $\dfrac{81u^{14}}{v^{22}}$ **29.** $3.8 \cdot 10^{-6}$ **31.** $4.057 \cdot 10^{13}$ **33.** $1.68 \cdot 10^8$

35. $3.7625 \cdot 10^{23}$ **37.** 83,000   **39.** $1.34 \cdot 10^8$ seconds   **41. a.** $-x^5 + 6x^4$ **b.** degree 5, binomial
43. a. $2w^2$ **b.** degree 2, monomial   **45.** $5z^2 - 5z - 1$ **47.** $4m^7 - m^6 + 14m^5 - 13m^4 - 9m^2 + 3m - 1$
49. $12u^3v - uv^2 - 3v^4$ **51.** $63h^8 - 21h^5 + 35h^3$ **53.** $3r^2 + 29r + 40$ **55.** $3x^2 - 5xy - 2y^2$
57. $4z^2 - 28z + 49$ **59.** $8w^3 - 44w^2 - 4w + 120$ **61.** $18u^7 - 12u^6 + 57u^5 - 42u^4 - 93u^3 + 48u^2 + 24u$

63. $8n^4 - 6n^2 + 10n$ **65.** $2h - 3$ **67.** $x - 1 - \dfrac{1}{4x - 3}$ **69.** $3q^2 + 2q - 5 + \dfrac{2q - 5}{q^2 - 2}$ **71.** $3z - 1$

Chapter 5

EXERCISES 5.1   *(page 229)*

1. $2^2 \cdot 3^2$ **3.** $3^2 \cdot 7$ **5.** $3 \cdot 5 \cdot 7$ **7.** $3^2 \cdot 5 \cdot 7$ **9.** 4   **11.** 12   **13.** 1   **15.** $2(u - 4)$ **17.** $7(3z - 2)$
19. $p(3p + 1)$ **21.** $2r^4(9r^5 + 1)$ **23.** $5h^3(4 - 5h)$ **25.** $2(x^2 + 8x + 14)$ **27.** $z(3z^2 - 4z + 6)$
29. $8v^3(4 + 2v - 3v^2)$ **31.** $7p^2(2p^2 + 3p - 1)$ **33.** $11(3m^2 - 11n^2)$ **35.** $16xy^2(xy + 5)$
37. $9pq(4p - 8q + 9)$ **39.** $7h^4k^2(7k^4 - 3hk^2 - 5h^2)$ **41.** $(y - 1)(3x + 7)$ **43.** $p(q^2 - 7)(p^2 - 3)$
45. $(3y + 4)(y^2 + 3)$ **47.** $(2z - 3)(5z^2 + 2)$ **49.** $(3p + 2)(6p - 5)$ **51.** $(r - 2)(7r - 6)$
53. $(2x - y)(4x + 3y)$ **55.** $(5p + q)(p - 10q)$ **57.** $2x(2h + x)$

Preparing for Section 5.2   *(page 229)*
59. $x^2 + 10x + 21$ **61.** $z^2 - 8z - 9$ **63.** $p^2 - 4p - 12$

EXERCISES 5.2   *(page 234)*

1. $(w + 3)(w + 5)$ **3.** $(u + 9)(u - 1)$ **5.** prime polynomial   **7.** prime polynomial   **9.** prime polynomial
11. $(h - 3k)(h - 4k)$ **13.** $(x + 4y)(x - 9y)$ **15.** $(s - t)(s - 12t)$ **17.** prime polynomial   **19.** $(p - 8q)(p - 8q)$
21. $2(q + 2)(q + 4)$ **23.** $r^4(r + 4)(r - 3)$ **25.** $x^2(x + 1)(x + 1)$ **27.** $z(z + 9)(z - 3)$ **29.** $4(u - 4)(u - 8)$
31. $2t^3(t + 5)(t - 3)$ **33.** $5h^6(h + 4)(h + 5)$ **35.** $9t^5(t - 1)(t - 1)$ **37.** $y(x^2 + 6xy - 18y^2)$
39. $6p^2q(q - 2)(q - 6)$ **41.** $3uv(u^2 + 5uv - 32v^2)$ **43.** $(w + z)(p + 8)(p + 9)$ **45.** $(r^2 + 2)(r^2 + 3)$

Preparing for Section 5.3   *(page 234)*
47. $2x^2 + 7x + 3$ **49.** $4w^2 + 31w - 45$ **51.** $15p^2 - 17p - 42$

EXERCISES 5.3   *(page 239)*

1. $(2w + 1)(w + 2)$ **3.** $(7n - 2)(n - 2)$ **5.** prime polynomial   **7.** $(4h - 3)(h + 5)$ **9.** $(5z + 2)(2z + 3)$
11. prime polynomial   **13.** $(7m - 5)(2m + 3)$ **15.** $(5x - 3y)(2x - 3y)$ **17.** $(7m - 3n)(2m + 5n)$
19. $x^2(x + 1)(x + 1)$ **21.** $w^3(w + 3)(w - 8)$ **23.** $4(m - 2)(m - 7)$ **25.** $8(h^2 - 2h + 5)$
27. $5r(r + 1)(r + 13)$ **29.** $2(2q + 3)(q - 1)$ **31.** $y^2(3y + 1)(3y + 1)$ **33.** $2k(3k + 5)(k - 4)$
35. $2s^4(5s + 2)(s - 1)$ **37.** $pq(4q + 1)(2q + 3)$ **39.** $4r^2s^7(2r - 3s)(4r - s)$

Preparing for Section 5.4   *(page 240)*
41. $x^2 + 10x + 25$ **43.** $z^2 - 36$ **45.** $49p^2 - 28p + 4$ **47.** $4m^2 + 36mn + 81n^2$

EXERCISES 5.4   *(page 244)*

1. $(w + 4)^2$ **3.** $(m + 8)^2$ **5.** prime polynomial   **7.** $(u - 11)^2$ **9.** $(3r - 2)^2$ **11.** $(7y + 1)^2$
13. $(9m + 1)^2$ **15.** $2(3h + 4)^2$ **17.** $2m^2(5m + 2)^2$ **19.** $4(2wz + 5)^2$ **21.** $(2t - 7r)^2$ **23.** $(4p + 5q)^2$
25. $3n(2m - 3n)^2$ **27.** $6s^4t^3(2s + t)^2$ **29.** $(w + 3)(w - 3)$ **31.** $(p + 7)(p - 7)$ **33.** $(k + 15)(k - 15)$

35. $(5z + 1)(5z - 1)$ **37.** $16(x + 2)(x - 2)$ **39.** $5v(v + 6)(v - 6)$ **41.** prime polynomial
43. $9(2p + 5q)(2p - 5q)$ **45.** $4(t + 7r)(t - 7r)$ **47.** $6y(2x + 3)(2x - 3)$ **49.** $8r^2s(2r + s)(2r - s)$
51. $6pq(p + 5q)(p - 5q)$ **53.** $(n^2 + 12)(n^2 - 12)$ **55.** $(4y^2 + 1)(2y + 1)(2y - 1)$ **57.** $3(r^2 + 9)(r + 3)(r - 3)$

Preparing for Section 5.5 *(page 245)*

59. $n = -5$ **61.** $p = \dfrac{3}{2}$ **63.** $u = -2$ **65.** $k = \dfrac{12}{7}$

EXERCISES 5.5 *(page 252)*

1. $x = 0, -3$ **3.** $w = -6, 4$ **5.** $p = \dfrac{1}{2}, 5$ **7.** $h = \dfrac{8}{5}, -\dfrac{3}{2}$ **9.** $r = 0, -4, -7$ **11.** $m = 0, 4, \dfrac{5}{2}$

13. $z = -2, -3$ **15.** $p = 6, -4$ **17.** $m = \dfrac{2}{3}, -\dfrac{1}{2}$ **19.** $r = \dfrac{5}{3}, -4$ **21.** $q = -4$ **23.** $y = \dfrac{5}{2}, -2$

25. $x = \dfrac{2}{5}$ **27.** $p = \dfrac{5}{4}, -\dfrac{5}{4}$ **29.** $n = 0, 8$ **31.** $h = 0, 12$ **33.** $t = 0, -20$ **35.** No solution by factoring

37. $x = 4, -\dfrac{5}{2}$ **39.** $q = \dfrac{5}{2}, -\dfrac{2}{3}$ **41.** $r = 5, -3$ **43.** No solution by factoring **45.** $s = 0, -11$

47. $y = 0, 4, -2$ **49.** 4, 7 or $-4, -7$ **51.** 6, 24 or $-6, -24$ **53. a.** After 3 seconds
b. After 6 seconds **55.** 100 or 200 videotapes **57.** $8 or $14 **59.** Length = 7 cm, width = 4 cm
61. Base = 20 in., height = 10 in. **63.** Length = 5 in., width = 3 in., height = 1 in. **65.** 3 ft **67.** 5

Preparing for Section 6.1 *(page 254)*

69. $-\dfrac{1}{2}$ **71.** $\dfrac{9}{2}$ **72.** $\dfrac{3y^3}{5}$ **75.** $\dfrac{2p^2}{3q^5}$

CHAPTER REVIEW *(page 255)*

1. $2 \cdot 3^2 \cdot 13$ **3.** 38 **5.** $7x(2x + 5)$ **7.** $3(3z^2 + z + 2)$ **9.** $9p^2(4p^2 - 3p + 2)$ **11.** $(v + 7)(v^2 + 1)$
13. $(3r + 4)(2r - 5)$ **15.** $(s + 1)(s + 13)$ **17.** $(k + 8)(k - 5)$ **19.** $(h + 12)(h - 7)$ **21.** $3(w + 3)(w - 18)$
23. $(2x - 5)(x - 6)$ **25.** $(y + 1)(4y - 3)$ **27.** $(3k + 5)(2k - 9)$ **29.** $s^2(3s + 2)(7s + 10)$ **31.** $(t - 9)^2$
33. $(2k + 7)^2$ **35.** $4(3x + 2y)^2$ **37.** $3m(3mn - 4)^2$ **39.** $(10 + u)(10 - u)$ **41.** $12t(2s + 3t)(2s - 3t)$

43. $n = -3, 2$ **45.** $m = 6, -8$ **47.** $k = -3, -\dfrac{1}{4}$ **49.** $s = \dfrac{9}{5}$ **51.** $q = 4, -4$ **53.** $x = 0, \dfrac{1}{10}, -\dfrac{1}{10}$

55. 11, 12 or $-11, -12$ **57. a.** After 1 second and after 2 seconds **b.** After 3 seconds
59. Length = 13 cm, width = 6 cm **61.** Width = 1 in.

Chapter 6

EXERCISES 6.1 *(page 263)*

1. a. $\dfrac{1}{2}$ **b.** $\dfrac{3}{2}$ **3. a.** $\dfrac{4}{5}$ **b.** $-\dfrac{4}{3}$ **5. a.** Not defined **b.** Not defined **7 a.** $\dfrac{4}{11}$ **b.** $\dfrac{8}{5}$

9. $x = 1$ **11.** $z = -\dfrac{3}{2}$ **13.** $p = 2, 4$ **15.** Defined for all real numbers **17.** $\dfrac{3}{h}$ **19.** $\dfrac{2x^3}{3}$

21. $\dfrac{4p^2q^2}{5}$ **23.** 4 **25.** $\dfrac{1}{3}$ **27.** $\dfrac{3}{7}$ **29.** $\dfrac{2k}{k + 6}$ **31.** $\dfrac{r - 2}{r(r - 1)}$ **33.** -1 **35.** $-(p + 5)$

37. $\dfrac{1 + w^2}{w^2 - 1} = \dfrac{1 + w^2}{(w + 1)(w - 1)}$ is in lowest terms   **39.** $\dfrac{1}{h - 4}$ **41.** $\dfrac{y - 2}{y + 6}$ **43.** $\dfrac{q - 3}{3q}$ **45.** $\dfrac{z + 1}{z + 5}$

47. $\dfrac{4m + 3}{3m + 4}$ **49.** $\dfrac{m^2 + 3mn + m + 3n}{3n^2 + 6mn} = \dfrac{(m + 1)(m + 3n)}{3n(n + 2m)}$ is in lowest terms   **51. a.** 93,750; 136,364; 176,471

b. 18 months

Preparing for Section 6.2   *(page 264)*

53. $\dfrac{4}{15}$ **55.** $\dfrac{3}{4}$ **57.** $\dfrac{9}{8}$

EXERCISES 6.2   *(page 270)*

1. $\dfrac{x}{6}$ **3.** -3 **5.** $\dfrac{p^3}{42}$ **7.** $\dfrac{54}{r^4}$ **9.** $\dfrac{16}{u^4}$ **11.** $-\dfrac{2}{5w^2}$ **13.** $\dfrac{2x^{10}}{31}$ **15.** $\dfrac{7r^4}{24}$ **17.** $\dfrac{14m}{3}$ **19.** 15

21. $2h$ **23.** $\dfrac{n}{5(n + 10)}$ **25.** $\dfrac{5(r + 5)}{6(r + 10)}$ **27.** $\dfrac{3(x - 2)}{4}$ **29.** $\dfrac{(q - 1)(q + 4)}{q(q - 2)}$ **31.** $\dfrac{h + 3}{4h}$ **33.** $\dfrac{r}{r + 1}$

35. $\dfrac{x - 2}{x + 2}$ **37.** $\dfrac{h^3}{(h + 7k)^2}$ **39.** $\dfrac{2(3m - 2)}{(m - 2)(3m + 2)}$ **41.** $-\dfrac{(2x + y)}{x + 4y}$

Preparing for Section 6.3   *(page 271)*

43. $\dfrac{3}{5}$ **45.** $\dfrac{5}{12}$ **47.** $\dfrac{13}{12}$ **49.** $\dfrac{1}{9}$

EXERCISES 6.3   *(page 281)*

1. $\dfrac{3}{x}$ **3.** $\dfrac{4}{w^2}$ **5.** $\dfrac{1}{2p}$ **7.** 1   **9.** -1 **11.** y **13.** $2(p + 3)$ **15.** $z + 5$ **17.** $35n$ **19.** $30s^3$

21. $12x(x - 2)$ **23.** $(w + 2)(w - 2)$ **25.** $4(q + 9)(q - 9)$ **27.** $k(k + 6)(k - 4)$ **29.** $(m + 6)(m - 8)^2$

31. $2(u - 3)(u - 4)$ **33.** $\dfrac{11}{12x}$ **35.** $\dfrac{8w - 7}{2w^2}$ **37.** $\dfrac{4p + 9}{24p^3}$ **39.** $\dfrac{v^2 + 2v + 6}{uv^2}$ **41.** $\dfrac{r - 2}{r(r + 1)}$ **43.** $-\dfrac{5}{m - 2}$

45. $\dfrac{10}{3(y + 4)}$ **47.** $\dfrac{h - 3}{2h(h + 3)}$ **49.** $\dfrac{5}{z - 5}$ **51.** $\dfrac{m + 10}{2(m + 2)(m - 6)}$ **53.** $\dfrac{r^2}{(r + 2)(r + 3)(r + 4)}$

55. $\dfrac{5x}{(x + 3)^2(x - 2)}$ **57.** $\dfrac{3}{u + 3}$ **59.** $\dfrac{2q + 1}{q - 6}$

Preparing for Section 6.4   *(page 282)*

61. 3   **63.** 1   **65.** $\dfrac{28}{3}$

EXERCISES 6.4   *(page 287)*

1. $\dfrac{1}{2}$ **3.** $\dfrac{5z}{4w}$ **5.** $\dfrac{h + 1}{h - 1}$ **7.** $\dfrac{r}{4(r - 2)}$ **9.** $\dfrac{3}{y}$ **11.** $-\dfrac{w(w + 1)}{2}$ **13.** $\dfrac{3m + 1}{3m - 1}$ **15.** $k - h$

17. $\dfrac{9}{9k - 1}$ **19.** $\dfrac{4m}{m - 2}$ **21.** $-\dfrac{t + 5}{5t}$ **23.** $\dfrac{x + 2}{2 - x}$ **25.** $2v - 1$ **27.** $\dfrac{3(2y - 5)}{2(y - 2)}$ **29.** $\dfrac{n - 7}{2(n - 8)}$

31. $-\dfrac{2(p - 3)(p - 2)}{5p(p - 1)}$ **33.** $\dfrac{k}{k - 2}$ **35.** $-\dfrac{mn}{m + n}$

Preparing for Section 6.5   *(page 288)*

37. $x = 5$ **39.** $w = 4$ **41.** $q = -2, -8$

EXERCISES 6.5 *(page 293)*

1. $u = -3$ **3.** $x = 12$ **5.** $w = -\dfrac{5}{2}$ **7.** $v = \dfrac{25}{16}$ **9.** $m = 10$ **11.** $r = \dfrac{1}{2}$ **13.** $y = 18, -2$ **15.** $p = 3$

17. No solution **19.** No solution **21.** $m = -2$ **23.** $s = 4, 7$ **25.** $x = 0$ **27.** $z = -1, -3$
29. $q = 5$ **31.** $k = -2$ **33.** $m = -3$

Preparing for Section 6.6 *(page 294)*

35. $600 - x$ **37.** $\dfrac{1}{3}$ **39.** $k = 6$ **41.** $k = 8$

EXERCISES 6.6 *(page 303)*

1. 6 mph **3.** 20 mph **5.** $3\dfrac{3}{7}$ h or 3.43 h **7.** 9 h **9.** $5\dfrac{1}{4}$ h or 5.25 h **11.** 36 min **13.** $D = kp$

15. $N = \dfrac{k}{d}$ **17.** $V = kz^3$ **19.** $y = 80$ **21.** $h = 11.25$ **23.** $m = 8$ **25.** $133\dfrac{1}{3}$ kg **27.** 0.25 s
29. 67.5 lb **31.** 22.5 foot candles

Preparing for Section 7.1 *(page 305)*
33. 1 **35.** 13 **37.** 14 **39.** -2

CHAPTER REVIEW *(page 306)*

1. 2 **3.** $v = 0$ **5.** $y = 6, -10$ **7.** $-\dfrac{8}{3w^2}$ **9.** $-\dfrac{s+1}{s-2}$ **11.** $\dfrac{q(q+3)}{q-3}$ **13.** $\dfrac{3x}{x+6}$

15. $\dfrac{6}{k^2(k+5)}$ **17.** $\dfrac{2(u+1)}{u(u+4)}$ **19.** $-\dfrac{n+2}{2(n+3)}$ **21.** $108v$ **23.** $8q^2(q+6)$ **25.** $(w+2)(w-2)^2$

27. $\dfrac{4}{h}$ **29.** $\dfrac{3}{k+3}$ **31.** $\dfrac{5r-1}{(2r+1)(2r-1)}$ **33.** $\dfrac{2(v-6)}{v(v+8)(v-4)}$ **35.** $\dfrac{1}{2(q-4)}$ **37.** $\dfrac{w^2}{3w+1}$

39. $\dfrac{3}{4h^2}$ **41.** $\dfrac{3k+1}{2k-3}$ **43.** p **45.** $\dfrac{m-10}{10m}$ **47.** $k = 36$ **49.** $h = -\dfrac{1}{3}$ **51.** $z = 5$

53. $q = 5, -1$ **55.** $m = 0, -3$ **57.** 15 mph **59.** $1\dfrac{7}{8}$ h or 1.875 h **61.** $p = ks$ **63.** $w = \dfrac{k}{l}$

65. $y = 30$ **67.** $q = 3.375$ **69.** 2500 calories per h

Chapter 7

EXERCISES 7.1 *(page 315)*

1. yes **3.** yes **5.** no **7.** yes **9.** yes **11.** no **13.** yes **15.** yes **17.** $(14, 3)$
19. $(-5, 4)$ **21.** $(0, 0.8)$ **23.** $(0.4, -4)$ **25.** $(1, 11)$ **27.** $(6, 9)$

29.

x	y
0	$\frac{1}{2}$
9	-1
-9	2
-27	5

31.

x	y
4	0
2	-3
0	-6
-8	-18

33.

x	y
2	6
$-\dfrac{2}{3}$	-2
5	15
-4	-12

35. a. $(-1, -3)$, $(0, -2)$, $(1, -1)$ **b.** $y = x - 2$ **37. a.** $(-1, 7)$, $(0, 6)$, $(1, 5)$ **b.** $x + y = 6$

39. a. $(-1, -12)$, $(0, 0)$, $(1, 12)$ **b.** $y = 12x$ **41. a.** $\left(-1, -\frac{1}{2}\right)$, $\left(0, -\frac{5}{12}\right)$, $\left(1, -\frac{1}{3}\right)$ **b.** $x = 12y + 5$

43. a. $(-1, -11)$, $(0, -11)$, $(1, -11)$ **b.** $y = -11$ **45.** $2l + 2w = 20$; linear relationship because it is of the form $ax + by = c$, where $x = l$, $y = w$, and $a = 2$, $b = 2$, $c = 20$

Preparing for Section 7.2   *(page 316)*

47. $y = 7$ **49.** $y = -\dfrac{11}{2}$ or -5.5 **51.** $x = 5$ **53.** $y = -\dfrac{22}{5}$ or -4.4

EXERCISES 7.2   *(page 324)*

1.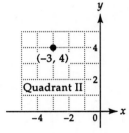

3.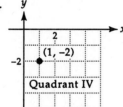

5.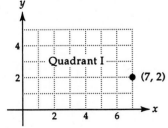

7.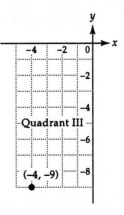

9. on *x* axis

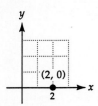

11. on *y* axis

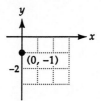

13. $(-2, -4)$ **15.** $(6, 2)$ **17.** $(4, 0)$ **19.** $(0, -2)$

21.

x	*y*
0	1
1	4
-2	-5

23.

x	*y*
2	2
6	0
-4	5

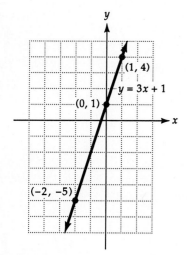

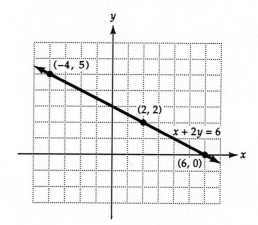

25.

x	*y*
0	0
8	-2
4	-1

27.

x	*y*
2	3
9	3
-1	3

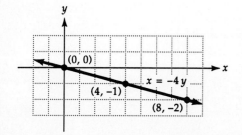

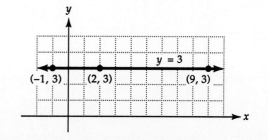

29. x intercept: 4
y intercept: -8

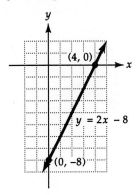

31. x intercept: 9
y intercept: -3

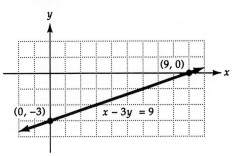

33. x intercept: -2
y intercept: 2

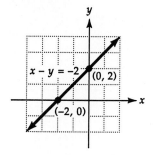

35. x intercept: 9
y intercept: 6

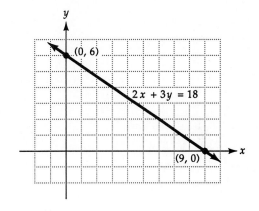

37. x intercept: 6
y intercept: -4

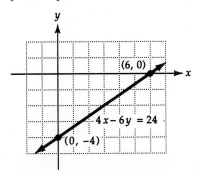

39. x intercept: 0
y intercept: 0

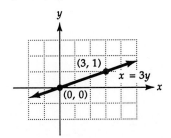

41. x intercept: 0
y intercept: 0

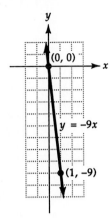

43. x intercept: 7.5
y intercept: -3

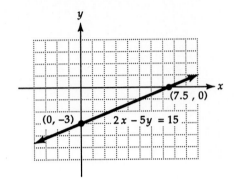

45. x intercept: $\dfrac{3}{5}$

y intercept: $\dfrac{3}{4}$

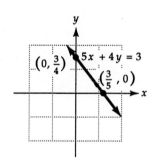

47. x intercept: 4
no y intercept

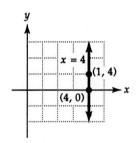

49. y intercept: -8
no x intercept

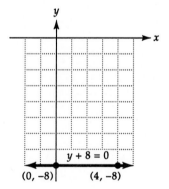

51. a. $55   **b.** $y = 30 + 0.20x$ **c.**   **d.** 149 miles

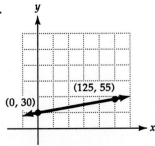

Preparing for Section 7.3 *(page 326)*

53. $\frac{5}{2}$ or 2.5   **55.** $\frac{5}{2}$ or 2.5   **57.** $-\frac{3}{4}$ or -0.75 **59.** $-\frac{3}{2}$ or -1.5

EXERCISES 7.3 *(page 333)*

1. 2   **3.** -1 **5.** -1 **7.** $\frac{13}{10}$ or 1.3   **9.** $\frac{1}{4}$ or 0.25   **11.** $-\frac{11}{5}$ or -2.2 **13.** 0   **15.** undefined

17. -32 **19.** -4 **21.** $\frac{2}{5}$ or 0.4   **23.** $-\frac{8}{5}$ or -1.6 **25.** undefined   **27.** 0

29. $m = 4$

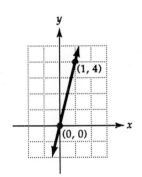

31. $m = -5$

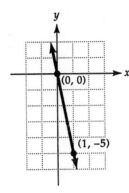

33. $m = 3$

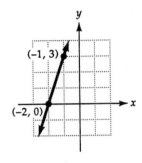

35. $m = -2$

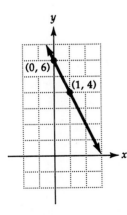

37. $m = -\frac{3}{4}$

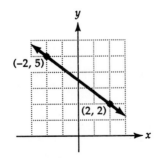

39. $m = \frac{3}{2}$

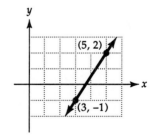

41. $m = -1.6$

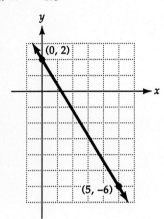

43. $m = 0$

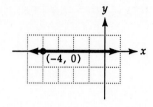

45. m is undefined

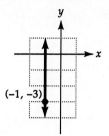

47. a. $y = 172 - 0.75x$; $m = -0.75$ **b.** 139.75

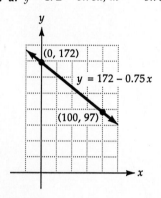

Preparing for Section 7.4   *(page 335)*

49. $y = 10 - x$ **51.** $y = -2x$ **53.** $y = 6x + 15$

EXERCISES 7.4   *(page 340)*

1. $m = 5, b = -3$ **3.** m is undefined, no y intercept   **5.** $m = -\dfrac{3}{2}, b = 5$ **7.** $m = \dfrac{2}{3}, b = 0$

9. $m = \dfrac{1}{6}, b = -3$ **11.** $m = -\dfrac{9}{8}, b = \dfrac{3}{2}$ **13.** $y = 7x + 4$ **15.** $y = \dfrac{2}{3}x$ **17.** $y = -4x - 3$ **19.** $y = 3$

21. $x = 6$ **23.** $y = -x + 3$ **25.** $y = \dfrac{15}{4}x + \dfrac{51}{4}$ **27.** $x = 5$ **29.** $y = -1$ **31.** $y = 15x - 6$

33. $y = -\dfrac{31}{13}x - \dfrac{157}{104}$ **35.** $y = -\dfrac{5}{3}x + 5$ **37.** $y = \dfrac{3}{2}x - 3$ **39. a.** $y = 0.75x + 320$ **b.** 320 m/s

c. $-427°C$

Preparing for Section 7.5 *(page 342)*

41. True **43.** False **45.** False **47.** True

EXERCISES 7.5 *(page 348)*

1.

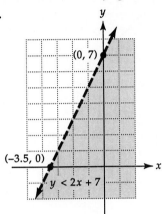

3.

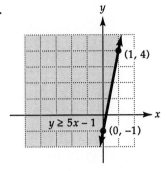

5.

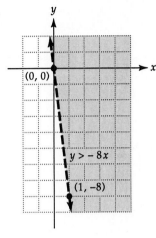

7.

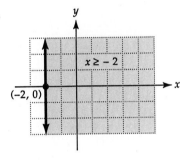

9.

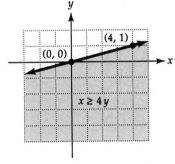

11.

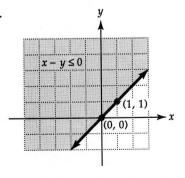

13.

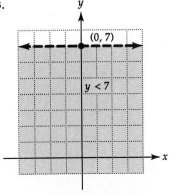

15.

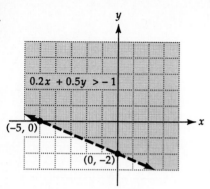

17.

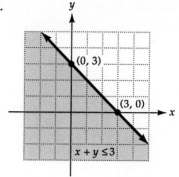

19.

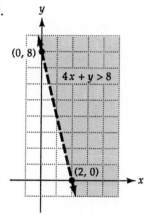

21.

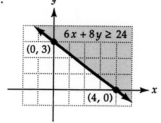

23.

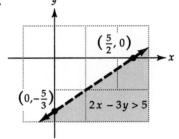

25. $x + y < 3$ **27.** $y \geq x$ **29.** $y \geq -4$ **31.** $x > 3$

33. a. $4x + 6y \leq 108$ **b.**

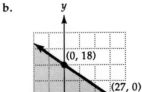

c. no; more than 108 labor hours are needed to make 10 regular and 18 scientific calculators.

Preparing for Section 8.1   *(page 350)*

35. $x + y = 7$ and $2x - y = 8$
$5 + 2 = 7$ $10 - 2 = 8$
$7 = 7$ $8 = 8$

37. $y = 3x - 13$ and $y = 0.4x$
$2 = 15 - 13$ $2 = 0.4(5)$
$2 = 2$ $2 = 2$

39.

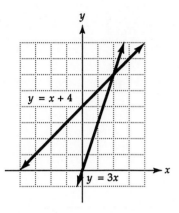

41.

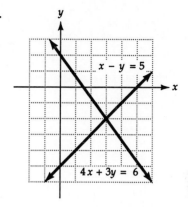

CHAPTER REVIEW   *(page 350)*

1. yes   **3.** no   **5.** $(-1, -2)$ **7.** $(-3, -5)$ **9.**

x	y
0	3
2	$\dfrac{3}{2}$
4	0
8	-3

11. a. $(-1, -3), (0, 3), (1, 9)$ **b.** $y = 6x + 3$

13. Quadrant IV

15. Quadrant III

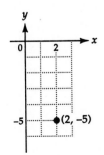

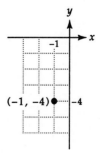

17. $(2, 3)$ **19.** $(0, -5)$

21.

x	y
0	8
3	20
−1	4

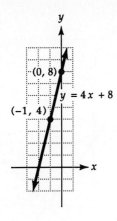

23. x intercept: 2
y intercept: −10

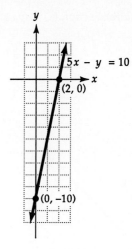

25. x intercept: 2
y intercept: 4

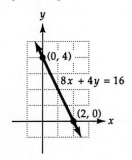

27. x intercept: 0
y intercept: 0

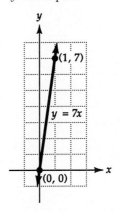

29. x intercept: −3
no y intercept

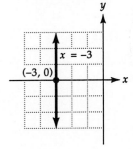

31. −1   **33.** undefined   **35.** 1   **37.** $m = \dfrac{3}{8}$

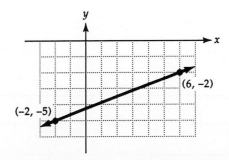

39. *m* is undefined

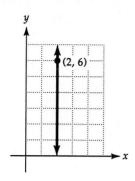

41. a. $m = -2$

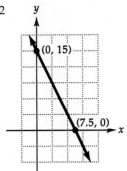

b. 2.5 cm

43. $m = -\dfrac{1}{2}$, $b = 4$ **45.** $m = \dfrac{1}{2}$, $b = 0$ **47.** $m = 0$, $b = 1$ **49.** $y = -2x + 4$ **51.** $y = -\dfrac{1}{2}x + 2$

53. $y = -3$ **55.** $y = 4x - 10$ **57.** $y = -x - 2$ **59.** $x = 6$ **61.** $y = \dfrac{4}{5}x + \dfrac{7}{5}$

63.

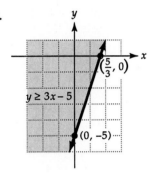

65.

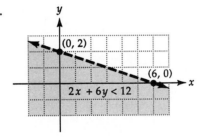

67.

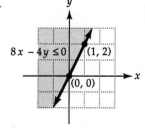

69.

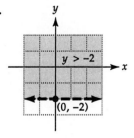

71. $y \geq -1.5x + 3$ **73.** $y < 4$

75. a. $120x = 180y > 3{,}600$ **b.**

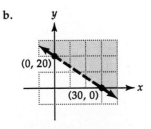

c. No; 14 women and 8 men weigh less than 3600 pounds.

Chapter 8

EXERCISES 8.1   *(page 363)*

1. $(8, 2)$ **3.** $(4, 0)$ **5.** $(-2, 6)$ **7.** $(-4, 5)$ **9.** $(4, -5)$ **11.** $(2, 0)$ **13.** $(-4, 1)$
15. Inconsistent system,; no solution   **17.** $(2, 0)$ **19.** Dependent system; infinite number
of solutions   **21.** $(0, 0)$ **23.** $(-2, -3)$ **25.** Inconsistent system; no solution
27. Dependent system; infinite number of solutions   **29.** $(-2, -18)$ **31.** $(2, 3)$
33. **a.** $3x + y = 3$: $m = -3, b = 3$ **b.** Inconsistent system
   $y = -3x + 6$: $m = -3, b = 6$

35. **a.** $4x - 3y = 12$: $m = \dfrac{4}{3}, \quad b = -4$ **b.** Consistent system

   $2x + 3y = 15$: $m = -\dfrac{2}{3}, \quad b = 5$

37. **a.** $x - 6y = 8$: $m = \dfrac{1}{6}, b = -\dfrac{4}{3}$ **b.** Dependent system

   $-3x + 18y = -24$: $m = \dfrac{1}{6}, b = -\dfrac{4}{3}$

39. **a.** 1st option: $y = 10 + 5x$ **b.**   **c.** 5 records
   2nd option: $y = 7x$

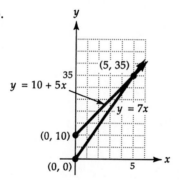

41. **a.** $y = 50x$ **b.** $y = 120,000 + 20x$ **c.** 4,000 pairs needed to break even

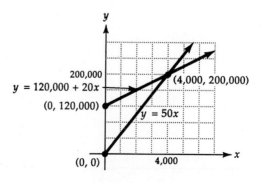

Preparing for Section 8.2   *(page 364)*

43. $3x + y$ **45.** $10x + 7y$ **47.** $2x$

EXERCISES 8.2   *(page 371)*

1. $(7, -2)$ **3.** $(4, 6)$ **5.** $(2, 5)$ **7.** Inconsistent system; no solution   **9.** $(13, 1)$ **11.** $(1, -1)$
13. $\left(\dfrac{1}{3}, 2\right)$ **15.** $(2, 4)$ **17.** $(2, -1)$ **19.** $(-1, 5)$ **21.** $(-6, -5)$ **23.** $(4, 3)$ **25.** $(0, 2)$
27. Inconsistent system; no solution   **29.** $(-3, 4)$ **31.** $(2, 0)$ **33.** $(-2, 5)$ **35.** $(-1, 3)$
37. $(11, -11)$ **39.** $(0, 0)$ **41.** Dependent system; infinite number of solutions   **43.** 4 and 5
45. \$25

Preparing for Section 8.3   *(page 372)*

47. $x = \dfrac{6}{5}$ **49.** $w = 3$ **51.** $p = -4$

EXERCISES 8.3   *(page 377)*

1. $(1, 11)$ **3.** $(2, 2)$ **5.** $(4, -2)$ **7.** $(1, 4)$ **9.** $(4, -2)$ **11.** $\left(\dfrac{1}{5}, \dfrac{3}{5}\right)$ **13.** $(5, -1)$
15. Dependent system; infinite number of solutions   **17.** $(0, 0)$ **19.** $(-4, -23)$ **21.** $(3, 6)$
23. Dependent system; infinite number of solutions   **25.** $(3, -1)$ **27.** $(5.2, 3.4)$
29. Inconsistent system; no solution   **31.** Dependent system; infinite number of solutions
33. \$2   **35. a.** Revenue: $y = 180x$ **b.** $(2{,}500, 450{,}000)$ **c.** 2,501 bikes
Total Cost: $y = 200{,}000 + 100x$

Preparing for Section 8.4   *(page 378)*

37. $x - y = 20$ **39.** $3x + 2y = 85$ **41.** $x = -1$ **43.** $w = 1{,}200$

EXERCISES 8.4   *(page 385)*

1. a   **3.** d   **5.** 13 and 5   **7.** Length: 15 cm; width: 5 cm   **9.** Wife: \$240,000; sister: \$60,000
11. 2000 orchestra tickets, 1500 balcony tickets   **13.** Buick: 400 mi; Honda: 500 mi
15. Brush: \$1.25; tube of paint: \$0.90   **17.** Speed in still air: 450 mph; speed of the wind: 50 mph
19. Walker: 3 mph; jogger: 6 mph   **21.** Enemy missile: 2,000 mi; anti-missile missile: 3,000 mi
23. 40% bluegrass: 300 lb; 70% bluegrass: 200 lb   **25.** Carbon monoxide: 500 mg; carbon dioxide: 1100 mg
27. Mix 1: 80 grams; mix 2: 60 grams   **29.** Kona: 5 lb; Dark Roast: 5 lb   **31.** \$6000 at 7%; \$4000 at 12%
33. A's: 4 h; B's: 10 h

Preparing for Section 8.5   *(page 388)*

35.

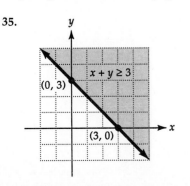

37.

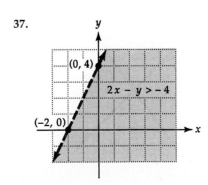

39.

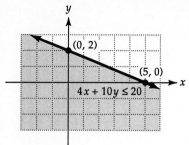

41.

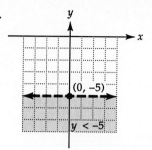

EXERCISES 8.5   *(page 392)*

1.

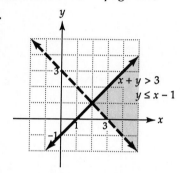

3.

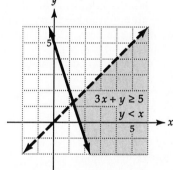

5.

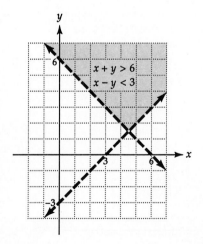

7.

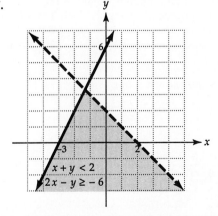

9. no solution

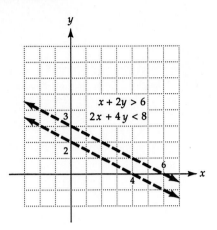

11.

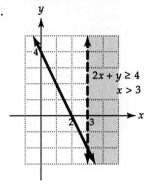

13.

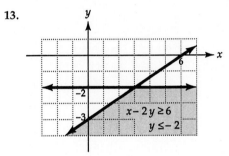

15.

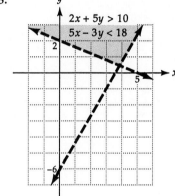

17.

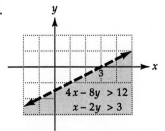

19.

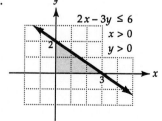

21.

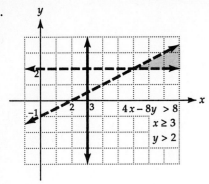

$4x - 8y > 8$
$x \geq 3$
$y > 2$

23.

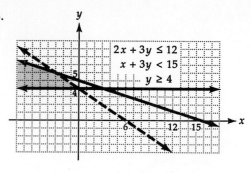

$2x + 3y \leq 12$
$x + 3y < 15$
$y \geq 4$

25.

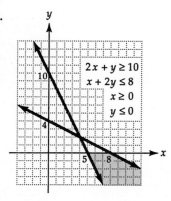

$2x + y \geq 10$
$x + 2y \leq 8$
$x \geq 0$
$y \leq 0$

27.

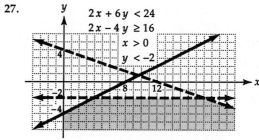

$2x + 6y < 24$
$2x - 4y \geq 16$
$x > 0$
$y < -2$

29. a. $x + y \leq 10$ **b.**
   $x \geq 4$
   $y \geq 3$

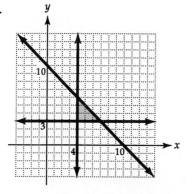

c. (4, 3), (5, 4) for example

31. a. $x + y \leq 5000$
   $45x + 60y \leq 270,000$
   $x \geq 0$
   $y \geq 0$

b.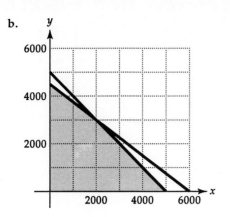

c. $(1,000, 3,000)$, $(2,000, 2,500)$
   for example

Preparing for Section 9.1   *(page 393)*
33. -243 **35.** $x = 10, -10$ **37.** $w = 16, -16$

CHAPTER REVIEW   *(page 394)*

1. yes   **3.** no   **5.** $(6, 2)$ **7.** Inconsistent system; no solution   **9.** $(8, -2)$ **11.** Dependent system;
infinite number of solutions   **13.** $(0, -2)$ **15.** $(-2, -3)$ **17.** $(8, -1)$ **19.** $(-2, 5)$ **21.** $(4, -1)$
23. $(3, 4)$ **25.** $(3, 1)$ **27.** $(3, 0)$ **29.** $(3, 5)$ **31.** $(-2, 4)$ **33.** $(-2, -6)$ **35.** $(1, 7)$ **37.** $(-22, -5)$
39. Larger: 15; smaller: -8 **41.** \$1,000 bonds: 1,000; \$500 bonds: 1,750
43. 10% acid: 12 liters; 15% acid: 8 liters

45.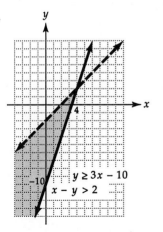

47.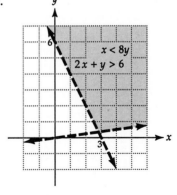

49.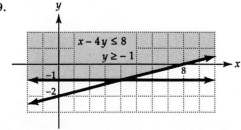

51.

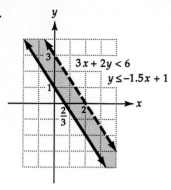

$3x + 2y < 6$
$y \leq -1.5x + 1$

53.

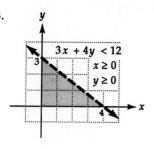

$3x + 4y < 12$
$x \geq 0$
$y \geq 0$

55. a. $x + y \leq 12,000$
$3x + 9y \geq 4,500$
$x > 0$
$y > 0$

b.

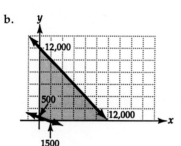

c. $(1,500, 2,000)$, $(5,000, 2,000)$ for example

Chapter 9

EXERCISES 9.1   *(page 403)*

1. a, c   **3.** 2.8   **5.** 4.1   **7.** 6.5   **9.** 8.1   **11.** 4.9   **13.** 3.1   **15.** 2.8   **17.** 8.6   **19.** 4.2
21. 13.2   **23.** 11.7   **25.** 25.7   **27.** 112.6   **29.** 12   **31.** $\dfrac{5}{8}$ **33.** $\dfrac{10}{3}$ or $3\dfrac{1}{3}$ **35.** $\dfrac{9}{13}$ **37.** 6
39. yes; $2^3 = 8$ **41.** no; $(3.4)^3 = 39.304 \neq 12$ **43.** yes; $5^3 = 125$ **45.** yes; $(-1)^3 = -1$ **47.** no;
$(-5)^3 = -125 \neq 120$ **49.** no; $(1.732)^2 = 2.999824 \neq 3$ **51.** no; $\sqrt{11} \approx 3.3166 > 3.3$ **53.** no; $(-5)^2 = 25$
55. yes; $\sqrt{135} = 11.6189$ and $11.6 < 11.6189 < 11.7$ **57.** True for $N = 16$ and all positive N. False for $N = -16$.
59. True for $N = 16$ and all positive N. False for $N = -16$.   **61.** False for $N = 16$. False for $N = -16$. Not true
for all positive N.   **63.** False for all N, except 0.   **65.** 74.40 in.

Preparing for Section 9.2   *(page 404)*
67. 7.0710678, equal   **69.** 2.6457513, equal   **71.** 1.4142135, equal

EXERCISES 9.2   *(page 409)*

1. x^2 **3.** x^2y^3 **5.** $2(x + 1)$ **7.** Equal   **9.** Unequal   **11.** Equal   **13.** Equal   **15.** $3\sqrt{2}$
17. $\sqrt{57}$ **19.** 8   **21.** $5\sqrt{6}$ **23.** $2\sqrt[3]{3}$ **25.** 15   **27.** 24   **29.** $3\sqrt{13}$ **31.** $4\sqrt{7}$ **33.** $8\sqrt{3}$
35. $9\sqrt{2}$ **37.** $2\sqrt[3]{4}$ **39.** $\sqrt{xy}$ **41.** $-10x$ **43.** $6xy^2\sqrt{xy}$ **45.** x **47. a.** $30\sqrt{5}$ cm^3
b. 80 cm^2 **c.** 110 cm^2 **49.** 7071 m

Preparing for Section 9.3   *(page 410)*

51. 2.4494897, equal   **53.** $\dfrac{3}{4}$ **55.** $\dfrac{4}{13}$

EXERCISES 9.3   *(page 414)*

1. $\dfrac{x}{y}$ **3.** $\dfrac{x^3}{4y}$ **5.** no   **7.** yes   **9.** yes   **11.** no   **13.** yes   **15.** $\dfrac{\sqrt{14}}{2}$ **17.** $\dfrac{3\sqrt{2}}{5}$ **19.** $\dfrac{\sqrt{6}}{4}$

21. $\dfrac{2\sqrt{10}}{9}$ **23.** $\dfrac{\sqrt{5}}{5}$ **25.** $2\sqrt{7}$ **27.** $\dfrac{\sqrt{34}}{2}$ **29.** $\dfrac{1}{2}$ **31.** $\dfrac{5\sqrt{15}}{3}$ **33.** $\dfrac{2\sqrt{3}}{9}$ **35.** $\dfrac{9\sqrt{2}}{2}$ **37.** $\dfrac{5\sqrt[3]{2}}{4}$

39. $\dfrac{\sqrt{2x}}{x}$ **41.** $\dfrac{\sqrt{5}}{x}$ **43.** $\dfrac{\sqrt{10xy}}{5x}$ **45.** $\dfrac{a^2\sqrt{6ab}}{4b}$ **47.** $\dfrac{\sqrt{5}}{y^2}$ **49.** $\dfrac{\sqrt{3}}{2}$

Preparing for Section 9.4   *(page 415)*
51. $13x$ **53.** $7\sqrt{3}$

EXERCISES 9.4   *(page 419)*

1. $11\sqrt{5}$ **3.** $5\sqrt{13}$ **5.** $-4\sqrt{23}$ **7.** $5\sqrt{2}$ **9.** $38\sqrt{3}$ **11.** 0 **13.** $10 + \sqrt[3]{100}$ **15.** $-3\sqrt[3]{5}$
17. $41\sqrt[4]{2}$ **19.** $-\sqrt{3m}$ **21.** $10\sqrt{5q}$ **23.** $7w\sqrt{3}$ **25.** $26p\sqrt{5}$ **27.** $8m\sqrt[3]{5}$ **29.** $(3-2k)\sqrt[3]{2k}$
31. $23x\sqrt[3]{3x^8}$ **33.** $8\sqrt{3}$ **35.** $26\sqrt{3} + 8\sqrt{2}$ **37.** $17\sqrt[3]{3}$ **39.** $4\sqrt{6w} - 6\sqrt{w}$ **41.** $-\sqrt{2} + 2\sqrt[3]{2}$
43. $4p\sqrt{7q} + (36 - 14q)\sqrt{7p}$

Preparing for Section 9.5   *(page 419)*
45. $x^2 - 30x + 225$ **47.** $16m^2 - 169$

EXERCISES 9.5   *(page 423)*

1. $6\sqrt{5} + 12$ **3.** 12 **5.** $-14\sqrt{6}$ **7.** $3x + 7\sqrt{x}$ **9.** $13 + 7\sqrt{3}$ **11.** $5 - 6\sqrt{2}$ **13.** $q - 3\sqrt{q} - 54$
15. $6r - 29\sqrt{r} + 28$ **17.** $2h - 4k\sqrt{2h} - 5k^2$ **19.** $15 + 4\sqrt{11}$ **21.** $y - 18\sqrt{y} + 81$ **23.** $14 + 8\sqrt{3}$
25. $3p + 10\sqrt{6pq} + 50q$ **27.** 1 **29.** $z - 8$ **31.** $24 - q$ **33.** $2r - 9t$ **35.** $3\sqrt{15} + 9$ **37.** $\sqrt{14} + \sqrt{10}$

39. $\dfrac{7 - \sqrt{33}}{4}$ **41.** $\dfrac{k + 4\sqrt{k} + 4}{k - 4}$ **43.** $\dfrac{r - 2\sqrt{rs} + s}{r - s}$ **45.** $\dfrac{49\sqrt{z+w} - 7z - 7w}{49 - z - w}$

Preparing for Section 9.6   *(page 424)*
47. $\dfrac{1}{x}$ **49.** $\dfrac{1}{m}$ **51.** $\dfrac{z^{14}}{4}$

EXERCISES 9.6   *(page 429)*

1. 8.246 **3.** 1.710 **5.** 0.585 **7.** 49.140 **9.** -0.091 **11.** -10 **13.** $\dfrac{3}{2}$ **15.** 6 **17.** 5

19. 16 **21.** 243 **23.** $-\dfrac{1}{6}$ **25.** $\dfrac{16}{9}$ **27.** 9 **29.** 3 **31.** 3 **33.** 3 **35.** $x^{7/5}$ **37.** $k^{1/4}$

39. $3p^{1/3}$ **41.** $\dfrac{1}{8m^{3/2}}$ **43.** $\dfrac{1}{w^{2/3}}$ **45.** $\dfrac{1}{z^{1/4}}$ **47.** $\dfrac{p^{1/4}}{2}$ **49.** $\sqrt[3]{h}$ **51.** $\sqrt[4]{m^3}$ **53.** $\dfrac{p^3}{q^2}$ **55.** 0.9
57. 70.71 beats per minute

Preparing for Section 9.7   *(page 430)*
59. 4 **61.** $6, -1$ **63.** $1, 2$

EXERCISES 9.7   *(page 435)*

1. $x = 34$ **3.** $p = 16$ **5.** $m = 25$ **7.** $S = \dfrac{48}{5}$ **9.** No solution   **11.** $y = 1$ **13.** No solution

15. $n = 4$ **17.** $t = 1, t = 2$ **19.** $h = 4, h = -4$ **21.** $v = 4$ **23.** $x = 9$ **25.** $z = -2$ **27.** $p = 5$
29. $n = 6$ **31.** $x = 9$ **33.** $h = 3$ **35.** $t = 1$ **37.** $u = 0, u = 1$ **39.** $d = 900$ ft   **41.** $L = 150$ ft

Preparing for Section 10.1   *(page 436)*

43. $h = 5$ **45.** $x = 11, x = -11$ **47.** $z = 6, z = 5$ **49.** $p = 7, p = -\dfrac{5}{3}$

CHAPTER REVIEW   *(page 437)*

1. 3.38   **3.** 127.35   **5.** -8.89 **7.** 15.36 cm   **9.** $2\sqrt{10}$ **11.** $6\sqrt{3}$ **13.** $15x\sqrt{x}$ **15.** $-3(x-1)^2$

17. $-15x^2y^2$ **19.** 15.6   **21.** $\dfrac{\sqrt{7}}{7}$ **23.** $\dfrac{\sqrt{6x}}{2x}$ **25.** $\dfrac{\sqrt{xy}}{y^2}$ **27.** 14.2 cm   **29.** $3x\sqrt{15}$ **31.** $-2\sqrt{3}$

33. $-5\sqrt{3y}$ **35.** 0   **37.** $2p\sqrt{6q} + 4p\sqrt{2q} - 20q\sqrt{2pq}$ **39.** $6 + 2\sqrt{2} - 3\sqrt{6} - 2\sqrt{3}$

41. $25 + 20\sqrt{y} + 4y$ **43.** $2k - 25h$ **45.** $\dfrac{\sqrt{10} - \sqrt{2}}{2}$ **47.** $\dfrac{2\sqrt{30} + 2\sqrt{21}}{3}$ **49.** $\dfrac{1}{2}$ **51.** 100

53. $\dfrac{27}{w^{3/2}}$ **55.** $\dfrac{1}{3^{1/5}}$ **57.** $p = 3$ **59.** $m = 3$ **61.** $k = -2$ **63.** $u = 8$ **65.** $d = 12{,}544$ ft

Chapter 10

EXERCISES 10.1   *(page 446)*

1. $(-3)^2 - 6(-3) - 27 = 9 + 18 - 27 = 0$ **3.** ± 6 **5.** 3, -1 **7.** 2, -6 **9.** 6, -5 **11.** $\pm\dfrac{7}{3}$

13. 2, $\dfrac{1}{2}$ **15.** $\dfrac{4}{3}$, -2 **17.** $\pm\dfrac{13}{6}$ **19.** $-2, -8$ **21.** 2, $-\dfrac{1}{5}$ **23.** 4, $-\dfrac{1}{3}$ **25.** 4, 8   **27.** ± 7

29. $\pm\dfrac{3}{4}$ **31.** ± 12 **33.** $\pm\dfrac{1}{11}$ **35.** $\pm 3\sqrt{5}$ **37.** $\pm\sqrt{30}$ **39.** ± 5 **41.** $\dfrac{-3 \pm \sqrt{6}}{2}$ **43.** 20 s

45. 7 and 8; or -8 and -7 **47.** $\dfrac{4}{3}$ and $-\dfrac{3}{4}$ **49.** 6 cm

Preparing for Section 10.2   *(page 447)*
51. Cannot be factored   **53.** 16   **55.** 0

EXERCISES 10.2   *(page 453)*

1. 1, 5   **3.** $\dfrac{-5 \pm \sqrt{17}}{2}$ **5.** $\pm\dfrac{3\sqrt{5}}{5}$ **7.** 5, -2 **9.** 0, $\dfrac{6}{5}$ **11.** 1, 3   **13.** No real roots

15. No real roots   **17.** $-3, \dfrac{4}{3}$ **19.** 0, $\dfrac{5}{3}$ **21.** 1, -3 **23.** $\dfrac{3}{2}, \dfrac{2}{3}$ **25.** $\dfrac{5 \pm \sqrt{37}}{2}$ **27.** $\dfrac{5 \pm \sqrt{31}}{2}$

29. $\dfrac{\pm\sqrt{10}}{2}$ **31.** $-1 \pm \sqrt{2}$ **33.** $\dfrac{-7 \pm \sqrt{69}}{2}$ **35.** 0, -11 **37.** 3.44, -0.44 **39.** 1.26, -0.89

41. 1.77, -2.5 **43. a.** $-11, \dfrac{5}{3}, \dfrac{43}{50}, \dfrac{33}{16}$, 420.6, 26.2144, 15.36   **b.** $37 - 42$, two each   **c.** none

d. 36, none   **45. a.** $18\dfrac{3}{4}$ gallons   **b.** 2.6 inches

Preparing for Section 10.3   *(page 454)*

47. A right angle is an angle that measures 90°.   **49.** $A = \dfrac{1}{2}pq$ **51.** 32 cm

EXERCISES 10.3   *(page 460)*

1. 13   **3.** 12   **5.** 5.7   **7.** 17.2   **9.** 9.8   **11.** 6.2   **13.** 50 cm², 34.1 cm   **15.** 960 cm², 132 cm
17. 4 m   **19.** 10.6 ft   **21.** 25 mph   **23.** 3 mph   **25.** 98.5 m   **27.** 28 seats, 22 rows

29. No, it is in front.   **31.** $\dfrac{1 + \sqrt{5}}{2}$ **33.** 39.5 pens   **35.** 16 pens

Preparing for Section 10.4   *(page 462)*

37.

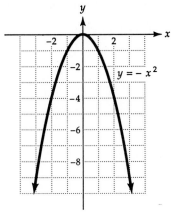

EXERCISES 10.4   *(page 470)*

1. a. upward   **b.** two   **3. a.** downward   **b.** two   **5. a.** upward   **b.** none
7. a. downward   **b.** two

9. (0, 0), (7, 0)

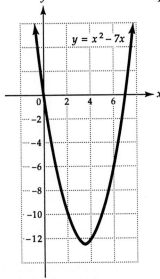

11. (1, 0), (−1.5, 0)

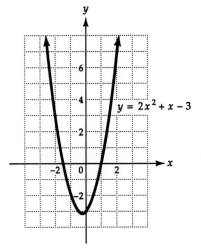

13. $(1, 0)$, $(-7, 0)$

15. $(1.4, 0)$, $(-0.4, 0)$

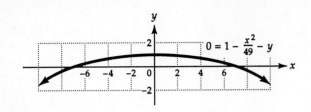

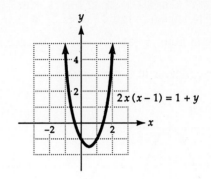

17. $0, 7$ **19.** $1, -1.5$ **21.** ± 7 **23.** $1.37, -0.37$ **25.** \$7500

27.

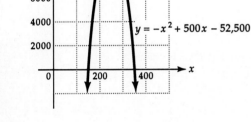

29. 250 **31.** no **33.** 50 mph

Preparing for Section 10.5 *(page 472)*

35. $3 + 2i$ **37.** $-3 - 10i - 7i^2$ **39.** 15 **41.** $6\sqrt{3}$

EXERCISES 10.5 *(page 480)*

1. 1; real; 1 **3.** -2; real; -2 **5.** 0; real; 0 **7.** -4; real; -4 **9.** -10; real; -10 **11.** $10i$
13. $-11 - 8i$ **15.** $7 - 5i$ **17.** $12i$ **19.** $6 + i$ **21.** $7 - 8i$ **23.** 6 **25.** $6 - 2i$ **27.** $6 + 18i$

29. 13 **31.** $\dfrac{5}{3} + 4i$ **33.** $-\dfrac{13}{25} + \dfrac{9}{25}i$ **35.** $-\dfrac{14}{25} + \dfrac{21}{50}i$ **37.** $\dfrac{37}{52} - \dfrac{21}{26}i$ **39.** $-7 + 9i$ **41.** $8 - 6i$

43. $6 + i$ **45.** $\pm 3i\sqrt{2}$ **47.** $0, -3$ **49.** $-1 \pm \sqrt{7}$ **51.** $\dfrac{3}{2} \pm \dfrac{\sqrt{11}}{2}i$ volts **53.** $\pm\sqrt{5}$

55. $\dfrac{1}{12} \pm \dfrac{\sqrt{23}}{12}i$ **57.** $-1 + 2i$; $c = 1 - 8i$ **59.** resistance: 15 ohms; reactance: 8 ohms **61.** $2 - 23i$

CHAPTER REVIEW   *(page 482)*

1. $-1, -4$ **3.** $3, 4$ **5.** 1 **7.** $2, 4$ **9.** $2, -\dfrac{1}{2}$ **11.** $2, -12$ **13.** $\dfrac{3}{4}$ or $\dfrac{4}{3}$ **15.** $\dfrac{1 \pm \sqrt{13}}{6}$

17. $\dfrac{-3 \pm \sqrt{5}}{2}$ **19.** $\dfrac{-1 \pm \sqrt{41}}{4}$ **21.** No real roots   **23.** $3, -2$ **25.** $\pm\sqrt{10}$ **27.** 9.7 ft high, 17.7 ft long   **29.** 10.8   **31.** 9.4   **33.** 8.9   **35.** 20 ft

37. about ± 1.4 **39.** no x intercepts

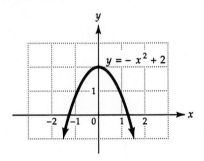

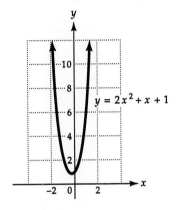

41. ± 1.4 **43.** no real solution   **45.** About 40 meters

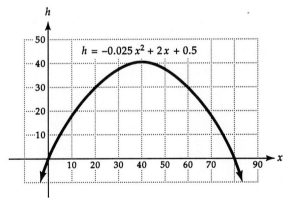

47. i; imaginary   **49.** $-2\sqrt{2}i$; imaginary   **51.** $-2i$ **53.** $-9 - 2i$ **55.** $3 + 4i$ **57.** $-20 + 35i$

59. $5 + 14i$ **61.** $-\dfrac{4}{5} - \dfrac{3}{5}i$ **63.** $(-3i)^2 + 12 = 3; 3i$
$-9 + 12 = 3$
$3 = 3$ **65.** $0, 3$ **67.** $1 \pm i$ **69.** $\dfrac{3}{5} \pm \dfrac{1}{5}i$ **71.** $\dfrac{9}{14} \pm \dfrac{5\sqrt{33}}{42}i$

Index

U.S. Measure

Weight	16 ounces (oz) = 1 pound (lb) 2000 pounds = 1 ton
Length	12 inches (in.) = 1 foot (ft) 3 feet = 1 yard (yd) 5280 feet = 1 mile (mi)
Area	144 square inches (in.^{2}) = 1 square foot (ft^2) 9 square feet = 1 square yard (yd^2) 43,560 square feet = 1 acre
Volume	1728 cubic inches (in.^{3}) = 1 cubic foot (ft^3) 27 cubic feet = 1 cubic yard (yd^3)
Capacity (liquids)	8 ounces (oz) = 1 cup (c) 2 cups = 1 pint (pt) 2 pints = 1 quart (qt) 4 quarts = 1 gallon (gal)
Time	60 seconds (s) = 1 minute (min) 60 minutes = 1 hour (h) 24 hours = 1 day

Metric Measure

Prefixes	milli (m) = 0.001 deka (da) = 10 centi (c) = 0.01 hecto (h) = 100 deci (d) = 0.1 kilo (k) = 1000
Weight	The base unit is the gram (g). *Example:* 0.01 gram = 1 centigram (1 cg)
Length	The base unit is the meter (m). *Example:* 1000 meters = 1 kilometer (1 km)
Area	Units are squared length units. *Example:* 100 square centimeters (100 cm^2) = 1 square decimeter (1 dm^2)
Volume	Units are cubed length units. *Example:* 1000 cubic decimeters (1000 dm^3) = 1 cubic meter (1 m^3)
Capacity (liquids)	The base unit is the liter (L). *Example:* 100 milliliters (100 ml) = 1 deciliter (1 dl)